✳

Understanding Society

An Introductory Reader

FOURTH EDITION

MARGARET L. ANDERSEN
University of Delaware

KIM A. LOGIO
Saint Joseph's University

HOWARD F. TAYLOR
Princeton University

WADSWORTH
CENGAGE Learning

Australia • Brazil • Japan • Korea • Mexico • Singapore • Spain • United Kingdom • United States

WADSWORTH
CENGAGE Learning™

Understanding Society: An Introductory Reader, Fourth Edition
Margaret L. Andersen,
Kim A. Logio, and Howard F. Taylor

Publisher/Executive Editor: Linda Schreiber-Ganster

Acquisitions Editor: Erin Mitchell

Assistant Editor: John Chelll

Editorial Assistant: Mallory Ortberg

Media Editor: Melanie Cregger

Marketing Manager: Andrew Keay

Marketing Assistant: Dimitri Hagnere

Marketing Communications Manager: Laura Locallio

Content Project Management: PreMediaGlobal

Creative Director: Rob Hugel

Art Director: Caryl Gorska

Print Buyer: Betsy Donaghey

Rights Acquisitions Account Manager, Text and Image: Roberta Broyer

Production Service: PreMediaGlobal

Copy Editor: PreMediaGlobal

Cover Designer: Riezebos Holzbaur/ Brie Hattey

Cover Image: Panoramic Images

Compositor: PreMediaGlobal

For product information and technology assistance, contact us at **Cengage Learning Customer & Sales Support, 1-800-354-9706.**

For permission to use material from this text or product, submit all requests online at **www.cengage.com/permissions.** Further permissions questions can be e-mailed to **permissionrequest@cengage.com**

Library of Congress Control Number: 2010940574

ISBN-13: 978-1-1111-8596-1

ISBN-10: 1-1111-8596-4

Wadsworth
20 Davis Drive
Belmont, CA 94002-3098
USA

Cengage Learning is a leading provider of customized learning solutions with office locations around the globe, including Singapore, the United Kingdom, Australia, Mexico, Brazil, and Japan. Locate your local office at **www.cengage.com/global**.

Cengage Learning products are represented in Canada by Nelson Education, Ltd.

To learn more about Wadsworth, visit **www.cengage.com/Wadsworth**

Purchase any of our products at your local college store or at our preferred online store **www.cengagebrain.com**.

Printed in the United States of America
2 3 4 5 6 7 8 13 12

Contents

Preface

"If you really acquire the sociological perspective, you can never be bored," writes June Jordan, a contemporary African American essayist. We agree; and we present these readings to help students see how fascinating the sociological perspective can be in interpreting human life. This anthology is intended for use in introductory sociology courses. Most of the students in these courses are first- or second-year students, many of whom are not majoring in sociology. We think this anthology will excite students about the sociological perspective and show what such a perspective can bring to understanding society.

This new edition keeps many of the same themes as the first three editions, but we added thirty-eight new articles that bring new material to the book and reflect some of the latest developments in society. The table of contents is modeled on the companion text, *Sociology The Essentials* by Andersen and Taylor, but it can easily be adapted for use with other introductory books or as a stand-alone text.

For this collection we selected articles that will engage students and show them what sociology can contribute to their understanding of the world. The readings include a variety of perspectives, research methodologies, and current topics. Most have been excerpted to make them more accessible to students. The anthology has a strong focus on diversity, both in the sections on class, race, gender, and age and in other selections throughout the book. We also included many articles that give students a global perspective on various sociological topics. The book features current research in sociology plus some classic readings from sociological theory. We developed this book with several themes in mind:

- **Contemporary research:** We wanted students to see examples of strong contemporary research, presented in a fashion that would be accessible to beginning undergraduates. The articles included here feature different styles of sociological research. For example, Nikki Jones's article, "Between Good and Ghetto" is an ethnographic study of African American girls living in impoverished neighborhoods. We have presented the part of her book that details her ethnographic methods, while also providing some parts of her argument about the young girls' lives. Contrasted with this is an article by Ann Meier and Gina Allen ("Romantic Relationships from Adolescence to Young Adulthood") that uses hypothesis testing to study the longitudinal development of adolescents' romantic relationships. These are but two examples of the diversity of sociological research found in this anthology, but they show students how research designs and methods differ, depending on the nature of the research question being asked.

- **Current events in society:** We want students to see how a sociological perspective can help them understand events in the contemporary world. BP oil spill. Earthquake in Haiti. Global warming. Economic recession. You wouldn't think sociologists would have a lot to say about these events that seem to be the province of other scientists. But students need to understand that even these phenomena have sociological dimensions. We want students to be able to use their sociological imaginations to interpret events happening in the world around them. For example, we include an article by Tanya Golash-Boza ("'Bandits Going Wild in Haiti!' and Other Post-Quake Myths") that explores how the media distorts the behavior that follows so-called natural disasters. And, Constance Lever-Tracy's article ("Global Warming and Sociology") brings a strong sociological argument to one of the most pressing issues of the day: climate change and global warming. Several articles included here discuss the sociological implications of the economic recession that has burdened so many people in recent times.

- **Diversity:** In keeping with our understanding that society is increasingly diverse, we selected articles that show the range of experiences people have by virtue of differences in race, gender, class, sexual orientation, and other characteristics (such as age and religion). Many of the selections bring a comprehensive analysis of race, class, and gender to the subject at hand, thus adding to students' understanding of how diverse groups experience the social structure of society. As an example, Shannon N. Zenk and her co-authors ("Neighborhood Racial Composition, Neighborhood Poverty, and the Spatial Accessibility of Supermarkets in Metropolitan Detroit") show how something as basic as the location of good supermarkets is related to the health of different racial-ethnic groups. Robert D. Bullard ("Environmental Justice in the 21st Century: Race Still Matters") shows the racial dimensions to the broad concern with sustainability and environmental degradation. And, several new articles look at different consequences of the current economic recession on various social classes (for example, John Bellamy Foster, "Aspects of Class in the United States").

- **Global perspective:** We also incorporated a global perspective into the reader, with many sections including articles that broaden students' world-view beyond the borders of the United States. Articles like Arlie Hochschild's essay "The Nanny Chain" will help students see how patterns of domestic help and contemporary immigration link the experiences of U.S. families to families from other nations, whose women are increasingly being employed to provide domestic help for professional workers in the United States. Similarly, Patricia Hill Collins's article, "New Commodities, New Consumers" shows how African American people are used in the capitalist quest for new global markets. Students will be able to see how their own consumption practices engage race and globalization in ways they might not have recognized.

- **Applying sociological knowledge:** Our students commonly ask, "What can you do with a sociological perspective?" We think this is an important question and one with many different answers. Sociologists use their knowledge in a variety of ways: to influence social policy formation, to interpret current events, and to educate people about common misconceptions and stereotypes, to name a few. Many of the new readings have a particularly contemporary appeal because the topics relate directly to changes in students' lives, such as the increasing use of the Internet for social interaction. For example, Maddy Coy's research ("Milkshakes, Lady Lumps and Growing Up To Want Boobies") details the increasing sexualization of popular culture and reviews research on the harm therefore done to young women. Also, Kathleen Gerson's research ("The Unfinished Revolution") relates directly to the ideals that the current generation holds for their own future families and work lives. We hope that students can see themselves in these and other articles throughout the anthology. The student exercises we have placed at the end of each section also enhance the theme of applying sociological knowledge to everyday life because they engage students in hands-on activities that will enrich their understanding of the course material.

- **Classical theory:** We think it is important that introductory students learn about the contributions of classical sociological theorists. Thus, we have kept articles by Weber, Marx and Engels, Du Bois, and Goffman that will showcase some of the most important classics. As with all of the selections, we include discussion questions at the end of each reading to help students think about how such classic pieces are reflected in contemporary issues. For example, Max Weber's argument about the Protestant ethic and the spirit of capitalism is fascinating to think about in the contemporary context of increased consumerism and increased class inequality. Students might ask whether contemporary patterns of wealth and consumption no longer reflect the asceticism and moral calling about which Weber wrote. W. E. B. Du Bois's reflections on double consciousness also continue to be relevant in discussions of race and group perceptions.

NEW TO THE FOURTH EDITION

The fourth edition of *Understanding Society* is organized in fourteen parts, following the outline of most introductory courses. The part titled "Social Institutions" is then subdivided into brief sections on six major social institutions: family, religion, education, work, government and politics, and health care. This organization allows faculty to focus on different institutions in any order they choose.

We replaced many of the articles in the thematic sections to reflect more current research and current events, particularly the sections on research, culture, class, sexuality, education, and health care.

We have continued the pedagogical features that we introduced in the third edition. Thus, at the conclusion of each article, we include a list of **key concepts** that are reflected in the article. These concepts are then defined in the glossary at the end of the book. We also include **student exercises at the end of each major section**, adding separate student exercises to each of the six institutions that we examine in Part XII. Through these exercises, students can—on their own or in group projects—apply what they have learned to their own observations of social behavior.

Brief introductions before each article place the article in context and help frame students' understanding of the selection. **Discussion questions** at the end of each reading help students think about the implications of what they have read.

The **thirty-eight new readings** in the book were selected to engage student interest, to reflect the richness of sociological thought, and to add articles that address issues that have emerged since the publication of the last edition (such as the economic recession, the Haiti earthquake, and the increasing racial segregation of schools, to name a few). In fact, we are always amazed at how, in a short period of time important new topics develop that influence our decisions about the content of this anthology. Thus, we have included a new piece on Mexican immigration (Edward Telles, "Mexican Immigrants and Immigrant Incorporation") which contrasts Mexican immigrant experiences with earlier European waves of immigration. The articles by Judith Treas ("The Great American Recession: Sociological Insights on Blame and Pain") and Elizabeth Warren ("America without a Middle Class") explore different sociological dimensions to the economic recession. We include an excellent article by John Brym ("Six Lessons of Suicide Bombers") that debunks some of common explanations of terrorist actions.

Our section on Culture also includes several new articles that explore popular culture—an increasingly important subject in the experiences of student readers. Thus, Ross Haenfler ("Games, Hackers, and Facebook: Computer Culture, Virtual Community, and Postmodern Identity") uses core sociological concepts about subcultures to examine cyberspace communities. Michael Messner ("Barbie Girls and Sea Monsters") also explores the dynamics of gender in play and popular culture.

Other new articles include those that examine the new politics of race, following the election of President Obama. Thus, Martell Teasley and David Ikard

("Barack Obama and the Politics of Race: The Myth of Post racism in America") question the idea that we now live in a post-racial society. Their article documents some of the persistent facts about structured racial inequality. Andrew J. Perrin's article on why people vote in presidential elections ("Why You Voted") also brings analytical insight to the contemporary politics of race and other characteristics that influence voting behavior.

And, in the institutions section, there are numerous new articles exploring diverse topics, such as myth of the absent Black father (Roberta L. Coles and Charles Green, "The Myth of the Missing Black Father"), religiosity among college students (Tim Clydesdale, "Abandoned, Pursued, or Safely Stowed?"), educational reform (Lisa Guernsey, "More Than "A is for Alligator': How to Ensure Early Childhood Systems Help Break the Cycle of Poverty"), and health care (T. R. Reid, "The Healing of America: A Global Quest for Better, Cheaper, and Fairer Health Care").

In sum, with this anthology we hope to capture student interest in sociology, provide interesting research and theory, incorporate the analysis of diversity into the core of the sociological perspective, analyze the increasingly global dimensions of society, and show students how what they learn about sociology can be applied to real issues and problems.

PEDAGOGICAL FEATURES

In addition to the sociological content of this reader, a number of pedagogical features enrich student learning and help instructors teach with the book. Each essay has a **brief introductory paragraph** that identifies the major themes and questions being raised in the article. A short list of **key concepts** at the end of each reading helps students understand how the reading is related to basic ideas in the field of sociology. All of the core concepts are defined in the **glossary** at the end of the book. This will especially help instructors who use this anthology as a stand-alone text. We also follow each article with **discussion questions** that students can use to improve their critical thinking and to reinforce their understanding of the article's major points. Many of these questions could also be used as the basis for class discussion, student papers, or research exercises and projects. And, new to this edition, are the **student exercises** included at the end of each major part. Finally, a **subject/name index** helps students and faculty locate specific topics and authors in the book.

ACKNOWLEDGMENTS

Many people helped in a variety of ways as we revised this anthology. We especially thank Katie Grunert for her assistance with some of the library work for this edition. We also appreciate the support of Dianna DiLorenzo, Sarah Hedrick, Linda Keen, Joan Stock, and Judy Watson, whose work makes a lot of things possible. And this edition is substantially improved based on the recommendations of our reviewers. We give special thanks to Wendy A. Moore, Texas A&M University; Kristin Marsh, University of Mary Washington; Suvarna Cherukun, Siena College; Catherine Marrone, SUNY Stony Brook; Ed-ward W. Morris, Ohio University; Gail Murphy-Geiss, Colorado College; and Patricia Gibbs, Foothill College.

We sincerely appreciate the enthusiasm and support provided by Wadsworth's outstanding editorial team. Thank you Erin Mitchell for your support for this work. And, most especially, we thank Richard, Jim, Nolan, Owen, and Isabel for all the love, smiles, and support they give us every day.

About the Editors

Margaret L. Andersen is the Edward F. and Elizabeth Goodman Rosenberg Professor of Sociology at the University of Delaware, where she also holds joint appointments in Women's Studies and Black American Studies. She is the author of *On Land and On Sea: A Century of Women in the Rosenfeld Collection; Thinking about Women: Sociological Perspectives on Sex and Gender; Living Art: The Life of African American Art Collector Paul R. Jones; Race, Class and Gender* (with Patricia Hill Collins); *Sociology Understanding a Diverse Society* (with Howard F. Taylor); and *Sociology The Essentials* (with Howard Taylor). She is a recipient of the American Sociological Association's Jessie Bernard Award and the Sociologists for Women in Society's Feminist Lecturer Award. She recently served as Vice President of the American Sociological Association and is the former president of the Eastern Sociological Society. She has won two teaching awards at the University of Delaware.

Kim A. Logio is Assistant Professor of Sociology at Saint Joseph's University in Philadelphia, where she teaches courses in research methods, childhood obesity, and women and health. Her research on adolescent body image has appeared in *Violence against Women* and has been presented at the American Sociological Association meetings. Her research on childhood obesity involves urban elementary school students and their understanding of nutrition and food choices. She is also the coauthor of *Adventures in Criminal Justice Research* with George Dowdall, Earl Babbie, and Fred Halley. She has been heard on National Public Radio discussing childhood obesity. She lives in Delaware with her husband and three children.

Howard F. Taylor is Professor of Sociology at Princeton University. He is the author of *Balance in Small Groups; The IQ Game; Sociology Understanding a Diverse Society* (with Margaret L. Andersen); and *Sociology The Essentials* (with Margaret

L. Andersen). He is the winner of the Du Bois-Johnson-Frazier Award, given by the American Sociological Association for distinguished research in race and ethnic relations, and has received the Princeton University President's Award for Distinguished Teaching. He is past president of the Eastern Sociological Society and is currently writing a book on *Race, Gender, Class, and the Big Test.*

✳

The Sociological Perspective

1

The Sociological Imagination

C. WRIGHT MILLS

First published in 1959, C. Wright Mills' essay, taken from his book The Sociological Imagination, *is a classic statement about the sociological perspective. Mills was a man of his times, and his sexist language intrudes on his argument, but the questions he poses about the connection between history, social structure, and peoples' biographies (or lived experiences) still resonate today. His central theme is that the task of sociology is to understand how social and historical structures impinge on the lives of different people in society.*

Nowadays men often feel that their private lives are a series of traps. They sense that within their everyday worlds, they cannot overcome their troubles, and in this feeling, they are often quite correct: What ordinary men are directly aware of and what they try to do are bounded by the private orbits in which they live; their visions and their powers are limited to the close-up scenes of job, family, neighborhood; in other milieux, they move vicariously and remain spectators. And the more aware they become, however vaguely, of ambitions and of threats which transcend their immediate locales, the more trapped they seem to feel.

Underlying this sense of being trapped are seemingly impersonal changes in the very structure of continent-wide societies. The facts of contemporary history are also facts about the success and the failure of individual men and women. When a society is industrialized, a peasant becomes a worker; a feudal lord is liquidated or becomes a businessman. When classes rise or fall, a man is employed or unemployed; when the rate of investment goes up or down, a man takes new heart or goes broke. When wars happen, an insurance salesman becomes a rocket launcher; a store clerk, a radar man; a wife lives alone; a child grows up without a father. Neither the life of an individual nor the history of a society can be understood without understanding both.

Yet men do not usually define the troubles they endure in terms of historical change and institutional contradiction. The well-being they enjoy, they do not usually impute to the big ups and downs of the societies in which they live. Seldom aware of the intricate connection between the patterns of their own lives and the course of world history, ordinary men do not usually know what this connection means for the kinds of men they are becoming and for the kinds of history-making in which they might take part. They do not possess the quality of mind essential to grasp the interplay of man and society, of biography and history, of self and world. They cannot cope with their personal troubles in such ways as to control the structural transformations that usually lie behind them....

The sociological imagination enables its possessor to understand the larger historical scene in terms of its meaning for the inner life and the external career of a variety of individuals. It enables him to take into account how individuals, in the welter of their daily experience, often become falsely conscious of their social positions. Within that welter, the framework of modern society is sought, and within that framework the psychologies of a variety of men and women are formulated. By such means the personal uneasiness of individuals is focused upon explicit troubles and the indifference of publics is transformed into involvement with public issues.

The first fruit of this imagination—and the first lesson of the social science that embodies it—is the idea that the individual can understand his own experience and gauge his own fate only by locating himself within his period, that he can know his own chances in life only by becoming aware of those of all individuals in his circumstances. In many ways it is a terrible lesson; in many ways a magnificent one. We do not know the limits of man's capacities for supreme effort or willing degradation, for agony or glee, for pleasurable brutality or the sweetness of reason. But in our time we have come to know that the limits of "human nature" are frighteningly broad. We have come to know that every individual lives, from one generation to the next, in some society; that he lives out a biography, and that he lives it out within some historical sequence. By the fact of his living he contributes, however minutely, to the shaping of this society and to the course of its history, even as he is made by society and by its historical push and shove.

The sociological imagination enables us to grasp history and biography and the relations between the two within society. That is its task and its promise. To recognize this task and this promise is the mark of the classic social analyst....

No social study that does not come back to the problems of biography, of history and of their intersections within a society has completed its intellectual journey. Whatever the specific problems of the classic social analysts, however limited or however broad the features of social reality they have examined, those who have been imaginatively aware of the promise of their work have consistently asked three sorts of questions:

1. What is the structure of this particular society as a whole? What are its essential components, and how are they related to one another? How does it differ from other varieties of social order? Within it, what is the meaning of any particular feature for its continuance and for its change?

2. Where does this society stand in human history? What are the mechanics by which it is changing? What is its place within and its meaning for the development of humanity as a whole? How does any particular feature

we are examining affect, and how is it affected by, the historical period in which it moves? And this period—what are its essential features? How does it differ from other periods? What are its characteristic ways of history-making?

3. What varieties of men and women now prevail in this society and in this period? And what varieties are coming to prevail? In what ways are they selected and formed, liberated and repressed, made sensitive and blunted? What kinds of "human nature" are revealed in the conduct and character we observe in this society in this period? And what is the meaning for "human nature" of each and every feature of the society we are examining?

Whether the point of interest is a great power state or a minor literary mood, a family, a prison, a creed—these are the kinds of questions the best social analysts have asked. They are the intellectual pivots of classic studies of man in society—and they are the questions inevitably raised by any mind possessing the sociological imagination. For that imagination is the capacity to shift from one perspective to another—from the political to the psychological; from examination of a single family to comparative assessment of the national budgets of the world; from the theological school to the military establishment; from considerations of an oil industry to studies of contemporary poetry. It is the capacity to range from the most impersonal and remote transformations to the most intimate features of the human self—and to see the relations between the two. Back of its use there is always the urge to know the social and historical meaning of the individual in the society and in the period in which he has his quality and his being.

That, in brief, is why it is by means of the sociological imagination that men now hope to grasp what is going on in the world, and to understand what is happening in themselves as minute points of the intersections of biography and history within society. In large part, contemporary man's self-conscious view of himself as at least an outsider, if not a permanent stranger, rests upon an absorbed realization of social relativity and of the transformative power of history. The sociological imagination is the most fruitful form of this self-consciousness. By its use men whose mentalities have swept only a series of limited orbits often come to feel as if suddenly awakened in a house with which they had only supposed themselves to be familiar. Correctly or incorrectly, they often come to feel that they can now provide themselves with adequate summations, cohesive assessments, comprehensive orientations. Older decisions that once appeared sound now seem to them products of a mind unaccountably dense. Their capacity for astonishment is made lively again. They acquire a new way of thinking, they experience a transvaluation of values: in a word, by their reflection and by their sensibility, they realize the cultural meaning of the social sciences.

Perhaps the most fruitful distinction with which the sociological imagination works is between "the personal troubles of milieu" and "the public issues of social structure." This distinction is an essential tool of the sociological imagination and a feature of all classic work in social science.

Troubles occur within the character of the individual and within the range of his immediate relations with others; they have to do with his self and with those limited areas of social life of which he is directly and personally aware.

Accordingly, the statement and the resolution of troubles properly lie within the individual as a biographical entity and within the scope of his immediate milieu—the social setting that is directly open to his personal experience and to some extent his willful activity. A trouble is a private matter: values cherished by an individual are felt by him to be threatened.

Issues have to do with matters that transcend these local environments of the individual and the range of his inner life. They have to do with the organization of many such milieux into the institutions of an historical society as a whole, with the ways in which various milieux overlap and interpenetrate to form the larger structure of social and historical life. An issue is a public matter: some value cherished by publics is felt to be threatened. Often there is a debate about what that value really is and about what it is that really threatens it. This debate is often without focus if only because it is the very nature of an issue, unlike even widespread trouble, that it cannot very well be defined in terms of the immediate and everyday environments of ordinary men. An issue, in fact, often involves a crisis in institutional arrangements, and often too it involves what Marxists call "contradictions" or "antagonisms."

In these terms, consider unemployment. When, in a city of 100,000, only one man is unemployed, that is his personal trouble, and for its relief we properly look to the character of the man, his skills, and his immediate opportunities. But when in a nation of 50 million employees, 15 million men are unemployed, that is an issue, and we may not hope to find its solution within the range of opportunities open to any one individual. The very structure of opportunities has collapsed. Both the correct statement of the problem and the range of possible solutions require us to consider the economic and political institutions of the society, and not merely the personal situation and character of a scatter of individuals.

Consider war. The personal problem of war, when it occurs, may be how to survive it or how to die in it with honor; how to make money out of it; how to climb into the higher safety of the military apparatus; or how to contribute to the war's termination. In short, according to one's values, to find a set of milieux and within it to survive the war or make one's death in it meaningful. But the structural issues of war have to do with its causes; with what types of men it throws up into command; with its effects upon economic and political, family and religious institutions, with the unorganized irresponsibility of a world of nation-states.

Consider marriage. Inside a marriage a man and a woman may experience personal troubles, but when the divorce rate during the first four years of marriage is 250 out of every 1,000 attempts, this is an indication of a structural issue having to do with the institutions of marriage and the family and other institutions that bear upon them.

Or consider the metropolis—the horrible, beautiful, ugly, magnificent sprawl of the great city. For many upper-class people, the personal solution to "the problem of the city" is to have an apartment with private garage under it in the heart of the city, and forty miles out, a house by Henry Hill, garden by Garrett Eckbo, on a hundred acres of private land. In these two controlled environments—with a small staff at each end and a private helicopter connection—most people could solve many of the problems of personal milieux caused by the facts of the city. But all this, however splendid, does not solve the public issues that the structural fact of the city poses. What should be done with this wonderful monstrosity? Break it all up into scattered units, combining residence and work? Refurbish it as it stands? Or, after evacuation, dynamite it and build new cities according to new plans in new places? What should those plans be? And who is to decide and to accomplish whatever choice is made? These are structural issues; to confront them and to solve them requires us to consider political and economic issues that affect innumerable milieux.

In so far as an economy is so arranged that slumps occur, the problem of unemployment becomes incapable of personal solution. In so far as war is inherent in the nation-state system and in the uneven industrialization of the world, the ordinary individual in his restricted milieu will be powerless—with or without psychiatric aid—to solve the troubles this system or lack of system imposes upon him. In so far as the family as an institution turns women into darling little slaves and men into their chief providers and unweaned dependents, the problem of a satisfactory marriage remains incapable of purely private solution. In so far as the overdeveloped megalopolis and the overdeveloped automobile are built-in features of the overdeveloped society, the issues of urban living will not be solved by personal ingenuity and private wealth.

What we experience in various and specific milieux, I have noted, is often caused by structural changes. Accordingly, to understand the changes of many personal milieux we are required to look beyond them. And the number and variety of such structural changes increase as the institutions within which we live become more embracing and more intricately connected with one another. To be aware of the idea of social structure and to use it with sensibility is to be capable of tracing such linkages among a great variety of milieux. To be able to do that is to possess the sociological imagination....

KEY CONCEPTS

issues   sociological imagination   troubles

DISCUSSION QUESTIONS

1. Using either today's newspaper or some other source of news, identify one example of what C. Wright Mills would call an issue. How is this issue reflected in the personal troubles of people it affects? Why would Mills call it a social issue?

2. What are the major historical events that have influenced the biographies of people in your generation? In your parents' generation? What does this tell you about the influence of society and history on biography?

2

Invitation to Sociology: A Humanistic Perspective

PETER BERGER

Peter Berger's classic book, Invitation to Sociology, *introduces the idea of debunking—that is, the process that sociologists use to see behind the taken-for-granted ways of thinking about social reality. As you read his essay, think about what it means to see behind the facades of common-sense explanations of society.*

… It can be said that the first wisdom of sociology is this—things are not what they seem. This too is a deceptively simple statement. It ceases to be simple after a while. Social reality turns out to have many layers of meaning. The discovery of each new layer changes the perception of the whole.

… People who like to avoid shocking discoveries, who prefer to believe that society is just what they were taught in Sunday School, who like the safety of the rules and the maxims of what Alfred Schuetz has called the "world-taken-for-granted," should stay away from sociology. People who feel no temptation before closed doors, who have no curiosity about human beings, who are content to admire scenery without wondering about the people who live in those houses on the other side of that river, should probably also stay away from sociology. They will find it unpleasant or, at any rate, unrewarding. People who are interested in human beings only if they can change, convert or reform them should also be warned, for they will find sociology much less useful than they hoped. And people whose interest is mainly in their own conceptual constructions will do just as well to turn to the study of little white mice. Sociology will be satisfying, in the long run, only to those who can think of nothing more entrancing than to watch men and to understand things human.

… The sociologist uses the term in a more precise sense, though, of course, there are differences in usage within the discipline itself. The sociologist thinks of "society" as denoting a large complex of human relationships, or to put it in more technical language, as referring to a system of interaction. The word "large" is difficult to specify quantitatively in this context. The sociologist may speak of a "society" including millions of human beings (say, "American society"), but he may also use the term to refer to a numerically much smaller

SOURCE: Peter Berger, Invitation to Sociology: A Humanistic Perspective. 1963. Garden City, NY: Doubleday.

collectivity (say, "the society of sophomores on this campus"). Two people chatting on a street corner will hardly constitute a "society," but three people stranded on an island certainly will. The applicability of the concept, then, cannot be decided on quantitative grounds alone. It rather applies when a complex of relationships is sufficiently succinct to be analyzed by itself, understood as an autonomous entity, set against others of the same kind.

… To ask sociological questions, then, presupposes that one is interested in looking some distance beyond the commonly accepted or officially defined goals of human actions. It presupposes a certain awareness that human events have different levels of meaning, some of which are hidden from the consciousness of everyday life. It may even presuppose a measure of suspicion about the way in which human events are officially interpreted by the authorities, be they political, juridical or religious in character.

The sociological perspective involves a process of "seeing through" the facades of social structures. We could think of this in terms of a common experience of people living in large cities. One of the fascinations of a large city is the immense variety of human activities taking place behind the seemingly anonymous and endlessly undifferentiated rows of houses. A person who lives in such a city will time and again experience surprise or even shock as he discovers the strange pursuits that some men engage in quite unobtrusively in houses that, from the outside, look like all the others on a certain street. Having had this experience once or twice, one will repeatedly find oneself walking down a street, perhaps late in the evening, and wondering what may be going on under the bright lights showing through a line of drawn curtains. An ordinary family engaged in pleasant talk with guests? A scene of desperation amid illness or death? Or a scene of debauched pleasures? Perhaps a strange cult or a dangerous conspiracy? The facades of the houses cannot tell us, proclaiming nothing but an architectural conformity to the tastes of some group or class that may not even inhabit the street any longer. The social mysteries lie behind the facades. The wish to penetrate to these mysteries is an analogon to sociological curiosity. In some cities that are suddenly struck by calamity, this wish may be abruptly realized. Those who have experienced wartime bombings know of the sudden encounters with unsuspected (and sometimes unimaginable) fellow tenants in the air-raid shelter of one's apartment building. Or they can recollect the startling morning sight of a house hit by a bomb during the night, neatly sliced in half, the facade torn away and the previously hidden interior mercilessly revealed in the daylight. But in most cities that one may normally live in, the facades must be penetrated by one's own inquisitive intrusions. Similarly, there are historical situations in which the facades of society are violently torn apart and all but the most incurious are forced to see that there was a reality behind the facades all along. Usually this does not happen and the facades continue to confront us with seemingly rocklike permanence. The perception of the reality behind the facades then demands a considerable intellectual effort.

… We would contend, then, that there is a debunking motif inherent in sociological consciousness. The sociologist will be driven time and again, by the very logic of his discipline, to debunk the social systems he is studying. This unmasking tendency need not necessarily be due to the sociologist's temperament or inclinations. Indeed, it may happen that the sociologist, who as an individual may be of a conciliatory disposition and quite disinclined to disturb the comfortable assumptions on which he rests his own social existence, is nevertheless compelled by what he is doing to fly in the face of what those around him take for granted. In other words, we would contend that the roots of the debunking motif in sociology are not psychological but methodological. The sociological frame of reference, with its built-in procedure of looking for levels of reality other than those given in the official interpretations of society, carries with it a logical imperative to unmask the pretensions and the propaganda by which men cloak their actions with each other. This unmasking imperative is one of the characteristics of sociology particularly at home in the temper of the modern era.

KEY CONCEPTS

debunking   social structure   society

DISCUSSION QUESTIONS

1. What does Berger mean by the "unmasking tendency" of sociology?
2. Pay attention to how a particular social issue is portrayed by a common news source (for example, job loss or a violent crime). What is the common explanation given for this phenomenon? How might a sociological explanation differ, and how does this illustrate Berger's concept of *debunking*?

3

"Bandits Going Wild in Haiti" and Other Post-Quake Myths

TANYA GALASH-BOZA

Tanya Golash-Boza uses a sociological perspective to challenge the usual way that disasters, such as the devastating Haiti earthquake, are routinely portrayed in the media. She argues that sociological evidence can debunk the suppositions that often frame media accounts of social problems such as disasters.

SOURCE: "Bandits Going Wild in Haiti", *Footnotes* 38 (February 2010). Washington, DC: American Sociological Association.

On January 25, 2010, I left for Haiti from the Dominican Republic with a team of five people from the Haitian non-governmental organization, Fondation Avenir, to meet with members of Haitian civil society to assess the possibilities for rebuilding the country in the aftermath of the devastating January 12 earthquake.

As we drove along the road from the border town of Malpasse to Port-au-Prince, the first major problem we encountered was a traffic jam in Croix-de–Bouquet, on the outskirts of Port-au-Prince. Closer to Port-au-Prince, we began to see more evidence of the destruction caused by the earthquake—flattened houses, tent cities, and lack of electricity. We saw few signs of the widespread civil unrest reported in the mainstream media. To the contrary, we found the city remarkably calm, with people selling goods on the streets, public transportation packed, and long lines outside money transfer outlets, cell phone stores, and waiting outside relief organizations.

As the electrical grid still was not functional, the city was quiet after dark. Many people slept in the streets. Some did this because they had lost their homes, others because their homes were unsafe, and still others because they feared there would be another earthquake. At these tent cities, despite the poor conditions, there was order and community. People arranged their tents into straight lines, left spaces for public use, and organized a security crew to watch over them at night and to ensure that cars did not trample people sleeping in the streets.

Press accounts of Haiti in the earthquake's aftermath emphasized the purported lack of public safety in Port-au-Prince. Many in the media reported that criminals were on the loose, rapes were commonplace, and banditry was omnipresent. As sociologists, we expect these sorts of reports after disasters, especially disasters involving people of African descent. It is our responsibility to insist on a more humane and accurate depiction of social life after disasters. There are three main points of contention that sociologists can address in terms of the popular representation of Haiti after the earthquake.

ADDRESSING FALSE REPORTING

The first point is the complete lack of historical context in media reports, especially of the role of the United States in Haiti over the course of the 20th century. For example, knowing about the U.S. occupations of Haiti contextualizes the current militarized response to the earthquake. The second point is that civil unrest and social violence are not common responses to disasters, yet typically are found in media portrayals of disasters. Disaster sociologist, Kathleen Tiemey, University of Colorado, and colleagues point out that not only do mass media consistently propagate the myth that lawlessness is a consequence of natural disasters, but that such myths justify a militarized response to these events (2006). The third point is that we can expect media representations of people of African descent to be influenced by "controlling images"—gendered and classed stereotypes about black people perpetuated by the media (Collins 2004).

As sociologists, one of our tasks is to educate the public on how to interpret the news and distinguish verifiable evidence from suppositions. We should participate in public debates and inform others how ideas about race, gender, and class influence perspectives. In the case of Haiti, preconceived notions about black men's sexuality have lent credibility to the idea that rapes are omnipresent, even with flimsy evidence. Ideas about black criminality also make it easier to believe that Haitians are looting and robbing. Many of the media reports of lawlessness are based on what most sociologists would consider flawed evidence (i.e., a woman hearing noises in a tent at night; a statement by a public official, and supposition by "experts" from afar).

Many of the reports that lead with headline about rampant rapes in Haiti are based almost entirely on one quote from Haiti's national police Chief Mario Andresol: "With the blackout that's befallen the Haitian capital, bandits are taking advantage to harass and rape women and young girls under the tents." This statement, the evidence for which is unclear, has picked up by many major media outlets. The diffusion of this statement has led to the widespread belief that rapes and banditry are omnipresent in Port-au-Prince following the earthquake.

FIELD OBSERVATIONS

I was in Port-au-Prince from January 25-28, 2010, and did not see any proof that social banditry reigned. Instead, I saw people in Port-au-Prince organizing themselves into groups and providing their own security. Of course, I do not have evidence that the news accounts are false. My perspective as a sociologist, however, inclines me to ask for the evidence, to consider the data journalists are citing, and to realize that realized notions of black criminality and sexuality make it likely for mass media outlets to pick up on these sorts of statements and to blow them out of proportion.

Rapes are widely underreported everywhere in the world, and it is not my intention to add to sexist contentions that rape is not a widespread problem. Instead, I refer to disaster researchers, such as Alice Fothergill, University of Vermont, who have confirmed that intimate partner violence often increases in the aftermath of disasters, yet less is known about sexual assault by strangers. John Barnshaw, University of Delaware, found in the case of Hurricane Katrina that reports of rapes tended to be based on rumors, not eyewitness accounts (2005).

Several mainstream media outlets stated that escaped prisoners from the destroyed jail are going on rampages and raping women. The animalistic discourse in headlines of these articles such as "Bandits going wild in Haiti" and "Escaped criminals raping, running wild in Haiti" are indicative of how Haitians are dehumanized and myths are spread. There is a tendency within popular discourse in the United States to associate blacks with unbridled sexuality and criminality. It is also worth noting that 80 percent of the escaped prisoners were in pre-trial detention, and thus that it is inaccurate to refer to them as criminals, as they had yet to be convicted of any crime. In actuality, the likelihood that escaped prisoners from the Haitian prison would randomly attack women in tent cities is ridiculous. Most rapes occur by people the victim knows.

These sensationalist headlines create the impression that Haitians are savages, and that a military response is the best response to the current disaster. As Tiemey et al. suggest, the portrayal of lawlessness justifies a militarized response to the disaster. The widespread fear that Haiti will descend further into lawlessness without a U.S. military presence prevents people in the United States from seeing that the military presence is doing little to alleviate the effects of the disaster, and that resources that could be used to provide Haitians with food and shelter are being misallocated to public safety.

For the mainstream media, however, headlines such as "Haitians removing, rubble with bare hands" or "There is nowhere for residents of tent cities to use, bathrooms" or "The military are great at setting up camps in Haiti—their own, that is" are not as likely to pull in the advertising dollars.

We sociologists should advise our students and our communities on the ways profit-oriented mass media corporations distort reality, and to direct people to alternative news outlets for a more balanced understanding of the world. The focus of disaster reports should instead be on the need and the cooperation that occurs.

REFERENCES

Collins, Patricia Hill. 2004. *Black Sexual Politics*. Routledge: New York.

Tierney, Kathleen, Christine Bevc, and Erica Kuligowski. 2006. "Metaphors Matter: Disaster Myths, Media Frames, and Their Consequences in Hurricane Katrina" *The ANNALS of the Americans Academy of Political and Social Science*. 604 (March): 57–81.

Barnshaw, John. 2005. "The Continuing Significance of Race and Class among Houston Hurricane Katrina Evacuees" *Natural Hazards Observer* 2: 11–12.

KEY CONCEPTS

controlling images   field research   social myths
stereotype

DISCUSSION QUESTIONS

1. Compare the media depictions of the Haiti earthquake to those that followed Hurricane Katrina. What common patterns do you see, and how might they be challenged by sociological research?

2. In what ways do portrayals of race and of gender frame media accounts of disasters? Are there stereotypes that are routinely invoked?

Applying Sociological Knowledge:
An Exercise for Students

One of the points in this section is how a sociological perspective differs from an individualistic or even psychological perspective. Pick one of the following topics: teen pregnancy, unemployment, child abuse. Compare and contrast what factors might be important to consider when explaining this social issue using a sociological perspective versus a psychological perspective.

PART II

✳

Sociological Research

4

Between Good and Ghetto: African American Girls and Inner-City Violence

NIKKI JONES

Nikki Jones uses the research method of participant observation to study how young African American girls frame their identities while growing up in a context of urban violence. Her research finds that these young women have to create an identity within a context that defines appropriate "feminine" behavior at the same time that they experience threats of violence.

Inner-city girls who live in distressed urban neighborhoods face a gendered dilemma: they must learn how to effectively manage potential threats of interpersonal violence—in most cases this means that they must work the code as boys and men do—at the risk of violating mainstream and local expectations regarding appropriate feminine behavior. This is a uniquely difficult dilemma for girls, since the gendered expectations surrounding girls' and women's use or control of violence are especially constraining. Conventional wisdom suggests that girls and women, whether prompted by nature, socialization, or a combination of the two, generally avoid physically aggressive or violent behavior: girls are expected to use relational aggression and fight with words and tears, not fists or knives. Inner-city girls, like most American girls, feel pressure to be "good," "decent," and "respectable." Yet, like some inner-city boys, they may also feel pressure to "go for bad" (Katz 1988) or to establish a "tough front" (Anderson 1999; Dance 2002) in order to deter potential challengers on the street or in the school setting. They too may believe that "sometimes you do got to fight"—and sometimes they do. In doing so, these girls, and especially those girls who become deeply invested in crafting a public persona as a tough or violent girl, risk evaluation by peers, adults, and outsiders as "street" or "ghetto."

Among urban and suburban adolescents, "ghetto" is a popular slang term that is commonly used to categorize a person or behavior as ignorant, stupid, or otherwise morally deficient. Inner-city residents use the term to describe the same kinds of actions and attitudes Elijah Anderson termed "street orientation." Analytically, the pairs "ghetto" and "good," or "street" and "decent," are used to

SOURCE: Nikki Jones. 2010. Between Good and Ghetto: African American Girls and Inner-City Violence. New Brunswick, NJ: Rutgers University Press.

represent "two poles of value orientation, two contrasting conceptual categories" that structure the moral order of inner-city life. In inner-city neighborhoods, the decent/street or good/ghetto distinctions are powerful. Community members use these distinctions as a basis for understanding, interpreting, and predicting their own and others' actions, attitudes, and behaviors, especially when it comes to interpersonal violence (Anderson 1999, 35). There is also a gendered dimension to these evaluative categories: good or decent girls are "young ladies" while "ghetto chicks" are adolescent girls whose "behaviors, dress, communication, and interaction styles" contrast with mainstream and Black middle class expectations of appropriate and respectable femininity (Thompson and Keith 2004, 58).

The branding of adolescent girls as ghetto is self-perpetuating, alienating the institutional forces that protect good girls and forcing adolescent girls who work the code of the street to become increasingly independent. Girls who are evaluated by adults or peers as ghetto, as opposed to those evaluated as good, ultimately may have the code as their only protection in too often violent inner-city world in which they live. Their efforts to protect themselves put them at risk of losing access to formal institutional settings like schools or the church, where girls who mirror normative gender expectations—girls who are perceived by others as good—can take some refuge. Yet, even for those good girls, this institutional protection is inadequate—they are aware that they may become targets in school or on the street and they too feel pressure to develop strategies that will help them successfully navigate their neighborhoods. Thus inner-city girls find themselves caught in what amounts to a perpetual dilemma, forced by violent circumstances to choose between two options, neither of which offers the level of security that is generally taken for granted in areas outside of urban poverty.

GAINING ACCESS: GIRLS IN THE VRP

I began systematically exploring the role of violence in the lives of inner-city girls after I was invited to work as an ethnographer for a city hospital–based violence reduction project (VRP). Each of the adolescent girls I discuss in this book had voluntarily enrolled in the project, which targeted Philadelphia youth aged twelve to twenty-four. All program participants were treated in the emergency department following an intentional violent incident, and all had been identified by VRP staff as either moderate or high risk for injury from similar incidents in the future. As a consequence of patterns of racial segregation within the city, almost all VRP enrollees were African American. Once enrolled in the VRP, a random selection of youth was assigned to receive intervention from a team of counselors who, over the course of several months, visited the young people in their homes, offered referrals, and provided mentoring with the aim of reducing the risk of subsequent violent injury.

A conversation with Tracey, a VRP project counselor, piqued my curiosity about girls' experiences with violence early on. Tracey, who is African American and was in her early twenties at the time, graduated from the same public city high school that some of the teenagers she now counseled attended. She lived in one of the neighborhoods included in the VRP's target area and could walk to some home visits. During our conversation, which took place in one of the hospital's conference rooms, I asked Tracey whether there were girls in the project. She said that there were. In fact, at that time, she said, her entire caseload was made up of girls. Most of the girls she counseled entered the emergency room with cuts or bruises from fights at school. I asked Tracey what the girls she worked with were fighting about.

"About being disrespected—that's about it," she replied.

"Being disrespected?"

"Yeah."

"So how's that look? What does that mean?" I asked.

"They're always saying, like, 'Nobody talks to me like that' and all. And I'm like, 'Yeah, but would you rather die over something somebody said?'"

"Do they see death as a real risk?" I asked her.

"No, no. They just see getting beat up and getting laughed at, that's all. And I try to tell them that life is too short to just do stupid stuff. You can't argue over dumb stuff. I don't expect you to go to school and not fight anymore because that would just be too unreal. I was like, 'But time will tell.' I don't know. I don't know. I don't know. Just crazy. I'm like, 'Okay, ya'll were fighting because she said your sneakers were ugly—okay… and [laughs] where does the argument start at?"

"Do they answer you? Do they tell you where the argument starts?" I asked.

"Yeah, they were, like, 'She said my sneakers were ugly, and I said this, and then she said this, and next thing you know this girl said this and we just all started fighting.'"

Tracey's claim that young women were fighting "about being disrespected— that's about it" foreshadowed the significant role that public displays of disrespect play in girls' accounts of how fights begin. Such an understanding of girls' fights challenges the popular assumption that girls fight *only* over the attention of boys. Tracey's admission, "I don't expect you to go to school and not fight anymore because that would just be *too unreal,*" also indicated a deep familiarity with the normative order of aggression in this setting. My conversation with Tracey encouraged me to focus my attention on uncovering the strategies that girls used to navigate inner-city settings where threats of interpersonal violence are encountered regularly, and the consequences of these strategies for girls in their everyday lives.

Looking back, my fieldwork, which included participant observation, direct observation, and formal and informal interviews with VRP participants, unfolded in three stages, over the period 2001 to 2003. Each stage logically built upon data collection and analysis undertaken during the previous stage(s). During the first phase, which lasted nearly a year and a half, I accompanied Tracey and other VRP counselors as they made home visits to meet individually with program participants. I also attended VRP events and interviewed members of the program's counseling staff—most of whom grew up in the city and were personally familiar with many of the neighborhoods in which the young people they counseled lived. In addition to observing these meetings and events, I also conducted formal and informal interviews with adolescent boys and girls enrolled in the VRP. In the second phase of the study, which lasted about a year, I visited program participants on my own conducted interviews with twenty-four adolescents, including fifteen African American girls and nine boys. I had met some of these participants during fieldwork conducted over the previous year, while I was meeting others for just the first time.

The third and final phase of the study lasted about a year and included observations and in-depth, open-ended interviews with three young women who were enrolled in the VRP: Terrie, a self-described "fighter," Danielle, a self-described "punk" (a non-"fighter"), and Amber, a young mother involved with a violent partner. I met each of them for the first time during the second phase of the study and then met with each of them several times throughout the final year. I visited Terrie and Danielle in their homes; my contact with Amber took me into a variety of settings, from a group home, to Planned Parenthood, to the city's Criminal Justice Center. I used the time I spent with Terrie, Danielle, and Amber to explore in greater depth some of the key concepts that had emerged during the previous phases of my data collection and analyses. These focused, in-depth interviews and observations helped me clarify the particular strategies adolescent girls used to negotiate conflict and violence.

Over time, I became an "observant participant" of interactions in the spaces and places that were significant in the lives of the young people I met (Anderson 2001). These spaces included trolley cars and buses, a neighborhood high school, the city's family and criminal courts, and various correctional facilities in the area, among others. I also intentionally engaged in extended conversations with grandmothers, mothers, sisters, brothers, cousins, and friends of the young people I visited and interviewed. I often recorded my observations and interactions in these settings in my field notes and used them to complement, supplement, test, and, at times, verify the information. I had collected during interviews or observations with VRP participants. This approach allowed me, during the first phase of the project, to develop a sound understanding of the physical, spatial, and symbolic context in which young people encountered various threats of violence; during the second and third phases, it helped me critically examine the strategies young people used to negotiate conflict and violence in their everyday lives and the various consequences of those strategies.

… I consider what the experiences of the girls discussed [here] reveal about contemporary conceptions of gender, strength, and survival among women and girls in poor, Black communities. Historically, the material circumstances of poor women's lives in general, and those of poor Black women in particular; have required a commitment to raising girls to become strong women. Whether they understand themselves as good girls or chiefly as fighters, the adolescent girls I came to know embraced locally held beliefs about the value of female strength. This positive embrace and unapologetic expression of female strength, which contrasts with traditional White, middle-class conceptions of femininity, and the gendered expectations embedded in Black respectability, was considered necessary for Black women's survival and for the survival of the Black community. The adolescent girls showed themselves to be no less concerned with survival than were strong Black women and girls in earlier periods. However, in today's inner city, where the culture of the code organizes much of social life, what a girl has got to do to survive has taken on new meanings.

REFERENCES

Anderson, Elijah. 2001. "Urban Ethnography". In *International encyclopedia of the social and behavioral sciences,* ed. Neil J. Smelser and Paul B. Baltes, Oxford: Pergamon Press.

_____1999. *Code of the Street: Decency, violence, and the moral life of the inner city.* New York: W.W. Norton.

Dance, Lory J. 2002. *Tough fronts: The impact of street culture on schooling.* New York: Routledge Press.

Katz, Jack. 1988. *Seductions of crime: Moral and sensual attractions in doing evil.* New York: Basic Books.

Thompson, Maxine, and Verna Keith. 2004. "Copper Brown and Blue Black: Colorism and Self Evaluation" pp. 45–64. In *Skin/Deep: How Race and Complexion Matter in the "Color Blind" Era*, edited by Cedric Herring, Verna M. Keith, and Hayward Derick Horton. Urbana: University of Illinois Press. Herring, Keith, and Horton.

KEY CONCEPTS

ethnography   open-ended interviews   participant observation

DISCUSSION QUESTIONS

1. What does Jones mean by saying that two "poles of value orientation"—good and ghetto—structure the moral order of inner-city life?
2. What specific research challenges might you face were you to do a participant observation study of adolescent girls in the neighborhood where you grew up?

5

Promoting Bad Statistics

JOEL BEST

In this article, Joel Best points out how numbers publicly used to describe social problems can be misleading. He shows how advocacy about a given problem can

SOURCE: Joel Best, "Promoting Bad Statistics." *Society*, March/April 2001: 11–15.

distort accurate, empirical observations. In addition, he emphasizes that statistical information is produced in a social context.

In contemporary society, social problems must compete for attention. To the degree that one problem gains media coverage, moves to the top of politicians' agendas, or becomes the subject of public concern, others will be neglected. Advocates find it necessary to make compelling cases for the importance of particular social problems. They choose persuasive wording and point to disturbing examples, and they usually bolster their case with dramatic statistics.

Statistics have a fetish-like power in contemporary discussions about social problems. We pride ourselves on rational policy making, and expertise and evidence guide our rationality. Statistics become central to the process: numbers evoke science and precision; they seem to be nodules of truth, facts that distill the simple essence of apparently complex social processes. In a culture that treats facts and opinions as dichotomous terms, numbers signify truth—what we call "hard facts." In virtually every debate about social problems, statistics trump "mere opinion."

Yet social problems statistics often involve dubious data. While critics occasionally call some number into question, it generally is not necessary for a statistic to be accurate—or even plausible—in order to achieve widespread acceptance. Advocates seeking to promote social problems often worry more about the processes by which policy makers, the press, and the public come to focus on particular problems, than about the quality of their figures. I seek here to identify some principles that govern this process. They are, if you will, guidelines for creating and disseminating dubious social problems statistics.

Although we talk about facts as though they exist independently of people, patiently awaiting discovery, someone has to produce—or construct—all that we know. Every social statistic reflects the choices that go into producing it. The key choices involve definition and methodology: Whenever we count something, we must first define what it is we hope to count, and then choose the methods by which we will go about counting. In general, the press regards statistics as facts, little bits of truth. The human choices behind every number are forgotten; the very presentation of a number gives each claim credibility. In this sense, statistics are like fetishes.

ANY NUMBER IS BETTER THAN NO NUMBER

By this generous standard, a number need not bear close inspection, or even be remotely plausible. To choose an example first brought to light by Christina Hoff Sommers, a number of recent books, both popular and scholarly, have repeated the garbled claim that anorexia kills 150,000 women annually. (The figure seems to have originated from an estimate for the total number of women who are anorexic; only about 70 die each year from the disease.) It should have been obvious that something was wrong with this figure. Anorexia typically affects *young* women. Each year, roughly 8,500 females aged 15–24 die from all causes; another 47,000 women aged 25–44 also die. What are the chances, then, that there could be 150,000 deaths from anorexia each year? But, of course, most of us have no idea how many young women die each year—("It must be a lot...."). When we hear that anorexia kills 150,000 young women per year, we assume that whoever cites the number must know that it is true. It is, after all, a number and therefore presumably factual.

Oftentimes, social problems statistics exist in splendid isolation. When there is only one number, that number has the weight of authority. It is accepted and repeated. People treat the statistic as authoritative because it is a statistic. Often, these lone numbers come from activists seeking to draw attention to neglected social phenomena. One symptom of societal neglect is that no one has bothered to do much research or compile careful records; there often are no official statistics or other sources for more accurate numbers. When reporters cover the story, they want to report facts. When activists have the only available figures, their numbers look like facts, so, in the absence of other numbers, the media simply report the activists' statistics.

Once a number appears in one news report, that story becomes a potential source for everyone seeking information about the social problem; officials, experts, activists, and other reporters routinely repeat figures that appear in press reports.

NUMBERS TAKE ON LIVES OF THEIR OWN

David Luckenbill has referred to this as "number laundering." A statistic's origin—perhaps simply as someone's best guess—is soon forgotten, and through repetition, the figure comes to be treated as a straightforward fact—accurate and authoritative. The trail becomes muddy, and people lose track of the estimate's original source, but they become confident that the number must be correct because it appears everywhere.

It barely matters if critics challenge a number, and expose it as erroneous. Once a number is in circulation, it can live on, regardless of how thoroughly it may have been discredited. Today's improved methods of information retrieval—electronic indexes, full-text databases, and the Internet—make it easier than ever to locate statistics. Anyone who locates a number can, and quite possibly will, repeat it. That annual toll of 150,000 anorexia deaths has been thoroughly debunked, yet the figure continues to appear in occasional newspaper stories. Electronic storage has given us astonishing, unprecedented access to information, but many people have terrible difficulty sorting through what's available and distinguishing good information from bad. Standards for comparing and evaluating claims seem to be wanting. This is particularly true for statistics that are, after all, numbers and therefore factual, requiring no critical evaluation. Why not believe and repeat a number that everyone else uses? Still, some numbers do have advantages.

BIG NUMBERS ARE BETTER THAN
LITTLE NUMBERS

Remember: social problems claims must compete for attention; there are many causes and a limited amount of space on the front page of the *New York Times.* Advocates must find ways to make their claims compelling; they favor melodrama—terrible villains, sympathetic, vulnerable victims, and big numbers. Big numbers suggest that there is a big problem, and big problems demand attention, concern, action. They must not be ignored.

Advocates seeking to attract attention to a social problem soon find themselves pressed for numbers. Press and policy makers demand facts ("You say it's a problem? Well, how big a problem is it?"). Activists believe in the problem's seriousness, and they often spend much of their time talking to others who share that belief. They know that the problem is much more serious, much more common than generally recognized ("The cases we know about are only the tip of the iceberg."). When asked for figures, they thus offer their best estimates, educated guesses, guesstimates, ballpark figures, or stabs in the dark. Mitch Snyder, the most visible spokesperson for the homeless in the early 1980s, explained on ABC's "Nightline" how activists arrived at the figure of three million homeless: "Everybody demanded it. Everybody said we want a number.... We got on the phone, we made a lot of calls, we talked to a lot of people, and we said, 'Okay, here are some numbers.' They have no meaning, no value." Because activists sincerely believe that the new problem is big and important, and because they suspect that there is a very large dark figure of unreported or unrecorded cases, activists' estimates tend to be high, and to err on the side of exaggeration.

This helps explain the tendency to estimate the scope of social problems in large, suspiciously round figures. There are, we are told, one million victims of elder abuse each year, two million missing children, three million homeless, 60 million functionally illiterate Americans; child pornography may be, depending on your source, a $1 billion or $46 billion industry, and so on. Often, these estimates are the only available numbers.

The mathematician John Allen Paulos argues that innumeracy—the mathematical counterpart to illiteracy—is widespread and consequential. He suggests that innumeracy particularly shapes the way we deal with large numbers. Most of us understand hundreds, even thousands, but soon the orders of magnitude blur into a single category: "It's a lot." Even the most implausible figures can gain widespread acceptance. When missing-children advocates charged that nearly two million children are missing each year, anyone might have done the basic math: there are about 60 million children under 18; if two million are missing, that would be one in 30; that is, every year, the equivalent of one child in every American schoolroom would be missing. A 900-student school would have 30 children missing from its student body each year. To be sure, the press debunked this statistic in 1985, but only four years after missing children became a highly publicized issue and the two-million estimate gained wide circulation.

And, of course, having been discredited, the number survives and can still be encountered on occasion.

It is remarkable how often contemporary discussions of social problems make no effort to define what is at issue. Often, we're given a dramatic, compelling example, perhaps a tortured, murdered child, then told that this terrible case is an example of a social problem—in this case, child abuse—and finally given a statistic: "There are more than three million reports of child abuse each year." The example, coupled with the problem's name, seems sufficient to make the definition self-evident. However, definitions cannot always be avoided.

DEFINITIONS: BETTER BROAD THAN NARROW

Because broad definitions encompass more kinds of cases, they justify bigger numbers, and we have already noted the advantages of big numbers. No definition is perfect; there are two principal ways definitions of social problems can be flawed. On the one hand, a definition might be too broad and encompass more than it ought to include. That is, broad definitions tend to identify what methodologists call false positives; they include some cases that arguably ought not to be included as part of the problem. On the other hand, a definition that is too narrow may exclude false negatives, cases that perhaps ought to be included as part of the problem.

In general, activists trying to promote a new social problem view false negatives as more troubling than false positives. Activists often feel frustrated trying to get people concerned about some social condition that has been ignored. The general failure to recognize and acknowledge that something is wrong is part of what the activists want to correct; therefore, they may be especially careful not to make things worse by defining the problem too narrowly. A definition that is too narrow fails to recognize a problem's full extent; in doing so, it helps perpetuate the history of neglecting the problem. Some activists favor definitions broad enough to encompass every case that ought to be included; that is, they promote broad definitions in hopes of eliminating all false negatives.

However, broad definitions may invite criticism. They include cases that not everyone considers instances of social problems; that is, while they minimize false negatives, they do so at the cost of maximizing cases that critics may see as false positives. The rejoinder to this critique returns us to the idea of neglect and the harm it causes. Perhaps, advocates acknowledge, their definitions may seem to be too broad, to encompass cases that seem too trivial to be counted as instances of the social problem. But how can we make that judgment? Here advocates are fond of pointing to terrible examples, to the victim whose one, brief, comparatively mild experience had terrible personal consequences; to the child who, having been exposed to a flasher, suffers a lifetime of devastating psychological consequences. Perhaps, advocates say, other victims with similar experiences suffer less or at least seem to suffer less. But is it fair to define a problem too narrowly to include everyone who suffers? Shouldn't our statistics measure the problem's full extent? While social problems statistics often go unchallenged, critics occasionally suggest that some number is implausibly large, or that a definition is too broad.

DEFENDING NUMBERS BY ATTACKING CRITICS

When activists have generated a statistic as part of a campaign to arouse concern about some social problem, there is a tendency for them to conflate the number with the cause. Therefore, anyone who questions a statistic can be suspected of being unsympathetic to the larger claims, indifferent to the victims' suffering, and so on. *Ad hominem* attack on the motives of individuals challenging numbers is a standard response to statistical confrontations. These attacks allow advocates to refuse to budge; making *ad hominem* arguments lets them imply that their opponents don't want to acknowledge the truth, that their statistics are derived from ideology, rather than methodology. If the advocates' campaign has been reasonably successful, they can argue that there is now widespread appreciation that this is a big, serious problem; after all, the advocates' number has been widely accepted and repeated, surely it must be correct. A fallback stance—useful in those rare cases where public scrutiny leaves one's own numbers completely discredited—is to treat the challenge as meaningless nitpicking. Perhaps our statistics were flawed, the advocates acknowledge, but the precise number hardly makes a difference ("After all, even one victim is too many.").

Similarly, criticizing definitions for being too broad can provoke angry reactions. For advocates, such criticisms seem to deny victim's suffering, minimize the extent of the problem, and by extension endorse the status quo. If broader definitions reflect progress, more sensitive appreciation of the true scope of social problems, then calls for narrowing definitions are retrograde, insensitive refusals to confront society's flaws.

Of course, definitions must be operationalized if they are to lead to statistics. It is necessary to specify how the problem will be measured and the statistic produced. If there is to be a survey, who will be sampled? And how will the questions be worded? In what order will they be asked? How will the responses be coded? Most of what we call social-scientific methodology requires choosing how to measure social phenomena. Every statistic depends upon these choices. Just as advocates' preference for large numbers leads them to favor broad definitions, the desirability of broad definitions shapes measurement choices.

MEASURES: BETTER INCLUSIVE THAN EXCLUSIVE

Most contemporary advocates have enough sociological sophistication to allude to the dark figure—that share of a social problem that goes unreported and unrecorded. Official statistics, they warn, inevitably underestimate the size of social problems. This undercounting helps justify advocates' generous estimates (recall all those references to "the tip of the iceberg"). Awareness of the dark figure also justifies measurement decisions that maximize researchers' prospects for discovering and counting as many cases as possible.

Consider the first federally sponsored National Incidence Studies of Missing, Abducted, Runaway, and Thrownaway Children (NISMART). This was an attempt to produce an accurate estimate for the numbers of missing children. To estimate family abductions (in which a family member kidnaps a child) researchers conducted a telephone survey of households. The researchers made a variety of inclusive measurement decisions: an abduction could involve moving a child as little as 20 feet; it could involve the child's complete cooperation; there was no minimum time that the abduction had to last; those involved may not have considered what happened an abduction; and there was no need that the child's whereabouts be unknown (in most family abductions identified by NISMART, the child was not with someone who had legal custody, but everyone knew where the child was). Using these methods of measurement, a non-custodial parent who took a child for an unauthorized visit, or who extended an authorized visit for an extra night, was counted as having committed a "family abduction." If the same parent tried to conceal the taking or to prevent the custodial parent's contact with the child, the abduction was classified in the most serious ("policy-focal") category. The NISMART researchers concluded that there were 163,200 of these more serious family abductions each year, although evidence from states -with the most thorough missing-children reporting systems suggests that only about 9,000 cases per year come to police attention. In other -words, the researchers' inclusive measurement choices led to a remarkably high estimate. Media coverage of the family-abduction problem coupled this high figure with horrible examples—cases of abductions lasting years, involving long-term sexual abuse, ending in homicide, and so on. Although most of the episodes identified by NISMART's methods were relatively minor, the press implied that very serious cases were very common ("It's a big number!").

There is nothing atypical about the NISMART example. Advocacy research has become an important source of social problems statistics. Advocates hope research will produce large numbers, and they tend to believe that broad definitions are justified. They deliberately adopt inclusive research measurements that promise to minimize false negatives and generate large numbers. These measurement decisions almost always occur outside public scrutiny and only rarely attract attention. When the media report numbers, percentages, and rates, they almost never explain the definitions and measurements used to produce those statistics.

While many statistics seem to stand alone, occasions do arise when there are competing numbers or contradictory statistical answers to what seems to be the same question. In general, the media tend to treat such competing numbers with a sort of even-handedness.

COMPETING NUMBERS ARE EQUALLY GOOD

Because the media tend to treat numbers as factual, and to ignore definitions and measurement choices, inconsistent numbers pose a problem. Clearly, both numbers cannot be correct. Where a methodologist might try to ask how different advocates arrived at different numbers (in hopes of showing that one figure is more accurate than another, or at least of understanding how the different numbers might be products of different methods), the press is more likely to account for any difference in terms of the competitors' conflicting ideologies or agendas.

Consider the case of the estimates for the crowd size at the 1995 Million Man March. The event's very name set a standard for its success: as the date for the March approached, its organizers insisted that it would attract a million people, while their critics predicted that the crowd would never reach that size. On the day of the March, the organizers announced success: there were, they said, 1.5 to 2 million people present. Alas, the National Park Service Park Police, charged by Congress with estimating the size of demonstrations on the Capitol Mall, calculated that the March drew only 400,000 people (still more than any previous civil rights demonstration). The Park Police knew the Mall's dimensions, took aerial photos, and multiplied the area covered by the crowd by a multiplier based on typical crowd densities. The organizers, like the organizers of many previous demonstrations on the Mall, insisted that the Park Police estimate was far too low. Enter a team of aerial photo analysts from Boston University who eventually calculated that the crowd numbered 837,000 plus or minus 25 percent (i.e., they suggested there might have been a million people in the crowd).

The press covered these competing estimates in standard "he said—she said" style. Few reporters bothered to ask why the two estimates were different. The answer was simple: the BU researchers used a different multiplier. Where the Park Police estimated that there was one demonstrator per 3.6 square feet (actually a fairly densely-packed crowd), the BU researchers calculated that there was a person for every 1.8 square feet (the equivalent of being packed in a crowded elevator). But rather than trying to compare or evaluate the processes by which people arrived at the different estimates, most press reports treated the numbers as equally valid, and implied that the explanation for the difference lay in the motives of those making the estimates.

The March organizers (who wanted to argue that the demonstration had been successful) produced a high number; the Park Police (who, the March organizers insisted, were biased against the March) produced a low one; and the BU scientists (presumably impartial and authoritative) found something in between. The BU estimate quickly found favor in the media: it let the organizers save face (because the BU team conceded the crowd might have reached one million); it seemed to split the difference between the high and low estimates; and it apparently came from experts. There was no effort to judge the competing methods and assumptions behind the different numbers, for example, to ask whether it was likely that hundreds of thousands of men stood packed as close together as the BU researchers imagined for the hours the demonstration lasted.

This example, like those discussed earlier, reveals that public discussions of social statistics are remarkably unsophisticated. Social scientists advance their careers by using arcane inferential statistics to interpret data. The standard introductory undergraduate statistics textbook tends to zip through descriptive statistics on the way to inferential statistics. But it is descriptive statistics—simple counts, averages, percentages, rates, and the like—that play the key role in public discussions over social problems and social policy. And the level of those discussions is not terribly advanced. There is too little critical thinking about social statistics. People manufacture, and other people repeat, dubious figures. While this can involve deliberate attempts to deceive and manipulate, this need not be the case. Often, the people who create the numbers—who, as it were, make all those millions—believe in them. Neither the advocates who create statistics, nor the reporters who repeat them, nor the larger public questions the figures.

What Paulos calls innumeracy is partly to blame—many people aren't comfortable with basic ideas of numbers and calculations. But there is an even more fundamental issue: many of us do not appreciate that every number is a social construction, produced by particular people using particular methods. The naïve, but widespread, tendency is to treat statistics as fetishes, that is, as almost magical nuggets of fact, rather than as someone's efforts to summarize, to simplify complexity. If we accept the statistic as a fetish, then several of the guidelines I have outlined make perfect sense. Any number is better than no number, because the number represents truth. Numbers take on lives of their own because they are true, and their truth justifies their survival. The best way to defend a number is to attack its critics' motives, because anyone who questions a presumably true number must have dubious reasons for doing so. And, when we are confronted with competing numbers, those numbers are equally good, because, after all, they are somehow equivalent bits of truth. At the same time, the guidelines offer those who must produce numbers justifications for favoring big numbers, broad definitions, and inclusive methods. Again, this need not be cynical. Often, advocates are confident that they know the truth, and they approach collecting statistics as a straightforward effort to generate the numbers needed to document what they, after all, know to be true.

Any effort to improve the quality of public discussion of social statistics needs to begin with the understanding that numbers are socially constructed. Statistics are not nuggets of objective fact that we discover; rather, they are people's creations. Every statistic reflects people's decisions to count, their choices of what to count and how to go about counting it, and so on. These choices inevitably shape the resulting numbers.

Public discussions of social statistics need to chart a middle path between naivete (the assumption that numbers are simply true) and cynicism (the suspicion that figures are outright lies told by people with bad motives). This middle path needs to be critical. It needs to recognize that every statistic has to be created, to acknowledge that every statistic is imperfect, yet to appreciate that statistics still offer an essential way of summarizing complex information. Social scientists have a responsibility to promote this critical stance in the public, within the press, and among advocates.

KEY CONCEPTS

anorexia nervosa   debunking   methodology

DISCUSSION QUESTIONS

1. What does Best mean by claiming that numbers are social constructions?
2. Find an example in the media in which someone is using numbers to promote concern about a particular social problem. Based on Best's article, what questions would you need to ask to find out if the numbers are accurate?

6

Romantic Relationships from Adolescence to Young Adulthood

ANN MEIER AND GINA ALLEN

This article is a good example of sociological research that is based on a large, nationally representative sample where the research design tests specific hypotheses. Because of the complexity of such a study and the introductory character of this book, the editors have eliminated much of the rigorous detail of this study, but present the background questions and the specific hypotheses tested with a brief summary of the findings. Paired with the Nikki Jones article, you can see the diverse ways that sociologists do research. In both cases the research design flows from the questions asked. You can see here how Meier and Allen derived their research questions from prior studies and theoretical issues.

Our study makes three important contributions. First, we use a large, nationally representative, and longitudinal sample to confirm findings on adolescent relationship progression from various past studies that were limited by small, nonrepresentative, cross-sectional, and age- or time-limited samples. Specifically, we find that nearly two-thirds of adolescents have some romantic relationship experience and that most follow normative patterns of relationship development as they age and gain relationship experience. Second, we find differences in adolescent relationship experiences by race, which may help us to better understand the divergent young-adult relationship experiences of blacks and whites. Finally, we show that adolescent relationships hold developmental currency for the more serious relationship commitments of young adulthood. While theories suggest

SOURCE: *The Sociological Quarterly*, 2009. Vol. 50: 308–335.

this should be the case, there is scant empirical evidence linking adolescent to young-adult relationships. With the changes of recent decades in prolonging the transition to adulthood and delayed entry into marriage, young adults now have more time and opportunity to gain valuable experience in romantic relationships before forming adult unions. Rigorous empirical testing of the foundational theories of romantic relationship development is imperative for the advancement of the field. Moreover, revealing the importance of early relationships can help inform policy and program efforts to strengthen the unions of young adults who are married or bearing children together.

EMPIRICAL LITERATURE

Patterns of Involvement

A few recent studies shed light on the normative patterning of adolescent romantic relationships. A study of fifth to eighth graders (approximately 10- to 14-year-olds) in Canada finds that while there is a substantial degree of stability in a phase over the course of one year, when they change, young adolescents generally move through romantic involvements sequentially and progressively. That is, they are more likely to *progress* than *regress* through phases of romantic relationships, and they do so mostly sequentially rather than by skipping a phase (Connolly et al. 2004). A study of slightly older adolescents (ages 15 and 16 years) at two points in time, one year apart, classified respondents into four relationship patterns defined at both points in time: (1) no dating relationships; (2) a single, casual dating relationship; (3) multiple, casual relationships; and (4) steady dating relationships (Davies and Windle 2000). The cross-classification of these four patterns of dating at times 1 and 2 reveals several patterns consistent with the relationship progression idea. Common transitions between the two time points are from (1) no dating to a single, casual relationship; (2) a single casual relationship to multiple casual relationships; (3) a single casual relationship to a steady dating relationship; and (4) multiple casual relationships to a steady dating relationship. Other research suggests that older adolescents (age 17 years) maintain relationships of longer durations or what might be classified as steady relationships (Seiffge-Krenke 2003). Thus, it appears most adolescents follow the orderly patterns predicted by theory—forward progress from fewer short and casual relationships to more relationships overall, often to a single steady relationship. Therefore, we offer the following hypothesis regarding relationship progression:

> Hypothesis 1: Most adolescents progress from one short relationship in early adolescence to multiple relationships in middle adolescence and a single, steady relationship in late adolescence.

Relationship Qualities

As patterns of relationship involvement change with age, so too do relationship qualities change from early to late adolescence and as youths gain more romantic experience. For example, some research indicates that younger adolescents do not differentiate in their ratings of the importance of support from parents, peers, or partners (Connolly and Johnson 1996). However, by late adolescence, as young people gain experience in more relationships and relationships of longer durations, partner support increases in ratings of importance in both an absolute sense and relative to peer and parental support (Seiffge-Krenke 2003), suggesting that longer relationships are characterized by more partner attachment (Miller and Hoicowitz 2004). Moreover, with age, partners report relationship behaviors that reflect higher levels of commitment, and emotional and sexual intensity (Carver et al. 2003). Thus, we hypothesize that,

> Hypothesis 2: As adolescents progress toward steady relationships, their relationships become more dyadic and more likely to include sexual and emotional intimacy than they were at lower stages of progression.

Sociodemographic Characteristics

While studies have shown race/ethnic, gender, and socioeconomic differences in adolescent sexual behavior (e.g., Upchurch et al. 1998; Giordano et al. 2005; Meier and Allen 2008), we still know relatively little or have mixed evidence about differences across these domains in adolescent romantic behavior. Connolly et al. (2004) find no gender differences in romantic behavior in their study of young adolescents, although several studies of older adolescents have shown that girls report more romantic involvement and more steady relationships than boys (Davies and Windle 2000; Carver et al. 2003). Thus, we hypothesize that,

> Hypothesis 3a: Girls are more likely than boys to be in a romantic relationship, especially a steady relationship.

Regarding race and ethnic differences in romantic relationships, Connolly et al. (2004) compared adolescents of European, Caribbean, and Asian descent and found that European and Caribbean adolescents follow the expected progression while Asian adolescents do not progress over the one-year period. Others have found lower rates of sexual activity among Asian-American teens (Upchurch et al. 1998). Together, these studies suggest that Asians are less romantically involved during adolescence. Compared with whites, black adolescents report longer but less emotionally intimate relationships (Carver et al. 2003; Giordano et al. 2005), and are less likely to make serious romantic commitments (Cohen et al. 2003), but are more likely to be sexually involved (Upchurch et al. 1998).

Some researchers suggest that it is low social class, not race, that inhibits family formation among blacks (Edin and Kefalas 2005). This claim is supported by findings that young adults not enrolled in postsecondary education are more likely than those in school to marry early, but they have lower marriage rates overall (Thornton, Axinn, and Teachman 1995). Yet, there is little research on class differences in adolescent relationships. One study using Add Health data found that adolescents in the highest family-income quintile were slightly more likely to report romantic involvement but found no consistent relationship between income and involvement over the full income distribution (Meier and Allen 2008).

With this past research in mind, we expect that Asian-American adolescents will be less likely than others to have romantic experience and that the relationships of black respondents will be longer and more sexually involved, but less emotionally intimate than those of their white counterparts. While empirical work is beginning to account for socioeconomic differences in adolescent romantic relationships (e.g., Meier and Allen 2008), the underdevelopment of the literature on class and adolescent romance precludes a clear hypothesis in the present analysis. Thus, we hypothesize that,

> Hypothesis 3b: Compared to whites, Asian-American adolescents are more likely to report no relationship experience and less likely to progress in involvement during adolescence.

> Hypothesis 3c: Compared to whites, blacks are more likely to have longer relationships that are characterized by more sexual intimacy but less emotional intimacy.

DATA, MEASURES, METHODS

Data

The Add Health includes interview data from respondents in grades 7 to 12 in 1995 who are followed up in a second interview approximately one year later in 1996 and in a third interview in 2001 to 2002. A sample of 80 high schools and 52 middle schools from the United States was selected with unequal probability of selection. Incorporating systematic sampling methods and implicit stratification into the Add Health study design ensured this sample is representative of U.S. schools with respect to region of the country, urbanicity, school size school type, and ethnicity (Udry 2003).

Questions on relationships were administered by Audio Computer-Assisted Self-Interview where respondents hear questions through headphones and see them on a computer screen. They enter responses into the computer without assistance from an interviewer. This method is used to get the most honest answers possible on potentially sensitive matters. All analyses are limited to heterosexual respondents.

Our findings offer support for our first hypothesis that early adolescents are more likely to have no relationships or to be initiating relationships for the first time, whereas older adolescents are more likely to progress to, or remain in, steady relationships. Many of the propositions in our third set of hypotheses regarding the sociodemographic correlates of relationship progression are also supported. Adolescent girls are less likely to have no relationships and more likely to have steady relationships across the one-year span (Hypothesis 3a). Asian adolescents are more likely to have no relationships (Hypothesis 3b). Compared with whites, black adolescents are more likely to have no relationships, but if they are romantically involved, then they are more likely to progress to steady relationships versus initiating or remaining in short-term relationships (Hypothesis 3c).

Perhaps their relatively quicker progression to steady relationships can partially explain their earlier age at first sex (Upchurch et al. 1998).

To summarize our findings regarding adolescent relationships, many adolescents remain in the same relationship state over the course of one year, but those who experience change are more likely to move forward in the idealized relationship progression. Girls and older adolescents are generally further along in the relationship progression and black and low-income adolescents are more likely to have no relationships. However, if they do have relationship experience, blacks and those from low-income families are more likely to move quickly to steady relationships. Being further along in the relationship progression is associated with more dyadic, sexual, and emotionally intimate relationships. Black and low-income adolescents report less dyadic mixing and emotional intimacy but more sexual intercourse with their partners.

DISCUSSION

Theories on romantic relationship development in adolescence posit a progression of involvement and a change in relationship quality to more emotional and sexual intensity, and more dyadic mixing with age, relationship duration, and experience in romantic relationships. In addition, theory suggests that adolescent romantic relationships should be an integral part of the social scaffolding on which young-adult romantic relationships rest. Furthermore, as the age at union formation increases in the United States and elsewhere, adolescent and young-adult relationships become ever more important, as they fill a longer span of time during which many people are not formally partnered—10 to 12 years or about half of a young adult's life span. In this study, we set out to integrate new theories and empirical studies on the development of romantic experiences during the transition to adulthood. To test these theories, we empirically assessed the types, qualities, and patterns of romantic relationships with a large, longitudinal, and representative data set that follows adolescents into early adulthood.

REFERENCES

Carver, Karen, Kara Joyner, and J. Richard Udry. 2003. "National Estimates of Adolescent Romantic Relationships." pp. 23–56 in *Adolescent Romantic Relationships and Sexual Behavior: Theory, Research, and Practical Implications*, edited by P. Florsheim. Mahwah, NJ: Lawrence Erlbaum.

Cohen, Patricia, Stephanie Kasen, Henian Chen, Claudia Hartmark, and Kathy Gordon. 2003. "Variations in Patterns of Developmental Transitions in the Emerging Adulthood Period." *Developmental Psychology* 39:657–69.

Connolly, Jennifer, Wendy Craig, Adele Goldberg, and Debra Pepler. 2004. "Mixed-Gender Groups, Dating, and Romantic Relationships in Early Adolescence." *Journal of Research on Adolescence* 14:185–207.

Connolly, Jennifer A. and Anne M. Johnson. 1996. "Adolescents Romantic Relationships and the Structure and Quality of Their Close Interpersonal Ties." *Personal Relationships* 3:185–95.

Davies, Patrick T. and Michael Windle. 2000. "Middle Adolescents' Dating Pathways and Psychosocial Adjustment." *Merrill-Palmer Quarterly* 46:90–118.

Edin, Kathryn and Maria Kefalas. 2005. *Promises I Can Keep: Why Poor Women Put Motherhood before Marriage.* Berkeley: University of California Press.

Giordano, Peggy C., Wendy D. Manning, and Monica A. Longmore. 2005. "The Romantic Relationships of African-American and White Adolescents." *The Sociological Quarterly* 46:545–68.

Meier, Ann and Gina Allen. 2008. "Intimate Relationship Development during the Transition to Adulthood: Differences by Social Class." *New Directions for Child and Adolescent Development* 119:25–40.

Miller, Judi Beinstein and Tova Hoicowitz. 2004. "Attachment Contexts for Adolescent Friendship and Romance," *Journal of Adolescence* 27:191–206.

Seiffge-Krenke, Inge. 2003. "Testing Theories of Romantic Development from Adolescence to Young Adulthood: Evidence of Developmental Sequence." *International Journal of Behavioral Development* 27:519–31.

Thornton, Arland, William G. Axinn, and Jay D. Teachman. 1995. "The Influence of Educational Aspirations and Experiences on Entrances into Cohabitation and Marriage." *American Sociological Review* 60:762–74.

Upchurch, Dawn, Lene Levy-Storms, Lea A. Sucoff, and Carol S. Anshensel 1998. "Gender & Ethnic Differences in Timing of First Sexual Intercourse." *Family Planning Perspectives* 30:121–7.

KEY CONCEPTS

hypothesis	longitudinal study	representative sample
research design	sample	survey

DISCUSSION QUESTIONS

1. What three hypotheses do Meier and Allen test? Were their hypothesis supported by the evidence they found?

2. Why was a longitudinal research design important for this study?

Applying Sociological Knowledge:
An Exercise for Students

This section includes examples of different methods of sociological research, one qualitative (Jones), one more quantitative (Meier and Allen), and one, critical of the misuse of statistics (Best). Were you doing research on a topic of interest to you, what would your major question be? Would you need to do qualitative or quantitative research to answer your question? What pitfalls would you need to avoid so as not to misuse statistics and numbers in reporting your research?

✳

Culture

7

Gamers, Hackers, and Facebook— Computer Cultures, Virtual Community, and Postmodern Identity

ROSS HAENFLER

Ross Haenfler uses cybercultures to explore basic sociological concepts pertaining to cultures and subcultures. Among other things, he shows how cyberspace interactions shape different cultural forms.

If you are of typical college-age people, you might have difficulty recalling a time before video games, downloaded music, and DVDs, let alone a time before personal computers. From cell phones and personal digital assistants to video games and the Internet, computers play a role in nearly every aspect of our lives. New technologies have spawned new subcultures and given established subcultures a new arena in which to interact. Immersive fantasy games such as *World of Warcraft* allow players to become heroes in fanciful realms, and *Second Life* enables residents to own virtual land, run virtual businesses, and even attend virtual concerts performed by real-life musicians. Message boards and listservs connect music fans help social activists network, and bring together people of every possible interest, from bird watching to sports. Blogs enable amateur journalists a forum to write (or rant) about politics, religion, and pop culture, and chat rooms and multi-user domains, or MUDs, serve as virtual cafes where people can socialize or cruise for a date. Auction sites, such as Ebay, make buying and selling nearly anything a mere mouse click away. Whether it's Xbox Online, iTunes, Facebook, or MySpace, many of us spend an increasing amount of our lives online, forging meaningful communities and online identities.

SOURCE: Haenfler, Ross. 2009. *Goths, Gamers, and Grrrls: Deviance and Youth Subcultures.*
New York: Oxford University Press.

... VIRTUAL SUBCULTURES AND SCENES

Hundreds of millions of people around the world use the Net, many of them for the pleasure and emotional support of socializing online (Hornsby 2005). Civic organizations, activists, political junkies, and a myriad of other groups form subcultures on the Net. Here we'll discuss one of the original online subcultures, hackers, and one of the fastest growing, players of online video games.

... Massive Multiplayer Online Games

Since the first video game, *Spacewar!* in 1961, the video game industry has grown exponentially, taking in billions of dollars a year in the United States and rivaling the film industry in profitability. Whether you grew up with Atari, Nintendo, Playstation, or Xbox, video games have become ubiquitous in contemporary youth culture. Some players have even gone professional, securing sponsorships and competing in tournaments as part of a pro circuit. The latest surge in video game popularity has been Massively Multiplayer Online Role Playing Games [MMORPG] such as *Everquest, Star Wars Galaxies, World of Warcraft, Ultima Online,* and *Dungeons and Dragons Online. World of Warcraft* alone has over 11.5 million subscriptions worldwide, including players in the United States, Korea, New Zealand, China, Australia, United Kingdom, Singapore, France, Germany, and Spain. In each of these games, players create a virtual persona called an avatar, character, or "toon." Unlike more conventional video games in which players assume a role created by the game designers (for example, you play Lara Croft in *Tomb Raider,* the Master Chief in *Halo,* or Marcus Fenix in *Gears of War*), MMORPG enthusiasts create and customize their own characters, including abilities, skills, appearance, profession, possessions or weapons, and names. Most games charge a subscription fee of around $15, making the most popular games big business.

... Players and game master use their imaginations to create a "shared fantasy" in which almost anything can happen, rolling dice to determine if they successfully accomplish tasks such as attacking a monster with a sword or sneaking past a guard undetected (Fine 2002). While MMORPGs require no dice, and replace the game master with faceless game developers, the notion of a party of adventurers questing in a fantasy world populated by mythical creatures remains.

After creating a character, players enter a vast virtual world of spectacular geographies and fantastic opponents. They explore the terrain, meet non-player (computer controlled) characters (NPCs), and undertake perilous quests. While players can play "solo," most band together forming groups in which each member performs a specialized role.

... Despite the lack of face-to-face contact, norms, values, and a sort of social order emerge in every MMORPG. Some players "role play" their characters, creating an in-game personality and speaking and acting accordingly. They craft a virtual self that may or may not reflect their own presentation of self. Just as in the non-virtual world, players exist in a status hierarchy and in organizational structures.

... VIRTUAL COMMUNITY—FACEBOOK, MYSPACE, AND MORE

If I were to ask you to tell me about your community; you would most likely describe the physical space and people in your neighborhood, town, or city. We tend to think of community in narrow terms and almost always tie community to geography, a physical place. In the information age, however, we need a broader definition. A **community** is a social network of people who somehow interact and have something in common such as geographic place (e.g., a university community), common interests (e.g., the poker community), distinct identity (e.g., the Latino and gay communities), or shared values (e.g., the Baptist community). Though we typically think of it in terms of place, a community is not explicitly tied to one location but is instead another way of identifying people who claim an identity, such as "the lesbian community," "the African American community," or "the Pagan community." Community can bring people with similar interests together, as in "the mountain biking community" or "the peace community." The Internet, especially, calls into question the idea that community is necessarily connected to a physical location.

Chances are you are one of the millions of people who have created a personal online profile page using MySpace, Facebook, or a similar social networking site. Facebook has over 200 million users worldwide. Each of these sites encourages users to post personal information such as favorite music, movies, and activities as well as pictures, blogs, videos, and songs. Members can customize their site and form and join groups based upon similar interests, anything from horror films to indie rock music. Many users are young adults in the United States and Europe, though the sites have spread to many countries and attracted people of all age groups. Users appreciate the opportunity to reconnect with old friends and to make new ones. In fact, given that you can immediately "screen" users' age, interests, and motivation (for example to make friends or find a date), profile sites are an *efficient* way to make friends.

In 1983 novelist William Gibson coined the term "**cyberspace**" which has since come to describe the virtual, computerized realm of Web pages, chat rooms, emails, video games, and blogs.[1] **Virtual communities** are communities of people who regularly interact and form ongoing relationships primarily via the Internet (Rhengold 2000). People who interact online are not automatically part of a community—virtual community entails more than surfing Web sites, making a Facebook page, reading emails, or engaging in brief chats on bulletin boards. Just because someone plays an MMORPG does not necessarily mean they are part of a virtual community; after all, you could theoretically play the game and never chat with another human player. On the other hand, someone who regularly plays an MMORPG with the same people (as part of a guild, for example), gets to know them a bit beyond playing the game, and develops personal relationships in the game is part of a community.

In postindustrial society we have more freedom to choose our communities. Before the advent of advanced communication technologies, affordable travel opportunities, and job mobility, people were more or less restricted to their local or regional communities—their hometown, with its churches, schools, and civic organizations. Now, to a certain extent, we can select the communities we are drawn to (Hewitt 2000). Our loyalties may be divided among many different communities, and we can leave and join communities relatively easily. Think about the myriad interest groups and subcultures that connect people online. You like goldfish? Chat with other enthusiasts at www.koivet.com. Enjoy bird watching? If you can't locate members of the Audubon Society (or if you can find them, but don't *like* them), join a bird-watching listserv. The main idea is this: community, now more than ever, is flexible and less tied to *geography*.

... The Web changes the nature of subcultures, potentially expanding community, but in a different form. If you can buy your favorite underground music for less money online, maybe you'll frequent your local independent record store less often, and eventually what was once perhaps a hub of a local scene might fade away.

Subcultures have typically relied upon physical spaces in which members can get together: clubs, record stores, skate parks, street corners, pubs, alternative fashion boutiques, and so on. You can think of these spaces as part of the **subcultural geography**, the terrain in which youth congregate and live the subculture day to day. Virtual subcultures have their own geographies, digital hangouts that bring together participants from all over the world. Kendall (2002) likens chat rooms/MUDs to virtual pubs, "neighborhood" hangouts where regulars meet and gossip (Kendall 2002). Correll (1995) claims that members of the Lesbian Café BBS (bulletin board system) talked about their virtual space as if it were a physical place in which they interacted. Instead of dropping by the neighborhood pub after work to have a laugh and catch up on news, many of us are logging in to virtual communities, often several times a day (and often *during* work!).

... Many people are skeptical of virtual communities, sometimes called "**computer-mediated communities**" to emphasize how interaction takes place via, or through, computers. In addition to concerns about online sexual predators, identity thieves, and other criminals they worry that computers will make us more isolated—the more we're "plugged in" the less we're interacting face to face (Nie and Ebring 2000). We all know the stereotype of the isolated computer nerd who substitutes virtual friendships for "real" ones, implying that virtual communities are less "authentic" than face-to-face relationships (Miller and Slater 2000). While the differences between face-to-face and virtual communities pose meaningful sociological questions, we should be cautious in assuming that new technologies automatically undermine community and wary of moral panics about "gaming addiction." After all, people initially had the same worries that the telephone would impede rather than help build genuine community.

ELEMENTS OF THE VIRTUAL SCENE

All of the standard elements of nonvirtual subcultures, from style and status hierarchy to jargon and gender ideology, have their equivalents in the virtual world. Although we do not have enough space to cover them all, a few are especially interesting in the way they transcend the virtual/nonvirtual divide.

... Geek and Gamer Language—Netspeak, 133t sp33k, txtspk

Just like any subculture, virtual cultures have produced their own languages. If I were to show you "/ooc 24 wiz lfg AQ3 pst," would you understand that I am communicating an "out of character" message, playing a "24th level wizard," am "looking for a group" to complete "armor quest 3," and would like you to "please send a tell" to me if I can join your group? Do the acronyms "ROFL," "brb," "afk," "lol," "mt," "pwn," or "lmao" mean anything to you? To online gamers and members of other virtual communities they mean "rolling on the floor laughing," "be right back," "away from keyboard," "laughing out loud," "mis-tell" (or "main tank," depending on context), "owned," and "laughing my ass off," respectively. Just as nonvirtual subcultures create and use their own vocabu-lary, virtual subcultures like hackers employ their own dialect consisting of short-hand words, acronyms, and computer jargon, sometimes called "133t" speak, short for elite where the numeral "1" substitutes for the letter "l" and "3" for "e." More commonly known as netspeak or text speak, abbreviated words and acronyms are now interwoven into daily language and text messaging. Words are symbols that convey meaning, and text becomes part of symbolic interaction. Knowledge of gaming-specific acronyms and ability to decipher text speak separate insiders from outsiders, contributing to a sense of community for those "in the know."

In real life, we use much more than words to communicate our intended meaning. Normal communication cues such as voice tone, facial expressions, ges-tures, and posture help us convey the meanings we want to accompany our words. Online talking with text alone leaves a lot of room for misinterpretation and mis-understanding, requiring other ways of conveying emotion and meaning. Many of you have probably used "emoticons" ("CONventions for expressing EMOTIons") to add feeling and emphasis to email messages (Hornsby 2005). Thus, as you surely know, :) and : (become smiley faces that can indicate a whole range of feelings, depending on the context: excitement, happiness, contentment. Likewise ;) is a winking smiley and connotes a shared joke, sarcasm, teasing, flirting, or similar meaning. In the MMORPG world, players can enact emotion and a presentation of self through their avatars. Toons dance, bow, smirk, threaten, flirt, scowl, cheer, clap, and blow kisses at the direction of their player-puppet masters.

Online language is no longer confined to virtual settings, as email shorthand and leet speak have moved to the nonvirtual world. In the last several years I have noticed (often with dismay) an increasing number of student papers including text message writing "u" substituted for "you" and "b4" for "before." The sheer numbers of peo-ple playing online games ensures the blending of texting and talking, with gamers exclaiming "Woot!!" (typed as w00t! in online gaming) to express joy in real life, "gee gee" (for gg, or good game) to congratulate, and calling each other "noobs" (for newbie, or newcomer) as a joking insult. Language, both virtual and nonvirtual, is fluid and will continue to evolve as the virtual and nonvirtual worlds overlap.

Virtual Gender and Nerd Masculinity

Long considered the domain of adolescent boys, video games often represent women in very sexualized ways. The hit series of *Tomb Raider* games is one of the few with a female protagonist; yet with her tiny waist and enormous breasts she hardly represents the typical female form. Another extremely popular series, *Grand Theft Auto,* has been maligned for the way players can direct the thuggish main character to commit violence against women. To regain health, players direct the main character to pay for sex with a female prostitute—afterwards players can beat and rob her to regain the money they just spent. To claim that video games are the source of real-life degradation of women is a simplistic attempt by moral entrepreneurs to create a moral panic or engage in symbolic politics; video games make an easy scapegoat for larger social problems, including sexism and violence against women. Nevertheless, in the context of a sexist culture, games that often depict violent men as tough and admirable heroes and women as seminaked sexual objects do perpetuate stereotypical gender representations and roles.

Despite the sexist depictions of women in games and the male-dominated tech world, computer culture has long been associated with geeks and nerds—the "computer geek" is a cultural icon. The "nerd" status serves a purpose in youth culture the as one of the identities that other groups define themselves against. Jocks, for example, are almost always defined in part against the stereotypical nerd—jocks are popular, strong, self-confident, and attractive to the opposite sex, while nerds are unpopular, weak, shy, and asexual. In a sense, nerd "connotes a lack of masculinity," particularly dating/sexual incompetence and little athletic ability (Kendall 2002, 80).

… Some self-described nerds proudly claim (or reclaim) their deviant status, embracing the nerd identity (Wright 1996). As with all deviance, meaning depends upon context—calling someone a nerd can be a demeaning slur or a show of affection and solidarity (among nerds). Technical expertise becomes a mark of superiority to more popular kids, particularly for male nerds. Female nerds face a more difficult situation; being stereotyped as unattractive has more negative consequences for women than for men. Computer programming has traditionally been thoroughly male dominated, making fitting into the "boys' club" a challenge for many young women. In addition, women who do manage to break into the boys' club risk being viewed as somehow less feminine or perhaps even intimidating to men because of their perceived intelligence and expertise (Seymour and Hewitt 1999).

Nerd masculinity encompasses both a critique and reinforcement of hegemonic masculinity. Like straight edgers, self-described nerds and geeks are sharply critical of the stereotypical young male bent on sexual conquest of women and domination of other men. Yet nerds, like skinheads, punks, and others who question what it means to be a man, do not fully resist hegemonic masculinity. For example, in MUDs and MMORPGs young men regularly talk trash as if they were on the basketball court or football field. They also talk about women as sexual objects even as they tease one another about their lack of experience with women and refuse to adopt the "asshole" persona necessary, in their minds, to be attractive to women (Kendall 2002). In virtual competitive games, players often express dominance and power, claiming to have "owned" (often typed pwned in-game) the other team and hurling insults at opposing players or less-skilled teammates. Many games reward players for high kill counts and set up online rankings—a kid who could never hold his own in gym class or on the football field could be the king of *Halo* rankings—virtual status, to be sure, but appealing nonetheless.

POSTMODERN IDENTITY AND THE VIRTUAL SELF

Most of us have, at one time or another, dreamed of being someone we are not—maybe a movie star, professional athlete, revolutionary, or supermodel. Virtual worlds may help us live out these dreams as, in a sense, we can all be rich, powerful, and beautiful, even achieving a measure of fame. We have the opportunity to recreate and remake, to an extent, who we are. In chat rooms we can express a different personality, in online games we can gender-bend by playing a toon of the opposite sex, and in the *Simms* we can pick up people in clubs and take them back to our virtual mansions. All of these possibilities raise questions about how we think of our "self" and our identities. Online games, especially, offer an opportunity to construct a self relatively free from some of the constraints of the material world (Crowe and Bradford 2007). Players choose their appearance, associations, professions, and so on. Yet it is important not to overemphasize the freedom offered in online forums. Participants bring with them their knowledge and experience from the material world. While opportunities to experiment with identity abound, status hierarchies emerge nonetheless. Players value some identities and expressions of self more than others. Having the most rare, expensive, or difficult-to-acquire armor becomes a status symbol in game, much as a sports car might in the material world—the difference is a players' assumption that anyone can have the armor with considerable effort.

We tend to think of our personal identity as coherent, ongoing, and stable—we might periodically abandon an identity (such as student) or adopt a new identity (such as becoming a parent), but ultimately we are who we are. Symbolic interactionists take a much more fluid and social view of the self. The **self** is *process* including one's thoughts, feelings, and choices as well as being something we *do* rather than simply *are* (Sandstrom, Martin, and Fine 2006). Our self emerges in interaction with others. Rather than packing a coherent self with us from time to time and place to place, we express our self (or many selves) depending upon the context of a particular interaction. Sociologists disagree to what extent the self continually changes, and virtual interaction has added another complex piece to the puzzle.

In diverse, mobile, technologically advanced, rapidly changing societies is it possible to construct a stable and coherent self? Rather than having one relatively stable identity, we have multiple identities, some only briefly. Think of your own life. You might have many different selves, one you express at work, one for your family, and another for your sorority. **Postmodern identity** is temporary, fragmented, unstable, and fluid. Rather than being deeply personal and unified, the self is relational and fragmented (Gergen 1991). People assume a variety of seemingly contradictory identities, such as an athlete who is also a band geek, a religious preacher who loves gory horror films, or a porn star happily married with a family. Rather than being tied to a consistent, stable "self," we bring a different, flexible self, so to speak, to each context (Zurcher 1977).

Changing computer and communication technologies are central to many theories of the self (see Agger 2004). Think about the ways you can communicate that differ from when your grandparents were your age: fax machines, email, cell phones, video conferencing, and answering machines. Now consider the technologies you may take for granted that people two generations ago could only dream of: satellite and cable TV, personal digital assistants, laptop computers. These technologies enable hackers, MMORPG players, and personal profile users to literally construct virtual selves unfettered by the same rules that apply to face-to-face relationships. The **virtual self** is "the person connected to the world and to others through electronic means such as the Internet, television, and cell phones" (Agger 2004, 1). It is a state of being, created and experienced through technology. Perhaps people feel more comfortable exploring and enacting taboo identities in the anonymity of cyberspace? Thus a shy, reserved, even socially awkward person can be outgoing, boisterous, and charming in the online world. You can never be sure when someone's online identity matches their off-line characteristics and when they are **masquerading,** or pretending to be something they are not (Turkle 1997; Kendall 2002). How do you determine someone's authenticity if you can't even reliably determine their age, sex, race, or real-life actions (Williams 2003)?

CONCLUSIONS

We are in a state of profound ambivalence about technology. We love the comforts and conveniences it affords us, but we are wary of its dangers, as films like *2001: A Space Odyssey, The Matrix,* and *Terminator* demonstrate. As more and more people "plug in" to the web, new moral panics arise: Internet stalkers, porn and video game addiction, and identity thieves. Yet for all the panic, many of us cannot wait to upgrade our cell phones, iPods, and Facebook pages, injecting a bit more of ourselves into the virtual universe.

The latest communication technologies are still so new that we continue to make a false distinction between "virtual" and "real" life, as if online experience is somehow secondary, less meaningful, and less real than face-to-face interaction (Chee, Vieta, and Smith 2006). Instead, we should be asking how (and *if*) virtual scenes differ from nonvirtual and, more importantly, how they overlap. Hackers, MMORPGs, chat rooms, blogs, and personal profile/networking sites force us to ask how the Internet might change our very conceptualization of subcultures/scenes. They show us how the boundaries between what we see as virtual and real are blurry (Taylor 2006; Castronova 2005). People meet online and then agree to meet offline, but they also meet offline and subsequently get together online. Our communities, and our identities, transcend the virtual-real divide.

NOTE

1.   William Gibson, *Neuromancer* (NewYork: Ace Books, 1983).

REFERENCES

Agger, Ben. 2004. *The Virtual Self: A Contemporary Sociology*. Oxford: Blackwell Publishing.

Correll, Shelley. 1995. "The Ethnography of an Electronic Bar: The Lesbian Café." *Journal of Contemporary Ethnography* 24 (3): 270–298.

Gergen, Kenneth. 1991. *The Saturated Self: Dilemmas of Identity in Contemporary Life*. New York: Basic Books.

Hornsby, Anne M. 2005. "Surfing the Net for Community: A Durkheimian Analysis of Electronic Gatherings." In *Illuminating Social Life: Classical and Contemporary Theory Revisited*, ed. Peter Kivisto, 59–91. Thousand Oaks, CA: Pine Forge Press.

Kendall, Lori. 2002. *Hanging Out in the Virtual Pub: Masculinities and Relationships Online*. Berkeley: University of California Press.

Miller, Daniel, and Don Slater. 2000. *The Internet: An Ethnographic Approach*. New York: Berg.

Nie, Norman H., and Lutz Erbriug 2000. "Our Shrinking Social Universe." *Public Perspective* 11 (3): 44–45.

Rheingold, Howard. 2000. *The Virtual Community: Homesteading on the Virtual Frontier*, rev. ed, Cambridge, MA: MIT Press.

Seymour, Elaine, and Nancy M. Hewitt. 1999. *Talking About Leaving: Why Undergraduates Leave the Sciences*. Boulder, CO: Westview Press.

Turkle, Sherry. 1997. *Life on the Screen: Identity in the Age of the Internet*. New York: Simon & Schuster.

Williams, J. Patrick. 2003. "The Straightedge Subculture on the Internet: A Case Study of Style-display Online." Media International Australia Incorporating Culture and Policy 107: 61–74.

Wright, R. 1996. The Occupational Masculinity of Computing." In *Masculinities in Organizations,* ed. C. Cheng, 77–96. Thousand Oaks, CA: Sage.

Zurcher, Louis. 1977. *The Mutable Self.* Beverly Hills, CA: Sage.

KEY CONCEPTS

community	social interaction	virtual communities
culture	subculture	

DISCUSSION QUESTIONS

1. What does Haenfler mean by "computer-mediated communities?" How is social interaction in such communities different from and similar to social interaction in face-to-face communities?

2. Do you participate in any online communities? If so, which one(s)? How would you describe the cultural characteristics of this community?

8

Milkshakes, Lady Lumps and Growing Up to Want Boobies: How the Sexualisation of Popular Culture Limits Girls' Horizons

MADDY COY

Using the British context, Coy explores the harmful impact of the increasing sexualization of young women in popular culture.

...The term 'sexualisation of culture' describes the current saturation of erotic imagery, particularly of women, in popular culture, for example, advertising and music videos (Gill, 2007). These sources disseminate several themes that Rosalind Gill (2007, p. 149) defines as prevailing in contemporary mass media: that women can use their bodies for profit as a means to power; the importance of individual choice; makeovers as reinventions of the self; and a focus on biological differences between men and women. All are stratified by social factors such as race, class and age, and all fundamentally reinforce heterosexual norms. An influential report by the American Psychological Association (APA) (2007) more specifically defines sexualisation as any one of the following: personal value based only on sex appeal; the equation of physical attractiveness with being sexy; construction as an object for others' sexual use; inappropriate imposition of sexuality.

SOURCE: *Child Abuse Review* Vol. 18: 372–383 (2009).

The same report concludes that sexualisation results in negative outcomes for girls and young women in terms of a lack of diminished educational achievement, as well as normalising abusive practices towards children.

While there is a tendency within some critiques of the sexualisation of culture to draw on notions of morality, particularly those influenced by abstinence discourses that seek to confine sexuality to the private realm of monogamous heterosexual relationships (Buckingham and Bragg, 2003), this article is based on a feminist perspective. Here, the sexualisation of culture is identified as a context that reinforces gender inequality by designating women as sexually available and objectified, perpetuates associations of masculinity and predatory sexual prowess, and justifies sexual violence. Research demonstrates that young people, particularly boys, who are exposed to sexualised media are likely to perceive women to be sex objects (Peter and Valkenburg, 2007).

… As this article will argue, sexualised media transmit messages about girlhood and womanhood that constrain the range of opportunities open to girls and young women, while the impact of sexualised cultures on boys and understandings of masculinity also affects young women's space for action.

While both girls and boys are often exposed to sexualised material aimed at adults, the marketing of products to children that are either sexualised (such as Bratz dolls) or implicitly associated with the sex industry (e.g. Playboy stationery and bedding and the sale of pole-dancing kits in supermarkets, high-street stores and children's toy departments) is a more recent development and is distinctly gendered (Rush and La Nauze, 2006). In December 2006, pole-dancing classes for children from 12-years old were introduced in Northumberland (BBC News, 2006). However, the 'background noise' of sexualisation that is directed at adults but available to, and absorbed by, children has also multiplied exponentially (Australian Senate Committee, 2008). The APA (2007, p. 13) review suggests that in this context young girls are 'adultified' and adult women are 'youthified'. It is this cumulative and multi-faceted sexualisation, referred to by Brown *et al.* (2006) as a 'sexy media diet', that is increasingly thought to be harmful to girls (APA, 2007; Australian Senate Committee, 2008). Some critical commentators have suggested that the media have replaced families and professionals as educators, assuming an authority of cool that renders it a 'super peer' (Levin and Kilbourne, 2008), thereby amplifying absorption by children and young people.

Children's levels of understanding are central to debates around the absorption and interpretation of socio-cultural messages, with very young children less developmentally equipped to recognise sexual meanings (Lamb and Brown, 2006). While beyond the scope of this article to discuss the differing significance of developmental stages, experiences of sexualised imagery from a very young age are nevertheless essential to its normalisation.

NARROWING 'SPACE FOR ACTION'

The concept of 'space for action' has been used by feminist researchers to explore how social norms, expectations and experiences shape possible routes for behaviour and thus constrain women's ability to act autonomously (Jeffner, 2000; Kelly, 2005; Lundgren, 1998). Here, it is used similarly to suggest that the ubiquity of sexualised images of women and the meanings attributed to them paint a picture of womanhood that narrows girls' horizons whilst appearing to stretch them into limitless possibilities.

Lamb and Brown (2006, p. 3) suggest that one of the key ways in which sexualisation of girls is harmful is by the false promises of choices—selling an identity story that they [marketers] call 'girl power', characterised by the consumption of beauty products and clothes, and social popularity. Girl power is a popular term for embracing a focus on appearance and femininity as emblems of empowerment. Lamb and Brown (2006) draw particular attention to the Disney version of girlhood based on princessdom, baring midriffs, arching backs and an obsession with mirrors, which they suggest parallels the fantasy woman of pornography. This is also reflected in a contemporary emphasis in advertising and popular culture which depicts women as actively celebrating sex-object status (Gill, 2007). Yet, pursuing these ideals often prevents girls developing other skills and activities, limiting their interests and opportunities (Rush and La Nauze, 2006). For younger girls, the preoccupation with shopping, crushes on boys and attention to physical appearance potentially limits creativity and identity exploration to a narrowly defined version of femininity (Levin and Kilbourne, 2008). As this sexualised 'identity story' is often regarded as not only gender appropriate but also an indicator of contemporary female empowerment, where expressing a sexualised self and selling femininity/beauty are manifestations that women can do and achieve anything they choose, there is little scope to engage in discussions of it as harmful to girls' personal and social development.

Some, perhaps many, young women may find that a sexualised identity gives them confidence and a sense that they are able to take control of their lives by defining themselves in a way that can be socially rewarding (Buckingham and Bragg, 2003; Reay, 2001). However, for girls who do not identify with WAGS (wives and girlfriends, typically used in relation to wealthy footballers) and Bratz dolls, there are few voices offering equivalent personal and social authority (Buckingham and Bragg, 2003). Young women's options for developing intellectual, athletic, creative identities that do not embrace hyper-femininity are stunted—a process that Lamb and Brown (2006, p. 20) refer to as 'girl typing'. This is empirically demonstrated by Diane Reay's (2001) research with seven-year olds in UK schools, which found that girls who adopted identities based on 'girl power' through overt heterosexuality were able to exercise more social power than those identifying as 'nice girls' or 'tomboys'. The 'tomboy' position itself was dependent on belittling 'feminine' characteristics and did not challenge the association of power with boys/masculinity. In other words, only by being sexualised could girls occupy a space to misbehave and/or be assertive (Ready, 2001).

… These values are encroaching further and further into childhood due to the marketing strategy of 'age compression', where previously adult/adolescent products are aimed at younger and younger children in order to guarantee more consumers (Lamb and Brown, 2006; Levin and Kilbourne, 2008). Perhaps the most extreme example of this is the recent launch of the 'Heelarious' range of soft high-heeled shoes for babies from birth to six months. Eight styles are available, some adorned with rhinestones and leopard print, marketed as offering babies 'fun, hilarity and glamour' (see www.heelarious.com). One result of this 'age compression' with respect to sexualised products is that the *transformation* of the sexualised self has become a *development* of the sexualised self. When six-year olds are wearing jeans with 'princess' across the rear, having hair extensions, or singing about their lady lumps and growing up to want boobies displaying a sexualised identity is no longer a milestone of impending adulthood, but an integral part of identity development itself. This is how sexualisation limits girls' space for action—at the same time as it seems to offer opportunities for material gain, personal achievement and sociocultural acceptance, and thus widen girls' choices, it fixes sexualisation as such a *normal* route that there is little space outside of it.

SEXUALISED AMBITIONS

In a recent BBC (2008) documentary, an 11-year-old girl seeking to be a 'beauty queen' cited the glamour model Jordan as her idol and showed off bleached hair, makeup, hair extensions and artificial tan. When asked about her self-image and future, she described herself as 'blonde, pretty, dumb—I don't need brains' (Sasha: Beauty Queen at 11, BBC, 2008). A 2005 UK online survey of almost 1000 girls aged 15-19 years found that 63 per cent considered 'glamour model' and 25 per cent 'lap dancer' their ideal profession from a list of choices including teacher, doctor (Deeley, 2008). Glamour-modelling agencies report dramatic increases in the numbers of young women wishing to register with them (Coy and Garner, in press). The 2008 BBC documentary series Glamour Girls followed young women hoping to establish their careers in topless modelling, primarily in search of a self-image of desirability and celebrity lifestyle (Coy and Garner, in press). Here, sexualisation was not viewed as harmful, but decadent and pleasurable: a means to a standard of living characterised by material consumption, pursuit of leisure and self-improvement.

… Sasha's comment about 'not needing brains' also illustrates another possible impact of sexualisation on girls' education. Young women aspiring to be footballers wives or glamour models (a girl's tee shirt sold on the high street in 2008 proclaimed 'When I grow up I want to be a WAG') are less likely to see the value or relevance of academic achievement.

… Similar messages a disseminated through the popular television show America's Next Top Model (ANTM) (The CW Television Network, 2003–2009), which offers young women the opportunity to become 'supermodels'. The show is regularly repeated on cable channels in the late afternoon and has just been launched as a video game. In previous cycles of the show, young women have been required to pose as if they are dead—murdered–while seminaked–'but still sexy'—described by one judge as 'broken down dolls… busted up, marionettes'. A frequently expressed sentiment from the show host Tyra Banks is 'sometimes modelling is about pretending you're a ho, but bringing it back to fashion'. The ANTM video game is rated suitable for players aged three years, promoting both the hyper-sexualised model 'look' and the ambition to model to very young girls.

Linked to these ambitions are ways in which young women pay close attention to how their selves and bodies match up to the airbrushed, surgically enhanced images of the media (Gill, 2007). Evaluating the female body against these glossy photographs cultivates a sense of personal shortfall that is regarded as part of femaleness, an ordinariness of anxiety about the body that Angela McRobbie (2007) terms 'normative discontent'. This affects girls' developing sense of self since 'in today's media it is possession of a 'sexy body' that is presented as women's key (if not sole) source of identity' (Gill, 2007, p. 149). The APA (2007) report concludes that sexualisation is linked to eating disorders, a lack of self-esteem and depression, as well as a sense of dissatisfaction with the body that leads to low self-confidence and anxiety.

… Young women are, as Bordo (1997) highlights, fixing, inadequacies (breast size, hair colour, skin tone, eyelash length) that popular culture has created, and perceiving that they have taken control of a personal, not social, situation. Girls' space for action is therefore shaped by expectations and ideals that limit their life, career and self-aspirations, and the relationship with their body, to those plastered over advertising billboards, music videos and reality television, with few alternatives presented as desirable.

SEXUALISED RELATIONSHIPS

Studies indicate that young people use pornography as an instruction manual for their own sexual relationships, with one finding that it influences young men's expectations of sexual relationships, 'lead[ing] to pressure on young women to comply' (Redgrave and Limmer, 2005, p. 22). This section considers the impact of what Lamb and Brown (2006 p. 151) refer to as 'watered down porn'—talk or visuals about sex that don't actually show body parts but imply sex other ways: stripping, moaning, pole dancing and jiggling'.

A recent survey found that that almost a quarter of 14-year-old girls have been coerced into sexual acts (WAFE/Bliss, 2008). The role of mass media in transmitting messages about appropriate and healthy sexual norms is significant. Researchers in the US found a connection between listening to sexualised music lyrics and early sexual activity (Martino et al., 2006). Given that early sexual activity is associated with teenage pregnancy and poor sexual health, media sources that influence young people to engage in early sexual relationships are at odds with [plans] to reduce teenage conception, be sexually healthy, and enable children to develop socially and emotionally.

… A crucial aspect of the connection between sexualised music lyrics and early sexual activity is that the causal factor is not just whether the music lyrics are sexual, but also 'degrading'—defined by the researchers as 'sexually insatiable men pursuing women valued only as sex objects' (Martino et al., 2006, p. 437).

These lyrics are only part of young people's 'sexy media diet' (Brown *et al.*, 2006) since they are also likely to be watching music videos with their frequent accompanying images of sexualised dancing (Martino *et al.*, 2006). Some recent examples in music videos include rap artist Nelly swiping a credit card through a young woman's buttocks (Tip Drill) and women being walked on leashes (P. I. M. P. by 50 cent) (Levin and Kilbourne, 2008).

The mainstreaming of 'pimp/ho chic' through rap and hip-hop music has sharpened attention to the portrayal of black girls and young women. Lamb and Brown's (2006) survey of girls in the US concluded that black girlhood is constructed in such a way that childhood innocence is disallowed; instead young women are 'bounced into the world of jiggling butts and cleavages' (p. 148). Here, there are opportunities for young men and young women to align themselves with the 'commercial hip hop trinity' of gangsta/pimp/ho as a form of developing personal and social power, but these figures reflect racialised stereotypes (Rose, 2008). For instance, masculinity is constructed so that young black men are depicted as predatory and young black women are limited to being hyper-sexual; power is restricted to attracting male attention and approval (Hill Collins, 2006; Rose, 2008). The recent launch of a black Disney princess may be an indicator of greater cultural diversity, but in terms of the 'girl power' values it carries the view that it is 'a great step... [and]could help black children see themselves more positively' (Adesioye, 2009) fails to address how it will reinforce messages of sexualisation for black girls.

Research with African-American girls aged 14-18 years concluded that where they viewed (hetero) sexual stereotypes of racialised gender subordination (defined as black women being fondled, controlled by black men, or using sex for material gain) in rap music videos they were more likely to have multiple sexual partners, use alcohol and drugs, and have a negative body image (Peterson *et al.*, 2007).

CONCLUSIONS

> 'Are there paths through the forest of sexy diva princess pink shopping hotties?' (Lamb and Brown, 2006, p. 263.)

The evidence base on the dimensions of harm from sexualisation of popular culture and links with early sexual activity, values that demean young women and lead to poor self-confidence, suggest a strong counter policy steer is necessary. Some countries have introduced measures with this aim. For instance, the investigation by the Australian Senate Committee (2008) culminated in a new Children's Code for advertising standards that has specific provision about the sexualisation of children. Some countries ban (Norway and Sweden), or restrict (Denmark, Greece and Belgium) advertising to children under 12 (Levin and Kilbourne, 2008). Given the diverse forms of media that children are exposed to, such measures have a limited (although welcome) reach.

The APA (2007) report suggests that harmful consequences of the sexualisation of girls can be mitigated by a range of strategies: media literacy programmes in schools to enable children to critically analyse media messages; increasing access to sport and other activities; more holistic sex education; campaigns to enable parents and careers to address the impact of sexualisation.

… The most significant challenge is to develop socio-cultural climates without sexualisation of girls and young women: 'to landscape the terrain in a slightly different way, so that certain topographical features will stand out against the ones that are currently so prominent to obscure all else' (Bordo, 1997, p. 68). Individual young women may find paths through the forest of pink shopping hotties; but the majority will find their paths navigated by the topographical features of sexualisation and a limited space for action, unless others 'trees' of positive ways to be girls and women are planted and nurtured.

REFERENCES

Adesioye L. 2009. All hail Princess Tiana. *Guardian*, 20 March.

American Psychological Association. 2007. *Report of the APA Task Force on the Sexualisation of Girls*. APA: Washington, DC.

Australian Senate Committee. 2008. *Sexualisation of Children in the Contemporary Media*. Parliament House: Canberra.

BBC. 2008. Sasha: Beauty Queen at 11. BBC One, 31 July.

BBC News. 2006. Children are taught pole dancing. Available: http://news.bbc.co.uk/l/hi/england/tyne/6173805.stm [29 May 2007].

Bordo S. 1997. *Twilight Zones: The Hidden Life of Cultural Images from Plato to O.J.* University of California Press: Berkeley.

Brown JD, L'Engle KL, Pardun CJ, Guo G, Kenneavy K, Jackson C. 2006. Sexy media matter: Exposure to sexual content in music, movies, television, and magazines predicts black and white adolescents' sexual behavior. *Pediatrics* **117**(4, April): 1018–1027. DOI: 10.1542/peds.2005-1406.

Buckingham D, Bragg S. 2003. *Young People, Media and Personal Relationships*. Advertising Standards Authority, British Board of Film Classification, BBC, Broadcasting Standards Commission, Independent Television Commission: London.

Coy M, Garner M. 2009. Glamour Modelling and the Marketing of Self-sexualisation: Critical Reflections. *International Journal of Cultural Studies*. In press.

Deeley L. 2008. I'm single, I'm sexy, and I'm only 13. The Times, 28 July.

Gill R. 2007. Postfeminist Media Culture: Elements of a Sensibility. *European Journal of Cultural Studies* **10**(2): 147–166. DOI: 10.1177/1367549407075898.

Hill Collins P. 2006. *From Black Power to Hip Hop: Racism, Nationalism, and Feminism*. Temple University Press: Philadelphia.

Jeffner S. 2000. Different space for action: The everyday meaning of young people's perception of rape. Paper at ESS Faculty Seminar, University of North London, May.

Kelly L. 2005. The Wrong Debate: Reflections on Why Force is Not the Key Issue with Respect to Trafficking in Women for Sexual Exploitation. *Feminist Review* 73: 139–144. DOI: 10.1057/palgrave.fr.9400086.

Lamb S, Brown L. 2006. *Packaging Girlhood: Rescuing Our Daughters from Marketers' Schemes*. St Martins Griffin: New York.

Levin D, Kilbourne J. 2008. *So Sexy So Soon: The New Sexualized Childhood and What Parents Can Do to Protect Their Kids*. Ballantine Books: New York.

Lundgren E. 1998. The hand that strikes and comforts: Gender construction and the tension between body and symbol. In *Rethinking Violence Against Women,* Dobash RE, Dobash RP (eds). Sage: London.

Martino SC, Collins RL, Elliott MN, Strachman A, Kanouse DE, Berry SH. 2006. Exposure to Degrading versus Nondegrading Music Lyrics and Sexual Behaviour among Youth. *Pediatrics* **118**(2): 430–441. DOI: 10.1542/peds.2006-0131.

McRobbie A. 2007. Illegible rage: Reflections on young women's post feminist disorders, Gender Institute, Sociology and ESRC New Femininities Series, 25 January. Available: http://www.lse.ac.uk/collections/LSEPublicLecturesAndEvents/pdf/20070125_McRobbie.pdf. 28 November 2008.

Peter J, Valkenburg PM. 2007. Adolescents' Exposure to a Sexualised Media environment and their Notions of Women as Sex Objects. *Sex Roles* 56: 318–395. DOI 10.1007/s11199-006-9176.

Peterson SH, Wingood GM, DiClemente R, Harrington K, Davies, S. 2007. Images of Sexual Stereotypes in Rap Videos and the Health of African American Female Adolescents. *Journal of Women's Health* **16**(8): 1157–1164. DOI: 10.1089/jwh.2007.0429.

Reay D. 2001. 'Spice Girls', 'Nice Girls', 'Girlies' and 'Tomboys': Gender Discourses, Girls' Cultures and Femininities in the Primary Classroom. *Gender and Education* **13**(2): 153-166. DOI: 10.1080/09540250120051178.

Redgrave K, Limmer M. 2005. *'It Makes You More Up For It': School Aged Young People's Perspectives on Alcohol and Sexual Health*. Rochdale Metropolitan Borough Council: Rochdale.

Rose T. 2008. *The Hip: Hop Wars: What We Talk About When We Talk About Hip-hop and Why It Matters*. Basic Books: New York.

Rush E, La Nauze A. 2006. *Letting Children be Children: Stopping the Sexualisation of Children in Australia*. The Australia Institute: Manuka.

KEY CONCEPTS

identity	popular culture
mass media	sexualization of culture

DISCUSSION QUESTIONS

1. What does Coy mean by the *sexualization of culture?* What evidence of this do you see in the cultural media you observe?

2. What specific consequences of the sexualization of culture are there for young women? Black women? Men?

9

Global Culture
Sameness or Difference?

MANFRED B. STEGER

Steger enters a debate here about the impact of globalization on local cultures. He shows some of the impact of global capitalism on local cultures, but also argues that local contexts have a role in whether or not cultures become the same or remain different.

Does globalization make people around the world more alike or more different? This is the question most frequently raised in discussions on the subject of cultural globalization. A group of commentators we all might call "pessimistic hyperglobalizers" argue in favour of the former. They suggest that we are not moving towards a cultural rainbow that reflects the diversity of the world's existing cultures. Rather, we are witnessing the rise of an increasingly homogenized popular culture underwritten by a Western "culture industry" based in New York, Hollywood, London, and Milan. As evidence for their interpretation, these commentators point to Amazonian Indians wearing Nike training shoes, denizens of the Southern Sahara purchasing Texaco baseball caps, and Palestinian youths proudly displaying their Chicago Bulls sweatshirts in downtown Ramallah. Referring to the diffusion of Anglo-American values and consumer goods as the "Americanization of the world," the proponents of this cultural homogenization thesis argue that Western norms and lifestyles are overwhelming more vulnerable cultures. Although there have been serious attempts by some countries to resist these forces of "cultural imperialism"—for example, a ban on satellite dishes in Iran, and the French imposition of tariffs and quotas on imported film and television—the spread of American popular culture seems to be unstoppable.

But these manifestations of sameness are also evident inside the dominant countries of the global North. American sociologist George Ritzer coined the term "McDonaldization" to describe the wide-ranging sociocultural processes by which the principles of the fast-food restaurant are coming to dominate more and more sectors of American society as well as the rest of the world. On the surface, these principles appear to be rational in their attempts to offer

SOURCE: "Global Culture: Sameness or Difference?" by Manfred B. Steger from *Globalization—A Very Short Introduction*. 2003. New York: Oxford University Press.

efficient and predictable ways of serving people's needs. However, looking behind the façade of repetitive TV commercials that claim to "love to see you smile," we can identify a number of serious problems. For one, the generally low nutritional value of fast-food meals—and particularly their high fat content—has been implicated in the rise of serious health problems such as heart disease, diabetes, cancer, and juvenile obesity. Moreover, the impersonal, routine operations of "rational" fast-service establishments actually undermine expressions of forms of cultural diversity. In the long run, the McDonaldization of the world amounts to the imposition of uniform standards that eclipse human creativity and dehumanize social relations.

Perhaps the most thoughtful analyst in this group of pessimistic hyperglobalizers is American political theorist Benjamin Barber. In his popular book on the subject, he warns his readers against the cultural imperialism of what he calls "McWorld"—a soulless consumer capitalism that is rapidly transforming the world's diverse populations into a blandly uniform market. For Barber, McWorld is a product of a superficial American popular culture assembled in the 1950s and 1960s, driven by expansionist commercial interests. Music, video, theatre, books, and theme parks are all constructed as American image exports that create common tastes around common logos, advertising slogans, stars, songs, brand names, jingles, and trademarks.

Barber's insightful account of cultural globalization also contains the important recognition that the colonizing tendencies of McWorld provoke cultural and political resistance in the form of "Jihad"—the parochial impulse to reject and repel the homogenizing forces of the West wherever they can be found.... Jihad draws on the furies of religious fundamentalism and ethnonationalism which constitute the dark side of cultural particularism. Fuelled by opposing universal aspirations, Jihad and McWorld are locked in a bitter cultural struggle for popular allegiance. Barber asserts that both forces ultimately work against a participatory form of democracy, for they are equally prone to undermine civil liberties and thus thwart the possibility of a global democratic future.

Optimistic hyperglobalizers agree with their pessimistic colleagues that cultural globalization generates more sameness, but they consider this outcome to be a good thing. For example, American social theorist Francis Fukuyama explicitly welcomes the global spread of Anglo-American values and lifestyles, equating the Americanization of the world with the expansion of democracy and free markets. But optimistic hyperglobalizers do not just come in the form of American chauvinists who apply the old theme of manifest destiny to the global arena. Some representatives of this camp consider themselves staunch cosmopolitans who celebrate the Internet as the harbinger of a homogenized "techno-culture." Others are free-market enthusiasts who embrace the values of global consumer capitalism.

It is one thing to acknowledge the existence of powerful homogenizing tendencies in the world, but it is quite another to assert that the cultural diversity existing on our planet is destined to vanish. In fact, several influential commentators offer a contrary assessment that links globalization to new forms of cultural expression. Sociologist Roland Robertson, for example, contends that global

The American Way Of Life

Number of types of packaged bread available at a Safeway in Lake Ridge, Virginia	104
Number of those breads containing no hydrogenated fat or diglycerides	0
Amount of money spent by the fast-food industry on television advertising per year	$3 billion
Amount of money spent promoting the National Cancer Institute's "Five A Day" programme, which encourages the consumption of fruits and vegetables to prevent cancer and other diseases	$1 million
Number of "coffee drinks" available at Starbucks, whose stores accommodate a stream of over 5 million customer per week, most of whom hurry in and out	26
Number of "coffee drinks" in the 1950s coffee houses of Greenwich Village, New York City	2
Number of new models of cars available to suburban residents in 2001	197
Number of convenient alternatives to the car available to most such residents	0
Number of U.S. daily newspapers in 2000	1,483
Number of companies that control the majority of those newspapers	6
Number of leisure hours the average American has per week	35
Number of hours the average American spends watching television per week	28

Sources: Eric Schossier, *Fast Food Nation* (Houghton & Mifflin, 2001), p. 47; www.naa.org/info/facts00/11.htm; *Consumer Reports Buying Guide 2001* (Consumers Union, 2001), pp. 147–163; Laurie Garrett, *Betrayal of Trust* (Hyperion, 2000), p. 353; www.roper.com/news/content/news169.htm; *The World Almanac and Book of Facts 2001* (World Almanac Books, 2001), p. 315; www.starbucks.com.

cultural flows often reinvigorate local cultural niches. Hence, rather than being totally obliterated by the Western consumerist forces of sameness, local difference and particularity still play an important role in creating unique local contexts. Robertson rejects the cultural homogenization thesis and speaks instead of "glocalization"—a complex interaction of the global and local characterized by cultural borrowing. The resulting expressions of cultural "hybridity" cannot be reduced to clear-cut manifestations of "sameness" or "difference." [S]uch processes of hybridization have become most visible in fashion, music, dance, film, food, and language.

In my view, the respective arguments of hyperglobalizers and sceptics are not necessarily incompatible. The contemporary experience of living and acting across cultural borders means both the loss of traditional meanings and the creation of new symbolic expressions. Reconstructed feelings of belonging coexist in uneasy tension with a sense of placelessness. Cultural globalization has contributed to a remarkable shift in people's consciousness. In fact, it appears that the old structures of modernity are slowly giving way to a new "postmodern" framework characterized by a less stable sense of identity and knowledge.

Given the complexity of global cultural flows, one would actually expect to see uneven and contradictory effects. In certain contexts, these flows might change traditional manifestations of national identity in the direction of a popular culture characterized by sameness; in others they might foster new expressions of cultural particularism; in still others they might encourage forms of cultural hybridity. Those commentators who summarily denounce the homogenizing effects of Americanization must not forget that hardly any society in the world today possesses an "authentic," self-contained culture. Those who despair at the flourishing of cultural hybridity ought to listen to exciting Indian rock songs, admire the intricacy of Hawaiian pidgin, or enjoy the culinary delights of Cuban-Chinese cuisine. Finally, those who applaud the spread of consumerist capitalism need to pay attention to its negative consequences, such as the dramatic decline of communal sentiments as well as the commodification of society and nature.

KEY CONCEPTS

cultural imperialism	global culture	homogenization
dominant culture	globalization	

DISCUSSION QUESTIONS

1. How would you answer Steger's central question, "Does globalization make people around the world more alike?"

2. What impacts does global capitalism have on diversity in world cultures? How do you see this in your particular environment?

Applying Sociological Knowledge: An Exercise for Students

Take a look around your campus and make note of any differences in the clothing that people are wearing that you would associate with cross-cultural differences. What issues would you face in your family, community, or friendship network were you to adopt that style of dress? What do these issues teach you about how cultural expectations shape social norms and group conformity?

✳

PART IV

Socialization and the Life Course

10

Leaving Home for College: Expectations for Selective Reconstruction of Self

DAVID KARP, LYNDA LYTLE HOLMSTROM, AND PAUL S GRAY

Many young adults leave home for the first time when they go away to college. This article addresses the changes young students go through when they leave high school and then family home. The authors discuss how personal changes in identity and perceptions of self are more significant than the geographical move to college.

In their important and much discussed critique of American culture, *Habits of the Heart* (1984), Robert Bellah and his colleagues remark that American parents are of two minds about the prospect of their children leaving home. The thought that their children will leave is difficult, but perhaps more troublesome is the thought that they might not. In contrast to many cultures, American parents place great emphasis on their children establishing independence at a relatively early age. Still, as Bellah's wry comment suggests, they are deeply ambivalent about their children leaving home. The data presented in this paper, part of a larger project on family dynamics during the year that a child applies for admission to college, show that such ambivalence is shared by the children. Our goal here is to document some of the social psychological complexities of achieving independence in America by analyzing the perspectives of 23 primarily upper-middle-class high school seniors as they moved through the college application process and contemplated leaving home.[1]

Of course, a great deal has been written about the internal conflict that surrounds any significant personal change (most obviously, Erik Enkson 1963, 1968, 1974, 1980; see also Manaster 1977; O'Mally 1995). Although researchers have attended to the phenomenon of "incompletely launched young adults" (Heer, Hodge, and Felson 1985; Grigsby and McGowan 1986; Schnaiberg and Goldenberg 1989), little has been written about how relatively sheltered, middle- to upper-middle-class children think about "leaving the nest." Leaving home for college is perhaps among the greatest changes that the economically comfortable students we interviewed have thus far encountered in their lives. For them, going to college carries great significance as a coming-of-age moment, in part because it

SOURCE: © 1998 by the Society for the Study of Symbolic Interaction.
Reprinted from *Symbolic Interaction.* Vol 21, No 3, pp 253–276 by permission.

has been long anticipated and not to do so would be unacceptable from a normative stand point. Literature on students who "beat the odds" by going to college suggests that this is also an important transition for them, but one carrying fundamentally different meanings. Unlike the middle- or upper-middle-class students we interviewed, who are trying, at the least, to retain their class position, students arriving at college from less privileged backgrounds must confront wholly new cultural worlds (Rodriguez 1982; Smith 1993; Hooks 1993).

The 23 students with whom we were able to complete interviews simply assumed, as did their parents, that they would go to college.[2] Among the 30 sets of parents, all but four individuals had attended college (and two received some different training beyond high school). All of the adults, however, felt strongly about the necessity of college attendance for their children. One of the four who did not go to college, a self-made and extraordinarily successful entrepreneur, did offer some reservation about the utility of an education in the rough and tumble "real world." Even so, both he and his wife were highly invested in getting their son into a prestigious college. While all of the children knew their parents' expectations and fully expected to meet them, we did speak with two students who had some misgivings about whether they really wanted to go to college. Like their counterparts in our sample, these students knew they would go, but still entertained private doubts about their interest in and motivation for college work.[3]

What does it mean to become independent of one's parents, family, and high school friendship groups? As Anna Freud noted, "few situations in life are more difficult to cope with than the attempts of adolescent children to liberate themselves" (Bassoff 1988, p.xi). Young people are ambivalent regarding independence; it is hard to break away. Their ambivalence embodies both symbolic and pragmatic dimensions. Symbolically, independence is the desired outcome of a necessary process of differentiation (Bios 1962). The task for adolescents is "to find their own way in the world and develop confidence that they are strong enough to survive outside the protective family circle" (Bassoff 1988, p. 3). To establish their own identity and sense of purpose, " … they need to wrench themselves away from those who threaten their developing selfhood" (Bassoff 1988, p. 3; see also Campbell, Adams, and Dobson 1984; Katchadounan and Boll 1994). However, independence also has a pragmatic side. In college, young people can "start over"; they can make new friends, establish intimate relationships, and develop the skills and knowledge to help them become self-supporting adults. "But the truth is that they are not sure they can take care of themselves or that they want to be left alone" (Bassoff 1988, p. 3)....

IDENTITY AFFIRMATION, IDENTITY RECONSTRUCTION, AND IDENTITY DISCOVERY

While the students in this study anticipated college as a time during which they would maintain, refine, build upon, and elaborate certain of their identities, they also anticipated negotiating some fundamental identity changes. The students saw college as the time for discovering who they *really* were. They anticipated finding wholly new and permanent life identities during the college years. In addition, they believed that going to college provides a unique opportunity to consciously establish some new identities. Repeatedly, students described the importance of going away to college in terms of an opportunity to discard disliked identities while making a variety of "fresh starts." Their words suggest that college-bound students look forward to re-creating themselves in a context far removed (often geographically, but always symbolically) from their family, high school, and community. The immediately following sections attend, in turn, to how upper-middle-class high school seniors (1) anticipate change, (2) strategize about solidifying certain identities, [and] (3) evaluate identities they wish to escape....

Anticipating Change

Along with such turning points as marriage, having children, and making an occupational commitment, it is plain that leaving for college is self-consciously understood as a dramatic moment of personal transformation. The students with whom we spoke all saw leaving home as a critical juncture in their lives. One measure of consensus in the way our 23 respondents interpreted the meaning of leaving home is the similarity of their words. Students used nearly identical phrases in describing the transition to college as the time to "move on," "discover who I really am," to "start over," to "become an adult," to "become independent," to "begin a new life." The students, moreover, explicitly saw going to college as the "next stage" of their lives....

While all the students interviewed recognized the need for change and were looking forward to it, their certainty about the appropriateness of moving on did not prevent them from feeling anxiety and ambivalence about the transition to college. Theirs is an anticipation composed of optimism, excitement, anxiety, and sometimes fear.

> [I'm] starting the rest of my life. I mean, deciding what I'm going to do and figuring out my future. I mean, that's one thing I'm looking forward to, but it's also one thing I'm not looking forward to. I have mixed feelings about that. It's exciting to figure out your future. In another sense it's scary to have all of the responsibility. (White male attending a public school)

These comments suggest that the prospect of leaving home generates an anticipatory socialization process characterized by multiple and sometimes contradictory feelings and emotions. Students long for independence, anticipate the excitement that accompanies all fresh starts, but worry about their ability to fully meet the challenge....

Affirming Who I Really Am

The one concrete and critical choice that college-bound students must make is which school, in fact, to attend. This decision is often an agonizing one for both students and their parents and involves very high levels of "emotion work" (Hochschild 1983). The significance of making the college choice and the anxiety that it occasions go well beyond questions of money, course curricula, or the physical amenities of the institutions themselves. What makes the decision so difficult is that the students know they are choosing the context in which their new identities will be established.... The fateful issue in the minds of the students is whether people with their identity characteristics and aspirations will be able to flourish. Consequently, it is not surprising that the most consistent and universal pattern in our data is the effort expended by students to find a school where "a person like me" will feel comfortable....

In the most global way, prospective students were searching for a place where the students seemed friendly. On several occasions, students remarked that they were turned on or off to a school because their "tour guide" was either really nice or not friendly enough....

In contrast to the students-like-me theme, an interesting sub-set of seniors expressed a strong interest in diversity. These students not only wanted to meet new people, but different kinds of new people. Students who wanted diversity were excited at the prospect of meeting people different from themselves as a critical learning experience. It is important to note that it was primarily the minority students we interviewed who looked for diversity as they contemplated colleges. An Asian student put it this way:

> The more mixed the better. I think interaction with other ethnic and
> racial groups is very healthy. If possible, I would not mind having,
> you know, like an Afro-American roommate. I'd love to. (Asian male
> attending a public school) ...

While the statements immediately above illustrate that students make careful assessments about the goodness of fit between certain aspects of themselves and the character of different colleges, a dominant theme in the interviews concerned change. Students repeatedly commented that, during their college years, they expected their identities to shift in two fundamental ways. First, they anticipated discovering "who I am" in the broadest sense. Second, they saw college as providing a fresh start because they could discard some of their disliked, sticky identities, often acquired as early as grade school.

Creating the Person I Want to Be

... Seen in terms of Erving Goffman's (1959) dramaturgical model of interaction, going to college provides a new stage and audience, together allowing for new identity performances. Goffman notes (1959, p. 6) that "When an individual appears before others his actions will influence the definition of the situation which they come to have. Sometimes the individual will act in a thoroughly calculating manner, expressing himself in a given way solely in order to give the kind of impression to others that is likely to evoke from them a specific response he is concerned to obtain." To the extent that such impression-management is most centrally dependent upon information control, leaving home provides an unparalleled opportunity to abandon labels that have most contributed to disliked and unshakable identities. When students speak of college as providing a fresh start, they have in mind the possibility of fashioning new roles and identities. Going to college promises the chance to edit, to revise, to re-write certain parts of their biographies.

> It's sort of like starting a new life. I'll have connections to the past, but I'm obviously starting with a clean slate.... Because no one cares how you did in your high school after you're in college. So everyone's equal now. (White male attending a public high school) ...

As students described their hopes about college, the theme of "fresh starts" was almost universally voiced.... Leaving home, friends, and community offers students the possibility to jettison identities which are the product of others' consistent definitions of them over many years. Going to college provides a unique opportunity to display new identities consistent with the person they wish to become.

The data presented thus far are meant to convey the symbolic weightiness of the transition from high school to college. Every student with whom we spoke saw leaving home as a critical biographical moment. They see it as a definitive life stage when their capacity for independence will be fully tested for the first time. Some have had a taste of independence at summer camps and the like, but the transition to college is viewed as the "real thing." Their words, we have been suggesting, indicate that they see strong connections among leaving home, gaining independence, achieving adult status, and transforming their identities. Students carefully attempt to pick a college where they will fit in, thus indicating the importance of retaining and consolidating certain parts of their identities (see Shreier 1991). In addition, they believe that they will discover, in a holistic sense, who they "really" are during the college years.

WILL THEY MISS ME?

... The family is a social system in which roles are interconnected and interdependent. When a child goes off to college, the system is disturbed and the family will try to adapt to the new circumstances. College-bound seniors worry about this process of adaptation. They speculate that their remaining siblings will miss them, or will be left to face the unremitting attentiveness and concern of parents. They also wonder about prospective changes in their parents' marital relationship. In particular, they are concerned for their mothers, whom they identify as being more invested than their fathers in keeping the family system *status quo ante*. Finally, and most significantly, these late adolescents manifest insecurity about their place in the family, especially now that they are leaving. Several of them remarked ruefully, "I should hope they feel some grief [laughter]." "I think they'll be lonelier. I hope they will." "They'll miss me, I hope.... I hope they feel my presence being gone.... They don't have to be, like, mourning my departure, but just a little bit would be nice." It's not that they actually want their parents and siblings to suffer, but missing them would be proof positive that their membership in the family was valued, and that their future place in the family system is assured, in spite of their changing addresses....

In many of our conversations, it appeared that the worst thing about going away to college was that the young people would no longer be able to participate in many aspects of family life. However, perhaps no issue symbolizes the worry associated with leaving home as powerfully as pending decisions over space in the household. How quickly one's bedroom is claimed by other members of the family is, for many of these students, a commentary on the fragility of their position. Although Silver (1996) points out that both the home room and college dorm room are used to symbolically affirm family relations, our conversations with students were more focused on the meanings they attached to their bedrooms at home. One senior said, "They always joke around and they say, 'Oh, we're going to make your room into a den.'" ...

Some of the seniors are beginning to understand that the nature of relations with their parents will be altered forever. They will have much more discretion concerning what to reveal about themselves, and therefore much more control over the impression they choose to give their parents. As one young woman put it, "I will experience a lot of things without them there, so that they won't know that they've happened ... [unless] I tell them or if they can see a difference in me." Others expressed shared anxieties about personal transformations and the consequent stability of their place in the family constellation....

What are we to make of these worries, speculations, and musings? College-bound young adults genuinely want to remain attached to their families, even as they are yearning for true independence. Getting into college is understood as a point of departure which has the potential to alter fundamentally their relationship with their family. However, in spite of their worries, most students see the transition to college as a good thing—a positive transformation with life-long consequences. They cannot predict precisely how their relations with parents and siblings will change, but they know for sure that they have initiated a process that will alter the character of these primary relationships. Such knowledge is plainly implicated in the calculus of ambivalence they feel about leaving home:

> It's like, if you want to be treated like an adult, you have to act like an adult. If you want to be treated like a child, act like a child. If you want to be treated like an adult the rest of your life, you've got to start sometime. (White male attending a public high school)

"You've got to start sometime." That, of course, is exactly what they are doing as they embark on their great adventure of self-discovery, into college first and hopefully, thereby, toward full adulthood.

NOTES

1. We used father's occupation as a proxy for social class. We characterized our sample as predominantly upper-middle class. A sampling of the types of father's occupations that warrant this description includes physician, lawyer, professor, administrator, and architect. A few occupations were either higher or lower in status.

2. Either because we could not reach them or because they declined to be interviewed, we did not speak to eight of the 31 students originally included in our sample. The number is 31 because one of the 30 families had twins.

3. One student, "who declined to be interviewed, did not complete the college application process during his senior year in high school. He "was the only student in our sample "who did not anticipate attending college in the year following high school graduation.

REFERENCES

Basoff, Evelyn. 1988. *Mothers and Daughters: Loving and Letting Go*. New York: Penguin Books.

Bellah, Robert, Richard Madsen, William Sullivan, Ann Swidler, and Steven Tipton. 1985. Habits of the Heart: Individualism and Commitment in American Life. Berkeley: University of California Press.

Blos, Peter. 1962. *On Adolescence: A Psychoanalytic Interpretation*. New York: Free Press.

Campbell, Eugene, Gerald Adams, and William Dobson. 1984. "Familial Correlates of Identity Formation in Late Adolescence: A Study of the Predictive Utility of Connectedness and Individuality in Family Relations." *Journal of Youth and Adolescence* 13: 509–525.

Erikson, Erik. 1963. Childhood and Society, 2nd ed. New York: W. W. Norton.

———. 1968. Identity: Youth and Crisis. New York: W. W. Norton.

———. 1974. Dimensions of a New Identity. New York: W. W. Norton.

———. 1980. Identity and the Life Cycle. New York: W. W. Norton.

Goffman, Erving. 1959. *The Presentation of Self in Everyday Life*. Garden City, NY: Doubleday Anchor.

Grigsby, Jill, and Jill McGowan. 1986. "Still in the Nest: Adult Children Living with Their Parents." *Sociology and Social Research* 70: 146–148.

Heer, David, Robert Hodge, and Marcus Felson. 1985. "The Cluttered Nest: Evidence That Young Adults Are More Likely to Live at Home Now Than in the Recent Past." *Sociology and Social Research* (69): 436–441.

Hochschild, Arlie. 1983. *The Managed Heart: Commercialization of Human Feeling*. Berkeley: University of California Press.

Hooks, Bell. 1993. "Keeping Close to Home: Class and Education." Pp. 99–111 in *Working-Class Women in the Academy*, edited by Michelle Tokarczyk and Elizabeth Fay. Amherst, MA: The University of Massachusetts Press.

Katchadourian, Herant, and John Boli. 1994. *Cream of the Crop: The Impact of Elite Education in the Decade After College*. New York: Basic Books.

Manaster, Guy. 1977. *Adolescent Development and the Life Tasks*. Boston: Allyn and Bacon.

O'Mally, Dawn. 1995. *Adolescent Development: Striking a Balance Between Attachment and Autonomy*. Ph.D. dissertation, Department of Psychology, Harvard University, Cambridge, MA.

Rodriguez, Richard. 1982. *Hunger of Memory: The Education of Richard Rodriguez*. Boston: David R. Godine.

Schnaiberg, Allan, and Sheldon Goldenberg. 1998. "From Empty Nest to Crowded Nest: The Dynamics of Incompletely-Launched Young Adults." *Social Problems* 36: 251–269.

Schreier, Barbara. 1991. *Fitting In: Four Generations of College Life*. Chicago: Chicago Historical Society.

Silver, Ira. 1996. "Role Transitions, Objects, and Identity." *Symbolic Interaction* 19: 1–20.

Smith, Patricia. 1993. "Grandma Went to Smith, All Right, But She Went from Nine to Five: A Memoir." Pp. 126–139 in *Working-Class Women in the Academy*, edited by Michelle Tokarczyk and Elizabeth Fay. Amherst, MA: The University of Massachusetts Press.

KEY CONCEPTS

anticipatory socialization   identity   rite of passage

DISCUSSION QUESTIONS

1. What changes did you go through (or are you going through) during your first year of college? How do these changes influence your self-identity and how others perceive you?

2. When you go home for vacations and visits, how does home feel differently now that you have lived away? What feels the same?

11

Barbie Girls versus Sea Monsters

Children Constructing Gender

MICHAEL A. MESSNER

In this article, Messner analyzes the gender differences among preschool soccer teams. His analysis uncovers how young boys and girls "do gender" when they

SOURCE: Michael A. Messner "Barbie Girls versus Sea Monsters: Children Constructing Gender" *Gender & Society* 14, no 6 (December 2000): 765–784.

interact and play with and among one another. Messner also discusses how the youth soccer league is structured in gendered ways. Finally, the research shows that popular culture icons, like Barbie dolls, provide symbols of gendered expectations for children.

In the past decade, studies of children and gender have moved toward greater levels of depth and sophistication (e.g., Jordan and Cowan 1995; McGuffy and Rich 1999; Thorne 1993). In her groundbreaking work on children and gender, Thorne (1993) argued that previous theoretical frameworks, although helpful, were limited: The top-down (adult-to-child) approach of socialization theories tended to ignore the extent to which children are active agents in the creation of their worlds—often in direct or partial opposition to values or "roles" to which adult teachers or parents are attempting to socialize them. Developmental theories also had their limits due to their tendency to ignore group and contextual factors while overemphasizing "the constitution and unfolding of *individuals* as boys or girls" (Thorne 1993, 4). In her study of grade school children, Thorne demonstrated a dynamic approach that examined the ways in which children actively construct gender in specific social contexts of the classroom and the playground. Working from emergent theories of performativity, Thorne developed the concept of "gender play" to analyze the social processes through which children construct gender. Her level of analysis was not the individual but *"group life*—with social relations, the organization and meanings of social situations, the collective practices through which children and adults create and recreate gender in their daily interactions" (Thorne 1993, 4).

A key insight from Thome's research is the extent to which gender varies in salience from situation to situation. Sometimes, children engage in "relaxed, cross sex play"; other times—for instance, on the playground during boys' ritual invasions of girls' spaces and games—gender boundaries between boys and girls are activated in ways that variously threaten or (more often) reinforce and clarify these boundaries. However, these varying moments of gender salience are not free-floating; they occur in social contexts such as schools and in which gender is formally and informally built into the division of labor, power structure, rules, and values (Connell 1987).

The purpose of this article is to use an observation of a highly salient gendered moment of group life among four- and five-year-old children as a point of departure for exploring the conditions under which gender boundaries become activated and enforced. I was privy to this moment as I observed my five-year-old son's first season (including weekly games and practices) in organized soccer. Unlike the long-term, systematic ethnographic studies of children conducted by Thorne (1993) or Adler and Adler (1998), this article takes one moment as its point of departure. I do not present this moment as somehow "representative" of what happened throughout the season; instead, I examine this as an example of what Hochschild (1994, 4) calls "magnified moments," which are "episodes of heightened importance, either epiphanies, moments of intense glee or unusual insight, or moments in which things go intensely but meaningfully wrong. In either case, the moment stands out; it is metaphorically rich, unusually elaborate and often echoes [later]." A magnified moment in daily life offers a window into the social construction of reality. It presents researchers with an opportunity to excavate gendered meanings and processes through an analysis of institutional and cultural contexts. The single empirical observation that serves as the point of departure for this article was made during a morning. Immediately after the event, I recorded my observations with detailed notes. I later slightly revised the notes after developing the photographs that I took at the event.

I will first describe the observation—an incident that occurred as a boys' four- and five-year-old soccer team waited next to a girls' four- and five-year-old soccer team for the beginning of the community's American Youth Soccer League (AYSO) season's opening ceremony. I will then examine this moment using three levels of analysis.

> *The interactional level:* How do children "do gender," and what are the contributions and limits of theories of performativity in understanding these interactions?
>
> *The level of structural context:* How does the gender regime, particularly the larger organizational level of formal sex segregation of AYSO, and the concrete, momentary situation of the opening ceremony provide a context that variously constrains and enables the children's interactions?
>
> *The level of cultural symbol:* How does the children's shared immersion in popular culture (and their differently gendered locations in this immersion) provide symbolic resources for the creation, in this situation, of apparently categorical differences between the boys and the girls?

Although I will discuss these three levels of analysis separately, I hope to demonstrate that interaction, structural context, and culture are simultaneous and mutually intertwined processes, none of which supersedes the others.

BARBIE GIRLS VERSUS SEA MONSTERS

It is a warm, sunny Saturday morning. Summer is coming to a close, and schools will soon reopen. As in many communities, this time of year in this small, middle and professional-class suburb of Los Angeles is marked by the beginning of another soccer season. This morning, 156 teams, with approximately 1,850 players ranging from 4 to 17 years old, along with another 2,000 to 3,000 parents, siblings, friends, and community digmtanes have gathered at the local high school football and track facility for the annual AYSO opening ceremonies. Parents and children wander around the perimeter of the track to find the assigned station for their respective teams. The coaches muster their teams and chat with parents. Eventually, each team will march around the track, behind their new team banner, as they are announced over the loudspeaker system and are applauded by the crowd. For now though, and for the next 45 minutes to an hour, the kids, coaches, and parents must stand, mill around, talk, and kill time as they await the beginning of the ceremony.

The Sea Monsters is a team of four- and five-year-old boys. Later this day, they will play their first-ever soccer game. A few of the boys already know each other from preschool, but most are still getting acquainted. They are wearing their new uniforms for the first time. Like other teams, they were assigned team colors—in this case, green and blue—and asked to choose their team name at their first team meeting, which occurred a week ago. Although they preferred "Blue Sharks," they found that the name was already taken by another team and settled on "Sea Monsters." A grandmother of one of the boys created the spiffy team banner, which was awarded a prize this morning. As they wait for the ceremony to begin, the boys inspect and then proudly pose for pictures in front of their new award-winning team banner. The parents stand a few feet away—some taking pictures, some just watching. The parents are also getting to know each other, and the common currency of topics is just how darned cute our kids look, and will they start these ceremonies soon before another boy has to be escorted to the bathroom?

Queued up one group away from the Sea Monsters is a team of four- and five-year-old girls in green and white uniforms. They too will play their first game later today, but for now, they are awaiting the beginning of the opening ceremony. They have chosen the name "Barbie Girls," and they also have a spiffy new team banner. But the girls are pretty much ignoring their banner, for they have created another, more powerful symbol around which to rally. In fact, they are the only team among the 156 marching today with a team float—a red Radio Flyer wagon base, on which sits a Sony boom box playing music, and a 3-foot-plus-tall Barbie doll on a rotating pedestal. Barbie is dressed in the team colors—indeed, she sports a custom-made green-and-white cheerleader-style outfit, with the Barbie Girls' names written on the skirt. Her normally all-blonde hair has been streaked with Barbie Girl green and features a green bow, with white polka dots. Several of the girls on the team also have supplemented their uniforms with green bows in their hair.

The volume on the boom box nudges up and four or five girls begin to sing a Barbie song. Barbie is now slowly rotating on her pedestal, and as the girls sing more gleefully and more loudly, some of them begin to hold hands and walk around the float, in sync with Barbie's rotation. Other same-aged girls from other teams are drawn to the celebration and, eventually, perhaps a dozen girls are singing the Barbie song. The girls are intensely focused on Barbie, on the music, and on their mutual pleasure.

As the Sea Monsters mill around their banner, some of them begin to notice, and then begin to watch and listen as the Barbie Girls rally around their float. At first, the boys are watching as individuals, seemingly unaware of each other's shared interest. Some of them stand with arms at their sides, slack-jawed, as though passively watching a television show. I notice slight smiles on a couple of their faces, as though they are drawn to the Barbie Girls' celebratory fun. Then, with side-glances, some of the boys begin to notice each other's attention on the Barbie Girls. Their faces begin to show signs of distaste. One of them yells out, "NO BARBIE!" Suddenly, they all begin to move—jumping up and down, nudging and bumping one other—and join into a group chant: "NO

BARBIE! NO BARBIE! NO BARBIE!" They now appear to be every bit as gleeful as the girls, as they laugh, yell, and chant against the Barbie Girls.

The parents watch the whole scene with rapt attention. Smiles light up the faces of the adults, as our glances sweep back and forth, from the sweetly celebrating Barbie Girls to the aggressively protesting Sea Monsters. "They are SO different!" exclaims one smiling mother approvingly. A male coach offers a more in-depth analysis: "When I was in college," he says, "I took these classes from professors who showed us research that showed that boys and girls are the same. I believed it, until I had my own kids and saw how different they are." "Yeah," another dad responds, "Just look at them! They are so different!"

The girls, meanwhile, show no evidence that they hear, see, or are even aware of the presence of the boys who are now so loudly proclaiming their opposition to the Barbie Girls' songs and totem. They continue to sing, dance, laugh, and rally around the Barbie for a few more minutes, before they are called to reassemble in their groups for the beginning of the parade.

After the parade, the teams reassemble on the infield of the track but now in a less organized manner. The Sea Monsters once again find themselves in the general vicinity of the Barbie Girls and take up the "NO BARBIE!" chant again. Perhaps put out by the lack of response to their chant, they begin to dash, in twos and threes, invading the girls' space, and yelling menacingly. With this, the Barbie Girls have little choice but to recognize the presence of the boys—some look puzzled and shrink back, some engage the boys and chase them off. The chasing seems only to incite more excitement among the boys. Finally, parents intervene and defuse the situation, leading their children off to their cars, homes, and eventually to their soccer games.

THE PERFORMANCE OF GENDER

In the past decade, especially since the publication of Judith Butler's highly influential *Gender Trouble* (1990), it has become increasingly fashionable among academic feminists to think of gender not as some "thing" that one "has" (or not) but rather as situationally constructed through the performances of active agents. The idea of gender as performance analytically foregrounds the agency of individuals in the construction of gender, thus highlighting the situational fluidity of gender: here, conservative and reproductive, there, transgressive and disruptive. Surely, the Barbie Girls versus Sea Monsters scene described above can be fruitfully analyzed as a moment of crosscutting and mutually constitutive gender performances: The girls—at least at first glance—appear to be performing (for each other?) a conventional four- to five-year-old version of emphasized femininity. At least on the surface, there appears to be nothing terribly transgressive here. They are just "being girls," together. The boys initially are unwittingly constituted as an audience for the girls' performance but quickly begin to perform (for each other?—for the girls, too?) a masculinity that constructs itself in opposition to Barbie, and to the girls, as not feminine. They aggressively confront—first through loud verbal chanting, eventually through bodily invasions—the girls' ritual space of emphasized femininity, apparently with the intention of disrupting its upsetting influence. The adults are simultaneously constituted as an adoring audience for their children's performances and as parents who perform for each other by sharing and mutually affirming their experience-based narratives concerning the natural differences between boys and girls.

In this scene, we see children performing gender in ways that constitute themselves as two separate, opposed groups (boys vs. girls) and parents performing gender in ways that give the stamp of adult approval to the children's performances of difference, while constructing their own ideological narrative that naturalizes this categorical difference. In other words, the parents do not seem to read the children's performances of gender as social constructions of gender. Instead, they interpret them as the inevitable unfolding of natural, internal differences between the sexes....

The parents' response to the Barbie Girls versus Sea Monsters performance suggests one of the main limits and dangers of theories of performativity. Lacking an analysis of structural and cultural context, performances of gender can all too easily be interpreted as free agents' acting out the inevitable surface manifestations of a natural inner essence of sex difference. An examination of structural and cultural contexts, though, reveals that there was nothing inevitable about the girls' choice of Barbie as their totem, nor in the boys' response to it.

THE STRUCTURE OF GENDER

In the entire subsequent season of weekly games and practices, I never once saw adults point to a moment in which boy and girl soccer players were doing the *same* thing and exclaim to each other, "Look at them! They are *so similar!*" The actual similarity of the boys and the girls, evidenced by nearly all of the kids' routine actions throughout a soccer season—playing the game, crying over a skinned knee, scrambling enthusiastically for their snacks after the games, spacing out on a bird or a flower instead of listening to the coach at practice—is a key to understanding the salience of the Barbie Girls versus Sea Monsters moment for gender relations. In the face of a multitude of moments that speak to similarity, it was this anomalous Barbie Girls versus Sea Monsters moment—where the boundaries of gender were so clearly enacted—that the adults seized to affirm their commitment to difference. It is the kind of moment—to use Lorber's (1994, 37) phrase—where "believing is seeing," where we selectively "see" aspects of social reality that tell us a truth that we prefer to believe, such as the belief in categorical sex difference. No matter that our eyes do not see evidence of this truth most of the rest of the time.

In fact, it was not so easy for adults to actually "see" the empirical reality of sex similarity in everyday observations of soccer throughout the season. That is due to one overdetermining factor: an institutional context that is characterized by informally structured sex segregation among the parent coaches and team managers, and by formally structured sex segregation among the children. The structural analysis developed here is indebted to Acker's (1990) observation that organizations, even while appearing "gender neutral," tend to reflect, re-create, and naturalize a hierarchical ordering of gender....

Adult Divisions of Labor and Power

There was a clear—although not absolute—sexual division of labor and power among the adult volunteers in the AYSO organization. The Board of Directors consisted of 21 men and 9 women, with the top two positions—commissioner and assistant commissioner—held by men. Among the league's head coaches, 133 were men and 23 women. The division among the league's assistant coaches was similarly skewed. Each team also had a team manager who was responsible for organizing snacks, making reminder calls about games and practices, organizing team parties and the end-of-the-year present for the coach. The vast majority of team managers were women. A common slippage in the language of coaches and parents revealed the ideological assumptions underlying this position: I often noticed people describe a team manager as the "team mom." In short, as Table 1 shows, the vast majority of the time, the formal authority of the head coach and assistant coach was in the hands of a man, while the backup, support role of team manager was in the hands of a woman.

These data illustrate Connell's (1987, 97) assertion that sexual divisions of labor are interwoven with, and mutually supportive of, divisions of power and authority among women and men. They also suggest how people's choices to volunteer for certain positions are shaped and constrained by previous institutional practices. There is no formal AYSO rule that men must be the leaders, women the supportive followers. And there are, after all, *some* women coaches and *some* men team managers. So, it may appear that the division of labor among adult volunteers simply manifests an accumulation of individual choices and preferences. When analyzed structurally, though, individual men's apparently free choices to volunteer disproportionately for coaching jobs, alongside individual women's apparently free choices to volunteer disproportionately for team manager jobs, can be seen as a logical collective result of the ways that the institutional structure of sport has differentially constrained and enabled women's and men's previous options and experiences (Messner 1992). Since boys and men

TABLE 1   **Adult Volunteers as Coaches and Team Managers, by Gender (in percentages) (*N* = 156 teams)**

	Head Coaches	Assistant Coaches	Team Managers
Women	15	21	86
Men	85	79	14

have had far more opportunities to play organized sports and thus to gain skills and knowledge, it subsequently appears rational for adult men to serve in positions of knowledgeable authority, with women serving in a support capacity (Boyle and McKay 1995). Structure—in this case, the historically constituted division of labor and power in sport—constrains current practice. In turn, structure becomes an object of practice, as the choices and actions of today's parents recreate divisions of labor and power similar to those that they experienced in their youth.

The Children: Formal Sex Segregation

As adult authority patterns are informally structured along gendered lines, the children's leagues are formally segregated by AYSO along lines of age and sex. In each age-group, there are separate boys' and girls' leagues. The AYSO in this community included 87 boys' teams and 69 girls' teams. Although the four- to five-year-old boys often played their games on a field that was contiguous with games being played by four- to five-year-old girls, there was never a formal opportunity for cross-sex play. Thus, both the girls' and the boys' teams could conceivably proceed through an entire season of games and practices in entirely homosocial contexts. In the all-male contexts that I observed throughout the season, gender never appeared to be overtly salient among the children, coaches, or parents. It is against this backdrop that I might suggest a working hypothesis about structure and the variable salience of gender: The formal sex segregation of children does not, in and of itself, make gender overtly salient. In fact, when children are absolutely segregated, with no opportunity for cross-sex interactions, gender may appear to disappear as an overtly salient organizing principle. However, when formally sex-segregated children are placed into immediately contiguous locations, such as during the opening ceremony, highly charged gendered interactions between the groups (including invasions and other kinds of border work) become more possible.

Although it might appear to some that formal sex segregation in children's sports is a natural fact, it has not always been so for the youngest age-groups in AYSO. As recently as 1995, when my older son signed up to play as a five-year-old, I had been told that he would play in a coed league. But when he arrived to his first practice and I saw that he was on an all-boys team, I was told by the coach that AYSO had decided this year to begin sex segregating all age-groups, because "during halftimes and practices, the boys and girls tend to separate into separate groups. So the league thought it would be better for team unity if we split the boys and girls into separate leagues." I suggested to some coaches that a similar dynamic among racial ethnic groups (say, Latino kids and white kids clustering as separate groups during halftimes) would not similarly result in a decision to create racially segregated leagues. That this comment appeared to fall on deaf ears illustrates the extent to which many adults' belief in the need for sex segregation—at least in the context of sport—is grounded in a mutually agreed-upon notion of boys' and girls' "separate worlds," perhaps based in ideologies of natural sex difference.

The gender regime of AYSO, then, is structured by formal and informal sexual divisions of labor and power. This social structure sets ranges, limits, and possibilities for the children's and parents' interactions and performances of gender, but it does not determine them. Put another way, the formal and informal gender regime of AYSO made the Barbie Girls versus Sea Monsters moment possible, but it did not make it inevitable. It was the agency of the children and the parents within that structure that made the moment happen. But why did this moment take on the symbolic forms that it did? How and why do the girls, boys, and parents construct and derive meanings from this moment, and how can we interpret these meanings? These questions are best grappled within in the realm of cultural analysis.

THE CULTURE OF GENDER

The difference between what is "structural" and what is "cultural" is not clear-cut. For instance, the AYSO assignment of team colors and choice of team names (cultural symbols) seem to follow logically from, and in turn reinforce, the sex segregation of the leagues (social structure). These cultural symbols such as team colors, uniforms, songs, team names, and banners often carried encoded gendered meanings that were then available to be taken up by the children in ways that constructed (or potentially contested) gender divisions and boundaries.

Team Names

Each team was issued two team colors. It is notable that across the various age-groups, several girls' teams were issued pink uniforms—a color commonly recognized as encoding feminine meanings—while no boys' teams were issued pink uniforms. Children, in consultation with their coaches, were asked to choose their own team names and were encouraged to use their assigned team colors as cues to theme of the team name (e.g., among the boys, the "Red Flashes," the "Green Pythons," and the blue-and-green "Sea Monsters"). When I analyzed the team names of the 156 teams by age-group and by sex, three categories emerged:

1. Sweet names: These are cutesy team names that communicate small stature, cuteness, and/or vulnerability. These kinds of names would most likely be widely read as encoded with feminine meanings (e.g., "Blue Butterflies," "Beanie Babes," "Sunflowers," "Pink Flamingos," and "Barbie Girls").

2. Neutral or paradoxical names: Neutral names are team names that carry no obvious gendered meaning (e.g., "Blue and Green Lizards," "Team Flubber," "Galaxy," "Blue Ice"). Paradoxical names are girls' team names that carry mixed (simultaneously vulnerable *and* powerful) messages (e.g., "Pink Panthers," "Flower Power," "Little Tigers").

3. Power names: These are team names that invoke images of unambiguous strength, aggression, and raw power (e.g., "Shooting Stars," "Killer Whales," "Shark Attack," "Raptor Attack," and "Sea Monsters").

... [A]cross all age-groups of boys, there was only one team name coded as a sweet name—"The Smurfs," in the 10- to 11-year-old league. Across all age categories, the boys were far more likely to choose a power name than anything else, and this was nowhere more true than in the youngest age-groups, where 35 of 40 (87 percent) of boys' teams in the four-to-five and six-to-seven age-groups took on power names. A different pattern appears in the girls' team name choices, especially among the youngest girls. Only 2 of the 12 four- to five-year-old girls' teams chose power names, while 5 chose sweet names and 5 chose neutral/ paradoxical names. At age six to seven, the numbers begin to tip toward the boys' numbers but still remain different, with half of the girls' teams now choosing power names. In the middle and older girls' groups, the sweet names all but disappear, with power names dominating, but still a higher proportion of neutral/ paradoxical names than among boys in those age-groups.

Barbie Narrative versus Warrior Narrative

How do we make sense of the obviously powerful spark that Barbie provided in the opening ceremony scene described above? Barbie is likely one of the most immediately identifiable symbols of femininity in the world. More conservatively oriented parents tend to happily buy Barbie dolls for their daughters, while perhaps deflecting their sons' interest in Barbie toward more sex-appropriate "action toys." Feminist parents, on the other hand, have often expressed open contempt—or at least uncomfortable ambivalence—toward Barbie. This is because both conservative and feminist parents see dominant cultural meanings of emphasized femininity as condensed in Barbie and assume that these meanings will be imitated by their daughters. Recent developments in cultural studies, though, should warn us against simplistic readings of Barbie as simply conveying hegemonic messages about gender to unwitting children (Attfield 1996; Seiter 1995). In addition to critically analyzing the cultural values (or "preferred meanings") that may be encoded in Barbie or other children's toys, feminist scholars of cultural studies point to the necessity of examining "reception, pleasure, and agency," and especially "the fullness of reception contexts" (Walters 1999, 246). The Barbie Girls versus Sea Monsters moment can be analyzed as a "reception context," in which differently situated boys, girls, and parents variously used Barbie to construct pleasurable intergroup bonds, as well as boundaries between groups....

... Indeed, as the Barbie Girls rallied around Barbie, their obvious pleasure did not appear to be based on a celebration of quiet passivity (as feminist parents might fear). Rather, it was a statement that they—the Barbie Girls—were here in this public space. They were not silenced by the boys' oppositional chanting. To the contrary, they ignored the boys, who seemed irrelevant to their celebration. And, when the boys later physically invaded their space, some of the girls responded by chasing the boys off. In short, when I pay attention to what the girls *did* (rather than imposing on the situation what I *think* Barbie "should" mean to the girls), I see a public moment of celebratory "girl power."

And this may give us better basis from which to analyze the boys' oppositional response. First, the boys may have been responding to the threat of displacement they may have felt while viewing the girls' moment of celebratory girl power. Second, the boys may simultaneously have been responding to the fears of feminine pollution that Barbie had come to symbolize to them. But why might Barbie symbolize feminine pollution to little boys? A brief example from my older son is instructive. When he was about three, following a fun day of play with the five-year-old girl next door, he enthusiastically asked me to buy him a Barbie like hers. He was gleeful when I took him to the store and bought him one. When we arrived home, his feet had barely hit the pavement getting out of the car before an eight-year-old neighbor boy laughed at and ridiculed him: "A *Barbie?* Don't you know that Barbie is a *girl's toy?*" No amount of parental intervention could counter this devastating peer-induced injunction against boys' playing with Barbie. My son's pleasurable desire for Barbie appeared almost overnight to transform itself into shame and rejection. The doll ended up at the bottom of a heap of toys in the closet, and my son soon became infatuated, along with other boys in his preschool, with Ninja Turtles and Power Rangers....

By kindergarten, most boys appear to have learned—either through experiences similar to my son's, where other male persons police the boundaries of gender-appropriate play and fantasy and/or by watching the clearly gendered messages of television advertising—that Barbie dolls are not appropriate toys for boys (Rogers 1999, 30). To avoid ridicule, they learn to hide their desire for Barbie, either through denial and oppositional/pollution discourse and/or through sublimation of their desire for Barbie into play with male-appropriate "action figures" (Pope et al. 1999). In their study of a kindergarten classroom, Jordan and Cowan (1995, 728) identified "warrior narratives ... that assume that violence is legitimate and justified when it occurs within a struggle between good and evil" to be the most commonly agreed-upon currency for boys' fantasy play. They observe that the boys seem commonly to adapt story lines that they have seen on television. Popular culture—film, video, computer games, television, and comic books—provides boys with a seemingly endless stream of Good Guys versus Bad Guys characters and stories—from cowboy movies, Superman and Spiderman to Ninja Turtles, Star Wars, and Pokémon—that are available for the boys to appropriate as the raw materials for the construction of their own warrior play....

A cultural analysis suggests that the boys' and the girls' previous immersion in differently gendered cultural experiences shaped the likelihood that they would derive and construct different meanings from Barbie—the girls through pleasurable and symbolically empowering identification with "girl power" narratives; the boys through oppositional fears of feminine pollution (and fears of displacement by girl power?) and with aggressively verbal, and eventually physical, invasions of the girls' ritual space. The boys' collective response thus constituted them differently, *as boys*, in opposition to the girls' constitution of themselves *as girls*. An individual girl or boy, in this moment, who may have felt an inclination to dissent from the dominant feelings of the group (say, the Latina Barbie Girl who, her mother later told me, did not want the group to be identified with Barbie, or a boy whose immediate inner response to the Barbie Girls' joyful celebration might be to join in) is most likely silenced into complicity in this powerful moment of border work.

What meanings did this highly gendered moment carry for the boys' and girls' teams in the ensuing soccer season? Although I did not observe the Barbie Girls after the opening ceremony, I did continue to observe the Sea Monsters' weekly practices and games. During the boys' ensuing season, gender never reached this "magnified" level of salience again—indeed, gender was rarely raised verbally or performed overtly by the boys. On two occasions, though, I observed the coach jokingly chiding the boys during practice that "if you don't watch out, I'm going to get the Barbie Girls here to play against you!" This warning was followed by gleeful screams of agony and fear, and nervous hopping around and hugging by some of the boys. Normally, though, in this sex-segregated, all-male context, if boundaries were invoked, they were not boundaries between boys and girls but boundaries between the Sea Monsters and other boys' teams, or sometimes age boundaries between the Sea Monsters and a small group of dads and older brothers who would engage them in a mock scrimmage during practice. But it was also evident that when the coach was having trouble getting the boys to act together, as a group, his strategic and humorous invocation of the dreaded Barbie Girls once again served symbolically to affirm their group status. They were a team. They were the boys.

CONCLUSION

The overarching goal of this article has been to take one empirical observation from everyday life and demonstrate how a multilevel (interactioinst, structural, cultural) analysis might reveal various layers of meaning that give insight into the everyday social construction of gender. This article builds on observations made by Thorne (1993) concerning ways to approach sociological analyses of children's worlds. The most fruitful approach is not to ask why boys and girls are so different but rather to ask how and under what conditions boys and girls constitute themselves as separate, oppositional groups. Sociologists need not debate whether gender is "there"—clearly, gender is always already there, built as it is into the structures, situations, culture, and consciousness of children and adults. The key issue is under what conditions gender is activated as a salient organizing principle in social life and under what conditions it may be less salient. These are important questions, especially since the social organization of categorical gender difference has always been so clearly tied to gender hierarchy (Acker 1990; Lorber 1994). In the Barbie Girls versus Sea Monsters moment, the performance of gendered boundaries and the construction of boys' and girls' groups as categorically different occurred in the context of a situation systematically structured by sex segregation, sparked by the imposing presence of a shared cultural symbol that is saturated with gendered meanings, and actively supported and applauded by adults who basked in the pleasure of difference, reaffirmed.

I have suggested that a useful approach to the study of such "how" and "under what conditions" questions is to employ multiple levels of analysis. At the most general level, this project supports the following working propositions.

Interactionist theoretical frameworks that emphasize the ways that social agents "perform" or "do" gender are most useful in describing how groups of people actively create (or at times disrupt) the boundaries that delineate seemingly categorical differences between male persons and female persons. In this case, we saw how the children and the parents interactively performed gender in a way that constructed an apparently natural boundary between the two separate worlds of the girls and the boys.

Structural theoretical frameworks that emphasize the ways that gender is built into institutions through hierarchical sexual divisions of labor are most useful in explaining under what conditions social agents mobilize variously to disrupt or to affirm gender differences and inequalities. In this case, we saw how the sexual division of labor among parent volunteers (grounded in their own histories in the gender regime of sport), the formal sex segregation of the children's leagues, and the structured context of the opening ceremony created conditions for possible interactions between girls' teams and boys' teams.

Cultural theoretical perspectives that examine how popular symbols that are injected into circulation by the culture industry are variously taken up by differently situated people are most useful in analyzing how the meanings of cultural symbols, in a given institutional context, might trigger or be taken up by social agents and used as resources to reproduce, disrupt, or contest binary conceptions of sex difference and gendered relations of power. In this case, we saw how a girls' team appropriated a large Barbie around which to construct a pleasurable and empowering sense of group identity and how the boys' team responded with aggressive denunciations of Barbie and invasions....

... The eventual interactions between the boys and the girls were made possible—although by no means fully determined—by the structure of the gender regime and by the cultural resources that the children variously drew on.

On the other hand, the gendered division of labor in youth soccer is not seamless, static, or immune to resistance. One of the few woman head coaches, a very active athlete in her own right, told me that she is "challenging the sexism" in AYSO by becoming the head of her son's league. As post—Title IX women increasingly become mothers and as media images of competent, heroic female athletes become more a part of the cultural landscape for children, the gender regimes of children's sports may be increasingly challenged (Dworkin and Messner 1999). Put another way, the dramatically shifting opportunity structure and cultural imagery of post—Title IX sports have created opportunities for new kinds of interactions, which will inevitably challenge and further shift institutional structures. Social structures simultaneously constrain and enable, while agency is simultaneously reproductive and resistant.

REFERENCES

Acker, Joan. 1990. Hierarchies, jobs, bodies: A theory of gendered organizations. *Gender & Society* 4:139–58.

Adler, Patricia A., and Peter Adler. 1998. *Peer power: Preadolescent culture and identity*. New Brunswick, NJ: Rutgers University Press.

Attfield, Judy. 1996. Barbie and Action Man: Adult toys for girls and boys, 1959–93. In *The gendered object*, edited by Pat Kirkham, 80–89. Manchester, UK, and New York: Manchester University Press.

Boyle, Maree, and Jim McKay. 1995. "You leave your troubles at the gate": A case study of the exploitation of older women's labor and "leisure" in sport. *Gender & Society* 9:556–76.

Butler, Judith. 1990. *Gender trouble: Feminism and the subversion of identity*. New York and London: Routledge.

Connell, R.W. 1987. *Gender and power*. Stanford, CA: Stanford University Press.

Dworkin, Shari L., and Michael A. Messner. 1999. Just do ... what?: Sport, bodies, gender. In *Revisioning gender*, edited by Myra Marx Ferree, Judith Lorber, and Beth B. Hess, 341–61. Thousand Oaks, CA: Sage.

Hochschild, Arlie Russell. 1994. The commercial spirit of intimate life and the abduction of feminism: Signs from women's advice books. *Theory, Culture & Society* 11:1–24.

Hooks, Bell. 1993. "Keeping Close to Home: Class and Education." Pp. 99–11 in *Working Class Women in the Academy*, edited by Michelle Tokarczyk and Elizabeth Fay. Amherst, MA: University of Massachusetts Press.

Jordan, Ellen, and Angela Cowan. 1995. Warrior narratives in the kindergarten classroom: Renegotiating the social contract? *Gender & Society* 9:727–43.

Lorber, Judith. 1994. *Paradoxes of gender*. New Haven, CT, and London: Yale University Press.

McGuffy, C. Shawn, and B. Lindsay Rich. 1999. Playing in the gender transgression zone: Race, class and hegemonic masculinity in middle childhood. *Gender & Society* 13:608–27.

Messner, Michael A. 1992. *Power at play: Sports and the problem of masculinity*. Boston: Beacon.

Pope, Harrison G., Jr., Roberto Olivarda, Amanda Gruber, and John Borowiecki. 1999. Evolving ideals of male body image as seen through action toys. *International Journal of Eating Disorders* 26:65–72.

Rogers, Mary F. 1999. *Barbie culture*. Thousand Oaks, CA: Sage.

Seiter, Ellen. 1995. *Sold separately: Parents and children in consumer culture*. New Brunswick, NJ: Rutgers University Press.

Thorne, Barrie. 1993. *Gender play: Girls and boys in school*. New Brunswick, NJ: Rutgers University Press.

Walters, Suzanna Danuta. 1999. Sex, text, and context: (In) between feminism and cultural studies. In *Revisioning gender*, edited by Myra Marx Ferree, Judith Lorber, and Beth B. Hess, 222–57. Thousand Oaks, CA: Sage.

KEY CONCEPTS

doing gender   gender segregation   gender socialization

DISCUSSION QUESTIONS

1. Think back to a time when you may have played in organized groups (sports, camps, or some other activity). How is the example of the soccer league described in this article similar to what you may have experienced? How is it different?

2. What are some popular children's toys today? How do they socialize children into gendered roles? Are there toys today that construct gender differently than when you were a child?

12

Gender and Ageism

LAURIE RUSSELL HATCH

Laurie Hatch details some of the age stereotypes that are prevalent in the media, arguing that such stereotypes are also entangled with gender stereotypes. She also reviews the biases that are thus present in various social policies for older people.

The "double standard of aging" is a phrase heard so frequently that it seems a truism. The social worth of women has been linked more closely with their physical appearance compared to the situation for men, and these social valuations decline more markedly with age for women than they do for men (e.g., Hurd, 2000). Looking "old" is viewed more harshly for women across diverse cultures and also, apparently, for people of different sexual orientations, extending beyond heterosexual bias (Harris, 1994). As encapsulated by Garner (1999), "women lose their social value simply by growing old. Men are more likely to be evaluated and rewarded for what they do" (p. 4).

This article looks into biased constructions of gender and aging, reflected in the mass media, embedded in social policies, and evidenced in differential treatment of older women and men in healthcare encounters. The main question is how we as individual women and men perceive ourselves as we grow older. Do we "buy into" cultural stereotypes of gender and aging, or do we resist such stereotypes? In addition to the potential for resistance at the level of individuals, what is the potential for reducing ageism in society more broadly? The aim is to

SOURCE: *Generations* (Fall 2005): 19–24.

explore these questions, recognizing that specific forms and relative virulence of ageism extend beyond gender to also encompass ethnicity, social class, sexuality, and other forms of social differentiation and inequality.

MEDIA PORTRAYALS

The role of mass media as cause or consequence of cultural values has long been contested. At minimum, the media reflect dominant values in a society: Whether, and how, older women and men are portrayed represents one measure of how ageism and sexism are embedded in the social fabric (e.g., Holbrook, 1987). Older adults of both sexes are underrepresented in U.S. popular films and television programs relative to younger age groups, and older women are even less likely to appear on screen than older men (Bazzini and McIntosh, 1997; Sanders, 2002). When older characters are portrayed, women are more frequently depicted in negative stereotypes and shown as less successful compared to older men. Similar findings regarding the portrayal of older women and men on screen have been reported in other western countries (e.g., Kessler, Rakoczy, and Staudinger, 2004).

Older adults also are underrepresented in advertisements relative to younger age groups, and negative portrayals increase with the age of individuals featured (e.g., Peterson and Ross, 1997). Studies of advertisements from the 1950s through the 1990s have found, however, that images of older people in U.S. television and print advertising have become more positive over time (Miller et al.,1999; Miller, Leyell, and Mazachek, 2004). The researchers concede that these "positive images" nevertheless convey stereotypes of aging. Older people featured in advertisements are portrayed most frequently as stereotypic "golden-agers" (either as adventurous or "productive" elders) or as "perfect grandparents." Further, these images are more often utilized for older women featured in advertisements than for older men.

A more nuanced consideration of media portrayal of women as golden-agers is found in Pedersen's (2002) in-depth study of a Kellogg's cereal print advertising campaign, with the slogan, "Look good on your own terms." These advertisements use direct messages as well as humor and exaggeration to "suggest that a healthy body is beautiful and that feeling healthy is more important than looking good according to media standards" (p. 170). At the same time, the advertisements also send the message that "anyone can become slim and beautiful by eating Special K[cereal]" (p. 181). Pedersen concludes that in so doing, this popular advertising campaign reinforces the very gender stereotypes and ageism that it purports to challenge.

With marketing efforts grounded in a market economy, it is not surprising that advertisers seek to mine "the gold in the gray." Miller and colleagues (2004) assert that because television advertisers have "just a few seconds to clearly convey a selling proposition" (p. 337), they should not necessarily be faulted in using (positive) stereotypical images of older women and men. A much wider and more varied set of images are utilized in advertisements featuring younger and middle-age groups, however, and these groups also are featured in media portrayals much more frequently than are older adults.

GENDER AND OLD-AGE SOCIAL POLICES

Although ageism typically is understood in terms of negative imagery, attitudes, and practices aimed at elders, old-age programs in the U.S. as well as in other countries increasingly are criticized as benefiting older age groups at the expense of children and younger adults (eg., Coombs and Dollery, 2004; Newacheck and Benjamin, 2004). A sharply contrasting perspective is that Social Security, Medicare, and other old-age programs reflect an intergenerational compact in which Americans of all ages share a long-term stake (Kingson, Hirshorn, and Cornman, 1986).

Whether viewed as biased in favor of older people or as promoting the common intergenerational good, extant old-age policies reflect the interests of men more closely than those of women (Hendricks, Hatch, and Cutler, 1999). The benefits afforded through Medicare represent a case in point. Medicare is geared primarily toward healthcare for acute illnesses and covers relatively few expenses incurred by chronic illnesses. Whereas older women are more likely than men to suffer from debilitating chronic illnesses, older men contract fewer long-term chronic illnesses but are more prone to have acute illnesses. Hence, Medicare meets the healthcare needs of older men more fully than it does those of older women.

Another example of this disparity is the differential effect of Medicare's failure to cover long-term care in a nursing home, a "last resort" residence option for more women than men. Primarily because of their longer life expectancies and a greater likelihood of being widowed and living alone (Steinbach, 1992; Wolinsky and Johnson, 1992), women outnumber men in nursing homes by a ratio of roughly three to one (U.S. Department of Health and Human Services, 2004). The public policy provision for nursing home coverage is through Medicaid, the means-tested health insurance program for poor Americans of all ages. Individuals who are not already poor when they enter a nursing home typically become so rather quickly by virtue of the high cost of nursing home care, which averages over $51,000 per year. After expending their personal assets, individuals may then become eligible for Medicaid benefits to cover nursing home care (see Schulz, 2001). "By default then, the United States has a poverty-based long-term care system" (Harrington Meyer, 1990, p. 9), in which older women predominate.

Social Security provisions, like those afforded through Medicare, are more congruent with a (primarily white, middle-class) male model of life experience. Social Security does not take into consideration the unpaid work that women frequently perform as family care providers. Social Security does provide spousal benefits, which some may assert implicitly recognize household and family work.

However, these benefits are linked solely with maintaining a marital relationship rather than with performing childcare, eldercare, or other family labor (Harrington Meyer, 1990). In contrast, public pensions in other countries including Britain, Germany, and Japan include provisions for homemaker credits (Schulz, 2001). In the U.S., spousal benefits are afforded only when marriage lasts a minimum of ten years, and to date have extended to heterosexual couples only. Recent legislation recognizing same-sex marriage in Massachusetts and pending in other states may put this benefit restriction to a legal test.

Social Security offers a dual-eligibility structure by which individuals can receive benefits either as wage earners or as spouses of wage earners. Although spousal benefits are paid out at the rate of just one-half the benefits of the primary earner, the payments that many women would receive from their own earnings histories would be even lower (Harrington Meyer, 1990). Because of their lower lifetime earnings and more often interrupted labor-force histories, women and racial/ethnic minorities generally receive lower average "primary earner" benefits from Social Security compared to white (primarily middle- and upper-class) men. More generally, old-age policies and programs perpetuate economic inequalities established prior to retirement, and these inequalities are linked closely with gender, race/ethnicity, and social class (see Hatch, 2000).

AGEISM AND SEXISM IN HEALTHCARE ENCOUNTERS

The nexus of ageism and sexism is not only embedded in societal institutions, but is also expressed in everyday interactions. Among those with the greatest consequence are interactions with healthcare providers, where ageism and sexism are especially problematic. In these encounters, stereotypes are often expressed in subtle ways and may not be held consciously, but they have important implications for health (Sharpe, 1995). For example, older adults interviewed by Minichiello and colleagues (2000) reported that healthcare workers typically did not behave in overtly ageist ways toward them but said that the workers conveyed a sense of the older patient as unimportant and often did not consult the person directly about healthcare decisions. Audiotaped interactions between internists and general medical outpatients found that the physicians were more egalitarian, patient, and engaged with younger patients than with older ones. In addition, the physicians were more likely to raise psychosocial concerns with younger patients than with older ones and to be more responsive to such concerns when expressed by younger patients (Sharpe, 1995).

Research suggests that the concerns of older women are especially likely to be trivialized in healthcare encounters (Sharpe, 1995). For example, a study of audiotaped interactions between male physicians and older women patients documented frequent interruptions from the physicians and devaluing of the women's concerns (Davis, 1988). More generally, healthcare providers have been more likely to label expressed health concerns as psychosomatic when the patient is a woman (Munch, 2004).

Such biases may result in differential health treatment for women and men. For example, Correa-de-Araujo (2004) indicates that men were more likely to receive a kidney transplant in every age category, but the gender gap was greatest in the group 46 to 60 years old. Women in this age group had only half the chance of receiving a transplant as men. The studies reviewed also suggest that cardiovascular disease is diagnosed and treated later for women than for men, resulting in higher mortality rates for coronary bypass surgery among women.

SEXISM, AGEISM, AND SELF-IDENTITIES

The most insidious ways in which sexist, ageist, and other biases can take shape and meaning is when they are internalized by the targeted groups. Consistent with a double standard of gender and physical attractiveness, Hurd's (2000) interviews with women ages 61 to 92 suggest that older women "exhibit the internalization of ageist beauty norms even as they assert that health is more important to them than physical attractiveness and comment on the 'naturalness' of the aging process" (p. 77). Although men also face pressures to dye their hair and appear more youthful (Thompson, 2004), women are far more likely than men to use cosmetic and other techniques to reduce the physical signs of aging. Harris (1994) reports that 34 percent of the women in her sample compared to just 6 percent of men dyed their hair, and that 24 percent of women compared to 1 percent of men used wrinkle cream. On the other hand, women apparently view physical signs of aging as less unattractive than do men (Harris, 1994).

Ageism no doubt takes somewhat different forms for women and men, with broad-ranging implications for self-identity. Thompson (2004) avers that the research literature has focused on the confluence of ageism and sexism for women and has neglected how older men's identities also can be profoundly influenced by ageism. According to Thompson, "growing older seems to be outside conceptualizations of masculinity.... Normative masculinity became and remains embodied by middle-aged and younger men" (p. 1). Drummond's (2003) interviews found that men who relied on their physicality for masculine identity felt inadequate as their bodies showed increasing signs of age, an observation that has been made more frequently for women.

COMBATING SEXISM AND AGEISM

This article has reviewed some of the biases reflected in media portrayals of older women and men, embedded in selected social policies, evidenced in interactions within healthcare settings, and potentially mirrored in the self-identities of older women and men. What is the potential for combating and reducing these biases?

Media portrayals. The evidence is that individuals can be made aware of media stereotyping and that they may take action on the basis of such recognition. Donlon, Ashman, and Levy (2005) studied the television viewing habits of older adults ages 60 to 92. Overall, the researchers found that greater exposure to television was associated with more negative views of aging. Study participants who were randomly assigned to an intervention group were asked to record diaries of their television viewing for one week, noting the frequency with which older characters appeared on screen and the ways in which older characters were portrayed. At the end of the study period, participants who were assigned to the intervention group were more aware of negative portrayals and infrequency of older characters on television than the control group, who had not maintained diaries of their television watching. Participants in the intervention group also expressed their intention to watch less television in the future. Donlon and colleagues concluded that "these findings suggest that the promotion of awareness provides a means of helping elders confront ageism" (p. 307).

Combating biased media representations is likely to require more than older people simply reducing television viewing. It is possible that the increasing size and presumed economic clout of the older population, combined with a greater awareness of media stereotyping of older women and men, will result in more varied (if not more balanced) portrayals of older adults.

Social policies. The increasing size and presumed clout of the older population also have important implications for old-age policies and programs. With the future of Social Security and other old-age programs increasingly in question, debates over gendered biases embedded in these programs and possible strategies to reduce gender inequities have receded further from scholarly and policy discussions in the U.S. Without doubt, Social Security, Medicare, Medicaid, and other programs have improved the lives of older Americans across gender, race, and ethnic groups. The trend toward retrenchment in these programs is likely to result in the impoverishment of more older people than are poor today, including the groups who are most in jeopardy for poverty: women and minorities (Hendricks, Hatch, and Cutler, 1999).

One aspect of the public discourse surrounding cutbacks in Social Security, Medicare, Medicaid, and other old-age programs that deserves more attention is the predomination of women in the older population, particularly the rapidly growing "oldest-old" group age 85 and older. There are almost three million women age 85-plus in the U.S., representing 70 percent of Americans in this age category (U.S. Census Bureau, 2004). The increasingly voiced concern that public monies cannot support the economic and healthcare needs of an aging population is linked implicitly with concerns regarding an aging female population. As Steckenrider (1998) notes, "Although age-related public policies, past and present, have not developed with the intent of deliberate gender bias, the male life cycle has been the standard for policy decisions" (p. 242). A question that is provoking to ponder but difficult to answer is, Might the increasingly heated debates surrounding intergenerational equity and old-age policy reform be framed differently if men, rather than women, represented the numeric majority in older age?

Healthcare encounters and the healthcare system. Ten years ago, Sharpe (1995) proposed a multifaceted approach to problems of ageism and sexism within the healthcare system, beginning with the healthcare encounter. He and others have called for educating health professionals about ageist and sexist stereotypes, attending also to cultural diversities. Yet, relatively few medical education programs to date have focused on communicating effectively with older patients (see Weir, 2004).

A recent study found that women receive better healthcare in National Centers of Excellence in Women's Health (initiated in 1996 through the Department of Health and Human Services and the Office on Women's Health) (Henderson et al., 2004). Women who received care in these comprehensive women's health centers were more likely to receive physical breast examinations and mammograms at age 50 and older than in other primary healthcare settings.

Self-identities of older women and men. Some scholars have asserted that while older adults exempt themselves from aging stereotypes, they apply these stereotypes to peers who are deemed "old" (e.g., Marshall and McPherson, 1994). Other scholars have argued that self-identities are greatly influenced by ageist stereotyping, and that women often internalize the "double standard" of aging and physical attractiveness (e.g., Hurd, 2000).

How do older women and men *themselves* understand ageism, sexism, and other biases, and what strategies do they use to negotiate their own experiences? Wray (2002) cautions that "conceptualizing acts of resistance [like refusing to dye gray hair] as something (older) people do to simply avoid ageist stereotypes reinforces the oppositional binary of youth and old age" (p. 523). In her study of older women of diverse ethnicities, Wray found that issues concerning racism were of greater concern to many of the women than remaining "youthful" (p. 523). Both individually and collectively, the women utilized culturally relevant strategies to maintain a sense of control in their lives. This study also highlights how older women took action in dealing with what many would consider to be debilitating health problems. The ability to "get on with life" and maintain a sense of autonomy in the face of racism, health disabilities, and other challenges is often overlooked, "as it is assumed that the primary desire of all older people is to resist ageing per se" (Wray, 2002, p. 518).

IN SUM

It is important to acknowledge that ageism and sexism are pervasive forces that underlie and shape the lives of older people. At the same time, these are of course not the only forces that individuals experience in particular contexts. Rather, there is a "multiplicity of points of resistance" (Foucault, 1980, p. 95) related to power relations that is grounded in the lived experiences of women and men at all ages.

REFERENCES

Bazzini, D. G., and McIntosh, W. D. 1997. "The Aging Women in Popular Film: Underrepresented, Unattractive, Unfriendly, and Unintelligent." *Sex Roles* 36(7/8): 531–43.

Coombs, G., and Dollety, B. 2004. "The Ageing of Australia: Fiscal Sustainability, Intergenerational Equity and Inter-temporal Fiscal Balance." *Australian Journal of Social issues* 39(4): 459–70.

Correa-de-Araujo, R. 2004. "A Wake-Up Call to Advance Women's Health." *Women's Health Issues* 14: 31–4.

Davis, K. 1988. "Paternalism Under the Microscope." In A. D. Todd and S. Fisher, eds., *Gender and Discourse: The power of Talk*. Norwood, N.J.: Ablex.

Donlon, M. M., Ashman, O., and Levy, B. R. 2005. "Re-Vision of Older Television Characters: A Stereotype-Awareness intervention." *Journal of Social Issues* 61(2): 307–19.

Garner, J. D. 1999. "Feminism and Feminist Gerontology." *Journal of Women & Aging* 11 (2/3): 3–12.

Harrington Meyer, M. 1990. "Family Status and Poverty Among Older Women: The Gendered Distribution of Retirement Income in the United States." *Social Problems* 37: 551–63.

Harris, M. B. 1994. "Growing Old Gracefully: Age Concealment and Gender." *Journals of Gerontology: Psychological Sciences* 49(4): 149–58.

Hatch, L. R. 2000. *Beyond Gender Differences: Adaptation to Aging in Life Course Perspective*. Amityville, N. Y.: Baywood.

Henderson, J. T., et al. 2004. "The Role of Physician Gender in the Evaluation of the National Centers of Excellence in Women's Health: Test of an Alternate Hypothesis." *Women's Health Issues* 14: 130–39.

Hendricks, J., Hatch, L. R, and Cutler, S. 1999. "Entitlement, Social Compacts, and the Trend Toward Retrenchment in U.S. Old-Age Programs." *Hallym International Journal of Aging* 1(1): 14–32.

Holbrook, M. B. 1987. "Mirror, Mirror, on the Wall, What's Unfair in the Reflections in Advertising?" *Journal of Marketing* 51: 95–103.

Hurd, L. C. 2000. "Older Women's Body Image and Embodied Experience: An Exploration." *Journal of Women & Aging* 12(3/4): 77–97.

Kessler, E. -M., Rakoczy, K., and Staudinger, U. M. 2004. "The Portrayal of Older People in Prime Time Television Series: The Match with Gerontological Evidence." *Ageing & Society* 24(4): 531–52.

Kingson, E. R, Hirshorn, B. A., and Cornman, J. M. 1986. *Ties That Bind: The Interdependence of Generations*. Washington, D.C.: Seven Locks Press.

Marshall, V. W., and McPherson, B. D. 1994. "Introduction." In V. W. Marshall and B. D. McPherson, eds., *Aging: Canadian Perspectives*. Peterborough, Ontario: Broadview.

Miller, D. W., Leyell, T.S., and Mazachek, J. 2004. "Stereotypes of the Elderly in U.S. Television Commercials from the 1950s to the 1990s." *International Journal of Aging and Human Development* 58(4): 315–40.

Miller, P. N., et al. 1999. "Stereotypes of the Elderly in Print Advertisements 1956–1996." *International Journal of Aging and Human Development* 49(4): 319–37.

Minichiello, V., Browne, J., and Kendig, H. 2000. "Perceptions and Consequences of Ageism: Views of Older People." *Ageing and Society* 20(3): 253–78.

Munch, S. 2004. "Gender-Biased Diagnosing of Women's Medical Complaints; Contributions of Feminist Thought, 1970–1995." *Women & Health* 40(1): 101–21.

Newacheck, P. W., and Benjamin, A. E. 2004. "Intergenerational Equity and Public Spending." *Health Affairs* 23(5): 142–6.

Pedersen, I. 2002. "Looking Good on Whose Terms? Ambiguity in Two Kellogg's Special K(r) Print Advertisements." *Social Semiotics* 12(2): 169–81.

Peterson, R. T., and Ross, D. T. 1997. "A Content Analysis of the Portrayal of Mature Individuals in Television Commercials." *Journal of Business Ethics* 15: 425–33.

Sanders, M. 2002. "Older Women and the Media." *Women in Action* 3(Sept.): 56.

Schulz, J. 2001. *The Economics of Aging*, 7th ed. Westport, Conn.: Auburn House.

Sharpe, P. A. 1995, "Older Women and Health Services: Moving from Ageism Toward Empowerment." *Women & Health* 22(3): 9–23.

Steckenrider, J. S. 1998. "Aging as a Female Phenomenon: The Plight of Older Women." In J. S. Steckenrider and T. M. Parrott, eds., *New Directions in Old-Age Policies*. Albany, N. Y.: State University of New York Press.

Steinbach, U. 1992. "Social Networks, Institutionalization, and Mortality Among Elderly People in the United States." *Journals of Gerontology: Social Sciences* 47(4): S183–90.

Thompson, E. H., Jr., 2004. "Editorial." *Journal of Men's Studies* 13(1): 1.

U.S. Census Bureau. 2004. *We the People: Aging in the United States*. Washington, D.C.: U.S. Department of Commerce.

U.S. Department of Health and Human Services. 2004. *Health, United States, 2004*. Washington, D.C.: National Center for Health Statistics.

Weir, E. C. 2004 "Identifying and Preventing Ageism Among Health-Care Professionals." *International Journal of Therapy and Rehabilitation* 11(2): 56–63.

Wolinsky, F. D., and Johnson, R. J. 1992. "Widowhood, Health Status, and the Use of Health Services by Older Adults: A Cross-Sectional and Prospective Approach." *Journals of Gerontology: Social Sciences* 47(1): S8–16.

KEY CONCEPTS

ageism	gerontology	Social Security
age stereotypes	Medicare	

DISCUSSION QUESTIONS

1. Why does Hatch argue that there is a "double standard of aging?" How does this affect the socialization process?

2. In what ways do social policies for older people embed some of the age and gender stereotypes that Hatch discusses?

Applying Sociological Knowledge:
An Exercise for Students

Socialization occurs in many different contexts. Pick one of the following situations that involve entry to a new role: becoming a college student, becoming a parent, or getting your first job. What changes do you experience in this new role? How are the expectations associated with your new status communicated to you? Are these formally or informally communicated? What does this teach you about the socialization process?

✳

Society and Social Interaction

13

The Presentation of Self in Everyday Life

ERVING GOFFMAN

Erving Goffman likens social interaction to a "congame," in which we are consistently trying to put forward a certain impression or "self" in order to get something from others. Although many will not see human behavior so cynically, Goffman's analysis sheds light on how people try to manage the impression that others have of them.

When an individual plays a part, he implicitly requests his observers to take seriously the impression that is fostered before them. They are asked to believe that the character they see actually possesses the attributes he appears to possess, that the task he performs will have the consequences that are implicitly claimed for it, and that, in general, matters are what they appear to be. In line with this, there is the popular view that the individual offers his performance and puts on his show "for the benefit of other people." It will be convenient to begin a consideration of performances by turning the question around and looking at the individual's own belief in the impression of reality that he attempts to engender in those among whom he finds himself.

At one extreme, one finds that the performer can be fully taken in by his own act; he can be sincerely convinced that the impression of reality which he stages is the real reality. When his audience is also convinced in this way about the show he puts on—and this seems to be the typical case—then for the moment at least, only the sociologist or the socially disgruntled will have any doubts about the "realness" of what is presented.

At the other extreme, we find that the performer may not be taken in at all by his own routine. This possibility is understandable, since no one is in quite as good an observational position to see through the act as the person who puts it on. Coupled with this, the performer may be moved to guide the conviction of his audience only as a means to other ends, having no ultimate concern in the conception that they have of him or of the situation. When the individual has no belief in his own act and no ultimate concern with the beliefs of his audience,

SOURCE: Erving Goffman. 1959. *The Presentation of Self in Everyday Life*. Garden City, NY: Anchor Doubleday, pp. 17–27.

we may call him cynical, reserving the term "sincere" for individuals who believe in the impression fostered by their own performance. It should be understood that the cynic, with all his professional disinvolvement, may obtain unprofessional pleasures from his masquerade, experiencing a kind of gleeful spiritual aggression from the fact that he can toy at will with something his audience must take seriously.

It is not assumed, of course, that all cynical performers are interested in deluding their audiences for purposes of what is called "self-interest" or private gain. A cynical individual may delude his audience for what he considers to be their own good, or for the good of the community, etc. For illustrations of this we need not appeal to sadly enlightened showmen such as Marcus Aurelius or Hsun Tzu. We know that in service occupations practitioners who may otherwise be sincere are sometimes forced to delude their customers because their customers show such a heartfelt demand for it. Doctors who are led into giving placebos, filling station attendants who resignedly check and recheck tire pressures for anxious women motorists, shoe clerks who sell a shoe that fits but tell the customer it is the size she wants to hear—these are cynical performers whose audiences will not allow them to be sincere. Similarly, it seems that sympathetic patients in mental wards will sometimes feign bizarre symptoms so that student nurses will not be subjected to a disappointingly sane performance. So also, when inferiors extend their most lavish reception for visiting superiors, the selfish desire to win favor may not be the chief motive; the inferior may be tactfully attempting to put the superior at ease by simulating the kind of world the superior is thought to take for granted.

I have suggested two extremes: an individual may be taken in by his own act or be cynical about it. These extremes are something a little more than just the ends of a continuum. Each provides the individual with a position which has its own particular securities and defenses, so there will be a tendency for those who have traveled close to one of these poles to complete the voyage. Starting with lack of inward belief in one's role, the individual may follow the natural movement described by Park:

> It is probably no mere historical accident that the word person, in its first meaning, is a mask. It is rather a recognition of the fact that everyone is always and everywhere, more or less consciously, playing a role.... It is in these roles that we know each other; it is in these roles that we know ourselves.[1]

In a sense, and in so far as this mask represents the conception we have formed of ourselves—the role we are striving to live up to—this mask is our truer self, the self we would like to be. In the end, our conception of our role becomes second nature and an integral part of our personality. We come into the world as individuals, achieve character, and become persons.[2] ...

Front, then, is the expressive equipment of a standard kind intentionally or unwittingly employed by the individual during his performance. For preliminary purposes, it will be convenient to distinguish and label what seem to be the standard parts of front.

First, there is the "setting," involving furniture, decor, physical layout, and other background items which supply the scenery and stage props for the spate of human action played out before, within, or upon it....

If we take the term "setting" to refer to the scenic parts of expressive equipment, one may take the term "personal front" to refer to the other items of expressive equipment, the items that we most intimately identify with the performer himself and that we naturally expect will follow the performer wherever he goes. As part of personal front we may include: insignia of office or rank; clothing; sex, age, and racial characteristics; size and looks; posture; speech patterns; facial expressions; bodily gestures; and the like. Some of these vehicles for conveying signs, such as racial characteristics, are relatively fixed and over a span of time do not vary for the individual from one situation to another. On the other hand, some of these sign vehicles are relatively mobile or transitory, such as facial expression, and can vary during a performance from one moment to the next....

In addition to the fact that different routines may employ the same front, it is to be noted that a given social front tends to become institutionalized in terms of the abstract stereotyped expectations to which it gives rise, and tends to take on a meaning and stability apart from the specific tasks which happen at the time to be performed in its name. The front becomes a "collective representation" and a fact in its own right.

When an actor takes on an established social role, usually he finds that a particular front has already been established for it. Whether his acquisition of the role was primarily motivated by a desire to perform the given task or by a desire to maintain the corresponding front, the actor will find that he must do both.

Further, if the individual takes on a task that is not only new to him but also unestablished in the society, or if he attempts to change the light in which his task is viewed, he is likely to find that there are already several well-established fronts among which he must choose. Thus, when a task is given a new front we seldom find that the front it is given is itself new.

NOTES

1. Robert Ezra Park, *Race and Culture* (Glencoe, IL: The Free Press, 1950), p. 249.
2. Ibid., p. 250.

KEY CONCEPTS

dramaturgical model   presentation of self

DISCUSSION QUESTIONS

1. How many "selves" do you think you could "play" or "do" in order to accomplish something with another person? Discuss two such selves and try to get them to be quite different from each other.

2. "All the world's a stage," wrote William Shakespeare. So might Goffman have said this. How does his analysis of the presentation of self in everyday life suggest that life is a drama where we all play our parts?

14

Code of the Street

ELIJAH ANDERSON

Elijah Anderson's study of interaction on the street shows the vast array of implicit "codes" of behavior or rules that guide street interaction. His analysis helps explain the complexity of street interaction and provides a sociological explanation of street violence.

In some of the most economically depressed and drug- and crime-ridden pockets of the city, the rules of civil law have been severely weakened, and in their stead a "code of the street" often holds sway. At the heart of this code is a set of prescriptions and proscriptions, or informal rules, of behavior organized around a desperate search for respect that governs public social relations, especially violence, among so many residents, particularly young men and women. Possession of respect—and the credible threat of vengeance—is highly valued for shielding the ordinary person from the interpersonal violence of the street. In this social context of persistent poverty and deprivation, alienation from broader society's institutions, notably that of criminal justice, is widespread. The code of the street emerges where the influence of the police ends and personal responsibility for one's safety is felt to begin, resulting in a kind of "people's law," based on "street justice." This code involves a quite primitive form of social exchange that holds would-be perpetrators accountable by promising an "eye for an eye," or a certain "payback" for transgressions. In service to this ethic, repeated displays of

SOURCE: Elijah Anderson. 1999. *Code of the Street*. New York: W. W. Norton, pp. 9–11, 32–34, 312–317. Reprinted with permission.

"nerve" and "heart" build or reinforce a credible reputation for vengeance that works to deter aggression and disrespect, which are sources of great anxiety on the inner-city street....

In approaching the goal of painting an ethnographic picture of these phenomena, I engaged in participant-observation, including direct observation, and conducted in-depth interviews. Impressionistic materials were drawn from various social settings around the city, from some of the wealthiest to some of the most economically depressed, including carryouts, "stop and go" establishments, laundromats, taverns, playgrounds, public schools, the Center City indoor mall known as the Gallery, jails, and public street corners. In these settings I encountered a wide variety of people—adolescent boys and young women (some incarcerated, some not), older men, teenage mothers, grandmothers, and male and female schoolteachers, black and white, drug dealers, and common criminals. To protect the privacy and confidentiality of my subjects, names and certain details have been disguised....

Of all the problems besetting the poor inner-city black community, none is more pressing than that of interpersonal violence and aggression. This phenomenon wreaks havoc daily on the lives of community residents and increasingly spills over into downtown and residential middle-class areas. Muggings, burglaries, carjackings, and drug-related shootings, all of which may leave their victims or innocent bystanders dead, are now common enough to concern all urban and many suburban residents.

The inclination to violence springs from the circumstances of life among the ghetto poor—the lack of jobs that pay a living wage, limited basic public services (police response in emergencies, building maintenance, trash pickup, lighting, and other services that middle-class neighborhoods take for granted), the stigma of race, the fallout from rampant drug use and drug trafficking, and the resulting alienation and absence of hope for the future. Simply living in such an environment places young people at special risk of falling victim to aggressive behavior. Although there are often forces in the community that can counteract the negative influences—by far the most powerful is a strong, loving, "decent" (as inner-city residents put it) family that is committed to middle-class values—the despair is pervasive enough to have spawned an oppositional culture, that of "the street," whose norms are often consciously opposed to those of mainstream society. These two orientations—decent and street—organize the community socially, and the way they coexist and interact has important consequences for its residents, particularly for children growing up in the inner city. Above all, this environment means that even youngsters whose home lives reflect mainstream values—and most of the homes in the community do—must be able to handle themselves in a street-oriented environment.

This is because the street culture has evolved a "code of the street," which amounts to a set of informal rules governing interpersonal public behavior, particularly violence. The rules prescribe both proper comportment and the proper way to respond if challenged. They regulate the use of violence and so supply a rationale allowing those who are inclined to aggression to precipitate violent encounters in an approved way. The rules have been established and are enforced mainly by the street-oriented; but on the streets the distinction between street and decent is often irrelevant. Everybody knows that if the rules are violated, there are penalties. Knowledge of the code is thus largely defensive, and it is literally necessary for operating in public. Therefore, though families with a decency orientation are usually opposed to the values of the code, they often reluctantly encourage their children's familiarity with it in order to enable them to negotiate the inner-city environment.

At the heart of the code is the issue of respect—loosely defined as being treated "right" or being granted one's "props" (or proper due) or the deference one deserves. However, in the troublesome public environment of the inner city, as people increasingly feel buffeted by forces beyond their control, what one deserves in the way of respect becomes ever more problematic and uncertain. This situation in turn further opens up the issue of respect to sometimes intense interpersonal negotiation, at times resulting in altercations. In the street culture, especially among young people, respect is viewed as almost an external entity, one that is hard-won but easily lost—and so must constantly be guarded. The rules of the code in fact provide a framework for negotiating respect. With the right amount of respect, individuals can avoid being bothered in public. This security is important, for if they *are* bothered, not only may they face physical danger, but they will have been disgraced or "dissed" (disrespected). Many of the forms dissing can take may seem petty to middle-class people (maintaining eye contact for too long, for example), but to those invested in the street code, these actions, a virtual slap in the face, become serious indications of the other person's intentions. Consequently, such people become very sensitive to advances and slights, which could well serve as a warning of imminent physical attack or confrontation.

The hard reality of the world of the street can be traced to the profound sense of alienation from mainstream society and its institutions felt by many poor inner-city black people, particularly the young. The code of the street is actually a cultural adaptation to a profound lack of faith in the police and the judicial system—and in others who would champion one's personal security. The police, for instance, are most often viewed as representing the dominant white society and as not caring to protect inner-city residents. When called, they may not respond, which is one reason many residents feel they must be prepared to take extraordinary measures to defend themselves and their loved ones against those who are inclined to aggression. Lack of police accountability has in fact been incorporated into the local status system: the person who is believed capable of "taking care of himself" is accorded a certain deference and regard, which translates into a sense of physical and psychological control. The code of the street thus emerges where the influence of the police ends and where personal responsibility for one's safety is felt to begin. Exacerbated by the proliferation of drugs and easy access to guns, this volatile situation results in the ability of the street-oriented minority (or those who effectively "go for bad") to dominate the public spaces....

The attitudes and actions of the wider society are deeply implicated in the code of the street. Most people residing in inner-city communities are not totally invested in the code; it is the significant minority of hard-core street youth who maintain the code in order to establish reputations that are integral to the extant social order. Because of the grinding poverty of the communities these people inhabit, many have—or feel they have—few other options for expressing themselves. For them the standards and rules of the street code are the only game in town.

And as was indicated above, the decent people may find themselves caught up in problematic situations simply by being at the wrong place at the wrong time, which is why a primary survival strategy of residents here is to "see but don't see." The extent to which some children—particularly those who through upbringing have become most alienated and those who lack strong and conventional social support—experience, feel, and internalize racist rejection and contempt from mainstream society may strongly encourage them to express contempt for the society in turn. In dealing with this contempt and rejection, some youngsters consciously invest themselves and their considerable mental resources in what amounts to an oppositional culture, a part of which is the code of the street. They do so to preserve themselves and their own self-respect. Once they do, any respect they might be able to garner in the wider system pales in comparison with the respect available in the local system; thus they often lose interest in even attempting to negotiate the mainstream system.

At the same time, many less alienated young people have assumed a street-oriented demeanor as way of expressing their blackness while really embracing a much more moderate way of life; they, too, want a nonviolent setting in which to live and one day possibly raise a family. These decent people are trying hard to be part of the mainstream culture, but the racism, real and perceived, that they encounter helps legitimize the oppositional culture and, by extension, the code of the street. On occasion they adopt street behavior; in fact, depending on the demands of the situation, many people attempt to code switch, moving back and forth between decent and street behavior....

In addition, the community is composed of working-class and very poor people since those with the means to move away have done so, and there has also been a proliferation of single-parent households in which increasing numbers of kids are being raised on welfare. The result of all this is that the inner-city community has become a kind of urban village, apart from the wider society and limited in terms of resources and human capital. Young people growing up here often receive only the truncated version of mainstream society that comes from television and the perceptions of their peers....

According to the code, the white man is a mysterious entity, a part of an enormous monolithic mass of arbitrary power, in whose view black people are insignificant. In this system and in the local social context, the black man has very little clout; to salvage something of value, he must outwit, deceive, oppose, and ultimately "end-run" the system.

Moreover, he cannot rely on this system to protect him; the responsibility is his, and he is on his own. If someone rolls on him, he has to put his body, and often his life, on the line. The physicality of manhood thus becomes extremely important. And urban brinksmanship is observed and learned as a matter of course....

Urban areas have experienced profound structural economic changes, as deindustrialization—the movement from manufacturing to service and high-tech—and the growth of the global economy have created new economic conditions. Job opportunities increasingly go abroad to Singapore, Taiwan, India, and Mexico, and to nonmetropolitan America, to satellite cities like King of Prussia, Pennsylvania. Over the last fifteen years, for example, Philadelphia has lost 102,500 jobs, and its manufacturing employment has declined by 53 percent. Large numbers of inner-city people, in particular, are not adjusting effectively to the new economic reality. Whereas low-wage jobs—especially unskilled and low-skill factory jobs—used to exist simultaneously with poverty and there was hope for the future, now jobs simply do not exist, the present economic boom notwithstanding. These dislocations have left many inner-city people unable to earn a decent living. More must be done by both government and business to connect inner-city people with jobs.

The condition of these communities was produced not by moral turpitude but by economic forces that have undermined black, urban, working-class life and by a neglect of their consequences on the part of the public. Although it is true that persistent welfare dependency, teenage pregnancy, drug abuse, drug dealing, violence, and crime reinforce economic marginality, many of these behavioral problems originated in frustrations and the inability to thrive under conditions of economic dislocation. This in turn leads to a weakening of social and family structure, so children are increasingly not being socialized into mainstream values and behavior. In this context, people develop profound alienation and may not know what to do about an opportunity even when it presents itself. In other words, the social ills that the companies moving out of these neighborhoods today sometimes use to justify their exodus are the same ones that their corporate predecessors, by leaving, helped to create.

Any effort to place the blame solely on individuals in urban ghettos is seriously misguided. The focus should be on the socioeconomic structure, because it was structural change that caused jobs to decline and joblessness to increase in many of these communities. But the focus also belongs on the public policy that has radically threatened the well-being of many citizens. Moreover, residents of these communities lack good education, job training, and job networks, or connections with those who could help them get jobs. They need enlightened employers able to understand their predicament and willing to give them a chance. Government, which should be assisting people to adjust to the changed economy, is instead cutting what little help it does provide....

The emergence of an underclass isolated in urban ghettos with high rates of joblessness can be traced to the interaction of race prejudice, discrimination, and the effects of the global economy. These factors have contributed to the profound social isolation and impoverishment of broad segments of the inner-city black population. Even though the wider society and economy have been experiencing accelerated prosperity for almost a decade, the fruits of it often miss the truly disadvantaged isolated in urban poverty pockets.

In their social isolation an oppositional culture, a subset of which is the code of the street, has been allowed to emerge, grow, and develop. This culture is essentially one of accommodation with the wider society, but different from past efforts to accommodate the system. A larger segment of people are now not simply isolated but ever more profoundly alienated from the wider society and its institutions. For instance, in conducting the fieldwork for this book, I visited numerous inner-city schools, including elementary, middle, and high schools, located in areas of concentrated poverty. In every one, the so-called oppositional culture was well entrenched. In one elementary school, I learned from interviewing kindergarten, first-grade, second-grade, and fourth-grade teachers that through the first grade, about a fifth of the students were invested in the code of the street; the rest are interested in the subject matter and eager to take instruction from the teachers—in effect, well disciplined. By the fourth grade, though, about three-quarters of the students have bought into the code of the street or the oppositional culture.

As I have indicated throughout this work, the code emerges from the school's impoverished neighborhood, including overwhelming numbers of single-parent homes, where the fathers, uncles, and older brothers are frequently incarcerated—so frequently, in fact, that the word "incarcerated" is a prominent part of the young child's spoken vocabulary. In such communities there is not only a high rate of crime but also a generalized diminution of respect for law. As the residents go about meeting the exigencies of public life, a kind of people's law results, ... Typically, the local streets are, as we saw, tough and dangerous places where people often feel very much on their own, where they themselves must be personally responsible for their own security, and where in order to be safe and to travel the public spaces unmolested, they must be able to show others that they are familiar with the code—that physical transgressions will be met in kind.

In these circumstances the dominant legal codes are not the first thing on one's mind; rather, personal security for self, family, and loved ones is. Adults, dividing themselves into categories of street and decent, often encourage their children in this adaptation to their situation, but at what price to the children and at what price to wider values of civility and decency? As the fortunes of the inner city continue to decline, the situation becomes ever more dismal and intractable....

KEY CONCEPTS

deindustrialization   norms   urban underclass

DISCUSSION QUESTIONS

1. List several ways that subtle or nonverbal behavior becomes important "on the street."

2. What specific ways does Anderson see street behavior as stemming from social structural conditions for African Americans?

15

The Impact of Internet Communications on Social Interaction

THOMAS WELLS BRIGNALL III AND THOMAS VAN VALEY

The increased use of the Internet has altered the way young people interact. This article applies Goffman's theory about the presentation of self to social interaction that takes place on the Internet. The "rules" for social interaction and the skill that develops when you engage in face-to-face interaction are changed when communicating online. The authors suggest that young people today are cyber-kids and experience social interaction and education differently because of the Internet.

INTRODUCTION

The Internet is clearly on the way to becoming an integral tool of business, communication, and popular culture in the United States and in other parts of the world. The Internet is also being presented as a pedagogical tool for much of public education. However, its extraordinary growth is not without concern. Of particular relevance is the issue of the potential impact of the Internet and especially computer-mediated communications on the nature and quality of social interaction among cyberkids.

According to NetValue, people who chat online are among the heaviest users of the Internet (2002, p. 1). Furthermore, 20% of female chatters are teenage girls. NetValue, Nielsen, and eMarketer (Ramsey 2000) all conclude that while teenagers are not yet the dominant demographic of Internet users, their usage is growing very rapidly. Regardless of country of origin, recent data on Internet usage indicates that young people are becoming some of the heaviest users of the Internet (Nielsen 2002, p. 1). According to Nielsen (2002), 74% of United States residents between the ages of 12 and 18 are now using the Internet.

SOURCE: Thomas Wells Brignall III and Thomas Van Valey. 2005. "The Impact of Internet Communications on Social Interaction." *Sociological Spectrum* 25: 335–348.

Among Internet users between 12 and 18, 35% spend 31–60 minutes per day online, and 44% spend more than an hour per day online. Indeed, almost 4% of these cyberkids spend four or more hours online each day. America Online reported in their national survey of more than 6,700 teens and parents of teens that "fifty-six percent of teens (aged 18 to 19) prefer Internet to the telephone" (Pastore 2002, p. 1). The survey also reported that in order to keep up their communications with friends, more than 81% of teens use e-mail, while 70% use instant messaging.

Certainly, one can argue that online communication is not yet the dominant form of communication among young people. However, there is little question that the phenomenon of online communication among teens and children is growing rapidly. Moreover, it is not hard to imagine a time in the near future when elementary and secondary school students may spend several hours a day doing schoolwork, communicating with friends, teachers, and family, and seeking entertainment, all via the computer and the Internet. It also follows, therefore, that substantial portions of students' experiences with interpersonal communications (especially with persons not already known to them) are likely to be computer-mediated. If children and teenagers are already using computers as a significant form of education, communication, and entertainment, it may well be that less time is being spent having face-to-face interactions with peers.

If the amount of time spent in face-to-face interactions among youth is shrinking, there may be significant consequences for their development of social skills and their presentation of self. Several decades ago, Goffman (1959, 1967) suggested that individuals who lack the normative communication, cultural, and civility skills in a society would find it difficult to interact with others successfully. At the time, Goffman was referring to the variety of visual and auditory cues that occur in face-to-face communications. Today, with computer-mediated communications, it is possible that none of the cues that Goffman wrote about may be present during online communication.

Some authors have already suggested that online behavior is different from offline behavior (Rheingold 1993; Postman 1992; Jones 1995; Miller 1996). Examples of such differences in behavior include individuals who are willing to misrepresent themselves by feigning a different gender, skin color, sexual orientation, physical condition, or age. Other differences in observed behaviors include the open display of group norm violations such as aggressive behavior, racism, sexism, homophobia, personal attacks, harassment, and a tendency for individuals to quickly abandon groups and conversations, refusing to deal with issues they find difficult to immediately resolve. If these authors are correct, it is critical that the specific elements of interpersonal communication involved in computer-mediated interactions be identified and compared with face-to-face interactions. This article examines the nature of Internet communications that take place among young people (particularly the elements of the presentation of self that do not occur, or occur with limited frequency), and suggests some potential consequences for the education and socialization of our youth.

THEORIES OF SOCIAL INTERACTION

In *Asylums* (1961), Goffman described how small rituals have replaced the big rituals that occurred in traditional interpersonal relations. "What remains are brief rituals one individual performs for and to another, attesting to civility and good will on the performer's part and to the recipient's possession of a small patrimony of sacredness" (Goffman 1961, p. 63). However, the choice of the particular set of rules an individual chooses to follow derives from requirements established in social encounters. Therefore, an individual's concept of self is shaped by the sum of the social interactions in which that individual engages.

Goffman (1967) further argues that the self is the actor in an ongoing play that responds to the judgments of others. While each individual has more than one role, he or she is saved from role strain by "audience segregation," that is, by employing multiple roles in their interactions with others. Because an individual can play a unique role in each social situation, that individual can effectively be a different person in each situation without contradiction. However, audience segregation regularly breaks down, and an individual can present a role incompatible with ones presented on other occasions. In these situations, role strain occurs.

Children learn how to cope with such role strains through their own social experience and by watching others navigate social interactions that involve contradictory or competing roles. Indeed, Goffman argues that coping with role strain and developing impression management are necessary skills for the success of individuals in everyday social interactions. Children also must learn how to manage "front stage" and "back stage" behavior. The front stage is open to judgment by an audience, where the back stage is a place where actors can discuss, polish, or refine their performances without revealing themselves to the same audience. However, because there are multiple layers of front stages and back stages, individuals learn how much they can reveal to other characters.

All front stage roles contain a number of elements that are visual or auditory in nature including physical appearance (e.g., demographic characteristics, and physical features, such as size, make up, hairstyle, posture), manner of speaking (e.g., the use of standard vs. slang dialect, accents, regional vocabulary choices, and voice inflections), and the use of various props (e.g., clothing, car, and food preferences). Together, these elements help to create the role that is presented to others. However, individuals in face-to-face interactions not only present these more obvious indicators of their roles, they also give out more subtle cues such as posture, hand gestures, tone of voice, movement in a conversation, eye contact, and levels of social formality. Moreover, many of these indicators of role (both the obvious and the subtle) vary widely across groups.

However, in order to maintain a positive on-going relationship in any difficult face-to-face circumstance, an individual must learn the appropriate socialization rituals. Knowing these rituals and being able to play a proper front stage role is crucial in order for an individual to get along with others. Indeed, the appearance of getting along with others is sometimes far more important than whether individuals actually like one another. Once again, it is difficult for individuals to succeed if they lack the proper social skills of the various groups with which they interact. Only with practice, will individuals develop and learn to improve their interaction skills, and develop a better presentation when communicating to individuals via their front and back stages....

ONLINE INTERACTION

From the beginning of the Internet, it was clear that the interaction taking place online was a new form of social interaction. What was not clear, and still is to be determined, are the consequences of this new form of social interaction. The Internet itself is neither negative nor positive. It is inanimate, an object or a tool that can be used in various ways. To reify the Internet and suggest that it is somehow inherently liberating or enslaving is misleading. Nevertheless, it is important to look at how interaction on the Internet differs from other forms of interaction. These differences may play havoc with traditional social interaction rituals....

Because online interaction is different, it is entirely possible that children with high levels of online interactions would adopt different techniques of social interaction. Several authors such as LaRose, Eastin, and Greeg (2001) and Schmitz (1997) cultivate the notion that for many individuals, online communication helps facilitate face-to-face communications with current relationships and sometimes with new relationships. Boyd and Walther (2002) even argue that Internet social support is superior to face-to-face social support. According to them, online social support offers benefits that face-to-face social networks cannot: anonymity, constant access to better quality expertise, and enhanced modes of expression, with less chance of embarrassment and without incurring an obligation to the support provider. However, Spears and Lea (1992) argue that the absence of social and contextual cues undermine the perception of leadership, status, and power, and leads to reduced impact of social norms and therefore to deregulated, anti-normative behavior.

THE IMPACT OF INTERNET COMMUNICATIONS

Our fundamental position is that online social interaction is one form of role-play, and thus an element in the development of the self. Moreover, if a substantial amount of communication is accomplished online, either at home, at school, or elsewhere, children and cyberkids are likely to develop the skills necessary for online interaction, but they are also likely to lack some of the skills that are involved in face-to-face interaction.

Tapscott (1998) claims that the children of the Internet generation are already way ahead of their parents and many other adults (who do not yet understand what the Internet is or how it works). However, Tapscott and others may have made a crucial mistake in their logic. They have forgotten that political and economic power are in the hands of adults. Therefore, the members of the cyberkid generation must understand, adapt, and modify their interactions in order to get along in society.

Moreover, if students develop unique interaction rituals based on online communication without enough experience or understanding of traditional face-to-face interaction rituals, the likelihood for friction and/or conflict to occur undoubtedly increases. Such youths may be perceived as rude, insolent, disconnected, spoiled, or apathetic. It can be argued that new cultural and social phenomena have typically produced tensions between the generations. However, not having the skills or the knowledge of how to communicate with people who have different values and attitudes complicates the traditional struggles between youth and their elders. Furthermore, such rifts will not only be between the young and the old, but between any groups that are different from one another....

The skills and lessons of socialization that students need to learn in order to cope in everyday life, however, cannot be manufactured by computer simulations or video games (at least not yet). Classrooms with computers hooked up to the Internet predispose students to work as individuals rather than as members of any social group. Although students can interact with others when they are on the Internet, they are often able to choose with whom they wish to interact and how they want to manage it. Even if the students are interacting with fellow classmates while online, is it not reasonable that the use of a computer in mediating those interactions will alter the interaction rituals in which they engage?...

DISCUSSION

Given that we know so little about the nature of the social interactions that take place over the Internet, not only but especially by youth, it is clear that much research is needed. One obvious arena would include studies of young children followed over time to determine if there are recognizable online interaction rituals and what consequences they may have on the social development of the children. There is a growing body of research on this general issue—whether computer-mediated communications "displace" more traditional face-to-face forms of communication. A special issue of *IT & Society* recently focused on it, reporting the results of a number of time-diary studies in the United States and elsewhere. However, the conclusions are mixed. Some studies have found reduced levels of time spent with friends and family (Nie and Hillygus 2002; Pronovost 2002; Nie and Erbring 2002). Others have either reported no difference (Robinson et al. 2002; Gershny 2002) or increased levels of sociability among Internet users (Neustadl and Robinson 2002; Cummings 2002).

We agree with Wellman and Gulia (1999) that "the internet … is not a separate reality. People bring to their online interactions such baggage as their gender, stage in the life cycle, cultural milieu, socioeconomic status, and offline connections with others" (p. 3). Online and face-to-face social interactions share some elements, and many individuals interact using both forms of communication.

This discussion is not about the members of the current generations who have grown up with the Internet. It is about future generations where the

Internet is used as a primary source of communication, the focal center of school-work and research, the main source of entertainment, and the primary medium for the development of contemporary issues. If the strength of the Internet of the future is the fact that individuals can choose with whom they want to interact, then it may also be one of the Internet's weaknesses when it comes to the development of social interaction skills. The demands of learning to get along with others are likely to become drowned out by self-interested pursuits. The possibility of a narrow world perspective seems certain for those individuals who choose to isolate themselves from people and ideas with whom they feel uncomfortable. If the easiest solution to avoid dissonance is to avoid situations that produce it, then the potential for an unrealistic social process is high. We are not opposing or supporting computer-mediated communications or the Internet in schools. We are simply suggesting that it is important to assess the social impacts of the Internet while the frequency of use is still relatively low. Once they become ubiquitous, the possibility of such research has been forever lost.

REFERENCES

Boyd, S. and Joseph, B. Walther. 2002. "Attraction to Computer-Mediated Social Support." Pp. 153–188 in *Communication Technology and Society: Audience Adoption and Uses*, edited by C. A. Lin and D. Atkin. Cresskill, NJ: Hampton Press.

Cummings, J., L. Sproull, and S. Kiesler. 2002. "Beyond Hearing: Where Real World and Online Support Meet." *Group Dynamics: Theory, Research, and Practices*, 6(1):78–88.

Gershney, J. 2002. "Mass Media, Leisure, and Home IT: A Panel Time-Diary Approach." Pp. 53–66 retrieved June 8, 2002 from http://www.IT and Society.org.

Goffman, E. 1959. *The Presentation of Self in Everyday Life*. New York: Anchor.

Goffman, E. 1961. *Asylums: Essays on the Social Situation of Mental Patients and Other Inmates*. New York: Anchor.

Goffman, E. 1967. Interaction Ritual: Essays on Face to Face Behavior. New York: Pantheon Books.

Jones, G. S. 1995. *CyberSociety: Computer-Mediated Communication and Community*. Thousand Oaks, CA: Sage Publications.

LaRose, R., M. S. Eastin, and J. Gregg. 2001. "Reformulating the Internet Paradox: Social Cognitive Explanations of Internet Use and Depression." Retrieved January 30, 2003 from www.behavior.net/JOB/v1n1/paradox.html.

Miller, E. S. 1996. *Civilizing Cyberspace: Policy, Power, and the Information Superhighway*. New York: ACM Press.

NetValue. (2002). "Internet Use Patterns." Edited by R. A. Cole. Retrieved December 5, 2002 from www.netvalue.com/corp/actionnaires/index.htm.

Neustadl, A. and J. P. Robinson. 2002. "Media Use Differences Between Internet Users and Nonusers in the General Social Survey." Pp. 100–120. Retrieved June 8, 2002 from http://www.IT and Society.org.

Nie, N. H. and L. Erbring. (2002). "Internet and Society: A Preliminary Report." *IT and Society* 1(1):275–283.

Nie, N. H. and D. S. Hillygus. (2002). "The Impact of Internet Use on Sociability: Time-diary Findings." *IT and Society* 1(1):1–20.

Nielsen Net Ratings. (2002). "Internet User Growth." Retrieved September 19, 2002 from http://www.nielsen-netratings.com/news.jsp.

Pastore, M. (2002). "Internet Key to Communication Among Youth." Retrieved January 30, 2003 from http://cyberatlas.internet.com/big_picture/demographics/article/0,5901_961881,00.html.

Postman, N. (1992). *Technopoly*. New York: Vintage Books.

Pronovost, G. (2002). "The Internet and Time Displacement: A Canadian Perspective." *IT and Society* 1(1):44–53.

Ramsey, G. (2000). *The E-demographics and Usage Patterns Report: September 2000*. Retrieved October 23, 2000 from http://www.emarketer.com/ereports/ecommerce_b2b/.

Rheingold, H. (1993). *The Virtual Community: Homesteading on the Electronic Frontier*. Cambridge, Massachusetts: Addison-Wesley Publishing Company Reading.

Robinson, J. P., M. Kestnbaum, A. Neustadl, and A. Alvavez. 2002. "The Internet and Other Uses of Time." Pp. 244–262 in *The Internet in Everyday Life*, edited by B. Wellman and C. Haythornthwaite. Malden, MA: Blackwell Publishing.

Schmitz, J. (1997). "Structural Relations, Electronic Media, and Social Change: The Public Electronic Network and the Homeless." Pp. 80–101 in *Virtual Culture: Identity and Communication in Cybersociety*, edited by S. G. Jones. London: Sage.

Spears, R. and M. Lea. (1992). "Social Influence and the Influence of the "Social" in Computer-mediated Communication." Pp. 30–65 in *Contexts of Computer-mediated Communication*, edited by M. Lea. London: Harvester-Wheatsheaf.

Tapscott, D. (1998). *Growing Up Digital: The Rise of the Net Generation*. New York: McGraw-Hill Trade.

Wellman, B. and M. Gulia. 1999. "Net Surfers Don't Ride Alone: Virtual Community as Community." Pp. 331–367 in *Networks in the Global Village*, edited by Barry Wellman. Boulder; Co: Westview Press.

KEY CONCEPTS

ritual   role strain   social interaction

DISCUSSION QUESTIONS

1. How have Facebook, MySpace, and other online communication sites changed the way you interact with friends? In your opinion, is the nature of the friendship changed when communicating online as opposed to in-person?

2. What "rules" do you think exist when e-mailing? For example, is there a difference in the way you address a professor in an e-mail compared to the way you address a friend or classmate? Consider e-mail etiquette. Is there such a thing? Should there be?

Applying Sociological Knowledge: An Exercise for Students

For a 48-hour period, keep a detailed log of every form of technology that you use for social interaction (including various forms of communication, networking, scheduling, etc.). How is this different from or similar to face-to-face interaction? Try to imagine life without some of these technologies. How do you think they have changed the character of social interaction? Can you also imagine what technologies might shape social interaction in the future?

✳

Groups and Organizations

16

Clique Dynamics

PATRICIA ADLER AND PETER ADLER

Patricia Adler and Peter Adler take a look at clique formation and friendship groupings in schools. In their study of children's friendship groups, they analyze how cliques can generate tremendous power and influence over clique members.

A dominant feature of children's lives is the clique structure that organizes their social world. The fabric of their relationships with others, their levels and types of activity, their participation in friendships, and their feelings about themselves are tied to their involvement in, around, or outside the cliques organizing their social landscape. Cliques are, at their base, friendship circles, whose members tend to identify each other as mutually connected. Yet they are more than that; cliques have a hierarchical structure, being dominated by leaders, and are exclusive in nature, so that not all individuals who desire membership are accepted. They function as bodies of power within grades, incorporating the most popular individuals, offering the most exciting social lives, and commanding the most interest and attention from classmates.... As such they represent a vibrant component of the preadolescent experience, mobilizing powerful forces that produce important effects on individuals.

The research on cliques is cast within the broader literature on elementary school children's friendship groups. A first group of such works examines independent variables that can have an influence on the character of children's friendship groups. A second group looks at the features of children's inter- and intra-group relations. A third group concentrates on the behavioral dynamics specifically associated with cliques. Although these studies are diverse in their focus, they identify several features as central to clique functioning without thoroughly investigating their role and interrelation: boundary maintenance and definitions of membership (exclusivity); a hierarchy of popularity (status stratification and differential power), and relations between m–groups and out-groups (cohesion and integration).

In this [essay] we look at these dynamics and their association, at the way clique leaders generate and maintain their power and authority (leadership, power/dominance), and at what it is that influences followers to comply so readily with clique leaders' demands (submission). These interactional dynamics are

SOURCE: Patricia Adler and Peter Adler. 1998. *Peer Power: Preadolescent Culture and Identity.* New Brunswick, NJ: Rutgers University Press, pp. 56–69. Reprinted with permission.

not intended to apply to all children's friendship groups, only those (populated by one-quarter to one-half of the children) that embody the exclusive and stratified character of cliques.

TECHNIQUES OF INCLUSION

The critical way that cliques maintained exclusivity was through careful membership screening. Not static entities, cliques irregularly shifted and evolved their membership, as individuals moved away or were ejected from the group and others took their place. In addition, cliques were characterized by frequent group activities designed to foster some individuals' inclusion (while excluding others). Cliques had embedded, although often unarticulated, modes for considering and accepting (or rejecting) potential new members. These modes were linked to the critical power of leaders in making vital group decisions. Leaders derived power through their popularity and then used it to influence membership and social stratification within the group. This stratification manifested itself in tiers and subgroups within cliques composed of people who were hierarchically ranked into levels of leaders, followers, and wannabes. Cliques embodied systems of dominance, whereby individuals with more status and power exerted control over others' lives.

Recruitment

Initial entry into cliques often occurred at the invitation or solicitation of clique members.... Those at the center of clique leadership were the most influential over this process, casting their votes for which individuals would be acceptable or unacceptable as members and then having other members of the group go along with them. If clique leaders decided they liked someone, the mere act of their friendship with that person would accord them group status and membership....

Potential members could also be brought to the group by established members who had met and liked them. The leaders then decided whether these individuals would be granted a probationary period of acceptance during which they could be informally evaluated. If the members liked them, the newcomers would be allowed to remain in the friendship circle, but if they rejected them, they would be forced to leave.

Tiffany, a popular, dominant girl, reflected on the boundary maintenance she and her best friend Diane, two clique leaders, had exercised in fifth grade:

Q: *Who defines the boundaries of who's in or who's out?*

TIFFANY: Probably the leader. If one person might like them, they might introduce them, but if one or two people didn't like them, then they'd start to get everyone up. Like in fifth grade, there was Dawn Bolton and she was new. And the girls in her class that were in our clique liked her, but Diane and I didn't like her, so we kicked her out. So then she went to the other clique, the Emily clique....

Application

A second way for individuals to gain initial membership into a clique occurred through their actively seeking entry.... Several factors influenced the likelihood that a person would be accepted as a candidate for inclusion, as Darla, a popular fourth-grade girl described: "Coming in, it's really hard coming in, it's like really hard, even if you are the coolest person, they're still like, 'What is *she* doing [exasperated]?' You can't be too pushy, and like I don't know, it's really hard to get in, even if you can. You just got to be there at the right time, when they're nice, in a nice mood."

According to Rick, a fifth-grade boy who was in the popular clique but not a central member, application for clique entry was more easily accomplished by individuals than groups. He described the way individuals found routes into cliques: "It can happen any way. Just you get respected by someone, you do something nice, they start to like you, you start doing stuff with them. It's like you just kind of follow another person who is in the clique back to the clique, and he says, 'Could this person play?' So you kind of go out with the clique for a while and you start doing stuff with them, and then they almost like invite you in. And then soon after, like a week or so, you're actually in. It all depends.... But you can't bring your whole group with you, if you have one. You have to leave them behind and just go in on your own."

Successful membership applicants often experienced a flurry of immediate popularity. Because their entry required clique leaders' approval, they gained associational status.

Friendship Realignment

Status and power in a clique were related to stratification, and people who remained more closely tied to the leaders were more popular. Individuals who wanted to be included in the clique's inner echelons often had to work regularly to maintain or improve their position.

Like initial entry, this was sometimes accomplished by people striving on their own for upward mobility. In fourth grade, Danny was brought into the clique by Mark, a longtime member, who went out of his way to befriend him. After joining the clique, however, Danny soon abandoned Mark when Brad, the clique leader, took an interest in him. Mark discussed the feelings of hurt and abandonment this experience left him with: "I felt really bad, because I made friends with him when nobody knew him and nobody liked him, and I put all my friends to the side for him, and I brought him into the group, and then he dumped me. He was my friend first, but then Brad wanted him.... He moved up and left me behind, like I wasn't good enough anymore."

The hierarchical structure of cliques, and the shifts in position and relationships within them, caused friendship loyalties within these groups to be less reliable than they might have been in other groups. People looked toward those above them and were more susceptible to being wooed into friendship with individuals more popular than they. When courted by a higher-up, they could easily drop their less popular friends....

Ingratiation

Currying favor with people in the group, like previous inclusionary endeavors, can be directed either upward (supplication) or downward (manipulation)…. Note that children often begin their attempts at entry into groups with low-risk tactics; they first try to become accepted by more peripheral members, and only later do they direct their gaze and inclusion attempts toward those with higher status. The children we observed did this as well, making friendly overtures toward clique followers and hoping to be drawn by them into the center.

The more predominant behavior among group members, however, involved currying favor with the leader to enhance their popularity and attain greater respect from other group members. One way they did this was by imitating the style and interests of the group leader. Marcus and Adam, two fifth-grade boys, described the way borderline people would fawn on their clique and its leader to try to gain inclusion:

MARCUS: Some people would just follow us around and say, "Oh yeah, whatever he says, yeah, whatever his favorite kind of music is, is my favorite kind of music."

ADAM: They're probably in a position then they want to be more in because if they like what we like, then they think more people will probably respect them. Because if some people in the clique think this person likes their favorite groups, say it's REM, or whatever, so it's say Bud's [the clique leader's], this person must know what we like in music and what's good and what's not, so let's tell him that he can come up and join us after school and do something.

Fawning on more popular people not only was done by outsiders and peripherals but was common practice among regular clique members, even those with high standing. Darla, a second-tier fourth-grade girl, … described how, in fear, she used to follow the clique leader and parrot her opinions: "I was never mean to the people in my grade because I thought Denise might like them and then I'd be screwed. Because there were some people that I hated that she liked and I acted like I loved them, and so I would just be mean to the younger kids, and if she would even say, 'Oh she's nice,' I'd say, 'Oh yeah, she's really nice!' " Clique members, then, had to stay abreast of the leader's shifting tastes and whims if they were to maintain status and position in the group. Part of their membership work involved a regular awareness of the leader's fads and fashions, so that they would accurately align their actions and opinions with the current trends in timely manner….

TECHNIQUES OF EXCLUSION

Although inclusionary techniques reinforced individuals' popularity and prestige while maintaining the group's exclusivity and stratification, they failed to contribute to other, essential, clique features such as cohesion and integration, the management of m-group and out-group relationships, and submission to clique leadership. These features are rooted, along with further sources of domination and power, in cliques' exclusionary dynamics.

Out-Group Subjugation

When they were not being nice to try to keep outsiders from straying too far from their realm of influence, clique members predominantly subjected outsiders to exclusion and rejection. They found sport in picking on these lower-status individuals. As one clique follower remarked, "One of the main things is to keep picking on unpopular kids because it's just fun to do." [Sociologist] Eder ... notes that this kind of ridicule, where the targets are excluded and not enjoined to participate in the laughter, contrasts with teasing, where friends make fun of each other in a more lighthearted manner but permit the targets to remain included in the group by also jokingly making fun of themselves. Diane, a clique leader in fourth grade, described the way she acted toward outsiders: "Me and my friends would be mean to the people outside of our clique. Like, Eleanor Dawson, she would always try to be friends with us, and we would be like, 'Get away, ugly.'"

Interactionally sophisticated clique members not only treated outsiders badly but managed to turn others in the clique against them. Parker and Gottman ... observe that one of the ways people do this is through gossip. Diane recalled the way she turned all the members of her class, boys as well as girls, against an outsider: "I was always mean to people outside my group like Crystal, and Sally Jones; they both moved schools.... I had this gummy bear necklace, with pearls around it and gummy bears. She [Crystal] came up to me one day and pulled my necklace off. I'm like, 'It was my favorite necklace,' and I got all of my friends, and all the guys even in the class, to revolt against her. No one liked her. That's why she moved schools, because she tore my gummy bear necklace off and everyone hated her. They were like, 'That was mean. She didn't deserve that. We hate you.'"...

In-Group Subjugation

Picking on people within the clique's confines was another way to exert dominance. More central clique members commonly harassed and were mean to those with weaker standing. Many of the same factors prompting the ill treatment of outsiders motivated high-level insiders to pick on less powerful insiders. Rick, a fifth-grade clique follower, articulated the systematic organization of downward harassment: "Basically the people who are the most popular, their life outside in the playground is picking on other people who aren't as popular, but are in the group. But the people just want to be more popular so they stay in the group, they just kind of stick with it, get made fun of, take it.... They come back everyday, you do more ridicule, more ridicule, more ridicule, and they just keep taking it because they want to be more popular, and they actually like you but you don't like them. That goes on a lot, that's the main thing in the group. You make fun of someone, you get more popular, because insults is what they like, they like insults."

The finger of ridicule could be pointed at any individual but the leader. It might be a person who did something worthy of insult, it might be someone who the clique leader felt had become an interpersonal threat, or it might be someone singled out for no apparent reason.... Darla, the second tier fourth grader discussed earlier, described the ridicule she encountered and her feelings of mortification when the clique leader derided her hair: "Like I remember, she embarrassed me so bad one day. Oh my God, I wanted to kill her! We were in music class and we were standing there and she goes, 'Ew! what's all that shit in your hair?' in front of the whole class. I was so embarrassed/cause, I guess I had dandruff or something."

Often, derision against insiders followed a pattern, where leaders started a trend and everyone followed it. This intensified the sting of the mockery by compounding it with multiple force. Rick analogized the way people in cliques behaved to the links on a chain: "Like it's a chain reaction, you get in a fight with the main person, then the person right under him will not like you, and the person under him won't like you, and et cetera, and the whole group will take turns against you. A few people will still like you because they will do their own thing, but most people will do what the person in front of them says to do, so it would be like a chain reaction. It's like a chain; one chain turns, and the other chain has to turn with them or else it will tangle."

Compliance

Going along with the derisive behavior of leaders or other high-status clique members could entail either active or passive participation. Active participation occurred when instigators enticed other clique members to pick on their friends. For example, leaders would often come up with the idea of placing phony phone calls to others and would persuade their followers to do the dirty work. They might start the phone call and then place followers on the line to finish it, or they might pressure others to make the entire call, thus keeping one step distant from becoming implicated, should the victim's parents complain.

Passive participation involved going along when leaders were mean and manipulative, as when Trevor submissively acquiesced in Brad's scheme to convince Larry that Rick had stolen his money. Trevor knew that Brad was hiding the money the whole time, but he watched while Brad whipped Larry into a frenzy, pressing him to deride Rick, destroy Rick's room and possessions, and threaten to expose Rick's alleged theft to others. It was only when Rick's mother came home, interrupting the bedlam, that she uncovered the money and stopped Larry's onslaught. The following day at school, Brad and Trevor could scarcely contain their glee. As noted earlier, Rick was demolished by the incident and cast out by the clique; Trevor was elevated to the status of Brad's best friend by his coconspiracy in the scheme....

Stigmatization

Beyond individual incidents of derision, clique insiders were often made the focus of stigmatization for longer periods of time. Unlike outsiders who commanded less enduring interest, clique members were much more involved in picking on their friends, whose discomfort more readily held their attention. Rick noted that the duration of this negative attention was highly variable: "Usually at certain times, it's just a certain person you will pick on all the time, if they do something wrong. I've been picked on for a month at a time, or a week, or a day, or just a couple of minutes, and then they will just come to respect you again." When people became the focus of stigmatization, as happened to Rick, they were rejected by all their friends. The entire clique rejoiced in celebrating their disempowerment. They would be made to feel alone whenever possible. Their former friends might join hands and walk past them through the play yard at recess, physically demonstrating their union and the discarded individual's aloneness.

Worse than being ignored was being taunted. Taunts ranged from verbal insults to put-downs to singsong chants. Anyone who could create a taunt was favored with attention and imitated by everyone. Even outsiders, who would not normally be privileged to pick on a clique member, were able to elevate themselves by joining in on such taunting....

The ultimate degradation was physical. Although girls generally held themselves to verbal humiliation of their members, the culture of masculinity gave credence to boys' injuring each other.... Fights would occasionally break out in which boys were punched in the ribs or stomach, kicked, or given black eyes. When this happened at school, adults were quick to intervene. But after hours or on the school bus boys could be hurt. Physical abuse was also heaped on people's homes or possessions. People spit on each other or others' books or toys, threw eggs at their family's cars, and smashed pumpkins in front of their house.

Expulsion

While most people returned to a state of acceptance following a period of severe derision ... this was not always the case. Some people became permanently excommunicated from the clique. Others could be cast out directly, without undergoing a transitional phase of relative exclusion. Clique members from any stratum of the group could suffer such a fate, although it was more common among people with lower status.

When Davey, mentioned earlier, was in sixth grade, he described how expulsion could occur as a natural result of the hierarchical ranking, where a person at the bottom rung of the system of popularity was pushed off. He described the ordinary dynamics of clique behavior:

Q: *How do clique members decide who they are going to insult that day?*

DAVEY: It's just basically everyone making fun of everyone. The small people making fun of smaller people, the big people making fun of the small people. Nobody is really making fun of people bigger than them because they can get rejected, because then they can say, "Oh yes, he did this and that, this and that, and we shouldn't like him anymore." And everybody else says, "Yeah, yeah, yeah," 'cause all the lower people like him, but all the higher people don't. So the lower-case people just follow the higher-case people. If one person is doing something wrong, then they will say, "Oh yeah, get out, good-bye."...

KEY CONCEPTS

clique   ingroup   outgroup

DISCUSSION QUESTIONS

1. Take a look at your own friendship group in school. Which of the processes of both inclusion and exclusion do you observe?
2. What forms of negative sanction, or punishment, do the more powerful high-status clique members deliver to others? List some, noting how they differ in severity.

17

Sexual Assault on Campus

A Multilevel, Integrative Approach to Party Rape

ELIZABETH A. ARMSTRONG, LAURA HAMILTON, AND BRIAN SWEENEY

In this article the authors discuss their research of a "party dorm" at a large university. The research uncovers patterns of gendered behavior that contribute to the risk of sexual assault during parties, specifically at fraternities. College women are struggling to find a balance between having fun and being in danger. The authors argue that parties are structured in such a way that puts men in control. Strategies that teach women to simply be careful fall short in their efforts to prevent sexual assault.

A 1991 National Institute of Justice study estimated that between one-fifth and one-quarter of women are the victims of completed or attempted rape while in college (Fisher, Cullen, and Turner 2000). College women "are at greater risk for rape and other forms of sexual assault than women in

SOURCE: Elizabeth A. Armstrong, Laura Hamilton, and Brian Sweeney 2006. "Sexual Assault on Campus: A Multilevel, Integrative Approach to Party Rape." *Social Problems* 53: 483–499.

the general population or in a comparable age group" (Fisher et al. 2000:iii). At least half and perhaps as many as three-quarters of the sexual assaults that occur on college campuses involve alcohol consumption on the part of the victim, the perpetrator, or both (Abbey et al. 1996; Sampson 2002). The tight link between alcohol and sexual assault suggests that many sexual assaults that occur on college campuses are "party rapes." A recent report by the U.S. Department of justice defines party rape as a distinct form of rape, one that "occurs at an off-campus house or on- or off-campus fraternity and involves ... plying a woman with alcohol or targeting an intoxicated woman" (Sampson 2002:6). While party rape is classified as a form of acquaintance rape, it is not uncommon for the woman to have had no prior interaction with the assailant, that is, for the assailant to be an in-network stranger (Abbey et al. 1996).

Colleges and universities have been aware of the problem of sexual assault for at least 20 years, directing resources toward prevention and providing services to students who have been sexually assaulted. Programming has included education of various kinds, support for *Take Back the Night* events, distribution of rape whistles, development and staffing of hotlines, training of police and administrators, and other efforts. Rates of sexual assault, however, have not declined over the last five decades (Adams-Curtis and Forbes 2004:95; Bachar and Koss 2001; Marine 2004; Sampson 2002:1).

Why do colleges and universities remain dangerous places for women in spite of active efforts to prevent sexual assault? While some argue that "we know what the problems are and we know how to change them" (Adams-Curtis and Forbes 2004:115), it is our contention that we do not have a complete explanation of the problem. To address this issue we use data from a study of college life at a large midwestern university and draw on theoretical developments in the sociology of gender (Connell 1987, 1995; Lorber 1994; Martin 2004; Risman 1998, 2004). Continued high rates of sexual assault can be viewed as a case of the reproduction of gender inequality—a phenomenon of central concern in gender theory.

We demonstrate that sexual assault is a predictable outcome of a synergistic intersection of both gendered and seemingly gender neutral processes operating at individual, organizational, and interactional levels. The concentration of homogenous students with expectations of partying fosters the development of sexualized peer cultures organized around status. Residential arrangements intensify students' desires to party in male-controlled fraternities. Cultural expectations that partygoers drink heavily and trust party-mates become problematic when combined with expectations that women be nice and defer to men. Fulfilling the role of the partier produces vulnerability on the part of women, which some men exploit to extract non-consensual sex. The party scene also produces fun, generating student investment in it. Rather than criticizing the party scene or men's behavior, students blame victims. By revealing mechanisms that lead to the persistence of sexual assault and outlining implications for policy, we hope to encourage colleges and universities to develop fresh approaches to sexual assault prevention.

APPROACHES TO COLLEGE SEXUAL ASSAULT

Explanations of high rates of sexual assault on college campuses fall into three broad categories. The first tradition, a psychological approach that we label the "individual determinants" approach, views college sexual assault as primarily a consequence of perpetrator or victim characteristics such as gender role attitudes, personality, family background, or sexual history (Flezzani and Benshoff 2003; Forbes and Adams-Curtis 2001; Rapaport and Burkhart 1984). While "situational variables" are considered, the focus is on individual characteristics (Adams-Curtis and Forbes 2004; Malamuth, Heavey, and Linz 1993). For example, Antonia Abbey and associates (2001) find that hostility toward women, acceptance of verbal pressure as a way to obtain sex, and having many consensual sexual partners distinguish men who sexually assault from men who do not. Research suggests that victims appear quite similar to other college women (Kalof 2000), except that white women, prior victims, first-year college students, and more sexually active women are more vulnerable to sexual assault (Adams-Curtis and Forbes 2004; Humphrey and White 2000).

The second perspective, the "rape culture" approach, grew out of second wave feminism (Brownmiller 1975; Buchwald, Fletcher, and Roth 1993; Lottes 1997; Russell 1975; Schwartz and DeKeseredy 1997). In this perspective, sexual assault is seen as a consequence of widespread belief in "rape myths," or ideas about the nature of men, women, sexuality, and consent that create an environment conducive to rape. For example, men's disrespectful treatment of women is normalized by the idea that men are naturally sexually aggressive. Similarly, the belief that women "ask for it" shifts responsibility from predators to victims (Herman 1989; O'Sullivan 1993). This perspective initiated an important shift away from individual beliefs toward the broader context. However, rape supportive beliefs alone cannot explain the prevalence of sexual assault, which requires not only an inclination on the part of assailants but also physical proximity to victims (Adams-Curtis and Forbes 2004:103).

A third approach moves beyond rape culture by identifying particular contexts—fraternities and bars—as sexually dangerous (Humphrey and Kahn 2000; Martin and Hummer 1989; Sanday 1990, 1996; Stombler 1994). Ayres Boswell and Joan Spade (1996) suggest that sexual assault is supported not only by "a generic culture surrounding and promoting rape," but also by characteristics of the "specific settings" in which men and women interact (p. 133). Mindy Stombler and Patricia Yancey Martin (1994) illustrate that gender inequality is institutionalized on campus by "formal structure" that supports and intensifies an already "high-pressure heterosexual peer group" (p. 180). This perspective grounds sexual assault in organizations that provide opportunities and resources.

We extend this third approach by linking it to recent theoretical scholarship in the sociology of gender. Martin (2004), Barbara Risman (1998, 2004), Judith Lorber (1994) and others argue that gender is not only embedded in individual selves, but also in cultural rules, social interaction, and organizational arrangements. This integrative perspective identifies mechanisms at each level that contribute to the reproduction of gender inequality (Risman 2004). Socialization processes influence gendered selves, while cultural expectations reproduce gender inequality in interaction. At the institutional level, organizational practices, rules, resource distributions, and ideologies reproduce gender inequality. Applying this integrative perspective enabled us to identify gendered processes at individual, interactional, and organizational levels that contribute to college sexual assault.

Risman (1998) also argues that gender inequality is reproduced when the various levels are "all consistent and interdependent" (p. 35). Processes at each level depend upon processes at other levels. Below we demonstrate how interactional processes generating sexual danger depend upon organizational resources and particular kinds of selves. We show that sexual assault results from the intersection of processes at all levels.

We also find that not all of the processes contributing to sexual assault are explicitly gendered. For example, characteristics of individuals such as age, class, and concern with status play a role. Organizational practices such as residence hall assignments and alcohol regulation, both intended to be gender neutral, also contribute to sexual danger. Our findings suggest that apparently gender neutral social processes may contribute to gender inequality in other situations.

METHOD

Data are from group and individual interviews, ethnographic observation, and publicly available information collected at a large midwestern research university. Located in a small city, the school has strong academic and sports programs, a large Greek system, and is sought after by students seeking a quintessential college experience. Like other schools, this school has had legal problems as a result of deaths associated with drinking. In the last few years, students have attended a sexual assault workshop during first-year orientation. Health and sexuality educators conduct frequent workshops, student volunteers conduct rape awareness programs, and *Take Back the Night* marches occur annually.

The bulk of the data presented in this paper were collected as part of ethnographic observation during the 2004–05 academic year in a residence hall identified by students and residence hall staff as a "party dorm." While little partying actually occurs in the hall, many students view this residence hall as one of several places to live in order to participate in the party scene on campus. This made it a good place to study the social worlds of students at high risk of sexual assault—women attending fraternity parties in their first year of college. The authors and a research team were assigned to a room on a floor occupied by 55 women students (51 first-year, 2 second-year, 1 senior, and 1 resident assistant [RA]). We observed on evenings and weekends throughout the entire academic school year. We collected in-depth background information via a detailed nine-page survey that 23 women completed and conducted interviews with 42 of the women (ranging from 1 1/4 to 2 1/2 hours). All but seven of the women on the floor completed either a survey or an interview.

With at least one-third of first-year students on campus residing in "party dorms" and one-quarter of all undergraduates belonging to fraternities or sororities, this social world is the most visible on campus. As the most visible scene on campus, it also attracts students living in other residence halls and those not in the Greek system. Dense pre-college ties among the many in-state students, class and race homogeneity, and a small city location also contribute to the dominance of this scene. Of course, not all students on this floor or at this university participate in the party scene. To participate, one must typically be heterosexual, at least middle class, white, American-born, unmarried, childless, traditional college age, politically and socially mainstream, and interested in drinking. Over three-quarters of the women on the floor we observed fit this description.

There were no non-white students among the first- and second-year students on the floor we studied. This is a result of the homogeneity of this campus and racial segregation in social and residential life. African Americans (who make up 3 to 5% of undergraduates) generally live in living-learning communities in other residence halls and typically do not participate in the white Greek party scene. We argue that the party scene's homogeneity contributes to sexual risk for white women. We lack the space and the data to compare white and African American party scenes on this campus, but in the discussion we offer ideas about what such a comparison might reveal.

We also conducted 16 group interviews (involving 24 men and 63 women) in spring 2004. These individuals had varying relationships to the white Greek party scene on campus. Groups included residents of an alternative residence hall, lesbian, gay, and bisexual students, feminists, re-entry students, academically-focused students, fundamentalist Christians, and sorority women. Eight group interviews were exclusively women, five were mixed in gender composition, and three were exclusively men. The group interviews covered a variety of topics, including discussions of social life, the transition to college, sexual assault, relationships, and the relationship between academic and social life. Participants completed a shorter version of the survey administered to the women on the residence hall floor. From these students we developed an understanding of the dominance of this party scene.

We also incorporated publicly available information about the university from informal interviews with student affairs professionals and from teaching (by all authors) courses on gender, sexuality, and introductory sociology. Classroom data were collected through discussion, student writings, e-mail correspondence, and a survey that included questions about experiences of sexual assault....

EXPLAINING PARTY RAPE

We show how gendered selves, organizational arrangements, and interactional expectations contribute to sexual assault. We also detail the contributions of processes at each level that are not explicitly gendered. We focus on each level in turn, while attending to the ways in which processes at all levels depend upon and reinforce others. We show that fun is produced along with sexual assault, leading students to resist criticism of the party scene.

Selves and Peer Culture in the Transition
from High School to College

Student characteristics shape not only individual participation in dangerous party scenes and sexual risk within them but the development of these party scenes. We identify individual characteristics (other than gender) that generate interest in college partying and discuss the ways in which gendered sexual agendas generate a peer culture characterized by high-takes competition over erotic status.

Non-Gendered Characteristics Motivate Participation in Party Scenes
Without individuals available for partying, the party scene would not exist. All the women on our floor were single and childless, as are the vast majority of undergraduates at this university; many, being upper-middle class, had few responsibilities other than their schoolwork. Abundant leisure time, however, is not enough to fuel the party scene. Media, siblings, peers, and parents all serve as sources of anticipatory socialization (Merton 1957). Both partiers and nonpartiers agreed that one was "supposed" to party in college. This orientation was reflected in the popularity of a poster titled "What I Really Learned in School" that pictured mixed drinks with names associated with academic disciplines. As one focus group participant explained,

> You see these images of college that you're supposed to go out and
> have fun and drink, drink lots, party and meet guys. [You are] supposed
> to hook up with guys, and both men and women try to live up to that.
> I think a lot of it is girls want to be accepted into their groups and guys
> want to be accepted into their groups.

Partying is seen as a way to feel a part of college life. Many of the women we observed participated in middle and high school peer cultures organized around status, belonging, and popularity (Eder 1985; Eder, Evans, and Parker 1995; Milner 2004). Assuming that college would be similar, they told us that they wanted to fit in, be popular, and have friends....

Peer Culture as Gendered and Sexualized   Partying was also the primary way to meet men on campus. The floor was locked to non-residents, and even men living in the same residence hall had to be escorted on the floor. The women found it difficult to get to know men in their classes, which were mostly mass lectures. They explained to us that people "don't talk" in class. Some complained they lacked casual friendly contact with men, particularly compared to the mixed-gender friendship groups they reported experiencing in high school.

Meeting men at parties was important to most of the women on our floor. The women found men's sexual interest at parties to be a source of self-esteem and status. They enjoyed dancing and kissing at parties, explaining to us that it proved men "liked" them. This attention was not automatic, but required the skillful deployment of physical and cultural assets (Stombler and Padavic 1997; Swidler 2001). Most of the party-oriented women on the floor arrived with appropriate gender presentations and the money and know-how to preserve and refine them. While some more closely resembled the "ideal" college party girl (white, even features, thin but busty, tan, long straight hair, skillfully made-up, and well-dressed in the latest youth styles), most worked hard to attain this presentation. They regularly straightened their hair, tanned, exercised, dieted, and purchased new clothes....

The psychological benefits of admiration from men in the party scene were such that women in relationships sometimes felt deprived. One woman with a serious boyfriend noted that she dressed more conservatively at parties because of him, but this meant she was not "going to get any of the attention." She lamented that no one was "going to waste their time with me" and that, "this is taking away from my confidence." Like most women who came to college with boyfriends, she soon broke up with him.

Men also sought proof of their erotic appeal. As a woman complained, "Every man I have met here has wanted to have sex with me!" ... The women found that men were more interested than they were in having sex. These clashes in sexual expectations are not surprising: men derived status from securing sex (from high-status women), while women derived status from getting attention (from high-status men). These agendas are both complementary and adversarial: men give attention to women en route to getting sex, and women are unlikely to become interested in sex without getting attention first.

University and Greek Rules, Resources, and Procedures

Simply by congregating similar individuals, universities make possible heterosexual peer cultures. The university, the Greek system, and other related organizations structure student life through rules, distribution of resources, and procedures (Risman 2004).

Sexual danger is an unintended consequence of many university practices intended to be gender neutral. The clustering of homogeneous students intensifies the dynamics of student peer cultures and heightens motivations to party. Characteristics of residence halls and how they are regulated push student partying into bars, off-campus residences, and fraternities. While factors that increase the risk of party rape are present in varying degrees in all party venues (Boswell and Spade 1996), we focus on fraternity parties because they were the typical party venue for the women we observed and have been identified as particularly unsafe (see also Martin and Hummer 1989; Sanday 1990). Fraternities offer the most reliable and private source of alcohol for first-year students excluded from bars and house parties because of age and social networks.

University Practices as Push Factors   The university has latitude in how it enforces state drinking laws. Enforcement is particularly rigorous in residence halls. We observed RAs and police officers (including gun-carrying peer police) patrolling the halls for alcohol violations. Women on our floor were "documented" within the first week of school for infractions they felt were minor. Sanctions are severe—a $300 fine, an 8-hour alcohol class, and probation for a year. As a consequence, students engaged in only minimal, clandestine alcohol consumption in their rooms. In comparison, alcohol flows freely at fraternities.

The lack of comfortable public space for informal socializing in the residence hall also serves as a push factor. A large central bathroom divided our floor. A sterile lounge was rarely used for socializing. There was no cafeteria, only a convenience store and a snack bar in a cavernous room furnished with big-screen televisions. Residence life sponsored alternatives to the party scene such as "movie night" and special dinners, but these typically occurred early in the evening. Students defined the few activities sponsored during party hours (e.g., a midnight trip to Wal-Mart) as uncool.

Intensifying Peer Dynamics   The residence halls near athletic facilities and Greek houses are known by students to house affluent, party-oriented students. White, upper-middle class, first-year students who plan to rush request these residence halls, while others avoid them. One of our residents explained that "everyone knows what [the residence hall] is like and people are dying to get in here. People just think it's a total party or something." Students of color tend to live elsewhere on campus. As a consequence, our floor was homogenous in terms of age, race, sexual orientation, class, and appearance....

The homogeneity of the floor intensified social anxiety, heightening the importance of partying for making friends. Early in the year, the anxiety was palpable on weekend nights as women assessed their social options by asking where people were going, when, and with whom. One exhausted floor resident told us she felt that she "needed to" go out to protect her position in a friendship group. At the beginning of the semester, "going out" on weekends was virtually compulsory. By 11 p.m. the floor was nearly deserted.

Male Control of Fraternity Parties   The campus Greek system cannot operate without university consent. The university lists Greek organizations as student clubs, devotes professional staff to Greek-oriented programming, and disbands fraternities that violate university policy. Nonetheless, the university lacks full authority over fraternities; Greek houses are privately owned and chapters answer to national organizations and the Interfraternity Council (IFC) (i.e., a body governing the more than 20 predominantly white fraternities).

Fraternities control every aspect of parties at their houses: themes, music, transportation, admission, access to alcohol, and movement of guests. Party themes usually require women to wear scant, sexy clothing and place women in subordinate positions to men. During our observation period, women attended parties such as "Pimps and Hos," "Victoria's Secret," and "Playboy Mansion"—the last of which required fraternity members to escort two scantily-clad dates. Other recent themes included: "CEO/Secretary Ho," "School Teacher/ Sexy Student," and "Golf Pro/Tennis Ho."

Some fraternities require pledges to transport first-year students, primarily women, from the residence halls to the fraternity houses. From about 9 to 11 p.m. on weekend nights early in the year, the drive in front of the residence hall resembled a rowdy taxi-stand, as dressed-to-impress women waited to be carpooled to parties in expensive late-model vehicles. By allowing party-oriented first-year women to cluster in particular residence halls, the university made them easy to find. One fraternity member told us this practice was referred to as "dorm-storming."

Transportation home was an uncertainty. Women sometimes called cabs, caught the "drunk bus," or trudged home in stilettos. Two women indignantly described a situation where fraternity men "wouldn't give us a ride home." The women said, "Well, let us call a cab." The men discouraged them from calling the cab and eventually found a designated driver. The women described the men as "just dicks" and as "rude."

Fraternities police the door of their parties, allowing in desirable guests (first-year women) and turning away others (unaffiliated men). Women told us of abandoning parties when male friends were not admitted. They explained that fraternity men also controlled the quality and quantity of alcohol. Brothers served themselves first, then personal guests, and then other women. Non-affiliated and unfamiliar men were served last, and generally had access to only the least desirable beverages. The promise of more or better alcohol was often used to lure women into private spaces of the fraternities.

Fraternities are constrained, though, by the necessity of attracting women to their parties. Fraternities with reputations for sexual disrespect have more success recruiting women to parties early in the year. One visit was enough for some of the women. A roommate duo told of a house they "liked at first" until they discovered that the men there were "really not nice."

The Production of Fun and Sexual Assault in Interaction

Peer culture and organizational arrangements set up risky partying conditions, but do not explain *how* student interactions at parties generate sexual assault. At the interactional level we see the mechanisms through which sexual assault is produced. As interactions necessarily involve individuals with particular characteristics and occur in specific organizational settings, all three levels meet when interactions take place. Here, gendered and gender neutral expectations and routines are intricately woven together to create party rape. Party rape is the result of fun situations that shift—either gradually or quite suddenly—into coercive situations. Demonstrating how the production of fun is connected with sexual assault requires describing the interactional routines and expectations that enable men to employ coercive sexual strategies with little risk of consequence....

Cultural expectations of partying are gendered. Women are supposed to wear revealing outfits, while men typically are not. As guests, women cede control of turf, transportation, and liquor. Women are also expected to be grateful for men's hospitality, and as others have noted, to generally be "nice" in ways that men are not (Gilligan 1982; Martin 2003; Phillips 2000; Stombler and

Martin 1994; Tolman 2002). The pressure to be deferential and gracious may be intensified by men's older age and fraternity membership. The quandary for women, however, is that fulfilling the gendered role of partier makes them vulnerable to sexual assault.

Women's vulnerability produces sexual assault only if men exploit it. Too many men are willing to do so. Many college men attend parties looking for casual sex. A student in one of our classes explained that "guys are willing to do damn near anything to get a piece of ass." A male student wrote the following description of parties at his (non-fraternity) house:

> Girls are continually fed drinks of alcohol. It's mainly to party but my roomies are also aware of the inhibition-lowering effects. I've seen an old roomie block doors when girls want to leave his room; and other times I've driven women home who can't remember much of an evening yet sex did occur. Rarely if ever has a night of drinking for my roommate ended without sex. I know it isn't necessarily and assuredly sexual assault, but with the amount of liquor in the house I question the amount of consent a lot.

Another student—after deactivating—wrote about a fraternity brother "telling us all at the chapter meeting about how he took this girl home and she was obviously too drunk to function and he took her inside and had sex with her." Getting women drunk, blocking doors, and controlling transportation are common ways men try to prevent women from leaving sexual situations. Rape culture beliefs, such as the belief that men are "naturally" sexually aggressive, normalize these coercive strategies. Assigning women the role of sexual "gatekeeper" relieves men from responsibility for obtaining authentic consent, and enables them to view sex obtained by undermining women's ability to resist it as "consensual" (e.g., by getting women so drunk that they pass out)....

We heard many stories of negative experiences in the party scene including at least one account of a sexual assault in every focus group that included heterosexual women. Most women who partied complained about men's efforts to control their movements or pressure them to drink. Two of the women on our floor were sexually assaulted at a fraternity party in the first week of school—one was raped. Later in the semester, another woman on the floor was raped by a friend. A fourth woman on the floor suspects she was drugged; she became disoriented at a fraternity party and was very ill for the next week.

Party rape is accomplished without the use of guns, knives, or fists. It is carried out through the combination of low level forms of coercion—a lot of liquor and persuasion, manipulation of situations so that women cannot leave, and sometimes force (e.g., by blocking a door, or using body weight to make it difficult for a woman to get up). These forms of coercion are made more effective by organizational arrangements that provide men with control over how partying happens and by expectations that women let loose and trust their party-mates. This systematic and effective method of extracting non-consensual sex is largely invisible, which makes it difficult for victims to convince anyone—even themselves—that a crime occurred. Men engage in this behavior with little risk of consequences.

Student Responses and the Resiliency of the Party Scene

The frequency of women's negative experiences in the party scene poses a problem for those students most invested in it. Finding fault with the party scene potentially threatens meaningful identities and lifestyles. The vast majority of heterosexual encounters at parties are fun and consensual. Partying provides a chance to meet new people, experience and display belonging, and to enhance social position. Women on our floor told us that they loved to flirt and be admired, and they displayed pictures on walls, doors, and websites commemorating their fun nights out.

The most common way that students—both women and men—account for the harm that befalls women in the party scene is by blaming victims. By attributing bad experiences to women's "mistakes," students avoid criticizing the party scene or men's behavior within it. Such victim-blaming also allows women to feel that they can control what happens to them. The logic of victim-blaming suggests that sophisticated, smart, careful women are safe from sexual assault. Only "immature," "naive," or "stupid" women get in trouble. When discussing the sexual assault of a friend, a floor resident explained that:

> She somehow got like sexually assaulted ... by one of our friends' old roommates. All I know is that kid was like bad news to start off with. So, I feel sorry for her but it wasn't much of a surprise for us. He's a shady character.

Another floor resident relayed a sympathetic account of a woman raped at knife point by a stranger in the bushes, but later dismissed party rape as nothing to worry about "'cause I'm not stupid when I'm drunk." Even a feminist focus group participant explained that her friend who was raped "made every single mistake and almost all of them had to with alcohol.... She got ridiculed when she came out and said she was raped." These women contrast "true victims" who are deserving of support with "stupid" women who forfeit sympathy (Phillips 2000). Not only is this response devoid of empathy for other women, but it also leads women to blame themselves when they are victimized (Phillips 2000).

Sexual assault prevention strategies can perpetuate victim-blaming. Instructing women to watch their drinks, stay with friends, and limit alcohol consumption implies that it is women's responsibility to avoid "mistakes" and their fault if they fail. Emphasis on the precautions women should take—particularly if not accompanied by education about how men should change their behavior—may also suggest that it is natural for men to drug women and take advantage of them. Additionally, suggesting that women should watch what they drink, trust party-mates, or spend time alone with men asks them to forgo full engagement in the pleasures of the college party scene....

Opting Out While many students find the party scene fun, others are more ambivalent. Some attend a few fraternity parties to feel like they have participated in this college tradition. Others opt out of it altogether. On our floor, 44 out of the 51 first-year students (almost 90%) participated in the party scene. Those on the floor who opted out worried about sexual safety and the consequences of engaging in illegal behavior. For example, an interviewee who did not drink was appalled by the fraternity party transport system. She explained that:

> All those girls would stand out there and just like, no joke, get into these big black Suburbans driven by frat guys, wearing like seriously no clothes, piled on top of each other. This could be some kidnapper taking you all away to the woods and chopping you up and leaving you there. How dumb can you be?

In her view, drinking around fraternity men was "scary" rather than "fun."

Her position was unpopular. She, like others who did not party, was an outsider on the floor. Partiers came home loudly in the middle of the night, threw up in the bathrooms, and rollerbladed around the floor. Socially, the others simply did not exist. A few of our "misfits" successfully created social lives outside the floor. The most assertive of the "misfits" figured out the dynamics of the floor in the first weeks and transferred to other residence halls.

However, most students on our floor lacked the identities or network connections necessary for entry into alternative worlds. Life on a large university campus can be overwhelming for first-year students. Those who most needed an alternative to the social world of the party dorm were often ill-equipped to actively seek it out. They either integrated themselves into partying or found themselves alone in their rooms, microwaving frozen dinners and watching television. A Christian focus group participant described life in this residence hall: "When everyone is going out on a Thursday and you are in the room by yourself and there are only two or three other people on the floor, that's not fun, it's not the college life that you want."

DISCUSSION AND IMPLICATIONS

We have demonstrated that processes at individual, organizational, and interactional levels contribute to high rates of sexual assault. Some individual level characteristics that shape the likelihood of a sexually dangerous party scene developing are not explicitly gendered. Party rape occurs at high rates in places that cluster young, single, party-oriented people concerned about social status. Traditional beliefs about sexuality also make it more likely that one will participate in the party scene and increase danger within the scene. This university contributes to sexual danger by allowing these individuals to cluster.

However, congregating people is not enough, as parties cannot be produced without resources (e.g., alcohol and a viable venue) that are difficult for underage students to obtain. University policies that are explicitly gender-neutral—such as the policing of alcohol use in residence halls—have gendered consequences. This policy encourages first-year students to turn to fraternities to party. Only fraternities, not sororities, are allowed to have parties, and men structure parties in ways that control the appearance, movement, and behavior of female guests.

Men also control the distribution of alcohol and use its scarcity to engineer social interactions. The enforcement of alcohol policy by both university and Greek organizations transforms alcohol from a mere beverage into an unequally distributed social resource.

Individual characteristics and institutional practices provide the actors and contexts in which interactional processes occur. We have to turn to the interactional level, however, to understand *how* sexual assault is generated. Gender neutral expectations to "have fun," lose control, and trust one's party-mates become problematic when combined with gendered interactional expectations. Women are expected to be "nice" and to defer to men in interaction. This expectation is intensified by men's position as hosts and women's as grateful guests. The heterosexual script, which directs men to pursue sex and women to play the role of gatekeeper, further disadvantages women, particularly when virtually *all* men's methods of extracting sex are defined as legitimate....

Our analysis also provides a framework for analyzing the sources of sexual risk in non-university partying situations. Situations where men have a home turf advantage, know each other better than the women present know each other, see the women as anonymous, and control desired resources (such as alcohol or drugs) are likely to be particularly dangerous. Social pressures to "have fun," prove one's social competency, or adhere to traditional gender expectations are also predicted to increase rates of sexual assault within a social scene.

This research has implications for policy. The interdependence of levels means that it is difficult to enact change at one level when the other levels remain unchanged.... Without change in institutional arrangements, efforts to change cultural beliefs are undermined by the cultural commonsense generated by encounters with institutions. Efforts to educate about sexual assault will not succeed if the university continues to support organizational arrangements that facilitate and even legitimate men's coercive sexual strategies. Thus, our research implies that efforts to combat sexual assault on campus should target all levels, constituencies, and processes simultaneously. Efforts to educate both men and women should indeed be intensified, but they should be reinforced by changes in the social organization of student life.

Researchers focused on problem drinking on campus have found that reduction efforts focused on the social environment are successful (Berkowitz 2003:21). Student body diversity has been found to decrease binge drinking on campus (Wechsler and Kuo 2003); it might also reduce rates of sexual assault. Existing student heterogeneity can be exploited by eliminating self-selection into age-segregated, white, upper-middle class, heterosexual enclaves and by working to make residence halls more appealing to upper-division students. Building more aesthetically appealing housing might allow students to interact outside of alcohol-fueled party scenes. Less expensive plans might involve creating more living-learning communities, coffee shops, and other student-run community spaces.

While heavy alcohol use is associated with sexual assault, not all efforts to regulate student alcohol use contribute to sexual safety. Punitive approaches sometimes heighten the symbolic significance of drinking, lead students to drink more hard liquor, and push alcohol consumption to more private and thus more dangerous spaces. Regulation inconsistently applied—e.g., heavy policing of residence halls and light policing of fraternities—increases the power of those who can secure alcohol and host parties. More consistent regulation could decrease the value of alcohol as a commodity by equalizing access to it.

Sexual assault education should shift in emphasis from educating women on preventative measures to educating both men and women about the coercive behavior of men and the sources of victim-blaming. Mohler-Kuo and associates (2004) suggest, and we endorse, a focus on the role of alcohol in sexual assault. Education should begin before students arrive on campus and continue throughout college. It may also be most effective if high-status peers are involved in disseminating knowledge and experience to younger college students.

Change requires resources and cooperation among many people. Efforts to combat sexual assault are constrained by other organizational imperatives. Student investment in the party scene makes it difficult to enlist the support of even those most harmed by the state of affairs. Student and alumni loyalty to partying (and the Greek system) mean that challenges to the party scene could potentially cost universities tuition dollars and alumni donations. Universities must contend with Greek organizations and bars, as well as the challenges of internal coordination. Fighting sexual assault on all levels is critical, though, because it is unacceptable for higher education institutions to be sites where women are predictably sexually victimized.

REFERENCES

Abbey, Antonia, Pam McAuslan, Tina Zawacki, A. Monique Clinton, and Philip Buck. 2001. "Attitudinal, Experiential, and Situational Predictors of Sexual Assault Perpetration." *Journal of Interpersonal Violence* 16:784–807.

Abbey, Antonia, Lisa Thomson Ross, Donna McDuffie, and Pam McAuslan. 1996. "Alcohol and Dating Risk Factors for Sexual Assault among College Women." *Psychology of Women Quarterly* 20:147–69.

Adams-Curtis, Leah and Gordon Forbes. 2004. "College Women's Experiences of Sexual Coercion: A Review of Cultural, Perpetrator, Victim, and Situational Variables." *Trauma, Violence, and Abuse: A Review Journal* 5:91–122.

Bachar, Karen and Mary Koss. 2001. "From Prevalence to Prevention: Closing the Gap between What We Know about Rape and What We Do." pp. 117–42 in *Sourcebook on Violence against Women*, edited by C. Renzetti, J. Edleson, and R. K. Bergen. Thousand Oaks, CA: Sage.

Berkowitz, Alan. 2003. "How Should We Talk about Student Drinking—And What Should We Do about It?" *About Campus* May/June:16–22.

Boswell, A. Ayres and Joan Z. Spade. 1996. "Fraternities and Collegiate Rape Culture: Why Are Some Fraternities More Dangerous Places for Women?" *Gender & Society* 10:133–47.

Brownmiller, Susan. 1975. *Against Our Will: Men, Women, and Rape*. New York: Bantam Books.

Buchwald, Emilie, Pamela Fletcher, and Martha Roth, eds. 1993. *Transforming a Rape Culture*. Minneapolis, MN: Milkweed Editions.

Connell, R. W. 1987. *Gender and Power*. Palo Alto, CA: Stanford University Press.

———. 1995. *Masculinities*. Berkeley, CA: University of California Press.

Eder, Donna. 1985. "The Cycle of Popularity: Interpersonal Relations among Female Adolescents." *Sociology of Education* 58:154–65.

Eder, Donna, Catherine Evans, and Stephen Parker. 1995. *School Talk: Gender and Adolescent Culture*. New Brunswick, NJ: Rutgers University Press.

Fisher, Bonnie, Francis Cullen, and Michael Turner. 2000. "The Sexual Victimization of College Women." Washington, DC: National Institute of Justice and the Bureau of Justice Statistics.

Flezzani, James and James Benshoff. 2003. "Understanding Sexual Aggression in Male College Students: The Role of Self-Monitoring and Pluralistic Ignorance." *Journal of College Counseling* 6:69–79.

Forbes, Gordon and Leah Adams-Curtis. 2001. "Experiences with Sexual Coercion in College Males and Females: Role of Family Conflict, Sexist Attitudes, Acceptance of Rape Myths, Self-Esteem, and the Big-Five Personality Factors." *Journal of Interpersonal Violence* 16:865–89.

Gilligan, Carol. 1982. *In a Different Voice: Psychological Theory and Women's Development*. Cambridge, MA: Harvard University Press.

Herman, Diane. 1989. "The Rape Culture." pp. 20–44 in *Women: A Feminist Perspective*, edited by J. Freeman. Mountain View, CA: Mayfield.

Humphrey, John and Jacquelyn White. 2000. "Women's Vulnerability to Sexual Assault from Adolescence to Young Adulthood." *Journal of Adolescent Health* 27:419–24.

Humphrey, Stephen and Arnold Kahn. 2000. "Fraternities, Athletic Teams, and Rape: Importance of Identification with a Risky Group." *Journal of Interpersonal Violence* 15:1313–22.

Kalof, Linda. 2000. "Vulnerability to Sexual Coercion among College Women: A Longitudinal Study." *Gender Issues* 18:47–58.

Lorber, Judith. 1994. *Paradoxes of Gender*. New Haven, CT: Yale University Press.

Lottes, Ilsa L. 1997. "Sexual Coercion among University Students: A Comparison of the United States and Sweden." *Journal of Sex Research* 34:67–76.

Malamuth, Neil, Christopher Heavey, and Daniel Linz. 1993. "Predicting Men's Antisocial Behavior against Women: The Interaction Model of Sexual Aggression." pp. 63–98 in *Sexual Aggression: Issues in Etiology, Assessment, and Treatment*, edited by G. N. Hall, R. Hirschman, J. Graham, and M. Zaragoza. Washington, D.C.: Taylor and Francis.

Marine, Susan. 2004. "Waking Up from the Nightmare of Rape." *The Chronicle of Higher Education*. November 26, p. B5.

Martin, Karin. 2003. "Giving Birth Like a Girl." *Gender & Society*. 17:54–72.

Martin, Patricia Yancey. 2004. "Gender as a Social Institution." *Social Forces* 82:1249–73.

Martin, Patricia Yancey and Robert A. Hummer. 1989. "Fraternities and Rape on Campus." *Gender & Society* 3:457–73.

Merton, Robert. 1957. *Social Theory and Social Structure*. New York: Free Press.

Milner, Murray. 2004. *Freaks, Geeks, and Cool Kids: American Teenagers, Schools, and the Culture of Consumption*. New York: Routledge.

Mohler-Kuo, Meichun, George W. Dowdall, Mary P. Koss, and Henry Weschler. 2004. "Correlates of Rape While Intoxicated in a National Sample of College Women." *Journal of Studies on Alcohol* 65:37–45.

O'Sullivan, Chris. 1993. "Fraternities and the Rape Culture." pp. 23–30 in *Transforming a Rape Culture*, edited by E. Buchwald, P. Fletcher, and M. Roth. Minneapolis, MN: Milkweed Editions.

Phillips, Lynn. 2000. *Flirting with Danger: Young Women's Reflections on Sexuality and Domination*. New York: New York University.

Rapaport, Karen and Barry Burkhart. 1984. "Personality and Attitudinal Characteristics of Sexually Coercive College Males." *Journal of Abnormal Psychology* 93:216–21.

Risman, Barbara. 1998. *Gender Vertigo: American Families in Transition*. New Haven, CT: Yale University Press.

———. 2004. "Gender as a Social Structure: Theory Wrestling with Activism." *Gender & Society* 18:429–50.

Russell, Diana. 1975. *The Politics of Rape*. New York: Stein and Day.

Sampson, Rana. 2002. "Acquaintance Rape of College Students." Problem-Oriented Guides for Police Series, No. 17. Washington, DC: U.S. Department of Justice, Office of Community Oriented Policing Services.

Sanday, Peggy. 1990. *Fraternity Gang Rape: Sex, Brotherhood, and Privilege on Campus*. New York: New York University Press.

———. 1996. "Rape-Prone versus Rape-Free Campus Cultures." *Violence against Women* 2:191–208.

Schwartz, Martin and Walter DeKeseredy. 1997. *Sexual Assault on the College Campus: The Role of Male Peer Support*. Thousand Oaks, CA: Sage Publications.

Stombler, Mindy. 1994. "'Buddies' or 'Slutties': The Collective Reputation of Fraternity Little Sisters." *Gender & Society* 8:297–323.

Stombler, Mindy and Patricia Yancey Martin. 1994. "Bringing Women In, Keeping Women Down: Fraternity 'Little Sister' Organizations." *Journal of Contemporary Ethnography* 23:150–84.

Stombler, Mindy and Irene Padavic. 1997. "Sister Acts: Resisting Men's Domination in Black and White Fraternity Little Sister Programs." *Social Problems* 44:257–75.

Swidler, Ann. 2001. *Talk of Love: How Culture Matters*. Chicago: University of Chicago Press.

Wechsler, Henry and Meichun Kuo. 2003. "Watering Down the Drinks: The Moderating Effect of College Demographics on Alcohol Use of High-Risk Groups." *American Journal of Public Health* 93:1929–33.

KEY CONCEPTS

acquaintance rape   doing gender   victimization

DISCUSSION QUESTIONS

1. Think about the most recent party you attended? Can you determine who was "in control" of the food and drink? Who controlled the guest list? What role, if any, did you have in the way the party was structured?

2. Consider the different perspectives on sexual assault presented in the article (individual determinants; rape culture; and particular contexts for sexual assault). Which do you believe offers the best explanation for sexual assault on college campuses?

18

The Social Organization of Toy Stores

CHRISTINE L. WILLIAMS

Christine Williams wrote a book about her ethnographic research working in toy stores. This article is excerpted from that book and talks about how retail toy stores are structured around race and gender. The social organization of toy stores typically places White women in positions to interact with customers and men of color in the back to work inventory. These patterns of work organization are not observed by the average customer, yet create inequality in the workplace.

Living in a consumer society means that we come into contact with retail workers almost every day. Over 22.5 million people work in this job, composing the largest sector of the service economy (Sandikci and Holt 1998, 305; U.S. Bureau of Labor Statistics 2004). But unless you have "worked retail," you probably know little about the working conditions of the job. At best, retail workers are taken for granted by consumers, noticed only when they aren't doing their job. At worst, they are stereotyped as either dim-witted or haughty, which is how they are often portrayed on television and in the movies.

Retail jobs, like other jobs in the service sector, have grown in number and changed dramatically over the past decades. Service jobs gradually have replaced manufacturing jobs as part of the general deindustrialization of the U.S. economy.

SOURCE: Christine L. Williams. 2006. *Inside Toyland: Working, Shopping, and Social Inequality.* Berkeley, CA: University of California Press, pp. 48–91.

This economic restructuring has resulted in boom times for wealthy American consumers as the prices for many commodities have dropped (a consequence of the movement of production overseas). It has also resulted in an erosion of working conditions for Americans in the bottom half of the economy, including service workers. Retail jobs have become increasingly "flexible," temporary, and part time. Over the past decades, workers in these jobs have experienced a loss of job security and benefits, a diminishment in the power of unions, and a lessening of the value of the minimum wage (McCall 2001). Yet while most retail workers have lost ground, the giant corporations they work for have enjoyed unprecedented prosperity and political clout.

George Ritzer (2002) aptly uses the term *Mcjobs* to describe the working conditions found in a variety of service industries today. The word is a pun on McDonald's, the fast-food giant that introduced and popularized this labor system. Mcjobs are not careers; they are designed to discourage long-term commitment. They have short promotion ladders, they provide few opportunities for advancement or increased earnings, and the technical skills they require are not transferable outside the immediate work environment. They target sectors of the labor force that presumably don't "need" money to support themselves or their families: young people looking for "fun jobs" before college; mothers seeking part-time opportunities to fit around their family responsibilities; older, retired people looking for the chance to get out of the house and to socialize. However, this image does not resonate with the increasing numbers of workers in these jobs who are struggling to support themselves and their families (Ehrenreich 2001; Talwar 2002). The marketing of Mcjobs on television commercials for Wal-Mart and fast-food restaurants obscures the harsh working conditions and low pay that contribute to the impoverished state of the working poor.

In addition to contributing to economic inequality, jobs in the retail industry are structured in ways that enhance inequality by gender and race. Although all retail workers are low paid, white men employed in this industry earn more money than any other group. Overall, about as many men as women work in retail trades, but they are concentrated in different kinds of stores. For example, men make up more than three-quarters of workers in retail jobs selling motor vehicles, lumber, and home and auto supplies, while women predominate in apparel, gift, and needlework stores (U.S. Bureau of Labor Statistics 2004).

In both stores where I worked, the gender ratio was about 60:40, with women outnumbering men. I was surprised that so many men worked in these toy stores. In my admittedly limited experience, I associated women with the job of selling toys. But I learned that because of the way that jobs are divided and organized, customers usually don't see the substantial numbers of men who are working there too.

Retail work is also organized by race and ethnicity. Ten percent of all employees in the retail trade industry are African American, and 12 percent are of Hispanic origin, slightly less than their overall representation in the U.S. population. But again, whites, African Americans, and Latinas/os are likely to work in different types of stores. For example, African Americans are under-represented (less than 5 percent) in stores that sell hardware, gardening equipment, and needle-work supplies and overrepresented (more than 15 percent) in department stores, variety stores, and shoe stores. Similarly Latinas/os are under-represented (less than 6 percent) in bookstores and gas stations and overrepresented (more than 16 percent) in retail florists and household appliance stores (U.S. Bureau of Labor Statistics 2004).

The two stores where I worked had radically different racial compositions. Sixty percent of the workers at the Toy Warehouse [a "big box" store] were African American, and 60 percent of those at Diamond Toys [a boutique, high-end store] were white. Only three African Americans, all women, worked at Diamond Toys. No black men worked at that store. In contrast, only four white women (including me) worked at the Toy Warehouse....

Sociologists have long recognized the workplace as a central site for the reproduction of social inequalities. Studies of factory work in particular have shown us how race and gender hierarchies are reproduced through the social organization of the work. I argue that the labor process in service industries is equally important for understanding social inequality, even though this sector has not come under the same degree of scrutiny by sociologists. I demonstrate how the working conditions at the two stores perpetuate inequality by class, gender, and race. The jobs are organized in such a way as to benefit some groups of workers and discriminate against others.

The stores where I worked represent a range of working conditions in large retail trade establishments. Although both were affiliated with national chains and both were in the business of selling toys, Diamond Toys was unionized and the Toy Warehouse was not. The union protected workers from some of the most egregious aspects of retail work. But ... the union could not overcome the race, gender, and class inequalities that are reproduced by the social organization of the industry.

STRATIFIED SELLING

Diamond Toys and the Toy Warehouse each employed about seventy workers. As in other large retail establishments, the workers were organized in an elaborate hierarchy. Each store was governed by a regional office, which in turn was overseen by the national corporate headquarters. There was no local autonomy in the layout or the merchandise sold in the stores. Within each store, directors were at the top, followed by managers, supervisors, and associates. Directors and managers were salaried employees; everyone else was hourly. Most directors and managers had a college degree. Candidates for these jobs applied to the regional headquarters and, once hired, were assigned to specific stores. These might not be the stores closest to where they lived. Olive, my manager at the Toy Warehouse, had a two-hour commute each way to work, even though there was a Toy Warehouse within five miles of her home.

The hierarchy of jobs and power within the stores was marked by race and gender. In both stores the directors and assistant directors were white men. Immediately below them were managers, who were a more diverse group, including men and women, whites and Latinas/os, and, at the Toy Warehouse, an African American woman (Olive). There were far more managers at Diamond Toys than at the Toy Warehouse; I met at least ten managers during my time there, versus only two at the Toy Warehouse.

The next layer of the hierarchy under managers were supervisors, who were drawn from the ranks of associates. They were among those who had the most seniority and thus the most knowledge of store procedures, and they had limited authority to do things like void transactions at the registers. All of the supervisors at Diamond Toys were white and most were men, while at the Toy Warehouse supervisors were more racially diverse and most were women. It took me a long time to figure out who the supervisors were at the Toy Warehouse. Many of those I thought were supervisors turned out to be regular employees. They had many of the same responsibilities as supervisors, but, as I came to find out, they were competing with each other for promotion to this position. When I asked why they were acting like supervisors, it was explained to me that the Toy Warehouse wouldn't promote anyone before he or she was proficient at the higher job. This policy justified giving workers more responsibilities without more pay. At Diamond Toys, in contrast, job descriptions were clearer and were enforced.

Associates were the largest group of workers at the stores (sometimes referred to as the staff). They included men and women of all races and ethnic groups and different ages, except at Diamond Toys, where I noted that there were no black men. Despite the apparent diversity among the staff, there was substantial segregation by race and gender in the tasks they were assigned. Employees of toy stores are divided between back- and front-of-house workers. The back-of-house employees and managers work in the storage areas, on the loading docks, and in the assembly rooms. In both stores where I worked, the back-of-house workers were virtually all men. The front-of-house workers, the ones who interacted with customers, included both men and women. But there, too, there was job segregation by gender and race, although, as I will discuss, it was harder to discern and on occasion it broke down.

There were two other jobs in the toy store: security guards and janitors, both of whom were subcontracted workers. Both the Toy Warehouse and Diamond Toys employed plainclothes security guards who watched surveillance monitors in their back offices and roamed the aisles looking for shoplifters. At the Toy Warehouse, the individuals who filled those jobs were mostly African American men and women, while only white men and women were hired for security at Diamond Toys. Finally, all of the cleaners at the two stores were Latinas. They were recent immigrants who didn't speak English.

What accounts for the race and gender segregation of jobs in the toy store? Conventional economic theory argues that job segregation is the product of differences in human capital attainment. According to this view, the marketplace sorts workers into jobs depending on their qualifications and preferences. Because men and women of different racial/ethnic groups possess different skills, aptitudes, and work experiences, they will be (and indeed should be) hired into different jobs. Economists generally see this process as benign, if not beneficial, in a society founded on meritocracy, individual liberty, and freedom of choice (Folbre 2001).

In contrast, when sociologists look at job segregation, they tend to see discrimination and structural inequality (Reskin and Roos 1990). Obtaining the right qualifications for a high-paying job is easier for some groups than others. Differential access to college education is an obvious example: society blocks opportunities for poor people to acquire this human capital asset while smoothing the path for the well-to-do. But the sociological critique of job segregation goes deeper than this. Sociologists argue that the definitions of who is qualified and what it means to be qualified for a job are linked to stereotypes about race and gender. Joan Acker (1990) argues that jobs are "gendered," meaning that qualities culturally associated with men (leadership, physical strength, aggression, goal orientation) are built into the job descriptions of the higher-status and higher-paid occupations in our economy. Qualities associated with women (dexterity, passivity, nurturing orientation) tend to be favored in low-paying jobs. In addition to being gendered, jobs are racialized. Black women have been subjected to a different set of gendered stereotypes than white women. Far from being seen as delicate and passive, they have been perceived as dominant, insubordinate, and aggressive (Collins 2000). Those who make hiring decisions draw upon these kinds of racialized stereotypes of masculinity and femininity when appointing workers to specific jobs....

This process may be exacerbated in interactive service work, where employers carefully pick workers who "look right" for the corporate image they attempt to project to the public. A recent court case against Abercrombie & Fitch illustrates this. A suit was brought against the retailer by Asian Americans and Latinas/os who said they were refused selling jobs because "they didn't project what the company called the A & F look" (Greenhouse 2003). Although the company denied the charge, the suit brings to light the common retail practice of matching employees with the image the company is seeking to cultivate. More egregious examples are found in sexualized service work, as in the case of Hooters restaurants (where only buxom young women are hired), and in theme parks like Disney, notorious for its resistance to hiring African Americans (Loe 1996; Project on Disney 1995).

This process of interpellation was apparent in the toy stores where I worked: managers imagined different kinds of people in each job, who came to see themselves in terms of these stereotypical expectations....

My experience illustrates this process of interpellation and resistance. I was hired to be a cashier at both toy stores. I didn't seek out this job, but this was how both managers who hired me envisioned my potential contribution. Only women were regularly assigned to work as cashiers at the Toy Warehouse, and I noticed that management preferred young or light-skinned women for this job. Some older African American women who wanted to work as cashiers had to struggle to get the assignment. Lazelle, for example, who was about thirty-five, had been asking to be put on register over the two months she had been working there. She had been assigned to be a merchandiser. Merchandisers retrieved items from the storeroom, priced items, and checked prices when the universal product codes (UPCs) were missing. Lazelle finally got her chance at the register the same day that I started. We set up next to each other, and I noticed with a bit of envy how much more competent and confident on the register she was compared to me. (Later she told me she had worked registers at other stores, including fast-food restaurants.) I told her that I had been hoping to get assigned to the job of merchandiser. I liked the idea of being free to walk around the store, engage with customers, and learn more about the toys. I had mentioned to Olive that I wanted that job, but she had made it clear that I was destined for cashiering and the service desk (and later, to my horror, computer accounting). Lazelle looked at me as if I were crazy. Merchandising was generally considered to be the worst job in the store because it was so physically taxing. From her point of view I had been assigned the better job, no doubt because of my race, and it seemed to her that I wanted to throw that advantage away.

The preference for whites in the cashier position reflected the importance of this job in the store's general operations. In discount stores like the Toy Warehouse, customers had few opportunities to interact and consult with salesclerks.... The cashier was the only human being that the customer was guaranteed to contact, giving the role enormous symbolic—and economic—importance for the organization. At the point of sale, transactions could break down if the customers were not treated in accordance with their expectations. The preference for white and light-skinned women as cashiers should be interpreted in this light: in a racist and sexist society, such women are generally believed to be the friendliest and most solicitous group and thus best able to inspire trust and confidence.

Personally, I hated working as a cashier. I thought it was a difficult, stressful, and thankless job. Learning to work a cash register is much like learning to use a new computer software package. Each store seems to use a different operating system. After working at these jobs I started to pay attention to every transaction that I made as a customer in a store, and I have yet to see the same computer system twice. The job looks simple from the outside, but because of the way it is organized cashiers have no discretionary power, making them completely dependent on others if anything out of the ordinary happens.... We couldn't even open up our registers to make change for the gum ball machine. Customers would often treat us like morons because we couldn't resolve these minor and routine situations on our own, but we were given no choice or autonomy.

We were, however, held accountable for everything in the register, which had to match the computer printout record of all transactions. At the Toy Warehouse, we were also responsible for requesting "pulls" (this was done automatically by the security personnel at Diamond Toys). A pull is when a manager removes large sums of cash from the register to protect the money in case of a robbery. We were told to request a pull whenever we accumulated more than $500 cash in our registers. If we didn't, and large sums of money were in the till when we closed out our registers at the end of the shift, we would be given a demerit. Interestingly, we were never given any instruction on what to do to protect ourselves in case of an actual robbery....

A few men were regularly assigned to work as cashiers at the Toy Warehouse, but this happened only in the electronics department. The electronics department was cordoned off from the rest of the store by a metal detector gate intended to curtail theft. All of the men with this regular assignment were Asian American. They had sought out this assignment because they were interested in computers and gaming equipment. Working a register in that section may have been more acceptable to them in part because the section was separated from the main registers and in part because Asian masculinity—as opposed to black or white masculinity—is often defined through technical expertise. My sense was that the stereotypical association of Asian American men with computers made these assignments desirable from management's perspective as well.

Occasionally men were assigned to work the registers outside the electronics department, but this happened only when there were staffing shortages or scheduling problems. Once I came to work to find Deshay, a twenty-five-year-old African American, and Shuresh, a twenty-one-year-old second-generation Indian American, both stationed at the main registers. I flew to the back of the store to clock in so I could take my station next to them, eager to observe them negotiating the demands of "women's work." But the minute I took my station they were relieved of cashiering and told to cash out their registers and return to their regular tasks. When a woman was available, the men didn't have to do the job....

Women also crossed over into the men's jobs, but this happened far less frequently. Management never assigned a woman to work in the back areas to make up for temporary staff shortfalls. At each store, only one woman worked in the back of the house, and both women were African American. At the Toy Warehouse, the only woman who worked in the back was Darlene, whom a coworker once described to me as "very masculine" (but also "really great"). Darlene, who worked in a contracting business on the side, took a lot of pride in her physical strength and stamina. She was also a lesbian, which made her the butt of mean-spirited joking (behind her back) but also probably made this assignment less dissonant in the eyes of management (women in nontraditional jobs are often stereotyped as lesbian). At Diamond Toys, the only woman in the back of the house was eighteen-year-old Chandnka. She started working in the back of the house but asked for and received a transfer to gift wrap. Chandnka, who was one of only three African Americans who worked at Diamond Toys, said she hated working in the storeroom because the men there were racist and "very misogynistic," telling sexist jokes and challenging her competence at the job.

Crossing over is a different experience for men and women. When a job is identified as masculine, men often will erect barriers to women, making them feel out of place and unwanted, which is what happened to Chandnka. In contrast, I never observed women trying to exclude men or marginalize men in "their" jobs. On the contrary, men tried to exclude themselves from "women's work." Job segregation by gender is in large part a product of men's efforts to establish all-male preserves, which help them to prove and to maintain their masculinity (Williams 1989). Management colludes in this insofar as they share similar stereotypes of appropriate task assignments for men and women or perceive the public to embrace such stereotypes. But they also insist on employee "flexibility," the widespread euphemism used to describe their fundamental right to hire, fire, and assign employees at will. At the Toy Warehouse, employees were often threatened that their hours would be cut if they were not "flexible" in terms of their available hours and willingness to perform any job. But in general managers shared men's preferences to avoid register duty unless no one else was available.

How and why a specific job comes to be "gendered" and "racialized," or considered appropriate only for women or for men, or for whites or nonwhites, depends on the specific context (which in the case of these toy stores was shaped—but not determined—by their national marketing strategies).... Thus, in contrast to the Toy Warehouse, Diamond Toys employed both men and women as cashiers, and only two of them were African American (both women). At the Toy Warehouse, most of the registers were lined up in the front of the store near the doors. Diamond Toys was more like a department store with cash registers scattered throughout the different sections. The preference for white workers seemed consistent with the marketing of the store's workers as "the ultimate toy experts." In retail service work, professional expertise is typically associated with whiteness, much as it is in domestic service (Wrigley 1995).

Although both men and women worked the registers, there was gender segregation by the type of toy we sold. Only women were assigned to work in the doll and stuffed animal sections, for example, and only men worked in sporting goods and electronics. Also, only women worked in gift wrap. Some sections, like the book department, were gender neutral, but most were as gender marked as the toys we sold....

The most firmly segregated job in the toy store was the job of cleaner. As I have noted, only Latinas filled these jobs. I never witnessed a man or a woman of different race/ethnicity in them....

HOURS, BENEFITS, AND PAY

When I started this project I thought I had the perfect career to combine with a part-time job in retail. As a college professor, I taught two courses per semester that met six hours per week. I thought I could pick a schedule for twenty hours a week that accommodated those teaching commitments. Wrong. To get a job in retail, workers must be willing to work weekends and to change their schedules from one week to another to meet the staffing needs of the store. This is the meaning of the word *flexible* in retail. It is exactly the kind of schedule that is incompatible with doing anything else....

Workers with seniority can gain some control over their schedules, but it takes years of "flexibility" to attain to this status. Moreover, in my experience, this control was guaranteed only at Diamond Toys, thanks to the union. The senior associates who had worked there more than a year had the same schedule from week to week. This didn't apply to the supervisors, though. They had to be willing to fill in as needed, since there had to be a supervisor on the floor at all times. Occasionally they even had to forgo breaks.

The fact that supervisors had less control over their schedules made the job less desirable, but to sweeten the pot the job paid $1 per hour more than what regular associates earned. From the perspective of at least two women senior associates I talked to, it just wasn't worth it to give up control over their schedules. But this effectively prevented them from rising in the hierarchy and making more money, and it contributed to the gender segregation of jobs.

What did people earn? At Diamond Toys we were instructed during our training session not to discuss our pay with anyone. Doing so was pointed out as an example of "unauthorized disclosure of confidential business information," a "serious willful violation" that could result in immediate discharge. So I wasn't about to ask anyone what he or she made. I made $8.75 per hour. I got a sense that that was about average but somewhat higher than what most new hires made (possibly because of my higher educational credentials).

To put this salary into perspective, a forty-hour, full-time, year-round worker making $8.75 per hour would earn about $17,500 before taxes. This was well above the median income of full-time cashiers in 2001, which was about $15,000 per year, and about average for retail sales workers in general, who earned a median income of $18,000 (U.S. Bureau of Labor Statistics 2003). Of course, most retail workers do not work full time and year round, so their incomes are much lower than this....

The unpredictability of scheduling presented a nightmare for many single mothers at the Toy Warehouse. Schedules were posted on Friday for the following week beginning on Sunday. The two-day notice of scheduling made it especially difficult to arrange child care. (In contrast, schedules came out on Tuesday for the following week at Diamond Toys, another benefit of the union.) Even worse, while I was working at the Toy Warehouse, management reduced everyone's hours, purportedly to make up for revenue shortfalls. We were all asked to fill in a form indicating our "availability" to work, from 6:00 A.M. until 10:00 P.M. This form, which was attached to our paychecks, warned that "associates with the flexible availability will get more hours than those who are limited." Part-timers who limited their availability were hit hard when schedules came out the following week, causing a great deal of anger and bad feelings. Angela, an experienced associate, was scheduled for only four hours, and she was so mad that no one could even talk to her. Some said that they were going to apply for unemployment....

One of the reasons management gave for cutting our hours was that the store had been experiencing major problems with "shrink," the retailer term for theft. It was insinuated that the workers were stealing, but I could never figure out how that could happen. In both stores, very elaborate surveillance systems were set up to monitor employees. Hidden cameras recorded activity throughout the store, including the areas around the emergency exits. Our bags, pockets, and purses were checked every time we left the store. Cashiers were monitored continuously via a back-room computer hooked up to every register. As I noted, the contents of the till had to match exactly with the register report. Being even slightly under was enough to cause a major panic....

TAKE THIS JOB AND SHOVE IT—OR NOT

Over the course of working at the two stores, I witnessed a great deal of employee turnover. I outlasted both of the others who were hired with me at the Toy Warehouse, and by the time I left Diamond Toys I was the third in seniority among the ten associates who worked in my section. On one of my last days there I was given the walkie-talkie, the direct line of communication to the storeroom and the managers' office, which indicated that I was at that time the most senior staff member in the section.

For most employees, retail work is a revolving door. Employers know this and expect and even cultivate it. Most new hires are not expected to last through the three-month probation period. One of my coworkers at the Toy Warehouse told me he rarely talked to the new people since they rarely lasted long. Kevin, a twenty-six-year-old African American supervisor, predicted I wouldn't stay very long because I didn't have the right personality for retail work. I asked him, "What kind of people stay at the store?" He said people who were quiet and didn't stress out easily (my major flaw), and then he whispered, "People who can kiss up to management." I asked him, did he do this? And he said yes, but in a different kind of way, not too obvious. I asked how, and he said that he didn't tell people what he "really" thought if he thought he was being screwed over.

Kevin was one of a handful of hourly workers who saw their jobs at the Toy Warehouse as their lifelong work. He told me that he had "grown up" at the Toy Warehouse and could never imagine leaving. There were also a half-dozen or so associates at Diamond Toys who had worked at the store for more than a year. It is much more understandable why workers might choose to stay there, given the union benefits. An extra incentive for staying was the possibility of moving into management. At least a third of the managers at Diamond Toys started working as regular associates. When they became managers they left the union and earned salaries, starting at about $35,000. In contrast, no associate ever moved into management at the Toy Warehouse, although one of my young coworkers maintained that it *could* happen. Vanme, a twenty-one-year-old first-generation immigrant from the Philippines, told me that she had met someone at another store in the chain (which had since closed down) who knew someone who had worked his way up from janitor to store director. I told her that I thought this seemed unlikely. Janitorial services were subcontracted, and director and management positions required a college degree and involved a completely separate application process through the regional office in another state. The story sounded like a Horatio Alger myth to me. But Vanme said that this man was a great inspiration to her, and I believed her.

Although acquiring a management position at Diamond Toys wasn't unprecedented, few of the long-term associates seemed interested in pursuing one. Some seemed resigned to keeping their associate position with its guaranteed hours, schedule, and benefits. Alyss, for example, had worked two years in the doll department and had a set schedule from 8:00 to 4:30, five days per week. She told me she couldn't believe she had worked at the store that long but that at this point she considered the job pretty easy. The only drawbacks were her dealings with Dorothy (the irritable section manager) and the occasional neurotic high-end Barbie collector. She had no interest in pursuing a supervisory position because that would mean losing her schedule and taking on more work....

Many longtime workers developed quasi-familial bonds at the stores, referring to each other as brother and sister, mother and grandmother. This happened at both stores. There were also real kin networks, most evident at the Toy Warehouse. Some of the older employees brought their children or grandchildren with them to the store during their shift. These kids hung out in the break room or played with the demonstration toys, including the video games that were set up in the electronics department. All the employees helped to keep an eye on them. Those with older children sometimes got them jobs in the store.

I was an outsider to these family and friendship networks, but I was often very touched by the mutual support and caring I witnessed among my coworkers, especially at the Toy Warehouse. For the longtime workers, the store was an extension of their family networks and responsibilities. They used their employee discount cards (10 percent at the Toy Warehouse, 30 percent at Diamond Toys) to buy toys for their kin and quasi-kin. But more important was the social and emotional support they experienced there.

I was struck by this once when I was in the break room at the Toy Warehouse and my coworkers Dwain and Lamomca walked in on a day they were not scheduled to work. They were an African American couple in their early twenties who had been dating for a short time. They had come in to pick up their paychecks and to talk to Selma, an African American woman in her forties who had worked at the store for six years. Dwain was upset with his mother, who was giving him a very hard time, telling him he was worthless and criticizing him for being too dependent. During the discussion Lamomca was sitting on a stack of chairs behind Dwain, sucking on her Jamba Juice. She smiled during the conversation or rolled her eyes, as if to say, "What are you going to do?" But she didn't really participate. Selma, on the other hand, was giving him reassurance and sympathy and moral support. I asked her if she knew his mother and she said no, just what Dwain had told her. I realized that Selma was a mentor for Dwain; he clearly cherished her advice and encouragement.

This is the backstage of stores that most shoppers don't see, but it is critical for understanding why people stay in crummy jobs with low pay. Barbara Ehrenreich (2001) considers this conundrum in her study of low-wage work. She wonders why workers don't leave when other, higher-paying opportunities arise. At the Toy Warehouse, most workers stayed no longer than three months. Those who stayed long term did so because that was where their family was.

WHAT A DIFFERENCE A UNION MAKES

The union at Diamond Toys helped to ameliorate some of the most egregious problems with working retail. It guaranteed hours and schedules for senior associates, mandated longer rest breaks, and provided health benefits, vacation pay, and a career track. I earned 17 percent more at the unionized workplace, which was in line with the national 20 percent wage premium that comes with union membership (McCall 2001, 181). We were always allowed to leave when we were scheduled, whereas in the Toy Warehouse we were kept up to an hour later than scheduled to finish cleaning up the store after closing (a practice that routinely resulted in our being scheduled for fewer breaks than we were lawfully due). I was also impressed by how the managers behaved professionally and respectfully toward the workers. They quickly responded to pages and patiently explained procedures to the new people.

The unfortunate exception was my area manager, Dorothy. She rushed around our section barking orders and shouting insults at us. The first time I met her she told me that my name was unacceptable, as there were already two others in the store named Chris or Christine, so she was going to call me by my middle name instead. (Luckily that didn't last long because I couldn't remember to respond to Louise.) Sometimes when the store was busy she would stand next to me and shout, "Hurry, hurry, hurry." She would roll her eyes and mutter about my stupidity whenever I had to ask her a question or get her help to solve a problem. When I paged her she would pick up the phone and say in an exasperated voice, "What is it *now*, Christine?" It was some consolation that she treated everyone this way. We all found different ways to cope. Carl dealt with it by keeping a happy song in his head, he said. Alyss became depressed and frustrated; she often looked on the verge of tears. Dennis dreamed of leaving retail altogether and becoming a full-time teacher. Chandnka told me that to deal with the abuse she prayed. She also claimed to have filed four formal complaints against Dorothy, but nothing ever came of that. I fantasized about making a principled scene on my last day. (I didn't.)

This is the part of the job that a union can't change. Managers are allowed to harass workers, and there is virtually no recourse unless that harassment targets a worker's race, gender, or other legally protected characteristic (Williams 2003). The other managers and the store directors knew about Dorothy's abusive behavior but oddly seemed to tolerate it. They attributed it to personal and family problems she was experiencing. Rumors abounded about the nature of these problems, but it was impossible to verify them. I desperately wanted to understand why she was so mean, but in the end it probably wouldn't have made a difference. Sadistic bosses are an unfortunate fact of life in many hierarchical work organizations (Gherardi 1995).

The union offered no protection from harassing and abusive customers either. Admittedly this was a bigger problem at the Toy Warehouse than at Diamond Toys, but I don't think that customers were better behaved at Diamond Toys because of the union. For many customers, part of the allure of shopping at Diamond Toys was the educated and solicitous, not to mention white, sales staff. The mixture of class, race, and gender frames the customer-server relationship, just as it does the social organization of retail work....

CONCLUSION

Most sociological research on retail stores looks at them as sites of consumption. But stores are also workplaces. Retail work makes up an increasing proportion of the jobs in our economy. Yet these are "bad" jobs. According to Frank Levy (1998), a "good job" is one that pays enough to support a family and provides benefits, security, and autonomy. In contrast, most jobs in retail pay low wages, offer few benefits, have high turnover, and restrict workers' autonomy.

... I have argued that the social organization of work in large toy stores also contributes to class, gender, and race inequalities. The Toy Warehouse, which had a predominately African American staff, paid extremely low wages, offered few benefits, and demanded "flexible" workers who made no scheduling demands. The store was segregated by race and gender, with white men in the director positions and African American women in managerial and supervisory roles. Among the staff, only white and light-skinned women and Asian American men were regularly assigned to cashiering positions, and only men (of all racial/ethnic groups) worked in the back room unloading and assembling the toys. African American men and women filled the positions of security guards, stockers, and gofers.

Because Diamond Toys was unionized, it offered better pay than the Toy Warehouse (but not a "living wage"), and its employees received health care and vacation benefits. Schedules were posted in advance, legally mandated breaks were honored, and career ladder promotions were available. For all of these reasons, a union does make a positive difference for workers....

... [T]he hierarchical and functional placement of workers according to managerial stereotypes results in advantages for white men and (to a lesser extent) white women and disadvantages for racial/ethnic minority men and women. These stereotypes are perhaps more deeply entrenched than low wages, based as they are on perceptions of customer preferences. Consumers therefore have a role in pressing for changes in these job assignments. But in my view, the struggle for equal access to "badjobs" is hardly worth an organized effort. There is little point in demanding equal access to jobs that don't support a family. Similarly, career ladders have to be created before equal opportunities for advancement are demanded. The fight against racism and sexism, then, should be folded into efforts to economically upgrade these jobs. The goal of restructuring jobs in toy stores, and in retail work in general, should be self-sufficiency—and hope—for all workers, regardless of race or gender.

REFERENCES

Acker, Joan. 1990. "Hierarchies, Jobs, Bodies: A Theory of Gendered Organizations." *Gender & Society* 4: 139–58.

Collins, Patricia Hill. 2000. *Black Feminist Thought.* New York: Routledge.

Ehrenreich, Barbara. 2001. *Nickel and Dimed: On (Not) Getting By in America.* New York: Metropolitan Books.

Folbre, Nancy. 2001. *The Invisible Heart: Economics and Family Values.* New York: New Press.

Greenhouse, Steven. 2003. "Abercrombie & Fitch Accused of Discrimination in Hiring." *New York Times,* June 17, A1.

Gherardi, Sylvia. 1995. *Gender, Symbolism and Organizational Culture.* Thousand Oaks, CA: Sage Publications.

Levy, Frank. 1998. *The New Dollars and Dreams: American Incomes and Economic Change.* New York: Russell Sage Foundation.

Loe, Meika. 1996. "Working for Men: At the Intersection of Power, Gender, and Sexuality." *Sociological Inquiry* 66: 399–421.

McCall, Leslie. 2001. *Complex Inequality: Gender, Class and Race in the New Economy.* New York: Routledge.

Project on Disney. 1995. *Inside the Mouse.* Durham, NC: Duke University Press.

Reskin, Barbara, and Patricia Roos. 1990. *Job Queues, Gender Queues.* Philadelphia: Temple University Press.

Ritzer, George. 2002. *McDonaldization: The Reader.* Thousand Oaks, CA: Pine Forge Press.

Sandikci, Ozlem, and Douglas Holt. 1998. "Malling Society: Mall Consumption Practices and the Future of Public Space." In *Servicescapes: The Concept of Place in Contemporary Markets*, ed. John Sherry, 305–36. Lincolnwood, IL: NTC Business Books.

Talwar, Jennifer Parker. 2002. *Fast Food, Fast Track: Immigrants, Big Business, and the American Dream.* Boulder, CO: Westview Press.

U.S. Bureau of Labor Statistics. 2003. "Household Data Annual Averages. 39. Median Weekly Earnings of Full-Time Wage and Salary Workers by Detailed Occupation and Sex." Retrieved March 22, 2005, from www.bls.gov/cps/cpsaat39.pdf.

U.S. Bureau of Labor Statistics. 2004. "Household Data Annual Averages. 18. Employed Persons by Detailed Industry, Sex, Race, and Hispanic or Latino Ethnicity." Retrieved April 6, 2005, from www.bls.gov/cps/cpsaat18.pdf.

Williams, Christine L. 1989. *Gender Differences at Work: Women and Men in Nontraditional Occupations.* Berkeley: University of California Press.

Williams, Christine L. 2003. "Sexual Harassment and Human Rights Law in New Zealand." *Journal of Human Rights* 2 (December) 573–84.

Wrigley, Julia. 1995. *Other People's Children.* New York: Basic Books.

KEY CONCEPTS

dual labor market   gendered institution   service sector

DISCUSSION QUESTIONS

1. In your most recent in-store shopping experience, did you notice the different roles of workers in the store? How was your shopping experience influenced by the salespeople working in the store?

2. Take the same analysis presented in this article and apply it to one of your work experiences. How did your gender and race-ethnicity influence the job(s) you were given? Was job segregation by gender and race?

Applying Sociological Knowledge:
An Exercise for Students

Think of an organization with which you are familiar (a college, a work organization, or a religious organization). What are the different groups that make up
this organization? Do the different groups that make up the organization have
different statuses within the organization? If so, describe each group's status. Is
there a hierarchy among these different groups and, if so, how does that affect
how they interact with each other?

PART VII

Deviance and Crime

19

The Functions of Crime

EMILE DURKHEIM

This classic essay, written in 1895 and translated many times since, points to crime as an inevitable part of society. Durkheim's main functionalist thesis that criminal behavior exists in all social settings is still the theoretical basis for many sociological inquiries into crime and deviance.

If there is a fact whose pathological nature appears indisputable, it is crime. All criminologists agree on this score. Although they explain this pathology differently, they nonetheless unanimously acknowledge it. However, the problem needs to be treated less summarily.

... Crime is not only observed in most societies of a particular species, but in all societies of all types. There is not one in which criminality does not exist, although it changes in form and the actions which are termed criminal are not everywhere the same. Yet everywhere and always there have been men who have conducted themselves in such a way as to bring down punishment upon their heads. If at least, as societies pass from lower to higher types, the crime rate (the relationship between the annual crime figures and population figures) tended to fall, we might believe that, although still remaining a normal phenomenon, crime tended to lose that character of normality. Yet there is no single ground for believing such a regression to be real. Many facts would rather seem to point to the existence of a movement in the opposite direction. From the beginning of the century statistics provide us with a means of following the progression of criminality. It has everywhere increased, and in France the increase is of the order of 300 percent. Thus there is no phenomenon which represents more incontrovertibly all the symptoms of normality, since it appears to be closely bound up with the conditions of all collective life. To make crime a social illness would be to concede that sickness is not something accidental, but on the contrary derives in certain cases from the fundamental constitution of the living creature. This would be to erase any distinction between the physiological and the pathological. It can certainly happen that crime itself has normal forms; this is what happens, for instance, when it reaches an excessively high level. There is no doubt that this excessiveness is pathological in nature. What is

SOURCE: Emile Durkheim. 1982. *The Rules of Sociological Method*, ed. Steven Lukes, trans. W. D. Halls. New York: The Free Press. A division of Macmillan, pp. 64–75.

normal is simply that criminality exists, provided that for each social type it does not reach or go beyond a certain level which it is perhaps not impossible to fix in conformity with the previous rules.

We are faced with a conclusion which is apparently somewhat paradoxical. Let us make no mistake: to classify crime among the phenomena of normal sociology is not merely to declare that it is an inevitable though regrettable phenomenon arising from the incorrigible wickedness of men; it is to assert that it is a factor in public health, an integrative element in any healthy society. At first sight this result is so surprising that it disconcerted even ourselves for a long time. However, once that first impression of surprise has been overcome it is not difficult to discover reasons to explain this normality and at the same time to confirm it.

In the first place, crime is normal because it is completely impossible for any society entirely free of it to exist.

Crime consists of an action which offends certain collective feelings which are especially strong and clear-cut. In any society, for actions regarded as criminal to cease, the feelings that they offend would need to be found in each individual consciousness without exception and in the degree of strength requisite to counteract the opposing feelings. Even supposing that this condition could effectively be fulfilled, crime would not thereby disappear; it would merely change in form, for the very cause which made the well-springs of criminality to dry up would immediately open up new ones.

Indeed, for the collective feelings, which the penal law of a people at a particular moment in its history protects, to penetrate individual consciousnesses that had hitherto remained closed to them, or to assume greater authority—whereas previously they had not possessed enough—they would have to acquire an intensity greater than they had had up to then. The community as a whole must feel them more keenly, for they cannot draw from any other source the additional force which enables them to bear down upon individuals who formerly were the most refractory....

In order to exhaust all the logically possible hypotheses, it will perhaps be asked why this unanimity should not cover all collective sentiments without exception, and why even the weakest sentiments should not evoke sufficient power to forestall any dissentient voice. The moral conscience of society would be found in its entirety in every individual, endowed with sufficient force to prevent the commission of any act offending against it, whether purely conventional failings or crimes. But such universal and absolute uniformity is utterly impossible, for the immediate physical environment in which each one of us is placed, our hereditary antecedents, the social influences upon which we depend, vary from one individual to another and consequently cause a diversity of consciences. It is impossible for everyone to be alike in this matter, by virtue of the fact that we each have our own organic constitution and occupy different areas in space. This is why, even among lower peoples where individual originality is very little developed, such originality does however exist. Thus, since there cannot be a society in which individuals do not diverge to some extent from the collective type, it is also inevitable that among these deviations some assume a criminal character. What confers upon them this character is not the intrinsic importance of the acts but the importance which the common consciousness ascribes to them. Thus if the latter is stronger and possesses sufficient authority to make these divergences very weak in absolute terms, it will also be more sensitive and exacting. By reacting against the slightest deviations with an energy which it elsewhere employs against those that are more weighty, it endues them with the same gravity and will brand them as criminal.

Thus crime is necessary. It is linked to the basic conditions of social life, but on this very account is useful, for the conditions to which it is bound are themselves indispensable to the normal evolution of morality and law.

Indeed today we can no longer dispute the fact that not only do law and morality vary from one social type to another, but they even change within the same type if the conditions of collective existence are modified. Yet for these transformations to be made possible, the collective sentiments at the basis of morality should not prove unyielding to change, and consequently should be only moderately intense. If they were too strong, they would no longer be malleable. Any arrangement is indeed an obstacle to a new arrangement; this is even more the case the more deep-seated the original arrangement. The more strongly a structure is articulated, the more it resists modification; this is as true for functional as for anatomical patterns. If there were no crimes, this condition would not be fulfilled, for such a hypothesis presumes that collective sentiments would have attained a degree of intensity unparalleled in history. Nothing is good indefinitely and without limits. The authority which the moral consciousness enjoys must not be excessive, for otherwise no one would dare to attack it and it would petrify too easily into an immutable form. For it to evolve, individual originality must be allowed to manifest itself. But so that the originality of the idealist who dreams of transcending his era may display itself, that of the criminal, which falls short of the age, must also be possible. One does not go without the other.

Nor is this all. Beyond this indirect utility, crime itself may play a useful part in this evolution. Not only does it imply that the way to necessary changes remains open, but in certain cases it also directly prepares for these changes. Where crime exists, collective sentiments are not only in the state of plasticity necessary to assume a new form, but sometimes it even contributes to determining beforehand the shape they will take on. Indeed, how often is it only an anticipation of the morality to come, a progression towards what will be! ... The freedom of thought that we at present enjoy could never have been asserted if the rules that forbade it had not been violated before they were solemnly abrogated. However, at the time the violation was a crime, since it was an offence against sentiments still keenly felt in the average consciousness. Yet this crime was useful since it was the prelude to changes which were daily becoming more necessary....

From this viewpoint the fundamental facts of criminology appear to us in an entirely new light. Contrary to current ideas, the criminal no longer appears as an utterly unsociable creature, a sort of parasitic element, a foreign, unassimilable body introduced into the bosom of society. He plays a normal role in social life. For its part, crime must no longer be conceived of as an evil which cannot be circumscribed closely enough. Far from there being cause for congratulation when it drops too noticeably below the normal level, this apparent progress assuredly coincides with and is linked to some social disturbance. Thus the number of crimes of assault never falls so low as it does in times of scarcity. Consequently, at the same time, and as a reaction, the theory of punishment is revised, or rather should be revised. If in fact crime is a sickness, punishment is the cure for it and cannot be conceived of otherwise; thus all the discussion aroused revolves round knowing what punishment should be to fulfill its role as a remedy. But if crime is in no way pathological, the object of punishment cannot be to cure it and its true function must be sought elsewhere....

KEY CONCEPTS

collective consciousness   deviance   functionalism
social facts

DISCUSSION QUESTIONS

1. According to Durkheim's theory, criminal behavior exists in all societies. Consider the possibility of a society without the ability to punish criminal behavior (no prisons, no courts, and so forth). How would individuals respond to crime? What informal social control mechanisms would help to maintain order?

2. How could you use Durkheim's theory as the basis for a research project on deviant behavior? What hypotheses could you test that would challenge or support the functionalist view of crime?

20

The Medicalization of Deviance

PETER CONRAD AND JOSEPH W. SCHNEIDER

This essay outlines the social construction of social deviance. The authors specifically refer to the medical profession as redefining certain deviant behaviors as

SOURCE: Peter Conrad and Joseph W. Schneider. 1992. *Deviance, and Medicalization: From Badness to Sickness*. Philadelphia: Temple University Press, pp. 28–37.

"illness," rather than as "badness." They argue that the "medicalization of deviance changes the social response to such behavior to one of treatment rather than punishment."

Consider the following situations. A woman rides a horse naked through the streets of Denver claiming to be Lady Godiva and after being apprehended by authorities, is taken to a psychiatric hospital and declared to be suffering from a mental illness. A well-known surgeon in a South-western city performs a psycho-surgical operation on a young man who is prone to violent outbursts. An Atlanta attorney, inclined to drinking sprees, is treated at a hospital clinic for his disease, alcoholism. A child in California brought to a pediatric clinic because of his disruptive behavior in school is labeled hyperactive and is prescribed methylphenidate (Ritalin) for his disorder. A chronically overweight Chicago housewife receives a surgical intestinal bypass operation for her problem of obesity. Scientists at a New England medical center work on a million-dollar federal research grant to discover a heroin-blocking agent as a "cure" for heroin addiction. What do these situations have in common? In all instances medical solutions are being sought for a variety of deviant behaviors or conditions. We call this "the medicalization of deviance" and suggest that these examples illustrate how medical definitions of deviant behavior are becoming more prevalent in modern industrial societies like our own. The historical sources of this medicalization, and the development of medical conceptions and controls for deviant behavior, are the central concerns of our analysis.

Medical practitioners and medical treatment in our society are usually viewed as dedicated to healing the sick and giving comfort to the afflicted. No doubt these are important aspects of medicine. In recent years the jurisdiction of the medical profession has expanded and encompasses many problems that formerly were not defined as medical entities.... There is much evidence for this general viewpoint—for example, the medicalization of pregnancy and childbirth, contraception, diet, exercise, child development norms—but our concern here is more limited and specific. Our interests focus on the medicalization of deviant behavior: the defining and labeling of deviant behavior as a medical problem, usually an illness and mandating the medical profession to provide some type of treatment for it. Concomitant with such medicalization is the growing use of medicine as an agent of social control, typically as medical intervention. Medical intervention as social control seeks to limit, modify, regulate, isolate, or eliminate deviant behavior with medical means and in the name of health....

Conceptions of deviant behavior change, and agencies mandated to control deviance change also. Historically there have been great transformations in the definition of deviance—from religious to state-legal to medical-scientific. Emile Durkheim (1893/1933) noted in *The Division of Labor in Society* that as societies develop from simple to complex, sanctions for deviance change from repressive to restitutive or, put another way, from punishment to treatment or rehabilitation. Along with the change in sanctions and social control agent there is a corresponding change in definition or conceptualization of deviant behavior. For example, certain "extreme" forms of deviant drinking (what is now called alcoholism) have been defined as sin, moral weakness, crime, and most recently illness.... In modern industrial society there has been a substantial growth in the prestige, dominance, and jurisdiction of the medical profession (Freidson, 1970). It is only within the last century that physicians have become highly organized, consistently trained, highly paid, and sophisticated in their therapeutic techniques and abilities.... The medical profession dominates the organization of health care and has a virtual monopoly on anything that is defined as medical treatment, especially in terms of what constitutes "illness" and what is appropriate medical intervention.... Although Durkheim did not predict this medicalization, perhaps in part because medicine of his time was not the scientific, prestigious, and dominant profession of today, it is clear that medicine is the central restitutive agent in our society.

EXPANSION OF MEDICAL JURISDICTION
OVER DEVIANCE

When treatment rather than punishment becomes the preferred sanction for deviance, an increasing amount of behavior is conceptualized in a medical framework as illness. As noted earlier, this is not unexpected, since medicine has always functioned as an agent of social control, especially in attempting to "normalize" illness and return people to their functioning capacity in society. Public health and psychiatry have long been concerned with social behavior and have functioned traditionally as agents of social control (Foucault, 1965; Rosen, 1972). What is significant, however, is the expansion of this sphere where medicine functions in a social control capacity. In the wake of a general humanitarian trend, the success and prestige of modern biomedicine, the technological growth of the 20th century, and the diminution of religion as a viable agent of control, more and more deviant behavior has come into the province of medicine. In short, the particular, dominant designation of deviance has changed; much of what was badness (i.e., sinful or criminal) is now sickness. Although some forms of deviant behavior are more completely medicalized than others (e.g., mental illness), recent research has pointed to a considerable variety of deviance that has been treated within medical jurisdiction: alcoholism, drug addiction, hyperactive children, suicide, obesity, mental retardation, crime, violence, child abuse, and learning problems, as well as several other categones of social deviance. Concomitant with medicalization there has been a change in imputed responsibility for deviance: with badness the deviants were considered responsible for their behavior, with sickness they are not, or at least responsibility is diminished (see Stoll, 1968). The social response to deviance is "therapeutic" rather than punitive. Many have viewed this as "humanitarian and scientific" progress; indeed, it often leads to "humanitarian and scientific" treatment rather than punishment as a response to deviant behavior....

A number of broad social factors underlie the medicalization of deviance. As psychiatric critic Thomas Szasz (1974) observes, there has been a major historical shift in the manner in which we view human conduct:

> With the transformation of the religious perspective of man into the
> scientific, and in particular the psychiatric, which became fully

articulated during the nineteenth century, there occurred a radical shift in emphasis away from viewing man as a *responsible agent acting in and on the world* and toward viewing him *as a responsive organism being acted upon* by biological and social "forces." (p. 149)

This is exemplified by the diffusion of Freudian thought, which since the 1920s has had a significant impact on the treatment of deviance, the distribution of stigma, and the incidence of penal sanctions.

Nicholas Kittrie (1971), focusing on decriminalization, contends that the foundation of the therapeutic state can be found in determinist criminology, that it stems from the *parens patnae* power of the state (the state's right to help those who are unable to help themselves), and that it dates its origin with the development of juvenile justice at the turn of the century. He further suggests that criminal law has failed to deal effectively (e.g., in deterrence) with criminals and deviants, encouraging a use of alternative methods of control. Others have pointed out that the strength of formal sanctions is declining because of the increase in geographical mobility and the decrease in strength of traditional status groups (e.g., the family) and that medicalization offers a substitute method for controlling deviance (Pitts, 1968). The success of medicine in areas like infectious disease has led to rising expectations of what medicine can accomplish. In modern technological societies, medicine has followed a technological imperative—that the physician is responsible for doing everything possible for the patient—while neglecting such significant issues as the patient's rights and wishes and the impact of biomedical advances on society (Mechanic, 1973). Increasingly sophisticated medical technology has extended the potential of medicine as social control, especially in terms of psychotechnology (Chorover, 1973). Psychotechnology includes a variety of medical and quasimedical treatments or procedures: psychosurgery, psychoactive medications, genetic engineering, disulfiram (Antabuse), and methadone. Medicine is frequently a pragmatic way of dealing with a problem (Gusfield, 1975). Undoubtedly the increasing acceptance and dominance of a scientific world view and the increase in status and power of the medical profession have contributed significantly to the adoption and public acceptance of medical approaches to handling deviant behavior.

THE MEDICAL MODEL AND "MORAL NEUTRALITY"

The first "victories" over disease by an emerging biomedicine were in the infectious diseases in which specific causal agents—germs—could be identified. An image was created of disease as caused by physiological difficulties located *within* the human body. This was the medical model. It emphasized the internal and biophysiological environment and deemphasized the external and social psychological environment.

There are numerous definitions of "the medical model."...We adopt a broad and pragmatic definition: the medical model of deviance locates the source of deviant behavior within the individual, postulating a physiological, constitutional, organic, or, occasionally, psychogenic agent or condition that is assumed to cause the behavioral deviance. The medical model of deviance usually, although not always, mandates intervention by medical personnel with medical means as treatment for the "illness." Alcoholics Anonymous, for example, adopts a rather idiosyncratic version of the medical model—that alcoholism is a chronic disease caused by an "allergy" to alcohol—but actively discourages professional medical intervention. But by and large, adoption of the medical model legitimates and even mandates medical intervention.

The medical model and the associated medical designations are assumed to have a scientific basis and thus are treated as if they were morally neutral (Zola, 1975). They are not considered moral judgments but rational, scientifically verifiable conditions.... Medical designations *are* social judgments, and the adoption of a medical model of behavior, a political decision. When such medical designations are applied to deviant behavior, they are related directly and intimately to the moral order of society. In 1851 Samuel Cartwright, a well-known Southern physician, published an article in a prestigious medical journal describing the disease "drapetomania," which only affected slaves and whose major symptom was running away from the plantations of their white masters (Cartwright, 1851). Medical texts during the Victorian era routinely described masturbation as a disease or addiction and prescribed mechanical and surgical treatments for its cure (Comfort, 1967; Englehardt, 1974). Recently many political dissidents in the Soviet Union have been designated mentally ill, with diagnoses such as "paranoia with counterrevolutionary delusions" and "manic reformism," and hospitalized for their opposition to the political order (Conrad, 1977). Although these illustrations may appear to be extreme examples, they highlight the fact that all medical designations of deviance are influenced significantly by the moral order of society and thus cannot be considered morally neutral....

Even after a social definition of deviance becomes accepted or legitimated, it is not evident what particular type of problem it is. Frequently there are intellectual disputes over the causes of the deviant behavior and the appropriate methods of control. These battles about deviance designation (is it sin, crime, or sickness?) and control are battles over turf: Who is the appropriate definer and treater of the deviance? Decisions concerning what is the proper deviance designation and hence the appropriate agent of social control are settled by some type of political conflict.

How one designation rather than another becomes dominant is a central sociological question. In answering this question, sociologists must focus on claims-making activities of the various interest groups involved and examine how one or another attains ownership of a given type of deviance or social problem and thus generates legitimacy for a deviance designation. Seen from this perspective, public facts, even those which wear a "scientific" mantle are treated as products of the groups or organizations that produce or promote them rather than as accurate reflections of "reality." The adoption of one deviance designation or another has consequences beyond settling a dispute about social control turf.

...When a particular type of deviance designation is accepted and taken for granted, something akin to a paradigm exists. There have been three major deviance paradigms: deviance as sin, deviance as crime, and deviance as sickness. When one paradigm and its adherents become the ultimate arbiter of "reality" in society, we say a hegemony of definitions exists. In Western societies, and American society in particular, anything proposed in the name of science gains great authority. In modern industrial societies, deviance designations have become increasingly medicalized. We call the change in designations from badness to sickness the medicalization of deviance....

REFERENCES

Cartwright, S. W. Report on the diseases and physical peculiarities of the negro race. *N. O. Med. Surg.* J., 1851, 7, 691–715.

Chorover, S. Big Brother and psychotechnology *Psychol. Today*, 1973, 7, 43–54 (Oct.).

Comfort, A. *The anxiety makers.* London: Thomas Nelson & Sons, 1967.

Conrad, P. Soviet dissidents, ideological deviance, and mental hospitalization. Presented at Midwest Sociological Society Meetings, Minneapolis, 1977.

Durkheim, E. *The division of labor in society.* New York: The Free Press, 1933. (Originally published 1893.)

Englehardt, H. T. Jr. The disease of masturbation: Values and the concept of disease. *Bull. Hist. Med.*, 1974, 48, 234–48 (Summer).

Foucault, M. *Madness and civilization.* New York: Random House, Inc. 1965.

Freidson, E. *Profession of medicine.* New York: Harper & Row Publishers Inc. 1970.

Gusfield, J. R. Categories of ownership and responsibility in social issues: Alcohol abuse and automobile use. *J. Drug Issues,* 1975, 5, 285–303 (Fall).

Kittrie, N. *The right to be different: Deviance and enforced therapy.* Baltimore: Johns Hopkins University Press, 1971.

Mechanic, D. Health and illness in technological societies. *Hastings Center Stud.,* 1973, 1(3), 7–18.

Pitts, J. Social control: The concept. In D. Sills (Ed.) *International Encyclopedia of Social Sciences.* (Vol. 14). New York: Macmillan Publishing Co., Inc. 1968.

Rosen, G. The evolution of social medicine. In H. E. Freeman, S. Levine, and L. Reeder (Eds.) *Handbook of medical sociology* (2nd ed.). Englewood Cliffs, NJ: Prentice-Hall, Inc. 1972.

Stoll, C. S. Images of man and social control. *Soc. Forces,* 1968, 47, 119–127 (Dec.).

Szasz, T. *Ceremonial chemistry.* New York: Anchor Books, 1974.

Zola, I. K. In the name of health and illness: On some socio-political consequences of medical influence. *Soc. Sci. Med.,* 1975, 9, 83–87.

KEY CONCEPTS

medicalization of deviance   social control

DISCUSSION QUESTIONS

1. Alcoholism is an example of a deviant behavior being medicalized. How has this altered the understanding and treatment of alcoholism? How does the involvement of health professionals in the treatment of alcoholism influence societal reaction to excessive drinking?

2. Some argue that rapists should be castrated. How does this illustrate the transformation of understanding rape as a move "from badness to sickness"? What assumptions guide the suggestion that rapists should be castrated as a way of stopping rape?

21

Six Lessons of Suicide Bombers

ROBERT J. BRYM

Robert Brym examines suicide bombings with an eye to understanding the sociological, as well as political, factors that influence such attacks. He debunks the idea that suicide bombings are the work of crazed religious zealots, arguing for a more sociological interpretation.

In October 1983, Shi'a militants attacked the military barracks, of American and French troops in Beirut, killing nearly 300 people. Today the number of suicide attacks worldwide has passed 1,000, with almost all the attacks concentrated in just nine countries: Lebanon, Sri Lanka, Israel, Turkey, India (Kashmir), Russia (Chechnya), Afghanistan, Iraq, and Pakistan. Israel, for example, experienced a wave of suicide attacks in the mid-1990s when Hamas and the Palestinian Islamic Jihad (PIJ) sought to undermine peace talks between Israel and the Palestinian Authority. A far deadlier wave of attacks began in Israel in October 2000 after all hope of a negotiated settlement collapsed. Altogether, between 1993 and 2005, 158 suicide attacks took place in Israel and the occupied Palestinian territories, killing more than 800 people and injuring more than 4,600.

SOURCE: Brym, Robert J. 2007. "Sex Lessons of Suicide Bombers." 6:40–45.

Over the past quarter century, researchers have learned much about the motivations of suicide bombers, the rationales of the organizations that support them, their modus operandi, the precipitants of suicide attacks, and the effects of counterterrorism on insurgent behavior. Much of what they have learned is at odds with conventional wisdom and the thinking of policymakers who guide counterterrorist strategy. This paper draws on that research, but I focus mainly on the Israeli/Palestinian case to draw six lessons from the carnage wrought by suicide bombers. In brief, I argue that (1) suicide bombers are not crazy, (2) nor are they motivated principally by religious zeal. It is possible to discern (3) a strategic logic and (4) a social logic underlying their actions. Targeted states typically react by repressing organizations that mount suicide attacks, but (5) this repression often makes matters worse. (6) Only by first taking an imaginative leap and understanding the world from the assailant's point of view can hope to develop a workable strategy for minimizing suicide attacks. Let us examine each of these lessons in turn.

LESSON 1: SUICIDE BOMBERS ARE NOT CRAZY

Lance Corporal Eddie DiFranco was the only survivor of the 1983 suicide attack on the U.S. Marine barracks in Beirut who saw the face of the bomber. DiFranco was on watch when he noticed the attacker speeding his truck full of explosives toward the main building on the marine base. "He looked right at me [and] smiled," DiFranco later recalled.

Was the bomber insane? Some Western observers thought so. Several psychologists characterized the Beirut bombers as "unstable individuals with a death wish." Government and media sources made similar assertions in the immediate aftermath of the suicide attacks on the United States on September 11, 2001. Yet these claims were purely speculative. Subsequent interviews with prospective suicide bombers and reconstructions of the biographies of successful suicide attackers revealed few psychological abnormalities. In fact, after examining many hundreds of cases for evidence of depression, psychosis, past suicide attempts, and so on, Robert Pape discovered only a single person who could be classified as having a psychological problem (a Chechen woman who may have been mentally retarded).

On reflection, it is not difficult to understand why virtually all suicide bombers are psychologically stable. The organizers of suicide attacks do not want to jeopardize their missions by recruiting unreliable people. A research report prepared for the Danish government a few years ago noted, "Recruits who display signs of pathological behavior are automatically weeded out for reasons of organizational security." It may be that some psychologically unstable people want to become suicide bombers, but insurgent organizations strongly prefer their cannons fixed.

LESSON 2: IT'S MAINLY ABOUT POLITICS,
NOT RELIGION

In May 1972, three Japanese men in business suits boarded a flight from Paris to Tel Aviv. They were members of the Japanese Red Army, an affiliate of the

Popular Front for the Liberation of Palestine. Eager to help their Palestinian comrades liberate Israel from Jewish rule, they had packed their carry-on bags with machine guns and hand grenades. After disembarking at Lod Airport near Tel Aviv, they began an armed assault on everyone in sight. When the dust settled, 26 people lay dead, nearly half of them Puerto Rican Catholics on a pilgrimage to the Holy Land.

Israeli guards killed one of the attackers. A second blew himself up, thus becoming the first suicide bomber in modern Middle Eastern history. The Israelis captured the third assailant, Kozo Okamoto.

Okamato languished in an Israeli prison until the mid-1980s, when he was handed over to Palestinian militants in Lebanon's Beka'a Valley in a prisoner exchange. Then, in 2000, something unexpected happened. Okamoto apparently abandoned or at least ignored his secular faith in the theories of Bakunin and Trotsky, and converted to Islam. For Okamoto, politics came first, then religion.

A similar evolution occurs in the lives of many people. Any political conflict makes people look for ways to explain the dispute and imagine a strategy for resolving it; they adopt or formulate an ideology. If the conflict is deep and the ideology proves inadequate, people modify the ideology or reject it for an alternative. Religious themes often tinge political ideologies, and the importance of the religious component may increase if analyses and strategies based on secular reasoning fail. When religious elements predominate, they may intensify the conflict.

For example, the Palestinians have turned to one ideology after another to explain their loss of land to Jewish settlers and military forces and to formulate a plan for regaining territorial control. Especially after 1952, when Gamal Abdel Nasser took office in Egypt, many Palestinians turned to Pan-Arabism, the belief that the Arab countries would unify and force Israel to cede territory. But wars failed to dislodge the Israelis. Particularly after the Six-Day War in 1967, many Palestinians turned to nationalism, which placed the responsibility for regaining control of lost territory on the Palestinians themselves. Others became Marxists, identifying wage-workers (and, in some cases, peasants) as the engines of national liberation. The Palestinians used plane hijackings to draw the world's attention to their cause, launched wave upon wave of guerilla attacks against Israel, and in the 1990s entered into negotiations to create a sovereign Palestinian homeland.

Yet Islamic fundamentalism had been growing in popularity among Palestinians since the late 1980s—ironically, without opposition from the Israeli authorities, who saw it as a conservative counterweight to Palestinian nationalism. When negotiations with Israel to establish a Palestinian state broke down in 2000, many Palestinians saw the secularist approach as bankrupt and turned to Islamic fundamentalism for political answers. In January 2006, the Islamic fundamentalist party, Hamas, was democratically elected to form the Palestinian government, winning 44 percent of the popular vote and 56 percent of the parliamentary seats. In this case, as in many others, secular politics came first. When secularism failed, notions of "martyrdom" and "holy war" gained in importance.

This does not mean that most modern suicide bombers are deeply religious, either among the Palestinians or other groups. Among the 83 percent of suicide attackers worldwide between 1980 and 2003 for whom Robert Pape found data on ideological background, only a minority—43 percent—were identifiably religious. In Lebanon, Israel, the West Bank, and Gaza between 1981 and 2003, fewer than half of suicide bombers had discernible religious inclinations. In its origins and at its core, the Israeli-Palestinian conflict is not religiously inspired, and suicide bombing, despite its frequent religious trappings, is fundamentally the expression of a territorial dispute. In this conflict, many members of the dominant group—Jewish Israelis—use religion as a central marker of identity. It is hardly surprising, therefore, that many Palestinian militants also view the struggle in starkly religious terms.

The same holds for contemporary Iraq. As Mohammed Hafez has recently shown, 443 suicide missions took place in Iraq between March 2003 and February 2006. Seventy-one percent of the identifiable attackers belonged to al-Qaeda in Iraq. To be sure, they justified their actions in religious terms. Members of al-Qaeda in Iraq view the Shi'a who control the Iraqi state as apostates. They want to establish fundamentalist, Sunni-controlled states in Iraq and other Middle Eastern countries. Suicide attacks against the Iraqi regime and its American and British supporters are seen as a means to that end.

But it is only within a particular political context that these ambitions first arose. After all, suicide attacks began with the American and British invasion of Iraq and the installation of a Shi'a-controlled regime. And it is only under certain political conditions that these ambitions are acted upon. Thus, Hafez's analysis shows that suicide bombings spike (1) in retaliation for big counterinsurgency operations and (2) as a strategic response to institutional developments which suggest that Shi'a-controlled Iraq is about to become more stable. So although communal identity has come to be religiously demarcated in Iraq, this does not mean that religion per se initiated suicide bombing or that it drives the outbreak of suicide bombing campaigns.

LESSON 3: SOMETIMES IT'S STRATEGIC

Suicide bombing often has a political logic. In many cases, it is used as a tactic of last resort undertaken by the weak to help them restore control over territory they perceive as theirs. This political logic is clear in statements routinely released by leaders of organizations that launch suicide attacks. Characteristically, the first communiqué issued by Hamas in 1987 stated that martyrdom is the appropriate response to occupation, and the 1988 Hamas charter says that jihad is the duty of every Muslim whose territory is invaded by an enemy.

The political logic of suicide bombing is also evident when suicide bombings occur in clusters as part of an organized campaign, often timed to maximize strategic gains. A classic example is the campaign launched by Hamas and the PIJ in the mid-90s. Fearing that a settlement between Israel and the Palestinian Authority would prevent the Palestinians from gaining control over all of Israel,

Hamas and the PIJ aimed to scuttle peace negotiations by unleashing a small army of suicide bombers.

Notwithstanding the strategic basis of many suicide attacks, we cannot conclude that strategic reasoning governs them all. More often than not, suicide bombing campaigns fail to achieve their territorial aims. Campaigns may occur without apparent strategic justification, as did the campaign that erupted in Israel after negotiations between Israel and the Palestinian Authority broke down in 2000. A social logic often overlays the political logic of suicide bombing.

LESSON 4: SOMETIMES IT'S RETALIATORY

On October 4, 2003, a 29-year-old lawyer entered Maxim restaurant in Haifa and detonated her belt of plastic explosives. In addition to taking her own life, Hanadi Jaradat killed 20 people and wounded dozens of others. When her relatives were later interviewed in the Arab press, they explained her motives as follows: "She carried out the attack in revenge for the killing of her brother and her cousin [to whom she had been engaged] by the Israeli security forces and in revenge for all the crimes Israel is perpetrating in the West Bank by killing Palestinians and expropriating their land." Strategic calculation did not inform Jaradat's attack. Research I conducted with Bader Araj shows that, like a majority of Palestinian suicide bombers between 2000 and 2005, Jaradat was motivated by the desire for revenge and retaliation.

Before people act, they sometimes weigh the costs and benefits of different courses of action and choose the one that appears to cost the least and offer the most benefits. But people are not calculating machines. Sometimes they just don't add up. Among other emotions, feelings of anger and humiliation can trump rational strategic calculation in human affairs. Economists have conducted experiments called "the ultimatum game," in which the experimenter places two people in a room, gives one of them $20, and tells the recipient that she must give some of the money—as much or as little as she wants—to the other person. If the other person refuses the offer, neither gets to keep any money. Significantly, in four out of five cases, the other person refuses to accept the money if she is offered less than $5. Although she will gain materially if she accepts any offer, she is highly likely to turn down a low offer so as to punish her partner for stinginess. This outcome suggests that emotions can easily override the rational desire for material gain. (Researchers at the University of Zürich have recently demonstrated the physiological basis of this override function by using MRI brain scans on people playing the ultimatum game.) At the political level, research I conducted with Bader Araj on the events precipitating suicide bombings, the motivations of suicide bombers, and the rationales of the organizations that support suicide bombings shows that Palestinian suicide missions are in most cases prompted less by strategic costbenefit calculations than by such human emotions as revenge and retaliation. The existence of these deeply human emotions also helps to explain why attempts to suppress suicide bombing campaigns sometimes do not have the predicted results.

LESSON 5: REPRESSION IS A BOOMERANG

Major General Doron Almog commanded the Israel Defense Forces Southern Command from 2000 to 2003. He tells the story of how, in early 2003, a wealthy Palestinian merchant in Gaza received a phone call from an Israeli agent. The caller said that the merchant's son was preparing a suicide mission, and that if he went through with it, the family home would be demolished, Israel would sever all commercial ties with the family, and its members would never be allowed to visit Israel again. The merchant prevailed upon his son to reconsider, and the attack was averted.

Exactly how many suicide bombers have been similarly deterred is unknown. We do know that of the nearly 600 suicide missions launched in Israel and its occupied territories between 2000 and 2005, fewer than 25 percent succeeded in reaching their targets. Israeli counterterrorist efforts thwarted three-quarters of them using violent means. In addition, Israel preempted an incalculable number of attacks by assassinating militants involved in planning them. More than 200 Israeli assassination attempts took place between 2000 and 2005, 80 percent of which succeeded in killing their main target, sometimes with considerable "collateral damage."

Common sense suggests that repression should dampen insurgency by increasing its cost. By this logic, when state organizations eliminate the people who plan suicide bombings, destroy their bomb-making facilities, intercept their agents, and punish the people who support them, they erode the insurgents' capabilities for mounting suicide attacks. But this commonsense approach to counterinsurgency overlooks two complicating factors. First, harsh repression may reinforce radical opposition and even intensify it. Second, insurgents may turn to alternative and perhaps more lethal methods to achieve their aims.

Consider the Palestinian case. Bader Araj and I were able to identify the organizational affiliation of 133 Palestinian suicide bombers between September 2000 and July 2005. Eighty-five of them (64 percent) were affiliated with the Islamic fundamentalist groups Hamas and the PIJ, while the rest were affiliated with secular Palestinian groups such as Fatah. Not surprisingly, given this distribution, Israeli repression was harshest against the Islamic fundamentalists, who were the targets of 124 Israeli assassination attempts (more than 60 percent of the total).

Yet after nearly five years of harsh Israeli repression—involving not just the assassination of leaders but also numerous arrests, raids on bomb-making facilities, the demolition of houses belonging to family members of suicide bombers, and so on—Hamas and PIJ leaders remained adamant in their resolve and much more radical than Palestinian secularist leaders. When 45 insurgent leaders representing all major Palestinian factions were interviewed in depth in the summer of 2006, 100 percent of those associated with Hamas and PIJ (compared to just 10 percent of secularist leaders) said they would never be willing to recognize the legitimacy of the state of Israel. That is, the notion of Israel as a Jewish state was still entirely unacceptable to each and every one of them. When asked how Israel's assassination policy had affected the ability of their organization to conduct suicide bombing operations, 42 percent of Hamas and PIJ respondents said that the policy had had no effect, while one-third said the policy had increased their organization's capabilities (the corresponding figures for secularist leaders were 5 percent and 9 percent, respectively).

And when asked how costly suicide bombing had been in terms of human and organizational resources, organizational damage, and so on, 53 percent of Hamas and PIJ leaders (compared to just 11 percent of secularist leaders) said that suicide bombing was less costly or at least no more costly than the alternatives. Responses to such questions probably tell us more about the persistent resolve of the Islamic fundamentalists than their actual capabilities. And that is just the point. Harsh Israeli repression over an extended period apparently reinforced the anti-Israel sentiments of Islamic fundamentalists.

Some counterterrorist experts say that motivations count for little if capabilities are destroyed. And they would be right if it were not for the substitutability of methods: increase the cost of one method of attack, and highly motivated insurgents typically substitute another. So, for example, Israel's late prime minister, Yitzhak Rabin, ordered troops to "break the bones" of Palestinians who engaged in mass demonstrations, rock throwing, and other nonlethal forms of protest in the late 1980s and early 1990s. The Palestinians responded with more violent attacks, including suicide missions. Similarly, after Israel began to crack down ruthlessly on suicide bombing operations in 2002, rocket attacks against Israeli civilians sharply increased in frequency. In general, severe repression can work for a while, but a sufficiently determined mass opposition can always design new tactics to surmount new obstacles, especially if its existence as a group is visibly threatened (and unless, of course, the mass opposition is exterminated in its entirety). One kind of "success" usually breeds another kind of "failure" if the motivation of insurgents is high.

LESSON 6: EMPATHIZE WITH YOUR ENEMY

In October 2003, Israeli Chief of Staff Moshe Ya'alon explicitly recognized this conundrum when he stated that Israel's tactics against the Palestinians had become too repressive and were stirring up potentially uncontrollable levels of hatred and terrorism. "In our tactical decisions, we are operating contrary to our strategic interests," he told reporters. Ya'alon went on to claim that the Israeli government was unwilling to make concessions that could bolster the authority of moderate Palestinian Prime Minister Mabmoud Abbas, and he expressed the fear that by continuing its policy of harsh repression, Israel would bring about the collapse of the Palestinian Authority, the silencing of Palestinian moderates, and the popularization of more radical voices like that of Hamas. The head of the General Security Service (Shabak), the defense minister, and Prime Minister Ariel Sharon opposed Ya'alon. Consequently, his term as chief of staff was not renewed, and his military career ended in 2005. A year later, all of Ya'alon's predictions proved accurate.

Ya'alon was no dove. From the time he became chief of staff in July 2002, he had been in charge of ruthlessly putting down the Palestinian uprising. He had authorized assassinations, house demolitions, and all the rest. But 15 months into the job, Ya'alon had learned much from his experience, and it seems that what he learned above all else was to empathize with the enemy—not to have warm and fuzzy feelings about the Palestinians, but to see things from their point of view in order to improve his ability to further Israel's chief strategic interest, namely, to live in peace with its neighbors.

As odd as it may sound at first, and as difficult as it may be to apply in practice, exercising empathy with one's enemy is the key to an effective counterterrorist strategy. Seeing the enemy's point of view increases one's understanding of the minimum conditions that would allow the enemy to put down arms. An empathic understanding of the enemy discourages counterproductive actions such as excessive repression, and it encourages tactical moves that further one's strategic aims. As Ya'alon suggested, in the Israeli case such tactical moves might include (1) offering meaningful rewards—for instance, releasing hundreds of millions of Palestinian tax dollars held in escrow by Israel, freeing selected Palestinians from Israeli prisons, and shutting down remote and costly Israeli settlements in the northern West Bank—in exchange for the renunciation of suicide bombing and (2) attributing the deal to the intercession of moderate Palestinian forces so as to buttress their popularity and authority. (From this point of view, Israel framed its unilateral 2005 withdrawal from Gaza poorly because most Palestinians saw it as a concession foisted on Israel by Hamas.) Once higher levels of trust and stability are established by such counterterrorist tactics, they can serve as the foundation for negotiations leading to a permanent settlement. Radical elements would inevitably try to jeopardize negotiations, as they have in the past, but Israel resisted the temptation to shut down peace talks during the suicide bombing campaign of the mid-1990s, and it could do so again. Empathizing with the enemy would also help prevent the breakdown of negotiations, as happened in 2000; a clear sense of the minimally acceptable conditions for peace can come only from an empathic understanding of the enemy.

CONCLUSION

Political conflict over territory is the main reason for suicide bombing, although religious justifications for suicide missions are likely to become more important when secular ideologies fail to bring about desired results. Suicide bombing may also occur for strategic or retaliatory reasons—to further insurgent aims or in response to repressive state actions.

Cases vary in the degree to which suicide bombers are motivated by (1) political or religious and (2) strategic or retaliatory aims. For example, research to date suggests that suicide bombing is more retaliatory in Israel than in Iraq, and more religiously motivated in Iraq than in Israel. But in any case, repression (short of a policy approaching genocide) cannot solve the territorial disputes that lie at the root of suicide bombing campaigns. As Zbigniew Brzezinski, President Jimmy Carter's national security adviser, wrote a few years ago in the *New York Times*, "to win the war on terrorism, one must … begin a political effort that focuses on the conditions that brought about [the terrorists'] emergence." These are wise words that Israel—and the United States in its own "war on terror"—would do well to heed.

KEY CONCEPTS

counterterrorism	fundamentalism	organizations

DISCUSSION QUESTIONS

1. In what ways does Brym use his analysis to challenge common-sense explanations of suicide bombings?
2. What does Brym's argument suggest for using a sociological perspective to think about ways to reduce terrorism?

22

The Rich Get Richer and the Poor Get Prison

JEFFREY H. REIMAN

This essay challenges the reader to view the criminal justice system from a radically different angle. Specifically, Jeffrey Reiman argues that the corrections system and broader criminal justice policy in the United States simply provide the illusion of fighting crime. In reality, he argues, criminal justice policies reinforce

SOURCE: Jeffrey H. Reiman. 2005. *The Rich Get Richer and the Poor Get Prison: Ideology, Class, and Criminal Justice,* 7th ed. Boston, MA: Allyn and Bacon, pp. 1–2.

public fears of crimes committed by the poor. These policies, in turn, help to maintain a "criminal class" of disadvantaged people.

A criminal justice system is a mirror in which a whole society can see the darker outlines of its face. Our ideas of justice and evil take on visible form in it, and thus we see ourselves in deep relief. Step through this looking glass to view the American criminal justice system—and ultimately the whole society it reflects—from a radically different angle of vision.

In particular, entertain the idea that the goal of our criminal justice system is not to eliminate crime or to achieve justice, *but to project to the American public a visible image of the threat of crime as a threat from the poor.* To do this, the justice system must present us with a sizable population of poor criminals. To do that, it must fail in the struggle to eliminate the crimes that poor people commit, or even to reduce their number dramatically. Crime may, of course, occasionally decline, as it has recently—*but largely because of factors other than criminal justice policies....*

In recent years, we have quadrupled our prison population and, in cities such as New York, allowed the police new freedom to stop and search people they suspect. No one can deny that if you lock up enough people, and allow the police greater and greater power to interfere with the liberty and privacy of citizens, you will eventually prevent some crime that might otherwise have taken place.... I shall point out just how costly and inefficient this means of reducing crime is, in money for new prisons, in its destructive effect on inner-city life, in reduced civil liberties, and in increased complaints of police brutality. I don't deny, however, that these costly means do contribute *in some small measure* to reducing crime. Thus, when I say... that criminal justice policy is failing, I mean that it is failing to eliminate our high crime rates. We continue to see a large population of poor criminals in our prisons and our courts, while our crime-reduction strategies do not touch on the social causes of crime. Moreover, our citizens remain fearful about criminal victimization, even after the recent declines....

Nearly 30 years ago, I taught a seminar for graduate students titled "The Philosophy of Punishment and Rehabilitation." Many of the students were already working in the field of corrections as probation officers, prison guards, or halfway-house counselors. Together we examined the various philosophical justifications for legal punishment, and then we directed our attention to the actual functioning of our correctional system. For much of the semester, we talked about the myriad inconsistencies and cruelties and the overall irrationality of the system. We discussed the arbitrariness with which offenders are sentenced to prison and the arbitrariness with which they are treated once there. We discussed the lack of privacy and the deprivation of sources of personal identity and dignity, the ever-present physical violence, as well as the lack of meaningful counseling or job training within prison walls. We discussed the harassment of parolees, the inescapability of the "ex-con" stigma, the refusal of society to let a person finish paying his or her "debt to society," and the absence of meaningful noncriminal opportunities for the ex-prisoner. We confronted time and again the bald irrationality of a society that builds prisons to prevent crime knowing full well that they do not, and one that does not seriously try to rid its prisons and postrelease practices of those features that guarantee a high rate of *recidivism,* the return to crime by prison alumni. How could we fail so miserably? We are neither an evil nor a stupid nor an impoverished people. How could we continue to bend our energies and spend our hard-earned tax dollars on cures we know are not working?

Toward the end of the semester, I asked the students to imagine that, instead of designing a criminal justice system to reduce and prevent crime, we designed one that would maintain a stable and visible "class" of criminals. What would it look like? The response was electrifying. Here is a sample of the proposals that emerged in our discussion.

First   It would be helpful to have laws on the books against drug use, prostitution, and gambling—laws that prohibit acts that have no unwilling victim. This would make many people "criminals" for what they regard as normal behavior and would increase their need to engage in *secondary crime* (the drug addict's need to steal to pay for drugs, the prostitute's need for a pimp because police protection is unavailable, and so on).

Second   It would be good to give police, prosecutors, and/or judges broad discretion to decide who got arrested, who got charged, and who got sentenced to prison. This would mean that almost anyone who got as far as prison would know of others who committed the same crime but were not arrested, were not charged, or were not sentenced to prison. This would assure us that a good portion of the prison population would experience their confinement as arbitrary and unjust and thus respond with rage, which would make them more antisocial, rather than respond with remorse, which would make them feel more bound by social norms.

Third   The prison experience should be not only painful but also demeaning. The pain of loss of liberty might deter future crime. But demeaning and emasculating prisoners by placing them in an enforced childhood characterized by no privacy and no control over their time and actions, as well as by the constant threat of rape or assault, is sure to overcome any deterrent effect by weakening whatever capacities a prisoner had for self-control. Indeed, by humiliating and brutalizing prisoners, we can be sure to increase their potential for aggressive violence.

Fourth   Prisoners should neither be trained in a marketable skill nor provided with a job after release. Their prison records should stand as a perpetual stigma to discourage employers from hiring them. Otherwise, they might be tempted *not* to return to crime after release.

Fifth   Ex-offenders' sense that they will always be different from "decent citizens," that they can never finally settle their debt to society, should be reinforced by the following means. They should be deprived for the rest of their lives of rights, such as the right to vote. They should be harassed by police as "likely suspects" and be subject to the whims of parole officers who can at any time threaten to send them back to prison for things no ordinary citizens could be arrested for, such as going out of town, or drinking, or fraternizing with the "wrong people."

And so on.

In short, *when asked to design a system that would maintain and encourage the existence of a stable and visible "class of criminals," we "constructed" the American criminal justice system*[1]...

... [T]he practices of the criminal justice system keep before the public the *real* threat of crime and the *distorted* image that crime is primarily the work of the poor. The value of this *to those in positions of power* is that it deflects the discontent and potential hostility of Middle America away from the classes above them and toward the classes below them. If this explanation is hard to swallow, it should be noted in its favor that it not only explains the dismal failure of criminal justice policy to protect us against crime but also explains why the criminal justice system functions in a way that is biased against the poor at every stage from arrest to conviction. Indeed, even at an earlier stage, when crimes are defined in law, the system concentrates primarily on the predatory acts of the poor and tends to exclude or deemphasize the equally or more dangerous predatory acts of those who are well off.

In sum, I will argue that *the criminal justice system fails in the fight against crime while making it look as if crime is the work of the poor.* This conveys the image that the real danger to decent, law-abiding Americans comes from below them, rather than from above them, on the economic ladder. This image sanctifies the status quo with its disparities of wealth, privilege, and opportunity, and thus serves the interests of the rich and powerful in America—the very ones who could change criminal justice policy if they were really unhappy with it.

Therefore, it seems appropriate to ask you to look at criminal justice "through the looking glass." On the one hand, this suggests a reversal of common expectations. Reverse your expectations about criminal justice and entertain the notion that the system's real goal is the very reverse of its announced goal. On the other hand, the figure of the looking glass suggests the prevalence of image over reality. My argument is that the system functions the way it does *because it maintains a particular image of crime' the image that it is a threat from the poor.* Of course, for this image to be believable, there must be a reality to back it up. The system must actually fight crime—or at least some crime—but only enough to keep it from getting out of hand and to keep the struggle against crime vividly and dramatically in the public's view, never enough to substantially reduce or eliminate crime.

I call this outrageous way of looking at criminal justice policy the *Pyrrhic defeat* theory. A "Pyrrhic victory" is a military victory purchased at such a cost in troops and treasure that it amounts to a defeat. The Pyrrhic defeat theory argues that the failure of the criminal justice system yields such benefits to those in positions of power that it amounts to success....

The Pyrrhic defeat theory has several components. Above all, it must provide an explanation of *how* the failure to reduce crime substantially could benefit anyone—anyone other than criminals, that is.... I argue there that the failure to reduce crime substantially broadcasts a potent *ideological* message to the American people, a message that benefits and protects the powerful and privileged in our society by legitimating the present social order with its disparities of wealth and privilege, and by diverting public discontent and opposition away from the rich and powerful and onto the poor and powerless.

To provide this benefit, however, not just any failure will do. It is necessary that the failure of the criminal justice system take a particular shape. *It must fail in the fight against crime while making it look as if serious crime and thus the real danger to society are the work of the poor.* The system accomplishes this both by what it does and by what it refuses to do…. I argue that the criminal justice system refuses to label and treat as crime a large number of acts of the rich that produce as much or more damage to life and limb as the crimes of the poor…. [E]ven among the acts treated as crimes, the criminal justice system is biased from start to finish in a way that guarantees that, *for the same crimes,* members of the lower classes are much more likely than members of the middle and upper classes to be arrested, convicted, and imprisoned—thus providing living "proof" that crime is a threat from the poor….

Our criminal justice system is characterized by beliefs about what is criminal, and beliefs about how to deal with crime, that predate industrial society. Rather than being anyone's conscious plan, the system reflects attitudes so deeply embedded in tradition as to appear natural. To understand why it persists even though it fails to protect us, all that is necessary is to recognize that, on the one hand, those who are the most victimized by crime are not those in positions to make and implement policy. Crime falls more frequently and more harshly on the poor than on the better off. On the other hand, there are enough benefits to the wealthy from the identification of crime with the poor and the system's failure to reduce crime that those with the power to make profound changes in the system feel no compulsion nor see any incentive to make them. In short, the criminal justice system came into existence in an earlier epoch and persists in the present because, even though it is failing—indeed, because of the way it fails—it generates no effective demand for change. When I speak of the criminal justice system as "designed to fail," I mean no more than this. I call this explanation of the existence and persistence of our failing criminal justice system the *historical inertia* explanation….

KEY CONCEPTS

labeling theory	social class	social institution

DISCUSSION QUESTIONS

1. What does Reiman mean in arguing that the current criminal justice system works to maintain a class of criminals? Do you agree or disagree that our corrections system fails to rehabilitate and fails to deter crime?

2. If you had the power to change the corrections system in the United States, what changes would you make to help reduce and prevent crime?

Applying Sociological Knowledge: An Exercise for Students

Become a norm breaker! Think of a norm we have in society that you can go out in public and violate. Make sure it is legal! How do people treat you when you stop doing something that is implicitly expected of you? How does it feel to go against what you feel you should be doing? Was this norm something you thought about doing before or is it something that you did without even thinking? Notice how hard it is to deviate from expected norms.

PART VIII

Social Class and Social Stratification

23

The Communist Manifesto

KARL MARX AND FREDERICH ENGELS

The analysis of the class system under capitalism, as developed by Marx and Engels, continues to influence sociological understanding of the development of capitalism and the structure of the class system. In this classic essay, first published in 1848, Marx and Engels define the class system in terms of the relationships between capitalism, the bourgeoisie, and the proletariat. Their analysis of the growth of capitalism and its influence on other institutions continues to provide a compelling portrait of an economic system based on the pursuit of profit.

BOURGEOIS AND PROLETARIANS

The history of all hitherto existing society is the history of class struggles.... Modern industry has established the world market, for which the discovery of America paved the way. This market has given an immense development to commerce, to navigation, to communication by land. This development has, in its turn, reacted on the extension of industry; and in proportion as industry, commerce, navigation, railways extended, in the same proportion the bourgeoisie developed, increased its capital, and pushed into the background every class handed down from the Middle Ages.

We see, therefore, how the modern bourgeoisie is itself the product of a long course of development, of a series of revolutions in the modes of production and of exchange....

The bourgeoisie has at last, since the establishment of modern industry and of the world market, conquered for itself, in the modern representative state, exclusive political sway. The executive of the modern state is but a committee for managing the common affairs of the whole bourgeoisie.

The bourgeoisie, historically, has played a most revolutionary part.

The bourgeoisie, wherever it has got the upper hand, has put an end to all feudal, patriarchal, idyllic relations. It has pitilessly torn asunder the motley feudal ties that bound man to his "natural superiors" and has left remaining no other

SOURCE: Karl Marx and Frederick Engels. 1998. *Manifesto of the Communist Party*. With introduction by Eric Hobsbawm. New York: Verso, pp. 33–51.

nexus between man and man than naked self-interest, than callous "cash payment." It has drowned the most heavenly ecstasies of religious fervour, of chivalrous enthusiasm, of philistine sentimentalism, in the icy water of egotistical calculation. It has resolved personal worth into exchange value, and in place of the numberless indefeasible chartered freedoms, has set up that single, unconscionable freedom—free trade. In one word, for exploitation, veiled by religious and political illusions, it has substituted naked, shameless, direct, brutal exploitation.

The bourgeoisie has stripped of its halo every occupation hitherto honoured and looked up to with reverent awe. It has converted the physician, the lawyer, the priest, the poet, the man of science, into its paid wage labourers.

The bourgeoisie has torn away from the family its sentimental veil, and has reduced the family relation to a mere money relation....

The need of a constantly expanding market for its products chases the bourgeoisie over the whole surface of the globe. It must nestle everywhere, settle everywhere, establish connections everywhere.

The bourgeoisie has through its exploitation of the world market given a cosmopolitan character to production and consumption in every country. To the great chagrin of reactionists, it has drawn from under the feet of industry the national ground on which it stood. All old, established national industries have been destroyed or are daily being destroyed. They are dislodged by new industries, whose introduction becomes a life and death question for all civilized nations, by industries that no longer work up indigenous raw material, but raw material drawn from the remotest zones; industries whose products are consumed, not only at home, but in every quarter of the globe. In place of the old wants, satisfied by the productions of the country, we find new wants, requiring for their satisfaction the products of distant lands and climes. In place of the old local and national seclusion and self-sufficiency, we have intercourse in every direction, universal interdependence of nations. And as in material, so also in intellectual production. The intellectual creations of individual nations become common property. National one-sidedness and narrow-mindedness become more and more impossible, and from the numerous national and local literatures, there arises a world literature.

The bourgeoisie, by the rapid improvement of all instruments of production, by the immensely facilitated means of communication, draws all, even the most barbarian, nations into civilization. The cheap prices of its commodities are the heavy artillery with which it batters down all Chinese walls, with which it forces the barbarians' intensely obstinate hatred of foreigners to capitulate. It compels all nations, on pain of extinction, to adopt the bourgeois mode of production; it compels them to introduce what it calls civilization into their midst, i.e., to become bourgeois themselves. In one word, it creates a world after its own image.

The bourgeoisie has subjected the country to the rule of the towns. It has created enormous cities, has greatly increased the urban population as compared with the rural, and has thus rescued a considerable part of the population from the idiocy of rural life. Just as it has made the country dependent on the towns, so it has made barbarian and semi-barbarian countries dependent on the civilized ones, nations of peasants on nations of bourgeois, the East on the West.

The bourgeoisie keeps more and more doing away with the scattered state of the population, of the means of production, and of property. It has agglomerated population, centralized means of production, and has concentrated property in a few hands. The necessary consequence of this was political centralization. Independent, or but loosely connected provinces, with separate interests, laws, governments and systems of taxation became lumped together into one nation, with one government, one code of laws, one national class interest, one frontier and one customs tariff....

The weapons with which the bourgeoisie felled feudalism to the ground are now turned against the bourgeoisie itself.

But not only has the bourgeoisie forged the weapons that bring death to itself; it has also called into existence the men who are to wield those weapons—the modern working class—the proletarians.

In proportion as the bourgeoisie, i.e., capital, is developed, in the same proportion is the proletariat, the modern working class, developed—a class of labourers, who live only so long as they find work, and who find work only so long as their labour increases capital. These labourers, who must sell themselves piecemeal, are a commodity, like every other article of commerce, and are consequently exposed to all the vicissitudes of competition, to all the fluctuations of the market.

Owing to the extensive use of machinery and to division of labour, the work of the proletarians has lost all individual character, and, consequently, all charm for the workman. He becomes an appendage of the machine, and it is only the most simple, most monotonous, and most easily acquired knack that is required of him. Hence, the cost of production of a workman is restricted, almost entirely, to the means of subsistence that he requires for his maintenance and for the propagation of his race. But the price of a commodity, and therefore also of labour, is equal to its cost of production. In proportion, therefore, as the repulsiveness of the work increases, the wage decreases. Nay more, in proportion as the use of machinery and division of labour increases, in the same proportion the burden of toil also increases, whether by prolongation of the working hours, by increase of the work exacted in a given time or by increased speed of the machinery, etc.

Modern industry has converted the little workshop of the patriarchal master into the great factory of the industrial capitalist. Masses of labourers, crowded into the factory, are organized like soldiers. As privates of the industrial army they are placed under the command of a perfect hierarchy of officers and sergeants. Not only are they slaves of the bourgeois class and of the bourgeois state; they are daily and hourly enslaved by the machine, by the overseer, and, above all, by the individual bourgeois manufacturer himself. The more openly this despotism proclaims gain to be its end and aim, the more petty, the more hateful and the more embittering it is.

The less the skill and exertion of strength implied in manual labour, in other words, the more modern industry becomes developed, the more is the labour of men superseded by that of women. Differences of age and sex have no longer any distinctive social validity for the working class. All are instruments of labour, more or less expensive to use, according to their age and sex....

But with the development of industry the proletariat not only increases in number; it becomes concentrated in greater masses, its strength grows, and it feels that strength more. The various interests and conditions of life within the ranks of the proletariat are more and more equalized, in proportion as machinery obliterates all distinctions of labour, and nearly everywhere reduces wages to the same low level. The growing competition among the bourgeois, and the resulting commercial crises, make the wages of the workers ever more fluctuating. The unceasing improvement of machinery, ever more rapidly developing, makes their livelihood more and more precarious; the collisions between individual workmen and individual bourgeois take more and more the character of collisions between two classes. Thereupon the workers begin to form combinations (trade unions) against the bourgeois....

This organization of the proletarians into a class, and consequently into a political party, is continually being upset again by the competition between the workers themselves. But it ever rises up again, stronger, firmer, mightier. It compels legislative recognition of particular interests of the workers, by taking advantage of the divisions among the bourgeoisie itself....

Altogether, collisions between the classes of the old society further, in many ways, the course of development of the proletariat. The bourgeoisie finds itself involved in a constant battle: at first with the aristocracy; later on, with those portions of the bourgeoisie itself, whose interests have become antagonistic to the progress of industry; at all times, with the bourgeoisie of foreign countries. In all these battles it sees itself compelled to appeal to the proletariat, to ask for its help, and thus to drag it into the political arena. The bourgeoisie itself, therefore, supplies the proletariat with its own elements of political and general education; in other words, it furnishes the proletariat with weapons for fighting the bourgeoisie.

Further, as we have already seen, entire sections of the ruling classes are, by the advance of industry, precipitated into the proletariat, or are at least threatened in their conditions of existence. These also supply the proletariat with fresh elements of enlightenment and progress.

Finally, in times when the class struggle nears the decisive hour, the process of dissolution going on within the ruling class, in fact within the whole range of old society, assumes such a violent, glaring character that a small section of the ruling class cuts itself adrift and joins the revolutionary class, the class that holds the future in its hands. Just as, therefore, at an earlier period, a section of the nobility went over to the bourgeoisie, so now a portion of the bourgeoisie goes over to the proletariat, and in particular, a portion of the bourgeois ideologists, who have raised themselves to the level of comprehending theoretically the historical movement as a whole.

Of all the classes that stand face to face with the bourgeoisie today, the proletariat alone is a really revolutionary class. The other classes decay and finally disappear in the face of modern industry; the proletariat is its special and essential product....

KEY CONCEPTS

capitalism   proletariat   working class

DISCUSSION QUESTIONS

1. What evidence do you see in contemporary society of Marx and Engels' claim that the need for a constantly expanding market means that capitalism "nestles everywhere"?

2. How do Marx and Engels depict the working class, and what evidence do you see of their argument in looking at the contemporary labor market?

24

Aspects of Class in the United States: An Introduction

JOHN BELLAMY FOSTER

John Bellamy Foster examines the increasing inequality that is characteristic of the contemporary United States. He shows that wealth is even more unevenly divided than is income and also discusses the decreased likelihood of social mobility for current generations.

If class war is continual in capitalist society, there is no doubt that in recent decades in the United States it has taken a much more virulent form. In a speech delivered at New York University in 2004 Bill Moyers pointed out that

> Class war was declared a generation ago in a powerful paperback polemic by William Simon, who was soon to be Secretary of the Treasury. He called on the financial and business class, in effect, to take back the power and privileges they had lost in the depression and the new deal. They got the message, and soon they began a stealthy class war against the rest of the society and the principles of our democracy. They set out to trash the social contract, to cut their workforces and wages, to scour the globe in search of cheap labor, and to shred the social safety net that was supposed to protect people from hardships beyond their control. *Business Week* put it bluntly at the time [in its October 12, 1974 issue]: "Some people will obviously have to do with less ... it will be a bitter

SOURCE: Foster, John Bellamy. 2006. "Aspects of Class in the United States: An Introduction." *Monthly Review* 58: 1-5.

pill for many Americans to swallow the idea of doing with less so that big business can have more."[1]

The effects of this relentless offensive by the vested interests against the rest of the society are increasingly evident. In 2005 the *New York Times* and the *Wall Street Journal* each published a series of articles focusing on class in the United States. This rare open acknowledgement of the importance of class by the elite media can be attributed in part to rapid increases in income and wealth inequality in U.S. society over the last couple of decades—coupled with the dramatic effects of the Bush tax cuts that have primarily benefited the wealthy. But it also grew out of a host of new statistical studies that have demonstrated that intergenerational class mobility in the United States is far below what was previously supposed, and that the United States is a more class-bound society than its major Western European counterparts, with the exception of Britain. In the words of *The Wall Street Journal* (May 13, 2005):

> Although Americans still think of their land as a place of exceptional opportunity—in contrast to class-bound Europe—the evidence suggests otherwise. And scholars have, over the past decade, come to see America as a less mobile society than they once believed. As recently as the later 1980s, economists argued that not much advantage passed from parent to child, perhaps as little as 20 percent. By that measure, a rich man's grand-child would have barely any edge over a poor man's grandchild.... But over the last 10 years, better data and more number-crunching have led economists and sociologists to a new consensus: The escalators of mobility move much more slowly. A substantial body of research finds that at least 45 percent of parents' advantage in income is passed along to their chil-dren, and perhaps as much as 60 percent. With the higher estimate, it's not only how much money your parents have that matters—even your great-great grandfather's wealth might give you a noticeable edge today.

As Paul Sweezy once observed, "self-reproduction is an *essential* characteristic of a class as distinct from a mere stratum."[2] What is clear from recent data is that the upper classes in the United States are extremely effective in reproducing them-selves—to a degree that invites no obvious historical comparison in modern capi-talist history. According to the *New York Times* (November 14, 2002), "Bhashkar Mazumber of the Federal Reserve Bank of Chicago ... found that around 65 per-cent of the earnings advantage of fathers was transmitted to sons." Tom Hertz, an economist at American University, states that "while few would deny that it is *possible* to start poor and end rich, the evidence suggests that this feat is more diffi-cult to accomplish in the United States than in other high-income nations."[3]

The fact that the rich are getting both relatively and absolutely richer, and the poor are getting relatively (if not absolutely) poorer, in the United States today is abundantly clear to all—although the true extent of this trend defies the imagination. Over the years 1950 to 1970, for each additional dollar made by those in the bottom 90 percent of income earners, those in the top 0.01 per-cent received an additional $162. In contrast, from 1990 to 2002, for every added dollar made by those in the bottom 90 percent, those in the uppermost 0.01 percent (today around 14,000 households) made an additional $18,000.[4]

Wealth is always far more unevenly divided than income. In 2001 the top 1 percent of wealth holders accounted for 33 percent of all net worth in the United States, twice the total net worth of the bottom 80 percent of the population. Measured in terms of financial wealth (which excludes equity in owner-occupied houses), the top 1 percent in 2001 owned more than four times as much as the bottom 80 percent of the population. Between 1983 and 2001, this same top 1 percent grabbed 28 percent of the rise in national income, 33 percent of the total gain in net worth, and 52 percent of the overall growth in financial worth.[5]

Nevertheless, a considerable portion of the population still seems willing to accept substantial differentials in economic rewards on the assumption that these represent returns to merit and that all children have a fighting chance to rise to the top. The United States, the received wisdom tells us, is still the "land of opportunity." The new data on class mobility, however, indicate that this is far from the case and that the barriers separating classes are hardening.

How class advantages are passed on from one generation to the next is of course enormously difficult to determine—if only because class privileges are so various. Class inequality manifests itself in wealth, income, and occupation, but also in education, consumption, and health—and each of these are among the means by which class advantages/disadvantages are transmitted. Class inequalities, Sweezy explained,

> are not only or perhaps even primarily a matter of income: [in certain social settings] a considerable range of income differentials would be compatible with all children having substantially equal life chances. More important are a number of other factors which are less well defined, less visible, and impossible to quantify: the advantages of coming from a more "cultured" home environment, differential access to educational opportunities, the possession of "connections" in the circles of those holding positions of power and prestige, and self-confidence which children absorb from their parents—the list could be expanded and elaborated.[6]

Such intangibles are difficult to measure, but in a capitalist society they tend to interact with large differentials in income and property ownership and hence leave their quantitative trace there. It is this whole constellation of class advantages roughly correlated with income and wealth, but not simply reducible to these elements, that allows the privileged to maintain their positions of economic status and power intergenerationally even in the context of a society that on the surface appears to have many of the characteristics of a meritocracy. The well-to-do get better education, enjoy better health, have more opportunities to travel, benefit from a wide array of personal services (derived from purchase of the labor services of others), etc.—all of which translates into class advantages passed on to their children.

The fact that strong barriers restricting upward class mobility exist is of course the first point to be considered in class analysis—since without this classes would be nonexistent. However, the real historical significance of class goes far beyond this. Class is not simply about the life chances of a given individual or a family; it is the prime mover in the constitution of modern society, governing both the distribution of power and the potential for social change. It therefore permeates all aspects of social existence.

At present there is no well-developed theory of class in all of its aspects, which remains perhaps the single biggest challenge facing the social sciences. Indeed, failure to advance in this area can be seen as symptomatic of the general stagnation of the social sciences over much of the twentieth century. Nevertheless, most Marxist analyses of class take their starting point from Lenin's famous definition of class:

> Classes are large groups of people differing from each other by the place they occupy in a historically determined system of social production, by their relation (in most cases fixed and formulated in law) to the means of production, by their role in the social organization of labour, and, consequently, by the dimensions of the share of social wealth of which they dispose and the mode of acquiring it.[7]

Like all brief definitions of class, this one has its weaknesses, since it is not able to take in the dynamic nature of class relations. As Sweezy argued, a systematic treatment of class and class struggle "needs also to encompass at least the following: the formation of classes in conflict with other classes, the character and degree of their self-consciousness, their internal organizational structures, the ways in which they generate and utilize ideologies to further their interests, and their modes of reproduction and self-perpetuation."[8] If we are speaking of a "ruling class" then the ways in which this class dominates the economy and the state need to be understood. Further, it is crucial to ascertain how class articulates itself in relation to other social relations and forms of oppression, such as race and gender.

An investigation of class thus leads to the analysis of society as a whole, its relationships of power, conflict, and change.

NOTES

1. Bill Moyers, "This is the Fight of Our Lives," keynote speech, Inequality Matters Forum, New York University, June 3, 2004, http://www.commondreams.org/views04/06l6-09.htm/.

2. Paul M. Sweezy, "Paul Sweezy Replies to Ernest Mandel," *Monthly Review* 31, no. 3 (July–August 1979), 82.

3. Tom Hertz, *Understanding Mobility in America*, Center for American Progress (April 26, 2006), i, 8, http://www.americanprogress.org/.

4. Correspondents of *The New York Times, Class Matters* (New York: Times Books, 2005), 186.

5. Edward N. Wolff, "Changes in Household Wealth in the 1980s and 1990s in the U.S.," (April 27, 2004, draft), forthcoming in Wolff, *International Perspectives on Household Wealth* (Brookfield, Vermont: Edward Elgar), http://www.econ.nyu.edu/user/wolffe/.

6. Sweezy's comments here were directed mainly at postrevolutionary societies, but he made it clear that the same issues related to the reproduction of class applied to capitalist societies. I have inserted a brief qualification in square brackets to avoid any misunderstanding related to the specific context in which he was writing. See Paul M. Sweezy, *Post-Revolutionary Society* (New York: Monthly Review Press, 1980), 79–80.

7. V. I. Lenin, *Selected Works* (Moscow: Progress Publishers, 1971), 486.

8. Sweezy, "Paul Sweezy Replies to Ernest Mandel," 79.

KEY CONCEPTS

capitalism	income	social mobility
contingent worker	social class (or class)	wealth
elites		

DISCUSSION QUESTIONS

1. What is the difference between *income* and *wealth,* and why is this important for the study of class inequality?

2. In what ways does Foster's analysis challenge the idea that the United States is a nation where anyone who works hard enough can get ahead?

25

America Without a Middle Class

ELIZABETH WARREN

Elizabeth Warren argues that the American tradition of having a strong middle class is now at risk because of the economic crisis that has beset America. She shows the increasingly fragile status of many middle-class families, who are now working harder than ever just to keep up with basic expenses.

Can you imagine an America without a strong middle class? If you can, would it still be America as we know it?

SOURCE: Warren, Elizabeth. 2009. "America Without a Middle Class." *Huffington Post,* December 3, 2009.

Today, one in five Americans is unemployed, underemployed, or just plain out of work. One in nine families can't make the minimum payment on their credit cards. One in eight mortgages is in default or foreclosure. One in eight Americans is on food stamps. More than 120,000 families are filing for bankruptcy every month. The economic crisis has wiped more than $5 trillion from pensions and savings, has left family balance sheets upside down, and threatens to put ten million homeowners out on the street.

Families have survived the ups and downs of economic booms and busts for a long time, but the fall-behind during the busts has gotten worse while the surge-ahead during the booms has stalled out. In the boom of the 1960s, for example, median family income jumped by 33% (adjusted for inflation). But the boom of the 2000s resulted in an almost-imperceptible 1.6% increase for the typical family. While Wall Street executives and others who owned lots of stock celebrated how good the recovery was for them, middle class families were left empty-handed.

The crisis facing the middle class started more than a generation ago. Even as productivity rose, the wages of the average fully-employed male have been flat since the 1970s.

But core expenses kept going up. By the early 2000s, families were spending twice as much (adjusted for inflation) on mortgages than they did a generation ago—for a house that was, on average, only ten percent bigger and 25 years older. They also had to pay twice as much to hang on to their health insurance.

To cope, millions of families put a second parent into the workforce. But higher housing and medical costs combined with new expenses for child care, the costs of a second car to get to work and higher taxes combined to squeeze families even harder. Even with two incomes, they tightened their belts. Families today spend less than they did a generation ago on food, clothing, furniture, appliances, and other flexible purchases—but it hasn't been enough to save them. Today's families have spent all their income, have spent all their savings, and have gone into debt to pay for college, to cover serious medical problems, and just to stay afloat a little while longer.

Through it all, families never asked for a handout from anyone, especially Washington. They were left to go on their own, working harder, squeezing nickels, and taking care of themselves. But their economic boats have been taking on water for years, and now the crisis has swamped millions of middle class families.

The contrast with the big banks could not be sharper. While the middle class has been caught in an economic vise, the financial industry that was supposed to serve them has prospered at their expense. Consumer banking—selling debt to middle class families—has been a gold mine. Boring banking has given way to creative banking, and the industry has generated tens of billions of dollars annually in fees made possible by deceptive and dangerous terms buried in the fine print of opaque, incomprehensible, and largely unregulated contracts.

And when various forms of this creative banking triggered economic crisis, the banks went to Washington for a handout. All the while, top executives kept their jobs and retained their bonuses. Even though the tax dollars that supported the bailout came largely from middle class families—from people already working hard to make ends meet—the beneficiaries of those tax dollars are now lobbying Congress to preserve the rules that had let those huge banks feast off the middle class.

Pundits talk about "populist rage" as a way to trivialize the anger and fear coursing through the middle class. But they have it wrong. Families understand with crystalline clarity that the rules they have played by are not the same rules that govern Wall Street. They understand that no American family is "too big to fail." They recognize that business models have shifted and that big banks are pulling out all the stops to squeeze families and boost revenues. They understand that their economic security is under assault and that leaving consumer debt effectively unregulated does not work.

Families are ready for change. According to polls, large majorities of Americans have welcomed the Obama Administration's proposal for a new Consumer Financial Protection Agency (CFPA). The CFPA would be answerable to consumers—not to banks and not to Wall Street. The agency would have the power to end tricks-and-traps pricing and to start leveling the playing field so that consumers have the tools they need to compare prices and manage their money. The response of the big banks has been to swing into action against the agency, fighting with all their lobbying might to keep business-as-usual. They are pulling out all the stops to kill the agency before it is born. And if those practices crush millions more families, who cares—so long as the profits stay high and the bonuses keep coming.

America today has plenty of rich and super-rich. But it has far more families who did all the right things, but who still have no real security. Going to college and finding a good job no longer guarantee economic safety. Paying for a child's education and setting aside enough for a decent retirement have become distant dreams. Tens of millions of once-secure middle class families now live paycheck to paycheck, watching as their debts pile up and worrying about whether a pink slip or a bad diagnosis will send them hurtling over an economic cliff.

America without a strong middle class? Unthinkable, but the once-solid foundation is shaking.

KEY CONCEPTS

American dream	economic crisis	social stratification
class consciousness	middle class	

DISCUSSION QUESTIONS

1. How does Warren's argument illustrate the problems of an increasing class divide?

2. What are the social forces that are threatening the status of middle-class families in the United States?

26

Making It by Faking It:

Working-Class Students in an Elite Academic Environment

ROBERT GRANFIELD

Robert Granfield uses Goffman's concept of stigma *to talk about the experience of working-class students in elite academic environments. In doing so, he illustrates that class is not just an economic status, but has strong social components as well.*

Research on stigma has generated significant insights into the complex relationship between self and society. The legacy of Goffman's (1963) seminal work on the subject can be found in studies on alcoholism, mental illness, homosexuality, physical deformities, and juvenile delinquency. Even the literature on gender and racial inequality has benefited from an emphasis on stigma.

... While women and minorities have been examined from the perspective of stigma (Schur 1984), little attention has been directed toward social class, despite Goffman's suggestion that lower social class was a potential stigma (p. 145). Individuals from the lower social classes often experience real or perceived devaluation and react in ways that are characteristic of stigma management. This is particularly evident within asymmetrical class interactions; that is, interactions that cross social class boundaries. Research on working-class youths suggest that the devaluation they feel as well as their related identity adjustments resemble the behavioral patterns associated with other stigmatized groups.

In this article, I focus on class stigma by examining a group of highly successful, upwardly mobile, working-class students who gained admission to a prestigious Ivy League law school in the East. While upward mobility from the working class occurs far less often within elite branches of the legal profession (Smigel 1969; Heinz and Laumann 1982) or corporate management (Useem and Karabel 1986), a certain amount of this type of mobility does take place. Working-class aspirants to the social elite, however, must accumulate cultural capital (Bourdieu and Passeron 1990; Cookson and Persell 1985) before they are able to transcend their status boundaries.

SOURCE: KL: Making it by Faking it: Working-Class Students in an Elite Academic Environment." *Journal of Contemporary Ethnography* 20: 331–351.

First, this article examines the ways in which working-class students experience a sense of differentness and marginality within the law school's elite environment. Next, I explore how these students react to their emerging class stigma by managing information about their backgrounds. I then demonstrate that these management strategies contribute to identity ambivalence and consider the secondary forms of adjustment students use to resolve this tension. Finally, I discuss why an analysis of social class can benefit from the insights forged by Goffman's work on stigma.

SETTING AND METHODOLOGY

The data analyzed for this article were collected as part of a much larger project associated with law school socialization (Granfield 1989). The subjects consist of students attending a prestigious, national law school in the eastern part of the United States. The school has had a long reputation of training lawyers who have become partners in major Wall Street law firms, Supreme Court judges, United States presidents and other politicians, heads of foundations, and an array of other eminent leadership positions. Throughout the school's history, it has drawn mostly on the talents of high-status males. It was not until the second half of the twentieth century that women, minorities, and members of the lower classes were allowed admission into this esteemed institution (Abel 1989).

Most of the students attending the university at the time the study was being conducted were White and middle class. The overwhelming majority are the sons and daughters of the professional-managerial class. Over 70% of those returning questionnaires had Ivy League or other highly prestigious educational credentials. As one would expect, fewer working-class students possessed such credentials.

A triangulated research design (Fielding and Fielding 1986) was used to collect the data. The first phase consisted of extensive fieldwork at the law school from 1985 to 1988, during which time I became a "peripheral member" (Adler and Adler 1987) in selected student groups. My activities while in the field consisted of attending classes with students, participating in their Moot Court preparations, studying with students on campus, and at times, in their apartments, lunching with them, becoming involved in student demonstrations over job recruiting and faculty hiring, attending extracurricular lectures presented on campus, and participating in orientation exercises for first-year students. Throughout the entire fieldwork phase, I assumed both overt and covert roles. During the observation periods in classrooms, I recorded teacher–student interactions that occurred.

To supplement these observations, I conducted in-depth interviews with 103 law students at various stages in their training. Both personal interviews and small-group interviews with three or four students were recorded. The interviews lasted approximately 2 hours each and sought to identify the lived process through which law students experience legal training.

Finally, I administered a survey to 50% of the 1,540 students attending the law school. The survey examined their backgrounds, motives for attending law school, subjective perceptions of personal change, expectations about future practice, and evaluations of various substantive areas of practice. Over half (391) of the questionnaires were returned—a high rate of response for a survey of six pages requiring approximately 30 minutes of the respondent's time.

For this article, a subset of working-class students was selected for extensive analysis. Of the 103 students interviewed for the larger study, 23 came from working-class backgrounds, none of these from either the labor aristocracy or the unstable sectors of the working class. Typical parental occupations include postal worker, house painter, factory worker, fireman, dock worker, and carpenter. Many of these students were interviewed several times during their law school career. Many of the students selected for interviews were identified through questionnaires, while others were selected through the process of snowball sampling (Chadwick, Bahr, and Albrecht 1984).

FEELING OUT OF PLACE

Working-class students entered this elite educational institution with a great deal of class pride. This sense of class pride is reflected in the fact that a significantly larger proportion of working-class students reported entering law school for the purposes of contributing to social change than their non-working-class counterparts (see Granfield and Koenig 1990). That these students entered law school with the desire to help the downtrodden suggests that they identified with their working-class kin. In fact, students often credited their class background as being a motivating factor in their decision to pursue a career in social justice. One third-year student, whose father worked as a postal worker, recalled her parental influence:

> I wanted a career in social justice. It seemed to me to be a good value for someone who wanted to leave this world a little better than they found it. My parents raised me with a sense that there are right things and wrong things and that maybe you ought to try to do some right things with your life.

... However, identification with the working class began to diminish soon after these students entered law school. Not long after arriving, most working-class law students encountered an entirely new moral career. Although initially proud of their accomplishments, they soon came to define themselves as different and their backgrounds a burden. Lacking the appropriate cultural capital (Bourdieu 1984) associated with their more privileged counterparts, working-class students began to experience a crisis in competency. Phrases such as "the first semester makes you feel extremely incompetent," "the first year is like eating humble pie," and "I felt very small, powerless and dumb" were almost universal among these working-class students. Some students felt embarrassed by their difficulty in using the elaborated speech codes (Bernstein 1977) associated with the middle class. One working-class woman said that she was very aware of using "proper" English, adding that "it makes me self-conscious when I use the wrong word or tense. I feel that if I had grown up in the middle class, I wouldn't have lapses. I have difficulty expressing thoughts while most other people here don't."

The recognition of their apparent differentness is perhaps best noted by examining the students' perception of stress associated with the first year of studies. Incoming working-class students reported significantly higher levels of personal stress than did their counterparts with more elite backgrounds. Much of this anxiety came from fears of academic inadequacy. Despite generally excellent college grades and their success in gaining admission to a nationally ranked law school, these students often worried that they did not measure up to the school's high standards. Nearly 62% of the first-year working-class students reported experiencing excessive grade pressure, compared to only 35% of those students from higher social class backgrounds.

In the words of Sennett and Cobb (1973) this lack of confidence is a "hidden injury of class," a psychological burden that working-class students experienced as they came to acquire the "identity beliefs" associated with middle-class society. While most students experience some degree of uncertainty and competency crisis during their first year, working-class students face the additional pressure of being cultural outsiders. Lacking manners of speech, attire, values, and experiences associated with their more privileged counterparts, even the most capable working-class student felt out of place:

> I had a real problem my first year because law and legal education are based on upper-middle class values. The class debates had to do with profit maximization, law and economics, atomistic individualism. I remember in class we were talking about landlords' responsibility to maintain decent housing in rental apartments. Some people were saying that there were good reasons not to do this. Well, I think that's bullshit because I grew up with people who lived in apartments with rats, leaks, and roaches. I feel really different because I didn't grow up in suburbia.

… Such experiences contributed to a student's sense of living in an alien world. The social distance these students experienced early in their law school career produced considerable discomfort.

This discomfort grew more intense as they became increasingly immersed into this new elite world.

… FAKING IT

… Initially, students who took pride in having accomplished upward mobility openly displayed a working-class presentation of self. Many went out of their way to maintain this presentation. One first-year student who grew up in a labor union family in New York explained that "I have consciously maintained my working class image. I wear work shirts or old flannel shirts and blue jeans every day." During his first year, this student flaunted his working-class background, frequently also donning an old army jacket, hiking boots, and a wool hat. Identifying himself as part of the "proletarian left," he tried to remain isolated from what he referred to as the "elitist" law school community.

This attempt to remain situated in the working class, however, not only separated these students from the entire law school community but alienated them from groups that shared their ideological convictions. While much of the clothing worn by non-working class law students suggests resistance to being identified as a member of the elite, working-class students become increasingly aware of their differentness. Although these students identify with the working class, others, despite their appearance, possess traits and life-styles that are often associated with more privileged groups (see Lurie 1983, Stone 1970). One first-year woman who described herself as "radical" complained that the other law school radicals were really "a bunch of upper-class White men." Subsequently, working-class students must disengage from their backgrounds if they desire to escape feeling discredited.

Working-class students disengaged from their previous identity by concealing their class backgrounds. Just as deviants seek to manage their identity by "passing" as nondeviants (Goffman 1963), these working-class law students often adopted identities that were associated with the more elite social classes. Concealment allowed students to better participate in the culture of eminence that exists within the law, and reap available rewards.

... AMBIVALENCE

Despite their maneuvers, these working-class students had difficulty transcending their previous identity. The attempt by these students to manage their stigma resulted in what Goffman (1963) termed "identity ambivalence" (p. 107). Working-class students who sought to exit their class background could neither embrace their group nor let it go. This ambivalence is often felt by working-class individuals who attain upward mobility into the professional-managerial class (Steinitz and Solomon 1986). Many experience the "stranger in paradise" syndrome, in which working-class individuals feel like virtual outsiders in middle-class occupations (Ryan and Sackrey 1984). Such experiences frequently lead to considerable identity conflict among working-class individuals who attempt to align themselves with the middle class.

The working-class law students in my sample typically experienced identity conflicts on their upward climb. Not only did they feel deceptive in their adjustment strategies, but many felt the additional burden of believing they had "sold out" their own class and were letting their group down. Like other stigmatized individuals who gain acceptance among dominant groups (Goffman 1963), these students often felt they were letting down their own group by representing elite interests. One third-year female student ruefully explained:

> My brother keeps asking me whether I'm a Republican yet. He thought
> that after I finished law school I would go to work to help people, not

work for one of those firms that do business. In a way, he's my conscience. Maybe he's right. I've got a conflict with what I'm doing. I
came from the working class and wanted to do public interest law. I
have decided not to do that. It's been a difficult decision for me. I'm not
completely comfortable about working at a large firm.

Another student, who grew up on welfare, expressed similar reservations
about his impending career in law:

I'm not real happy about going to a large firm. I make lots of apologies.
I'm still upset about me fact that my clients are real wealthy people, and
it's not clear as to what the social utility of that will be.

Like the previous example, this student experienced a form of self-alienation
as a result of his identity ambivalence. Students often experience a sense of guilt
as they transcend their working-class backgrounds. Such guilt, however, needs to
be abated if these students are to successfully adjust to their new reference group
and reduce the status conflict they experience. For these working-class students,
making the primary adjustment to upward mobility required strategies of accommodation in personal attitudes regarding their relationship to members of less
privileged social classes. Secondary identity adjustments were therefore critical
in helping students mitigate the ambivalence they experienced over their own
success and subsequent separation from the working class.

CONCLUSION

This article has demonstrated that a focus on stigma holds considerable promise
in analyzing social class relations and particularly the difficulties that upwardly
mobile working-class youths face as they ascend the status hierarchy.
    … The recognition that social class is an "experienced" and constructed reality
offers insights into responses to stigma. Over the years, various stigmatized groups
have directly combated attempts to relegate them to a secondary status within society. Certainly, women, racial minorities, gays, and those with disabilities have
fought against the unjust system of devaluation which restricted their opportunities, reduced their humanity, and forced them to make adjustments, such as cov-
ering, passing, and careful disclosure, for the benefit of dominant groups. The
willingness of these groups to confront critical social typifications came directly
from their growing realization of the arbitrariness of such evaluations.
    However, in regard to social class, the ideology of meritocracy serves to le-
gitimate devaluation of the lower classes. Because social class position is frequently seen as the outcome of individual talent and effort, the assignment of
stigma to lower socioeconomic groups is not seen as being based on arbitrary
evaluation. Given the legitimacy of the meritocratic ideology, is it any wonder
that upwardly mobile working-class students choose not to directly confront the
devaluation they experience but rather to forge a new identity which effectively
divorces them from the working class? It is not surprising to find, for example,
that the movements to reform law and make it more accessible to persons of

lower economic status emanated not from working-class intellectuals but from elites who were sympathetic to their plight (Katz 1984).

Upwardly mobile working-class students in this study, as well as in others, interpret and experience their social class from the perspective of stigma. However, since the stigma of being a member of the lower classes is thought to be just, upwardly mobile working-class students frequently construct identities in which they seek to escape the taint associated with their affiliation. Overcoming this stigma is therefore considered an individual rather than a collective effort. As was demonstrated in this study, such efforts often involve managing in one's identity in the ways that Goffman outlined.

REFERENCES

Abel, R. 1989. *American lawyers*. New York: Oxford University Press.

Adler, P., and P. Adler. 1987. *Membership roles in field research*. Newbury Park, CA: Sage.

Bernstein, B. 1977. *Class codes and control*, vol. 3: *Towards a theory of educational transmission*. London: Routledge & Kegan Paul.

Bourdieu. P. 1984. *Distinction: A social critique of the judgment of taste*. Cambridge, MA: Harvard University Press.

Bourdieu, P., and J. C. Passeron. 1990. *Reproduction in education, society and culture*. London: Routledge & Kegan Paul.

Chadwick, B., H. Bahr, and S. Albrecht. 1984. *Social science research methods*. Englewood Cliffs, NJ: Prentice-Hall.

Cookson, P., and C. Persell. 1985. *Preparing for power: American's elite boarding schools*. New York: Basic Books.

Fielding, N. and J. Fielding. 1986. *Linking data*. Beverly Hills, CA: Sage.

Goffman, E. 1963. *Stigma: Notes on the management of spoiled identity*. Englewood Cliffs, NJ: Prentice-Hall.

Granfield, R. 1989. Making the elite lawyer: Culture and ideology in legal education. Ph.d. diss., Northeastern University, Boston.

Granfield, R., and T. Koenig. 1990. From activism to pro bone: The redirection of working class altruism at Harvard Law School. *Critical Sociology* 17: 57–80.

Heinz, J., and E. Laumann. 1982. *Chicago lawyers: The social structure of the bar*. New York: Russell Sage.

Katz, J. 1984. *Poor people's lawyers in transition*. New Brunswick, NJ: Rutgers University Press.

Lurie, A. 1983. *The language of clothes*. New York: Vintage.

Ryan, J., and C. Sackrey. 1984. *Strangers in paradise: Academics from the working class*. Boston: South End Press.

Schur, E. 1984. *Labeling women deviant: Gender, stigma, and social control*. New York: Random House.

Sennett, R., and R. Cobb. 1973. *The hidden injuries of class*. New York: Random House.

Smigel, E. 1969. *The Wall Street lawyer*. Bloomington: Indiana University Press.

Steinitz, V., and E. Solomon. 1986. *Starting out: Class and community in the lives of working class youth*. Philadelphia: Temple University Press.

Stone, G. 1970. Appearance and the self. In *Social psychology through symbolic interaction*, edited by G. Stone and H. Farberman, 394–414. New York: Wiley.

Useem, M., and J. Karabel. 1986. Paths to corporate management. *American Sociological Review* 51: 184–200.

KEY CONCEPTS

identity management	stigma	working class
prestige		

DISCUSSION QUESTIONS

1. What is meant by "the hidden injuries of class," and how are they shown by Granfield's research?

2. Granfield's research was done with working-class students in an elite law school. What might you find if you studied the experience of working-class students in your school?

27

The Great American Recession
Sociological Insights on Blame and Pain

JUDITH TREAS

Judith Treas provides a sociological analysis of the impact of the recent economic recession on American families. She criticizes the tendency for people to use individualistic explanations of this crisis, rather than examining social structural causes.

SOURCE: Treas, Judith. 2010. "The Great American Recession: Sociological Insights on Blame and Pain." *Sociological Perspectives*, 53: 13–17.

In the fall of 2008, the shifting foundations of inequality—long the focus of sociological study—became a social, political, and economic earthquake of enormous magnitude. For sociologists, The Big One, a global economic melt-down, had finally arrived. We were riveted by the most serious economic calamity of our lifetimes—a worldwide economic pandemic, financial institutions in seeming freefall, and the crushed dreams of millions of Americans. As the crisis passed, the outlines of an explanation for these frightening developments began to emerge, but it offered little comfort to the many individuals and families who were the casualties of this economic debacle.

Sociologists will be writing about these historic times for decades to come, and our theoretical frameworks and empirical perspectives will offer an invaluable counterpoint to the interpretations of journalists, politicians, economists, and the Federal Reserve. One place for us to begin is with the lives of ordinary Americans who have been struggling. The focus is on the casualties of the recession on Main Street and Elm Street, not Wall Street. Although this article has little to say about investment bankers brought low, recent events make it clear that there is also a need for sociologists to "study up," to investigate the elites who have a disproportionate influence on our lives.

In this article, my objective is to illustrate how basic sociological concepts and data speak to the lived experience of Americans who struggled to pay their bills, went without needed medical care, or lost their homes in the recession. First, I invite an analysis of one set of popular and political narratives that were invoked in the recession, explanations for our economic difficulties that sociologists will recognize as "blaming the victim." Second, I refute many of the assumptions in these blame game narratives by showing how money problems affected the lives of Americans, unleashing a cascade of hardships. This analysis draws on 1991 and 2004 General Social Survey (GSS) data, which underscore the serious systemic vulnerabilities that existed for Americans even *before* the current recession.

BLAMING THE VICTIM

Anybody who has read an introductory sociology textbook has probably been captivated by that intuitive paradigm of injustice, the process of "blaming the victim." According to this perspective, we deflect attention from the real—usually structural—causes of misery by pinning responsibility on those individuals who suffer the problem. Because Americans favor individual explanations over structural ones (Huber and Form 1973; Kluegel and Smith 1986), we gravitate to victim blaming. Many popular analyses of the nation's financial troubles are also based on blaming the victim.

Consider, for example, the accusation that U.S. automakers were pushed to the brink of bankruptcy by the wage demands of their workers. The *Wall Street Journal* encapsulated this view when it published a reader's letter under the peculiar headline "UAW Sociology May Bring Down Detroit" (Maisonneuve 2009). When a $14 billion emergency bailout for U.S. automakers collapsed in the Senate in December 2008, CBS News (2008) reported the view that the United Auto Workers were to blame for not accepting Republican demands for wage cuts.

Although management came in for its share of criticism (Leonhardt 2008), the auto workers hard hit by the recession were faulted for their success in negotiating decent wages and good benefits in better times. Typically unreported were the painful sacrifices that workers had already made to keep a flailing industry afloat.

Another big blame game pins responsibility for the nation's economic woes on individual borrowers. With the subprime mortgage crisis, the defaulting homebuyer came in for contempt from some quarters. Even an industry analysis, however, concludes that "broker-facilitated fraud represents the most serious mortgage fraud risk to lenders" (BasePoint 2007). Homebuyers needed the collusion of brokers and appraisers to fib on their loan applications (Farrell 2008). With an eye on their own commissions and market share, brokers did not turn down risky loan applications. Instead, they seduced naïve borrowers with shady lending practices and winked at overvalued houses and inflated incomes (Schmidt and Tamman 2009).

Of course, whether they blame the borrower or the lender, these individualistic accounts serve mostly to obscure the real structural causes of the subprime problem. Sensible government regulation could have kept both borrowers and lenders from the practices that were unsustainable. For instance, because financial organizations that loaned money could quickly sell the mortgages to others, they had no incentive to worry about the financial viability of borrowers. Regulations could have set limits on the investment banks that bundled sub-prime mortgages, the rating companies that pronounced the derivatives to be safe investments, and the universities, pension funds, too-big-to-fail financial organizations, and even nations that bought the now toxic debt. They could have reined in the adjustable (and sometimes deceptive) mortgages with low teaser rates, which well-meaning borrowers found themselves unable to pay or refinance in a recession.

Of course, research shows that major setbacks, not opportunism, are what lead to bankruptcy court. In 2001, about two-thirds of bankrupt families had suffered a loss of income due to a business failure, a stint of unemployment, etc. (Sullivan, Warren, and Westbrook 2006). Behind half of all bankruptcy filings in 2007 were medical expenses (Giron 2009; Himmelstein et al. 2005). Nonetheless, blaming the victim allowed banks and credit card companies to rally support for their lender-friendly, bankruptcy legislation. Setting the stage for current economic problems, this legislation made it much harder for struggling Americans to get out of trouble by getting their debts wiped out in court.

RUNNING WILD OR MAKING ENDS MEET

Any clear-eyed assessment of family economic problems has to contend with the fact that we live in a consumer culture. The status competition that Veblen identified in the consumption of the 19th century "leisure class" has permeated American society.

Sophisticated marketing and omnipresent media have introduced everyone to the allure of pricy products and luxury labels. How many women even knew there were $500 shoes before *Sex and the City* introduced them to Manolo Blahnik's 5-inch heels? The ready availability of credit permitted Americans to indulge—with predictable hits to their debt levels and savings rates. Americans received six *billion* credit card solicitations in 2005 (Garcia 2007). On our college campuses, credit card pitches have become as much a part of freshman orientation as free pizza. Bombarded with this assurance of their credit-worthiness, is there any wonder that Americans succumb to borrowing? The industry they are in league with, however, has evolved to exploit them. Faced with inflation in 1980, Congress enacted legislation that took the teeth out of state usury laws that had protected consumers from extraordinary interest rates (Manning 2000). Meanwhile, the average credit card contract ballooned from a single page to more than thirty pages of complex technical provisions too deviously worded for a Harvard law professor to understand (Herbert 2009). How many new credit card holders understand enough of the fine print to realize that the industry's business model courted poor credit risks in order to profit from ballooning balances, late fees, and punitive rate hikes?

Other analyses emphasize that the downfall of Americans has not been keeping up with the fictional free-spending Carrie Bradshaw, but just making ends meet. They note that the escalating cost of necessities such as housing and health care—not a taste for upscale goods—has weighed heavily on personal finances (Warren 2007). At one time, people used credit to front-load the purchase of a big ticket item (say, a washing machine) that they could not afford right away. Grocery stores and gas stations did not accept American Express. Some Americans held out. They had to capitulate when a credit card became a necessity for anyone who needed to buy an airline ticket on the Internet. In fact, people who pay cash for airline tickets today run the risk of being profiled as terrorists. Americans started to use credit cards to pay for discretionary purchases as well as necessities. Aggressively marketed plastic became a way to cope with economic adversity and to smooth out consumption when income fell. And, of course, borrowing became a widely acceptable means for young people to invest in a better future. More students graduated from college with daunting debts, including unsubsidized student loans, which swelled due to unpaid interest while the borrower was in school (Chaker 2009).

Victim blamers argue that people with credit problems have ignored the inconvenient truth about their budgetary limits. While the siren call of big-screen TVs and other luxury goods may be the downfall of some Americans, others use credit to buy the children's school clothes, pay the utilities, and maintain some semblance of a decent living standard when they are out of work.

Sociologists trace the origins of these crises to the shifting foundations of inequality. A neo-liberal philosophy of market fundamentalism led the government to trim the public safety net that workers counted on for protection against poverty, unemployment, disability, and old age (Somers and Block 2005). These cutbacks came at a time when businesses—citing the need to compete in a global economy—were outsourcing jobs to other countries and reducing pay and benefits for U.S. workers who remained. More Americans found themselves in non-standard employment—part-time jobs, temporary work, and contract labor that offered less security, lower wages, and fewer benefits (Kalleberg, Reskin, and Hudson 2000). The percent of college graduates who thought it likely they would lose their job in the next year drifted upwards from 1977 to 2004 (Jacobs and Newman 2008). The economic crisis came early for some people and pre-dates the latest recession, which the National Bureau of Economic Research dates to the end of 2007.

CONCLUSION

The recession stands as an especially troubling episode in the history of this na-tion. For many Americans, the recession is a bewildering tragedy that has pushed them to the brink of economic despair and tested their faith in our social institu-tions. Against this backdrop of confusion and uncertainty, it is important to point out how much the discipline of sociology has had to offer in times like these. As even some economists have acknowledged, models that stress precise estimation over realistic assumptions and adequate data have clearly failed (Krugman 2009). So have the optimistic theories that assume a sustainable societal equilibrium growing out of individual self-interest.

Sociological perspectives and research have been vindicated by the events of the recession. It is sociologists who saw social networks and social structures rather than atomized actors. It is sociologists who pulled back the curtain to show how social pressures lead to overvalued stocks (Davis 2003) and how insiders loot banks (Calavita, Pontell, and Tillman 1997). It is sociologists who documented the con-sequences for ordinary Americans of free market failures and unjust social policies (Seccombe and Hoffman 2007). Our commitment to the disadvantaged might lead some people to conclude that sociology is the true dismal science. Nonethe-less, sociologists have offered an uplifting message by showing how collective action can right societal wrongs. On this point, it is important to note that narra-tives blaming the victims for our current economic crisis coexist with narratives that place responsibility on elites. Targeting bankers, lobbyists, mortgage brokers, politicians, and economists, this angry populist pushback embodies a withering sociological critique of power and a demand for accountability and social justice.

Drawing on sociological theory and sociological data has allowed me to demonstrate our sociological understanding of the recession in the lives of Amer-icans. Blaming the victim alerts us to the stigmatizing narratives that can add to the suffering of those who are selling off their prized possessions, foregoing needed health care, and losing their homes. While there are undoubtedly un-scrupulous borrowers who cheated a willing system, it is unfair to tar the victims of fraudulent brokers and a souring economy. Individualistic explanations, whether they blame the homeowner or the broker, divert attention from the structural reasons for the subprime mortgage crisis and the rising wave of delin-quencies and bankruptcies. Succumbing to an ideology of free markets and minimal government, Americans lacked commonsense regulatory controls. We also lacked the will to address rising income inequality and economic insecurity.

The growing troubles families faced getting health insurance or paying bills were unheeded warnings of the cascading hardships that Americans encountered when the economy turned south.

REFERENCES

BasePoint. 2007. "Broker-Facilitated Fraud—The Impact on Lenders." *Mortgage Solutions White Paper*. Retrieved October 20, 2009 from http://counterecon.files.wordpress.com/2008/11/white-paper-broker-facilitated-mortgage-fraud-2007.pdf.

Calavita, Kitty, Henry N. Pontell, and Robert Tillman. 1997. *Big Money Crime: Fraud and Politics in the Savings and Loan Crisis*. Berkeley: University of California Press.

Chaker, Anne-Marie. 2009. "Students Borrow More Than Ever for College." *Wall Street Journal*, September 3, pp. D1, D10.

Davis, Gerald F. 2003. "American Cronyism: How Executive Networks Inflated the Corporate Bubble." *Contexts* 2(3):34–40.

Farrell, Greg. 2008. "Las Vegas Called 'Mortgage Fraud Ground Zero.'" *U.S.A. Today*, June 3. Retrieved January 25, 2009 from http://www.usatoday.com/money/economy/housing/2008-06-02-mortage-fraud-las-vegas_N.htm.

Garcia, Jose A. 2007. *Borrowing to Make Ends Meet: The Rapid Growth of Credit Card Debt in America*. New York: Demos.

Giron, Lisa. 2009. "Medical Bills Tied to More Bankruptcies." *Los Angeles Times*, June 4, pp. B1.

Himmelstein, David U., Elizabeth Warren, Deborah Thorne, and Steffie Woolhandler. 2005. "MarketWatch: Illness and Injury as Contributors to Bankruptcy." *Health Affairs* 25:w84–w88.

Huber, Joan and William Form. 1973. *Income and Ideology*. New York: Free Press.

Jacobs, Elizabeth and Katherine Newman. 2008. "Rising Angst? Change and Stability in Perceptions of Instability." Pp. 74–101 in *Laid Off, Laid Low: Political and Economic Consequences of Employment Insecurity*, by K. S. Newman. New York: Columbia/SSRC.

Kalleberg, Arne L., Barbara F. Reskin, and Ken Hudson. 2000. "Bad Jobs in America: Standard and Nonstandard Employment Relations and Job Quality in the United States." *American Sociological Review* 65:256–78.

Kluegel, James R. and Eliot R. Smith. 1986. *Beliefs about Inequality*. New York: Aldine de Gruyter.

Krugman. Paul. 2009. "How Did Economists Get It So Wrong?" *New York Times Magazine*, September 2. Retrieved October 2, 2009 from http://www.nytimes.com/2009/09/06/magazine/06Economic-t.html.

Leonhardt, David. 2008. "$73 an Hour: Adding It Up." *New York Times*, December 10. Retrieved September 19, 2009 from http://www.nytimes.com/2008/12/10/business/worldbusiness/10iht-10leonhardt.18542483.html.

Maisonneuve, Mark. 2009. "UAW Sociology May Bring Down Detroit." *Wall Street Journal*, January 3, p. A8.

Manning, Robert D. 2000. *Credit Card Nation*. New York: Basic Books.

Schmidt, Susan and Maurice Tamman. 2009. "Housing Push for Hispanics Spawns Wave of Foreclosures." *Wall Street Journal*, January 5. p. A1.

Seccombe, Karen and Kim Hoffman. 2007. *Just Don't Get Sick: Access to Health Care in the Aftermath of Welfare Reform*. New Brunswick, NJ: Rutgers University Press.

Somers, Margaret R. and Fred Block. 2005. "From Poverty to Perversity: Ideas, Markets, and Institutions over 200 Years of Welfare Debate." *American Sociological Review* 70(2):260–87.

Sullivan, Teresa A., Elizabeth Warren, and Jay Lawrence Westbrook. 2006. "Less Stigma or More Financial Distress: An Empirical Analysis of the Extraordinary Increase in Bankruptcy Filings." *Stanford Law Review* 59(2):213–55.

Warren, Elizabeth. 2007. "The Vanishing Middle Class." Pp. 38–52 in *Ending Poverty in America: How to Restore the American Dream*, edited by J. Edwards, M. Crain, and A. L. Kalleberg. New York: The New Press.

KEY CONCEPTS

blaming the victim	neoliberalism	redlining
labor market	recession	

DISCUSSION QUESTIONS

1. How has "blaming the victim" been pervasive in common-sense accounts of the economic recession? In what ways does this reveal the American tendency toward individualism?

2. What are the social structural factors that Treas identifies as leading to the recent economic crisis?

Applying Sociological Knowledge: An Exercise for Students

Brand labels in a class-based society communicate our status to others. Make a list of all of the clothing labels you can think of and then match each label to a ranking in the class system (working class, middle class, upper-middle class, etc.). What class images are projected by each? Are there class stereotypes suggested by different labels? Whom do they affect and how? In what ways do these labels reproduce our class identities? Do they do any harm?

✳

PART IX

Global Stratification

28

Myth of the Global Safety Net

JAN BREMAN

Jan Breman examines the impact of global changes in the economy on the world's most vulnerable people—those who survive through an informal economy of low-wage work. This field research in India shows how the global economic crisis is affecting the lives of those already disadvantaged by working in some of the most devalued labor markets.

Media reports on the economic meltdown have mainly concentrated on the impact of the crisis on the rich nations, with little concern for the mass of the population living in what used to be called the Third World. The current view seems to be that the setbacks in these 'emerging economies' may be less severe than expected. China's and India's high growth rates have slackened, but the predicted slump has not materialized. This line of thought, however, analyzes only the effects of the crisis on countries as a whole, masking its differential impact across social classes. If one considers income distribution, and not just macro-calculations of GDP, the global downturn has taken a disproportionately higher toll on the most vulnerable sectors: the huge armies of the poorly paid, under-educated, resourceless workers that constitute the overcrowded lower depths of the world economy.

To the extent that these many hundreds of millions are incorporated into the production process, it is as informal labour, characterized by casualized and fluctuating employment and piece-rates, whether working at home, in sweatshops, or on their own account in the open air; and in the absence of any contractual or labour rights, or collective organization. In a haphazard fashion, still little understood, work of this nature has come to predominate within the global labour force at large. The International Labour Organization estimates that informal workers comprise over half the workforce in Latin America, over 70 per cent in Sub-Saharan Africa and over 80 per cent in India; an Indian government report suggests a figure of more than 90 per cent.[1] Cut loose from their original social moorings, the majority remain stuck in the vast shanty towns ringing city outskirts across the global South.

Recently, however, the life of street hawkers in Cairo, tortilla vendors in Mexico City, rickshaw drivers in Calcutta or scrap mongers in Jakarta has been

SOURCE: Breman, Jan. 2009. "Myth of the Global Safety Net." *New Left Review* 59: 29–36.

cast in a much rosier light. The informal sector, according to the *Wall Street Journal*, is 'one of the last safe havens in a darkening financial climate' and 'a critical safety net as the economic crisis spreads'.[2] Thanks to these jobs, former IMF Chief Economist Simon Johnson is quoted as saying, 'the situation in desperately poor countries isn't as bad as you'd think'. On this view, an admirable spirit of self-reliance enables people to survive in the underground circuits of the economy, unencumbered by the tax and benefit systems of the 'formal sector'. These streetwise operators are able to get by without expensive social provisions or unemployment benefit. World Bank economist W. F. Maloney assures the *WSJ* that the informal sector 'will absorb a lot of people and offer them a source of income' over the next year.

The *WSJ* draws its examples from Ahmedabad, the former mill city in Gujarat where I conducted fieldwork in the 1990s. Here, in the Manek Chowk market—'a row of derelict stalls', where 'vendors peddle everything from beans to brass pots as monkeys scramble overhead'–Surajben 'Babubhai' Patni sells tomatoes, corn and nuts from a makeshift shelter: 'She makes as much as 250 rupees a day, or about $5, but it's enough to feed her household of nine, including her son, who recently lost his job as a diamond polisher.' Enough: really? Five dollars for nine people is less than half the amount the World Bank sets as the benchmark above extreme poverty: one dollar per capita per day. Landless households in villages to the south of Ahmedabad have to make do with even less than that—on the days they manage to find work.[3]

Earlier this year I returned to the former mill districts of the city to see how the economic crisis was affecting people there. By 2000, these former working-class neighbourhoods had already degenerated into pauperized quarters. But the situation has deteriorated markedly even since then. Take the condition of the garbage pickers—all of them women, since this is not considered to be man's work. They are now paid half what they used to get for the harvest of paper, rags and plastic gleaned from the waste dumps on their daily rounds. To make up the loss, they now begin their work at 3am instead of at 5 am, bringing along their children to provide more hands. The Self-Employed Women's Association, which organizes informal-sector workers in the city, reports that 'incomes have declined, days of work decreased, prices have fallen and livelihoods disappeared'.[4]

A SEWA activist based in Ahmedabad reports on the anguish she met when visiting local members. One of these, Ranjanben Ashokbhai Parmar, started to cry: 'Who sent this recession! Why did they send it?'

> I was speechless. Her situation is very bad, her husband is sick, she has 5 children, she stays in a rented house, she has to spend on the treatment of her husband and she is the sole earner in the family, how can she meet her ends? When she goes to collect scrap she takes along her little daughter, while her husband sits at home and makes wooden ice-cream spoons, from which he can earn not more than 10 rupees a day.

In the industrial city of Surat, 120 miles south of Ahmedabad, half the informal labour force of the diamond workshops was laid off overnight at the end of 2008, with the collapse of worldwide demand for jewels. Some 200,000 diamond cutters and polishers found themselves jobless, while the rest had to contend with drastic reductions in hours and piece-rates. A wave of suicides swept the dismissed workers, who—with a monthly income of little more than $140—were reputed to belong to the most skilled and highest paid ranks of the informal economy. These bitter experiences of the recession-struck informal economy in Gujarat can be repeated for region after region across India, Africa and much of Latin America. Confronted with such misery it is impossible to concur with the World Bank's and *Wall Street Journal's* optimism about the sector's absorptive powers. As for their praise for the 'self-reliance' of those struggling to get by in these conditions: living in a state of constant emergency saps the energy to cope and erodes the strength to endure. To suggest that these workers constitute a 'vibrant' new class of self-employed entrepreneurs, ready to fight their way upward, is as misleading as portraying children from the *chawls* of Mumbai as slumdog millionaires.

RURAL ROPE'S END

The second option currently being touted by the Western media as a 'cushion for hard times' is a return to the countryside. As an Asian Development Bank official in Thailand recently informed the *International Herald Tribune*, 'returning to one's traditional village in the countryside is a sort of "social safety net"'. The complacent assumption is that large numbers of rural migrants made redundant in the cities can retreat to their families' farms and be absorbed in agricultural work, until they are recalled to their urban jobs by the next uptick of the economy. The *IHT* evokes a paradisial rural hinterland in northeast Thailand. Even in the dry season,

> there are still plenty of year-round crops—gourds, beans, coconuts and bananas among them—that thrive with little rainwater. Farmers raise chickens and cows, and dig fish ponds behind their homes…. Thailand's king, Bhumibol Adulyadej, has long encouraged such self-sufficiency.[5]

Similar views were published at the time of the Asian financial crisis in 1997. Then, World Bank consultants assumed that agriculture could act as a catchment reservoir for labour made redundant in other sectors, based on the notion that the army of migrants moving back and forth between the country and urban-growth poles had never ceased their primary occupation. The myth persisted that Southeast Asian countries were still essentially peasant societies. These tillers of the land might go to the city to earn extra wages for cash expenditure, but if they lost their jobs they were expected to reintegrate into the peasant economy with no difficulty. This was far from the case, as I wrote then.

Returning to the localities of my fieldwork in Java this summer, I listened to the latest stories of men and women who had come back to the village, having lost their informal-sector jobs elsewhere, and find no work here, either. Of course not: they were driven out of the village economy in the first place because of lack of land or other forms of capital. There is no family farm to fall back on. The departure of the landless and the land-poor was a flight, part of a coping strategy. Now that the members of this rural proletariat have become redundant in Jakarta or Bangkok, or as contract workers in Taiwan or Korea for that matter, they are back to square one, due to an acute and sustained lack of demand for their labour power in their place of origin. A comparable drama is taking place in China. Out of the 120 to 150 million migrants who made the trek from the rural interior to the rapidly growing coastal cities during the last twenty-five years, official sources report that about 10 to 15 million are now unemployed. For these victims of the new economy, there is no alternative but to go back 'home' to a deeply impoverished countryside.

The Asian village economy is not capable of accommodating all those who possess no means of production; nor has the urban informal sector the elasticity to absorb all those eager to drift into it. According to policymakers' notions of cross-sectoral mobility, the informal economy should swallow up the labour surplus pushed out of higher-paid jobs, enabling the displaced workforce to stick it out through income-sharing arrangements until the economic tide turned again. I have never found any evidence that such a horizontal drift has taken place. Street vendors do not turn into *becak* drivers, domestic servants or construction workers overnight. The labour market of the informal sector is highly fragmented; those who are laid off in their branch of activity have no alternative but to go back 'home', because staying on in the city without earnings is next to impossible. But returning to their place of origin is not a straightforward option, given the lack of space in the rural economy. Nevertheless, my informants do not simply lay the blame for their predicament on the economic meltdown. From the perspective of the world's underclasses, what looks like a conjunctural crisis is actually a structural one, the absence of regular and decent employment. The massive army of reserve labour at the bottom of the informal economy is entrapped in a permanent state of crisis which will not be lifted when the Dow Jones Index goes up again.

NEW ECONOMIC ORDER

The transformation that took place in nineteenth-century Western Europe, as land-poor and landless peasants migrated to the towns, is now being repeated on a truly global scale. But the restructuring that would create an industrial-urban order, of the sort which vastly improved the lot of the former peasants of the Northern hemisphere, has not materialized. The ex-peasants of the South have failed to find secure jobs and housing on their arrival in the cities. Struggling to gain a foothold there, they have become mired for successive generations in the deprivation of the shanties, a vast reserve army of informal labour.

In the 1960s and 70s, Western policymakers viewed the informal sector as a waiting room or temporary transit zone: newcomers could find their feet there and learn the ways of the urban labour market. Once savvy to these, they would increasingly be able to qualify for higher wages and more respectable working conditions. In fact the trend went in the opposite direction, due in large part to the onslaught of market-driven policies, the retreat of the state in the domain of employment and the decisive weakening of organized labour. The small fraction that made their way to the formal sector was now accused of being a labour aristocracy, selfishly laying claim to privileges of protection and security. At the same time, the informal sector began to be heralded by the World Bank and other transnational agencies as a motor of economic growth. Flexibilization became the order of the day—in other words, dismantling of job security and a crackdown on collective bargaining. The process of informalization that has taken shape over the last twenty years saw, among other things, the end of the large-scale textile industry in South Asia. In Ahmedabad itself, more than 150,000 mill workers were laid off at a stroke. This did not mean the end of textile production in the city. Cloth is now produced in power-loom workshops by operators who work twelve-hour days, instead of eight, and at less than half the wages they received in the mill; garment manufacture has become home-based work, in which the whole family is engaged day and night. The textile workers' union has all but disappeared. Sliding down the labour hierarchy has plunged these households into a permanent social and economic crisis.

It is not only that the cost of labour at the bottom of the world economy has been scaled down to the lowest possible level; fragmentation also keeps the under-employed masses internally compartmentalized. These people are competitors in a labour market in which the supply side is now structurally larger than the—constantly fluctuating—demand for labour power. They react to this disequilibrium by trying to strengthen their ties along lines of family, region, tribe, caste, religion, or other primordial identities which preclude collective bargaining on the basis of work status and occupation. Their vulnerability is exacerbated by their enforced rootlessness: they are pushed off the land, but then pushed back onto it again, roaming around in an endless search for work and shelter.

The emergence of the early welfare state in the Western hemisphere at the end of the nineteenth century has been attributed to the bourgeoisie's fear that the policy of excluding the lower ranks of society could end in the collapse of the established order.[6] The propertied part of mankind today does not seem to be frightened by the presence of a much more voluminous *classe dangereuse*. Their appropriation of ever-more wealth is the other side of the trend towards informalization, which has resulted in the growing imbalance between capital and labour. There are no signs of a change of direction in this economic course. Promises of poverty reduction by global leaders are mere lip service, or photo-opportunities. During his campaign, Obama would once in a while air his appreciation for Roosevelt's New Deal. Since his election the idea of a broad-based social-welfare scheme has been shelved without further ado. The global crisis is being tackled by a massive transfer of wealth from poor to rich. The logic suggests a return to nineteenth-century beliefs in the principle and practice of natural inequality. On this view, it is not poverty that needs to be eradicated. The problem is the poor people themselves, who lack the ability to pull themselves up out of their misery. Handicapped by all kinds of defects, they constitute a useless residue and an unnecessary burden. How to get rid of this ballast?

NOTES

1. 'Decent Work and the Informal Economy', International Labour Organization, Geneva 2002; *Report on the Conditions of Work and Promotion of Livelihoods in the Unorganised Sector*, National Commission for Enterprises in the Unorganised Sector, Government of India, New Delhi 2008.

2. Patrick Barta, 'The Rise of the Underground', *Wall Street Journal*, 14 March 2009.

3. Breman, *The Poverty Regime in Village India*, Delhi 2007.

4. Self-Employed Women's Association newsletter, *We the Self Employed*, no. 18, 15 May 2009. SEWA began organizing informal-sector workers in Ahmedabad in the 1970s, and has subsequently expanded its activities across India, and even beyond.

5. Thomas Fuller, 'In Southeast Asia, Unemployed Abandon Cities for Their Villages,' iht, 28 February 2009.

6. Abram de Swaan, *In Care of the State: Health Care, Education and Welfare in Europe and the USA in the Modern Era*, Cambridge 1988.

KEY CONCEPTS

extreme poverty	informal economy (or informal sector)	world economy

DISCUSSION QUESTIONS

1. Who are some of the informal sector workers whom Breman identifies? How do they try to put together a subsistence?

2. What has been the impact of global economic transformations on the lives of these informal workers?

29

The Nanny Chain

ARLIE RUSSELL HOCHSCHILD

Arlie Hochschild identifies the "nanny chain" as a global system of work in which women workers from poor nations provide the "care work" for more privileged workers in other parts of the world. This pattern of labor is transforming social relations of care worldwide and, according to Hochschild, makes care and love a commodity that is transferred and exchanged in the world market.

Vicky Diaz, a 34-year-old mother of five, was a college-educated schoolteacher and travel agent in the Philippines before migrating to the United States to work as a housekeeper for a wealthy Beverly Hills family and as a nanny for their two-year-old son. Her children, Vicky explained to Rhacel Parrenas,

> were saddened by my departure. Even until now my children are trying to convince me to go home. The children were not angry when I left because they were still very young when I left them. My husband could not get angry either because he knew that was the only way I could seriously help him raise our children, so that our children could be sent to school. I send them money every month.

In her book *Servants of Globalization*, Parrenas, an affiliate of the Center for Working Families at the University of California, Berkeley, tells an important and disquieting story of what she calls the "globalization of mothering." The Beverly Hills family pays "Vicky" (which is the pseudonym Parrenas gave her) $400 a week, and Vicky, in turn, pays her own family's live-in domestic worker back in the Philippines $40 a week. Living like this is not easy on Vicky and her family. "Even though it's paid well, you are sinking in the amount of your work. Even while you are ironing the clothes, they can still call you to the kitchen to wash the plates. It ... [is] also very depressing. The only thing you can do is give all your love to [the two-year-old American child]. In my absence from my children, the most I could do with my situation is give all my love to that child."

Vicky is part of what we could call a global care chain: a series of personal links between people across the globe based on the paid or unpaid work of caring. A typical global care chain might work something like this: An older

SOURCE: Arlie Russell Hochschild. 2000. "The Nanny Chain" *The American Prospect*, 3 (January 2000), pp. 33–36.

daughter from a poor family in a third world country cares for her siblings (the first link in the chain) while her mother works as a nanny caring for the children of a nanny migrating to a first world country (the second link) who, in turn, cares for the child of a family in a rich country (the final link). Each kind of chain expresses an invisible human ecology of care, one care worker depending on another and so on. A global care chain might start in a poor country and end in a rich one, or it might link rural and urban areas within the same poor country. More complex versions start in one poor country and extend to another slightly less poor country and then link to a rich country.

Global care chains may be proliferating. According to 1994 estimates by the International Organization for Migration, 120 million people migrated—legally or illegally—from one country to another. That's 2 percent of the world's population. How many migrants leave loved ones behind to care for other people's children or elderly parents, we don't know. But we do know that more than half of legal migrants to the United States are women, mostly between ages 25 and 34. And migration experts tell us that the proportion of women among migrants is likely to rise. All of this suggests that the trend toward global care chains will continue.

How are we to understand the impact of globalization on care? If, as globalization continues, more global care chains form, will they be "good" care chains or "bad" ones? Given the entrenched problem of third world poverty—which is one of the starting points for care chains—this is by no means a simple question. But we have yet to fully address it, I believe, because the world is globalizing faster than our minds or hearts are. We live global but still think and feel local.

FREUD IN A GLOBAL ECONOMY

Most writing on globalization focuses on money, markets, and labor flows, while giving scant attention to women, children, and the care of one for the other. Most research on women and development, meanwhile, draws a connection between, say, World Bank loan conditions and the scarcity of food for women and children in the third world, without saying much about resources expended on caregivmg. Much of the research on women in the United States and Europe focuses on a chainless, two-person picture of "work-family balance" without considering the child care worker and the emotional ecology of which he or she is a part. Fortunately, in recent years, scholars such as Ernestine Avila, Evelyn Nakano Glenn, Pierette Hondagneu-Sotelo, Mary Romero, and Rhacel Parrenas have produced some fascinating research on domestic workers. Building on this work, we can begin to focus on the first world end of the care chain and begin spelling out some of the implications of the globalization of love.

One difficulty in understanding these implications is that the language of economics does not translate easily into the language of psychology. How are we to understand a "transfer" of feeling from one link in a chain to another? Feeling is not a "resource" that can be crassly taken from one person and given to another. And surely one person can love quite a few people; love is not a resource limited the same way oil or currency supply is. Or is it?

Consider Sigmund Freud's theory of displacement, the idea that emotion can be redirected from one person or object to another. Freud believed that if, for example, Jane loves Dick but Dick is emotionally or literally unavailable, Jane will find a new object (say, John, Dick and Jane's son) onto which to project her original feeling for Dick. While Freud applied the idea of displacement mainly to relations within the nuclear family, the concept can also be applied to relations extending far outside it. For example, immigrant nannies and au pairs often divert feelings originally directed toward their own children toward their young charges in this country. As Sau-ling C. Wong, a researcher at the University of California, Berkeley, has put it, "Time and energy available for mothers are diverted from those who, by kinship or communal ties, are their more rightful recipients."

If it is true that attention, solicitude, and love itself can be "displaced" from one child (let's say Vicky Diaz's son Alfredo, back in the Philippines) onto another child (let's say Tommy, the son of her employers in Beverly Hills), then the important observation to make here is that this displacement is often upward in wealth and power. This, in turn, raises the question of the equitable distribution of care. It makes us wonder, is there—in the realm of love—an analogue to what Marx calls "surplus value," something skimmed off from the poor for the benefit of the rich?

Seen as a thing in itself, Vicky's love for the Beverly Hills toddler is unique, individual, private. But might there not be elements in this love that are borrowed, so to speak, from somewhere and someone else? Is time spent with the first world child in some sense "taken" from a child further down the care chain? Is the Beverly Hills child getting "surplus" love, the way immigrant farm workers give us surplus labor? Are first world countries such as the United States importing maternal love as they have imported copper, zinc, gold, and other ores from third world countries in the past?

This is a startling idea and an unwelcome one, both for Vicky Diaz, who needs the money from a first world job, and for her well-meaning employers, who want someone to give loving care to their child. Each link in the chain feels she is doing the right thing for good reasons—and who is to say she is not?

But there are clearly hidden costs here, costs that tend to get passed down along the chain. One nanny reported such a cost when she described (to Rhacel Parrenas) a return visit to the Philippines: "When I saw my children, I thought, 'Oh children do grow up even without their mother.' I left my youngest when she was only five years old. She was already nine when I saw her again but she still wanted for me to carry her [weeps]. That hurt me because it showed me that my children missed out on a lot."

Sometimes the toll it takes on the domestic worker is overwhelming and suggests that the nanny has not displaced her love onto an employer's child but rather has continued to long intensely for her own child. As one woman told Parrenas, "The first two years I felt like I was going crazy.... I would catch myself gazing at nothing, thinking about my child. Every moment, every second of the day, I felt like I was thinking about my baby. My youngest, you have to understand, I left when he was only two months old.... You know, whenever

I receive a letter from my children, I cannot sleep. I cry. It's good that my job is more demanding at night."

Despite the anguish these separations clearly cause, Filipina women continue to leave for jobs abroad. Since the early 1990s, 55 percent of migrants out of the Philippines have been women; next to electronic manufacturing, their remittances make up the major source of foreign currency in the Philippines. The rate of female emigration has continued to increase and includes college-educated teachers, businesswomen, and secretaries. In Parrenas's study, more than half of the nannies she interviewed had college degrees and most were married mothers in their 30s.

Where are men in this picture? For the most part, men—especially men at the top of the class ladder—leave child-rearing to women. Many of the husbands and fathers of Parrenas's domestic workers had migrated to the Arabian peninsula and other places in search of better wages, relieving other men of "male work" as construction workers and tradesmen, while being replaced themselves at home. Others remained at home, responsible fathers caring or helping to care for their children. But some of the men tyrannized their wives. Indeed, many of the women migrants Parrenas interviewed didn't just leave; they fled. As one migrant maid explained:

> You have to understand that my problems were very heavy before I left the Philippines. My husband was abusive. I couldn't even think about my children, the only thing I could think about was the opportunity to escape my situation. If my husband was not going to kill me, I was probably going to kill him.... He always beat me up and my parents wanted me to leave him for a long time. I left my children with my sister.... In the plane ... I felt like a bird whose cage had been locked for many years.... I felt free.... Deep inside, I felt homesick for my children but I also felt free for being able to escape the most dire problem that was slowly killing me.

Other men abandoned their wives. A former public school teacher back in the Philippines confided to Parrenas: "After three years of marriage, my husband left me for another woman. My husband supported us for just a little over a year. Then the support was stopped.... The letters stopped. I have not seen him since." In the absence of government aid, then, migration becomes a way of coping with abandonment.

Sometimes the husband of a female migrant worker is himself a migrant worker who takes turns with his wife migrating. One Filipino man worked in Saudi Arabia for 10 years, coming home for a month each year. When he finally returned home for good, his wife set off to work as a maid in America while he took care of the children. As she explained to Parrenas, "My children were very sad when I left them. My husband told me that when they came back home from the airport, my children could not touch their food and they wanted to cry. My son, whenever he writes me, always draws the head of Fido the dog with tears on the eyes. Whenever he goes to Mass on Sundays, he tells me that he misses me more because he sees his friends with their mothers. Then he comes home and cries."

THE END OF THE CHAIN

Just as global capitalism helps create a third world supply of mothering, it creates a first world demand for it. The past half-century has witnessed a huge rise in the number of women in paid work—from 15 percent of mothers of children aged 6 and under in 1950 to 65 percent today. Indeed, American women now make up 45 percent of the American labor force. Three-quarters of mothers of children 18 and under now work, as do 65 percent of mothers of children 6 and under. In addition, a recent report by the International Labor Organization reveals that the average number of hours of work per week has been rising in this country.

Earlier generations of American working women would rely on grandmothers and other female kin to help look after their children; now the grandmothers and aunts are themselves busy doing paid work outside the home. Statistics show that over the past 30 years a decreasing number of families have relied on relatives to care for their children—and hence are compelled to look for nonfamily care. At the first world end of care chains, working parents are grateful to find a good nanny or child care provider, and they are generally able to pay far more than the nanny could earn in her native country. This is not just a child care problem. Many American families are now relying on immigrant or out-of-home care for their *elderly* relatives. As a Los Angeles elder-care worker, an immigrant, told Parrenas, "Domestics here are able to make a living from the elderly that families abandon." But this often means that nannies cannot take care of their own ailing parents and therefore produce an elder-care version of a child care chain—caring for first world elderly persons while a paid worker cares for their aged mother back in the Philippines.

My own research for two books, *The Second Shift* and *The Time Bind*, sheds some light on the first world end of the chain. Many women have joined the law, academia, medicine, business—but such professions are still organized for men who are free of family responsibilities. The successful career, at least for those who are broadly middle class or above, is still largely built on some key traditional components: doing professional work, competing with fellow professionals, getting credit for work, building a reputation while you're young, hoarding scarce time, and minimizing family obligations by finding someone else to deal with domestic chores. In the past, the professional was a man and the "someone else to deal with [chores]" was a wife. The wife oversaw the family, which—in pre-industrial times, anyway—was supposed to absorb the human vicissitudes of birth, sickness, and death that the workplace discarded. Today, men take on much more of the child care and housework at home, but they still base their identity on demanding careers in the context of which children are beloved impediments; hence, men resist sharing care equally at home. So when parents don't have enough "caring time" between them, they feel forced to look for that care further down the global chain.

The ultimate beneficiaries of these various care changes might actually be large multinational companies, usually based in the United States. In my research on a Fortune 500 manufacturing company I call Amerco, I discovered a disproportionate number of women employed in the human side of the company:

public relations, marketing, human resources. In all sectors of the company, women often helped others sort out problems—both personal and professional—at work. It was often the welcoming voice and "soft touch" of women workers that made Amerco seem like a family to other workers. In other words, it appears that these working mothers displace some of their emotional labor from their children to their employer, which holds itself out to the worker as a "family." So, the care in the chain may begin with that which a rural third world mother gives (as a nanny) the urban child she cares for, and it may end with the care a working mother gives her employees as the vice president of publicity at your company.

HOW MUCH IS CARE WORTH?

How are we to respond to the growing number of global care chains? Through what perspective should we view them?

I can think of three vantage points from which to see care chains: that of the primordialist, the sunshine modernist, and (my own) the critical modernist. The primordialist believes that our primary responsibility is to our own family, our own community, our own country. According to this view, if we all tend our own primordial plots, everybody will be fine. There is some logic to this point of view. After all, Freud's concept of displacement rests on the premise that some original first object of love has a primary "right" to that love, and second and third comers don't fully share that right. (For the primordialist—as for most all of us—those first objects are members of one's most immediate family.) But the primordialist is an isolationist, an antiglobalist. To such a person, care chains seem wrong—not because they're unfair to the least-cared-for children at the bottom of the chain, but because they are global. Also, because family care has historically been provided by women, primordialists often believe that women should stay home to provide this care.

The sunshine modernist, on the other hand, believes care chains are just fine, an inevitable part of globalization, which is itself uncritically accepted as good. The idea of displacement is hard for the sunshine modernists to grasp because in their equation—seen mainly in economic terms—the global market will sort out who has proper claims on a nanny's love. As long as the global supply of labor meets the global demand for it, the sunshine modernist believes, everything will be okay. If the primordialist thinks care chains are bad because they're global, the sunshine modernist thinks they're good for the very same reason. In either case, the issue of inequality of access to care disappears.

The critical modernist embraces modernity but with a global sense of ethics. When the critical modernist goes out to buy a pair of Nike shoes, she is concerned to learn how low the wage was and how long the hours were for the third world factory worker making the shoes. The critical modernist applies the same moral concern to care chains: The welfare of the Filipino child back home must be seen as some part, however small, of the total picture. The critical modernist sees globalization as a very mixed blessing, bringing with it new opportunities—such as the nanny's access to good wages—but also new problems, including emotional and psychological costs we have hardly begun to understand.

From the critical modernist perspective, globalization may be increasing inequities not simply in access to money—and those inequities are important enough—but in access to care. The poor maid's child may be getting less motherly care than the first world child. (And for that matter, because of longer hours of work, the first world child may not be getting the ideal quantity of parenting attention for healthy development because too much of it is now displaced onto the employees of Fortune 500 companies.) We needn't lapse into primordialism to sense that something may be amiss in this.

I see no easy solutions to the human costs of global care chains. But here are some initial thoughts. We might, for example, reduce the incentive to migrate by addressing the causes of the migrant's economic desperation and fostering economic growth in the third world. Thus one obvious goal would be to develop the Filipino economy.

But it's not so simple. Immigration scholars have demonstrated that development itself can *encourage* migration because development gives rise to new economic uncertainties that families try to mitigate by seeking employment in the first world. If members of a family are laid off at home, a migrant's monthly remittance can see them through, often by making a capital outlay in a small business or paying for a child's education.

Other solutions might focus on individual links in the care chain. Because some women migrate to flee abusive husbands, a partial solution would be to create local refuges from such husbands. Another would be to alter immigration policy so as to encourage nannies to bring their children with them. Alternatively, employers or even government subsidies could help nannies make regular visits home.

The most fundamental approach to the problem is to raise the value of caring work and to ensure that whoever does it gets more credit and money for it. Otherwise, caring work will be what's left over, the work that's continually passed on down the chain. Sadly, the value ascribed to the labor of raising a child has always been low relative to the value of other kinds of labor, and under the impact of globalization, it has sunk lower still. The low value placed on caring work is due neither to an absence of demand for it (which is always high) nor to the simplicity of the work (successful caregiving is not easy) but rather to the cultural politics underlying this global exchange.

The declining value of child care anywhere in the world can be compared to the declining value of basic food crops relative to manufactured goods on the international market. Though clearly more essential to life, crops such as wheat, rice, or cocoa fetch low and declining prices while the prices of manufactured goods (relative to primary goods) continue to soar in the world market. And just as the low market price of primary produce keeps the third world low in the community of nations, the low market value of care keeps low the status of the women who do it.

One way to solve this problem is to get fathers to contribute more to child care. If fathers worldwide shared child care labor more equitably, care would spread laterally instead of being passed down a social-class ladder, diminishing in value along the way. Culturally, Americans have begun to embrace this idea—but they've yet to put it into practice on a truly large scale [see Richard Weissbourd, "Redefining Dad," *TAP*, December 6, 1999]. This is where norms and policies established in the first world can have perhaps the greatest influence on reducing costs along global care chains.

According to the International Labor Organization, half of the world's women between ages 15 and 64 are working in paid jobs. Between 1960 and 1980, 69 out of 88 countries for which data are available showed a growing proportion of women in paid work (and the rate of increase has skyrocketed since the 1950s in the United States, Scandinavia, and the United Kingdom). If we want developed societies with women doctors, political leaders, teachers, bus drivers, and computer programmers, we will need qualified people to help care for children. And there is no reason why every society cannot enjoy such loving paid child care. It may even remain the case that Vicky Diaz is the best person to provide it. But we would be wise to adopt the perspective of the critical modernist and extend our concern to the potential hidden losers in the care chain. These days, the personal is global.

KEY CONCEPTS

emotional labor   global care chain   theory of displacement

DISCUSSION QUESTIONS

1. What does Hochschild mean by the "globalization of love," and how is this phenomenon linked to the status of women in the United States? In other parts of the world?

2. What different perspectives on the care chain does Hochschild identify? What solutions to the problem does each perspective suggest? What would you recommend?

30

New Commodities, New Consumers

Selling Blackness in a Global Marketplace

PATRICIA HILL COLLINS

Patricia Hill Collins examines how the process of globalization requires ever-expanding consumer markets, thus ensuring greater profits. In this context, she argues, Black men and women become highly commodified—their bodies used for the interests of a global capitalist economy. She also shows how sexuality in a global marketplace is intertwined with racial and gender inequality.

In the eyes of many Americans, African American youth such as hip hop legend Tupac Shakur constitute a threatening and unwanted population. No longer needed for cheap, unskilled labor in fields and factories, poor and working-class black youth find few job opportunities in the large, urban metropolitan areas where most now reside. Legal and undocumented immigrants now do the dirty work in the hotels, laundries, restaurants and construction sites of a growing service economy. Warehoused in inner city ghettos that now comprise the new unit of racial segregation, poor black youth face declining opportunities and an increasingly punitive social welfare state. Because African American youth possess citizenship rights, social welfare programs legally can no longer operate in racially discriminatory ways. Yet, rather than providing African American youth with educational opportunities, elites chose instead to attack the social welfare state that ensured benefits for everyone. Fiscal conservatives have cut funding for public schools, public housing, public health clinics, and public transportation that would enable poor and working-class black youth to get to burgeoning jobs in the suburbs. Hiding behind a rhetoric of colorblindness, elites claim that these policies lack racial intentionality (Guinier and Torres, 2002; Bonilla-Silva, 2003). Yet when it comes to who is affected by these policies, African American youth constitute a sizable segment of the 'truly disadvantaged' (Wilson, 1987).

... African American youth are often conceptualized as a marginalized, powerless and passive population within macroeconomic policies of globalization. They serve as examples of an economic analysis that only rarely examines intersections of class and race. In contrast, I suggest that because African American

SOURCE: Patricia Hill Collins. 2006. "New Commodities, New Consumers: Selling Blackness in a Global Marketplace." *Ethnicities* 6, no. 3: 297–317.

youth are in the belly of the beast of the sole remaining world superpower, they present an important local location for examining new configurations of social class that is refracted through the lens of race, gender, age and sexuality. Stated differently, because they are centrally located within the United States, Black American youth constitute one important population of social actors who negotiate the contradictions of a racialized globalization as well as the new social class relations that characterize it.

This article asks, what might the placement of poor and working-class African American youth in the global political economy, both as recipients of social outcomes of globalization as well as social agents who respond to those outcomes, tell us about the new racialized class formations of globalization? Conversely, what light might the experiences of poor and working-class African American women and men shed on new global forms of racism? These are very large questions, and I briefly explore them by sketching out a two-part argument. First, I investigate how ideas of consumption, commodification and control situate black youth within a global political economy. I suggest that shifting the focus of class analysis from production to consumption provides a better understanding of black youth. Second, I develop a framework for understanding the commodification of black bodies that ties this process more closely to social class relations. In particular, I use the status of African American youth to explore how the literal and figurative commodification of blackness fosters new strategies of control.

NEW COMMODITIES: ADVANCED CAPITALISM AND BLACK BODY POLITICS

... African American youth are a hot commodity in the contemporary global marketplace and global media. Their images have catalyzed new consumer markets for products and services. The music of hip hop culture, for example, follows its rhythm and blues predecessor as a so-called crossover genre that is very popular with whites and other cultural groups across the globe. Circulated through film, television, and music, news and advertising, mass media constructs and sells a commodified black culture from ideas about class, gender and age. Through a wide array of genres ranging from talk shows to feature-length films, television situation comedies to CDs, video rentals to cable television, the images produced and circulated within this area all aim to entertain and amuse a highly segmented consumer market. This market is increasingly global and subject to the contradictions of global marketplace phenomena.

One implication of the significance of consumption for understanding social class relations of black youth concerns the constant need to stimulate consumer markets. Contemporary capitalism relies not just on cutting the costs attached to production, but also on stimulating consumer demand. Just as sustaining relations of production requires a steady supply of people to do the work, sustaining relations of consumption needs ever-expanding consumer markets. Moreover, just as people do not naturally work and must be encouraged or compelled to do so, people do not engage in excess consumption without prompting. In this context, advertising constitutes an important site that creates demand for commodities of all sorts. Marketing and advertising often create demand for things that formerly were not seen as commodities, for example, the rapid growth of the bottled water industry, as well as for intangible entities that seem difficult to commodify. In this regard, the rapid growth of mass media and new informational technologies has catalyzed a demand for black culture as a commodity....

Under this ever-expanding impetus to create new consumer markets, nothing is exempt from commodification and sale, including the pain that African American youth experience with poverty and powerlessness. Nowhere is this more evident than in the contradictions of rap.... In this context, rap becomes the only place where black youth have public voice, yet it is a public voice that is commodified and contained by what hip hop producers think will sell. Despite these marketplace limitations, rap remains a potential site of contestation, a place where African American youth can rebel against the police brutality, lack of jobs, and other social issues that confront them (Kelley, 1994). Thus, work on the black culture industry illustrates how images of black culture function to catalyze consumption.

The actual bodies of young African Americans may also be commodified as part of a new black body politics.... New forms of commodification within the constant pressure to expand consumer markets catalyze a new black body politics where social class relations rest not solely on exploiting labor power and/or mystifying exploitation through images, but also on the appropriation of bodies themselves. Whereas young black bodies were formerly valued for their labor power, under advanced capitalism, their utility lies elsewhere....

... The growth of the punishment industry also illustrates how black male bodies are objectified, commodified and incorporated in service to maintaining prisons as consumer markets. In essence, Black men's commodified bodies become used as raw materials for the growing prison industry. It is very simple—no prisoners means no jobs for all of the ancillary industries that service this growth industry. Because prisons express little interest in rehabilitating prisoners, they need a steady supply of bodies. The focus is less on appropriating the labor of incarcerated black men (although this does happen) than in finding profitable uses for their bodies while the state absorbs the costs of incarceration. If Kentucky Fried Chicken found chickens in short supply, they would close and their profitability would shrink. The Kentucky Fried Chicken Corporation has little interest in extracting labor from its chickens or in coaxing them to change their ways. Rather, the corporation needs a constant supply of cheap, virtually identical chickens to ensure that their business will remain profitable. In this way, prisons made use of the bodies of unemployed, unskilled young black men, the virtually indistinguishable young black men who populate corners of American cities.

The vast majority of young black people who are incarcerated by the punishment industry are male, yet it is important to remember that disproportionately high numbers of young black women are also incarcerated and thus are subject to this form of commodification. Moreover, young black women may also encounter an additional bodily commodification of their sexuality. The majority of sex workers may be female, obscuring the minority of males who also perform sex work as well as the objectification and commodification of black male bodies within mass media as an important component of the sex work industry. In essence, the bodies and images of young African Americans constitute new commodities that are central to global relations of consumption, not marginalized within them.

NEW CONSUMERS: SEX WORK AND HIP HOP CAPITALISM

Here, I want to take a closer look at this process by exploring how sexuality has grown in importance in the commodification of the bodies and images of black American youth and how this sex work in turn articulates with black agency in responding to advanced capitalism. In essence, black youth are now caught up in a burgeoning sex work industry, one that is far broader than commercial sex work as depicted in the media. Young African Americans participate in the sex work industry, not primarily as commercial workers as is popularly imagined, but rather as representations of commodified black sexuality as well as potential new consumer markets eager to consume their own images.

Racialized images of pimps and prostitutes may be the commercial sex workers who are most visible in the relations of production, yet the industry itself is much broader. A broader definition of sex work suggests how the sex work industry has been a crucial part of the expansion of consumer markets. The sex work industry encompasses a set of social practices, many of which may not immediately be recognizable as sex work, as well as a constellation of representations that create demand for sexual services, attach value to such services, identify sexual commodities with race, gender and age-specific individuals, and rules that regulate this increasingly important consumer market.

… Sex work is permeating the very fabric of African American communities in ways that resemble how sex work has changed the societies of developing countries. In essence, poor and working-class African American youth increasingly encounter few opportunities for jobs in urban neighborhoods while the mass marketing of sexuality permeates consumer markets. In this sense, their situation resembles that of black youth globally who confront similar pressures in response to globalization. At the same time, the situation of African American youth is unique in that the sexualized images that they encounter are of themselves. In essence, their own bodies often serve as symbols of this sexualized culture, placing African American youth in the peculiar position claiming and rejecting themselves. How might this happen?

… I investigate how reconfiguration of the sex work industry within the United States has shaped the domestic relations within African American communities generally and for poor and working-class African American youth in particular.

Nigeria, the most populous nation state on the African continent, provides an important case for building such an analysis. Reporting on patterns of trafficking in Italy, Eshohe Aghatise describes differential mechanisms used to traffic women from Eastern Europe and Nigeria as well as the differential value placed on women within Italian sex markets. The trafficking of Nigerian women and young girls into Italy for prostitution began in the 1980s in response to Nigeria's economic problems caused by structural adjustment policies of the International Monetary Fund (IMF) (Emeagwali, 1995). As Agahtise points out:

> women and girls started leaving Nigeria for Europe on promises of fantastic well-paying jobs to be obtained in factories, offices, and farms. They arrived in Italy only to find themselves lured into prostitution and sold into sexual slavery to pay off debts, which they were told they incurred in being 'helped' to come to Europe. (Aghatise, 2004: 1129)

Most Nigerian victims of trafficking are illiterate and lacked any exposure to urban life.

The shifting patterns of economic and social change within Nigerian society also contributed to the patterns of trafficking. Traffickers preyed not only on the poverty of Nigerian victims, but also on the breakdown of social and cultural values within Nigerian society, in particular, the disintegration of family structures and a weak social welfare state. For many families, sending female children abroad became a status symbol:

> Subscribing to a consumerist model that is widely publicized on television and in magazines with messages of high living in the West, and in the oral reports of 'been-tos' (a popular name in Nigeria given to those who have been to Western Europe, Canada or the United States), many families believe that it is easy to obtain wealth abroad, and that earning money, in whatever way, will be quick. (Aghatise, 2004: 1132)

Aghatise offers an especially harsh criticism of a society that embraces consumerism and sells its daughters to pay for it:

> The beginning years of Nigeria's economic boom from petrol dollars left the legacy of a people who had acquired a taste for a high standard of living and a consumer society that no longer had the means to satisfy its purchasing habits but was not ready to admit or accept it. (Aghatise, 2004: 1133)

Trafficking of women and girls to Italy demonstrates the fraying social fabric of Nigerian society, especially the ways in which women absorb the pressures placed on families under changing public policies. Most of the women trafficked to Italy are from polygamous families from the Edo ethnic group where wives are in a continuous struggle for a share of the family resources for themselves and their children. Even if men have jobs, their earnings are rarely enough to provide for the needs of the entire family.... The worsening conditions within the Nigerian economy, the weak welfare state, and cultural expectations of women meant that women who were trafficked in the 1990s were mainly much younger girls who set out on a job search to help their families....

... With no jobs for its large youth population, poor Nigerian families learned to look the other way when traffickers commodified and exported its girls and women for the international sex industry and/or when girls saw domestic sex work as their only option. They learned to accommodate a changing set of social norms that pushed young girls toward sex work, for some for reasons of basic survival, yet for others as part of the costs of upward social mobility. Poor and working-class African American girls seemingly confront a similar set of challenges in the context of a different set of circumstances. In this regard, the continuum of sex work from sugar daddies, night brides, floating prostitutes, call girls and trafficked women also applies, yet in a different constellation that reflects the political and economic situation of African Americans as well as cultural values of American society.

Two important features may shape young African American women's participation in the sex work industry. For one, because African American girls are American citizens, they cannot be as easily trafficked as other groups of poor women who lack US citizenship. Girls are typically trafficked into the United States, not out of it. African American girls do enjoy some protections from these forms of exploitation, yet expanding the definition of sex work itself suggests that their patterns of participation have changed. For another, commercial sex work is not always a steady activity but may occur simultaneously with other forms of income-generating work. In the global context, women sex workers also engage in domestic service, informal commercial trading, market-vending, shining shoes, or office work (Kempadoo and Doezama, 1998: 3). In a similar fashion, African American girls may have multiple sources of income, one of which is sex work. The 'night brides' and 'call girls' of Nigeria may find a domestic counterpart among black American adolescent girls, yet this activity would not be labeled 'sex work', nor would it be seen as prostitution. Restricting the concept of sex work and prostitute to the image of the streetwalker thus obscures the various ways that young black women's bodies and images are commodified and then circulated within the sex work industry....

The consequences of this sex-for-material-goods situation can be tragic. The pressures for young black women to engage in sex work have affected the rapid growth of HIV/AIDS among poor black women in the Mississippi Delta and across the rural South. Between 1990 and 2000, Southern states with large African American populations experienced a dramatic increase in HIV infections among African American women. For example, in Mississippi, 28.5 percent of those reporting new HIV infections in 2000 were black women, up from 13 percent in 1990. In Alabama, the number rose to 31 percent, from 13 percent, whereas in North Carolina, it rose to 27 percent, from 18 percent (Sack, 2001). Most of the women contracted HIV through heterosexual contact, and most found out that they were HIV positive when they became pregnant. The women took risks that may at first seem nonsensical. Yet in the context of their lives there was a sense that because they had so little control over other aspects of their lives, they felt that if God wanted them to get AIDS, then they resigned themselves to getting it.

These examples suggest that many young African American women resign themselves to commodifying their bodies as a necessary source of income. They may not be streetwalkers in the traditional sense, but they also view commodified black sexuality as the commodity of value that they can exchange. These relations also become difficult to disrupt in the context of a powerful mass media that defines and sells images of sexualized black women as one icon of seemingly authentic black culture. Young African American women encounter a set of representations that naturalizes and normalizes social relations of sex work. Whether she sleeps with men for pleasure, drugs, revenge, or money, the sexualized bitch constitutes a modern version of the Jezebel, repackaged for contemporary mass media. In discussing this updated Jezebel image, cultural critic Lisa Jones distinguishes between gold diggers/skeezers, namely, women who screw for status, and crack 'hos', namely, women who screw for a fix (1994: 79). Some women are the 'hos' who trade sexual favors for jobs, money, drugs and other material items. The female hustler, a materialist woman who is willing to sell, rent, or use her sexuality to get whatever she wants constitutes this sexualized variation of the bitch. This image appears with increasing frequency, especially in conjunction with trying to catch an African American man with money. Athletes are targets, and having a baby with an athlete is a way to garner income. Black women who are sex workers, namely, those who engage in phone sex, lap dancing, and prostitution for compensation, also populate this universe of sexualized bitches. The prostitute who hustles without a pimp and who keeps the compensation is a bitch who works for herself.

Black male involvement in the sex work industry may not involve the direct exploitation of black men's bodies as much as the objectification and commodification of sexualized black male images within hip hop culture. The prevalence of representations of black men as pimps speaks to this image of black men as sexual hustlers who use their sexual prowess to exploit women, both black and white. Ushered in by a series of films in the "Blaxploitation" era, the ubiquitous black pimp seems here to stay. Kept alive through HBO produced quasi-documentaries such as *Pimps Up, Hos Down*, African American men feature prominently in mass media. Despite these media constructions, actual pimps see themselves more as businessmen than as sexual predators. For example, the men interviewed in the documentary *American Pimp* all discuss the skills involved in being a successful pimp. One went so far as to claim that only African American men made really good pimps. Thus, the controlling image of the black pimp combines all of the elements of the more generic hustler, namely, engaging in illegal activity, using women for economic gain, and refusing to work.

Representations of black women and men as prostitutes and pimps permeate music videos, film and television. In the context of a powerful global mass media, black men's bodies are increasingly objectified within popular culture in ways that resemble the treatment of all women. Violence and sexuality sell, and associating black men with both is virtually sure to please. Yet the real struggle is less about the content of black male and black female images and more about the treatment of black people's bodies as valuable commodities within advertising and entertainment. Because this new constellation of images participates in commodified global capitalism, in all cases, representations of black people's bodies are tied to structures of profitability. Athletes and criminals alike are profitable, not for the vast majority of African American men, but for the people who own the teams, control the media, provide food, clothing and telephone services to the prisons, and who consume seemingly endless images of pimps, hustlers, rapists, and felons. What is different, however, is how these images of authentic blackness generate additional consumer markets beyond the selling of these specific examples of cultural production....

REFERENCES

Aghatise, E. (2004) 'Trafficking for Prostitution in Italy', *Violence Against Women* 10(10): 1126–55.

Emeagwali, G.T. (1995) *Women Pay the Price: Structural Adjustment in Africa and the Caribbean.* Trenton, N.J.: Africa World Press.

Guinier, L. and G. Torres (2002) *The Miner's Canary: Enlisting Race, Resisting Power, Transforming Democracy.* Cambridge, MA: Harvard University Press.

Jones, Lisa (1994) *Bulletproof Diva: Tales of Race, Sex, and Hair.* New York: Doubleday.

Kelley, R.D.G. (1994) *Race Rebels: Culture, Politics, and the Black Working Class.* New York: Free Press.

KEY CONCEPTS

globalization   sex trafficking   sex work

DISCUSSION QUESTIONS

1. What does Hill Collins mean by saying that Black men and women become a new commodity in the current reality of a global economy?

2. What evidence do you see in popular culture for Hill Collins's argument that Black men and women are sexualized in race- and gender-specific ways?

3. What parallels does Hill Collins draw regarding the social conditions in Nigeria and in the United States that drives Black women into sex work?

Applying Sociological Knowledge: An Exercise for Students

Take a look at the clothes in your closet. Where are they made? Do some research online into the living environment in some of these countries. Go to the CIA World Fact Book (www.cia.gov). Using the links to these different countries, check under "Economy," "People," and "Government" to answer these questions:

1. What is the life expectancy of people in this nation?
2. What is the infant mortality rate?
3. What percent of people live below the poverty line?
4. What is the unemployment rate?
5. What percentage of household income is held by the highest and lowest income groups?
6. What transnational issues (see the bottom of the page) does the nation face?

Having answered these questions, what would you now say about global stratification?

PART X

Race and Ethnicity

31

The Souls of Black Folk

W. E. B. DU BOIS

W.E. B. Du Bois, the first African American Ph.D. from Harvard University, is a classic sociological analyst. In this well-known essay, he develops the idea that African Americans have a "double consciousness"—one that they must develop as a protective strategy to understand how Whites see them. Originally writing this essay in 1903, Du Bois also reflects on the long struggle for Black freedom.

Between me and the other world there is ever an unasked question: unasked by some through feelings of delicacy; by others through the difficulty of rightly framing it. All, nevertheless, flutter round it. They approach me in a half-hesitant sort of way, eye me curiously or compassionately, and then, instead of saying directly, How does it feel to be a problem? they say, I know an excellent colored man in my town; or, I fought at Mechanicsville; or, Do not these Southern outrages make your blood boil? At these I smile, or am interested, or reduce the boiling to a simmer as the occasion may require. To the real question, How does it feel to be a problem? I answer seldom a word....

After the Egyptian and Indian, the Greek and Roman, the Teuton and Mongolian, the Negro is a sort of seventh son, born with a veil, and gifted with second-sight in this American world,—a world which yields him no true self-consciousness, but only lets him see himself through the revelation of the other world. It is a peculiar sensation, this double-consciousness, this sense of always looking at one's self through the eyes of others, of measuring one's soul by the tape of a world that looks on in amused contempt and pity. One ever feels his twoness,—an American, a Negro; two souls, two thoughts, two unreconciled strivings: two warring ideals in one dark body, whose dogged strength alone keeps it from being torn asunder.

The history of the American Negro is the history of this strife—this longing to attain self-conscious manhood, to merge his double self into a better and truer self. In this merging he wishes neither of the older selves to be lost. He would not Africanize America, for America has too much to teach the world and Africa. He would not bleach his Negro soul in a flood of white Americanism, for he knows that Negro blood has a message for the world. He simply wishes to

SOURCE: W. E. B. Du Bois. 1989. *The Souls of Black Folk*, edited and with an introduction by Donald B. Gibson. New York: Penguin, pp. 3–12.

make it possible for a man to be both a Negro and an American, without being cursed and spit upon by his fellows, without having the doors of Opportunity closed roughly in his face.

This, then, is the end of his striving: to be a co-worker in the kingdom of culture, to escape both death and isolation, to husband and use his best powers and his latent genius. These powers of body and mind have in the past been strangely wasted, dispersed, or forgotten. The shadow of a mighty Negro past flits through the tale of Ethiopia the Shadowy and of Egypt the Sphinx. Throughout history, the powers of single black men flash here and there like falling stars, and die sometimes before the world has rightly gauged their brightness. Here in America, in the few days since Emancipation, the black man's turning hither and thither in hesitant and doubtful striving has often made his very strength to lose effectiveness, to seem like absence of power, like weakness. And yet it is not weakness—it is the contradiction of double aims. The double-aimed struggle of the black artisan—on the one hand to escape white contempt for a nation of mere hewers of wood and drawers of water, and on the other hand to plough and nail and dig for a poverty-stricken horde—could only result in making him a poor craftsman, for he had but half a heart in either cause. By the poverty and ignorance of his people, the Negro minister or doctor was tempted toward quackery and demagogy; and by the criticism of the other world, toward ideals that made him ashamed of his lowly tasks. The would-be black *savant* was confronted by the paradox that the knowledge people needed was a twice-told tale to his white neighbors, while the knowledge which would teach the white world was Greek to his own flesh and blood. The innate love of harmony and beauty that set the ruder souls of his people a-dancing and a-singing raised but confusion and doubt in the soul of the black artist; for the beauty revealed to him was the soul-beauty of a race which his larger audience despised, and he could not articulate the message of another people. This waste of double aims, this seeking to satisfy two unreconciled ideals, has wrought sad havoc with the courage and faith and deeds often thousand of thousands people,—has sent them often wooing false gods and invoking false means of salvation, and at times has even seemed about to make them ashamed of themselves....

The Nation has not yet found peace from its sins; the freedman has not yet found in freedom his promised land. Whatever of good may have come in these years of change, the shadow of a deep disappointment rests upon the Negro people—a disappointment all the more bitter because the unattained ideal was unbounded save by the simple ignorance of a lowly people....

... Merely a concrete test of the underlying principles of the great republic is the Negro Problem, and the spiritual striving of the freedmen's sons is the travail of souls whose burden is almost beyond the measure of their strength, but who bear it in the name of an historic race, in the name of this the land of their fathers' fathers and in the name of human opportunity.

KEY CONCEPTS

caste system   double consciousness

DISCUSSION QUESTIONS

1. What does Du Bois mean by "double consciousness," and how does this affect how African American people see themselves and others?

2. In the contemporary world, what examples do you see that Black people are still defined as "a problem," as Du Bois notes? How does this affect the Black experience?

32

Color-Blind Privilege

The Social and Political Functions of Erasing the Color Line in Post Race America

CHARLES A. GALLAGHER

Charles A. Gallagher discusses the problem of a color-blind approach to race and race relations in this country. By denying race as a structural basis for inequality, we fail to recognize the privilege of Whiteness. With the blurring of racial lines, White college students lack a clear understanding of how the existing social, political, and economic systems advantage or privilege Whites.

INTRODUCTION

An adolescent white male at a bar mitzah wears a FUBU shirt while his white friend preens his tightly set, perfectly braided corn rows. A black model dressed in yachting attire peddles a New England yuppie boating look in Nautica advertisements. It is quite unremarkable to observe white, Asian or African-Americans with dyed purple, blond or red hair. White, black and Asian students decorate their bodies with tattoos of Chinese characters and symbols. In

SOURCE: Gallagher, Charles A. 2003. "Color-Blind Privilege: The Social and Political Functions of Erasing the Color Line in Post Race America." *Race, Gender and Class* (June). Reprinted with permission of the author.

cities and suburbs young adults across the color line wear hip-hop clothing and listen to white rapper Eminem and black rapper Jay-Z. A north Georgia branch of the NAACP installs a white biology professor as its president. The music of Jimi Hendrix is used to sell Apple Computers. Du-Rag kits, complete with bandana headscarf and elastic headband, are on sale for $2.95 at hip-hop clothing stores and family centered theme parks like Six Flags. Salsa has replaced ketchup as the best selling condiment in the United States. Companies as diverse as Polo, McDonalds, Tommy Hilfiger, Walt Disney World, Master Card, Skechers sneakers, IBM, Giorgio Armani and Neosporin antibiotic ointment have each crafted advertisements that show a balanced, multiracial cast of characters interacting and consuming their products in a post-race, color-blind world....

Americans are constantly bombarded by depictions of race relations in the media which suggest that discriminatory racial barriers have been dismantled. Social and cultural indicators suggest that America is on the verge, or has already become, a truly color-blind nation. National polling data indicate that a majority of whites now believe discrimination against racial minorities no longer exists. A majority of whites believe that blacks have "as good a chance as whites" in procuring housing and employment or achieving middle class status while a 1995 survey of white adults found that a majority of whites (58%) believed that African Americans were better off finding jobs than whites.[1] Much of white America now see a level playing field, while a majority of black Americans see a field which is still quite uneven.... The color-blind or race neutral perspective holds that in an environment where institutional racism and discrimination have been replaced by equal opportunity, one's qualifications, not one's color or ethnicity, should be the mechanism by which upward mobility is achieved. Whites and blacks differ significantly, however, in their support for affirmative action, the perceived fairness of the criminal justice system, the ability to acquire the "American Dream," and the extent to which whites have benefited from past discrimination.[2]

This article examines the social and political functions color-blindness serves for whites in the United States. Drawing on interviews and focus groups with whites from around the country, I argue that color-blindness maintains white privilege by negating racial inequality. Embracing post-race, color-blind perspective provides whites with a degree of psychological comfort by allowing them to imagine that being white or black or brown has no bearing on an individual's or a group's relative place in the socio-economic hierarchy. My interviews included seventeen focus group and thirty individual interviews with whites around the country. While my sample is not representative of the total white population, I used personal contacts and snowball sampling to purposively locate respondents raised in urban, suburban and rural environments. Twelve of the seventeen focus groups were conducted in a university setting, one in a liberal arts college in the Rocky Mountains and the other at a large urban university in the Northeast. Respondents in these focus groups were selected randomly from the student population. The occupational range for my individual interviews was quite eclectic and included a butcher, construction worker, hair stylist, partner in a prestigious corporate law firm, executive secretary, high school principal, bank president from a small town, retail workers, country lawyer and custodial workers. Twelve of the thirty individual interviews were with respondents who were raised in rural and/or agrarian settings. The remaining respondents lived in suburbs of large cities or in urban areas.

What linked this rather disparate group of white individuals together was their belief that race-based privilege had ended. As a majority of my respondents saw it, color-blindness was now the norm in the United States. The illusion of racial equality implicit in the myth of color-blindness was, for many whites, a form of comfort. This aspect of pleasure took the form of political empowerment ("what about whites' rights") and moral gratification from being liberated from "oppressor" charges ("we are not responsible for the past"). The rosy picture that color-blindness presumes about race relations and the satisfying sense that one is part of a period in American history that is morally superior to the racist days of the past is, quite simply, a less stressful and more pleasurable social place for whites to inhabit.

THE NORM OF COLOR-BLINDNESS

The perception among a majority of white Americans that the socio-economic playing field is now level, along with whites' belief that they have purged themselves of overt racist attitudes and behaviors, has made color-blindness the dominant lens through which whites understand contemporary race relations. Color-blindness allows whites to believe that segregation and discrimination are no longer an issue because it is now illegal for individuals to be denied access to housing, public accommodations or jobs because of their race.... Individuals from any racial background can wear hip-hop clothing, listen to rap music (both purchased at Wal-Mart) and root for their favorite, majority black, professional sports team. Within the context of racial symbols that are bought and sold in the market, color-blindness means that one's race has no bearing on who can ... live in an exclusive neighborhood, attend private schools or own a Rolex.

The passive interaction whites have with people of color through the media creates the impression that little, if any, socio-economic difference exists between the races. Research has found that whites who are exposed to images of upper-middle class African Americans ... believe that blacks have the same socioeconomic opportunities as whites. Highly visible and successful racial minorities like Secretary of State Colin Powell and National Security Advisor Condoleezza Rice are further proof to white America that the state's efforts to enforce and promote racial equality has been accomplished. Reflecting on the extent to which discrimination is an obstacle to socio-economic advancement and the perception of seeing African-Americans in leadership roles, Tom explained:

> If you look at some prominent black people in society today and I don't really see [racial discrimination]. I don't understand how they can keep bringing this problem onto themselves. If they did what society would want them to I don't see that society is making problems for them. I don't see it.

... The new color-blind ideology does not, however, ignore race; it acknowledges race while ignoring racial hierarchy by taking racially coded styles and products and reducing these symbols to commodities or experiences which whites and racial minorities can purchase and share. It is through such acts of shared consumption that race becomes nothing more than an innocuous cultural signifier. Large corporations have made American culture more homogeneous through the ubiquitousness of fast food, television, and shopping malls but this trend has also created the illusion that we are all the same through consumption. Most adults eat at national fast food chains like McDonalds, shop at mall anchor stores like Sears and J. C. Penney's and watch major league sports, situation comedies or television drama. Defining race only as cultural symbols that are for sale allows whites to experience and view race as nothing more than a benign cultural marker that has been stripped of all forms of institutional, discriminatory or coercive power. The post-race, color-blind perspective allows whites to imagine that depictions of racial minorities working in high status jobs and consuming the same products, or at least appearing in commercials for products whites desire or consume, is the same as living in a society where color is no longer used to allocate resources or shape group outcomes. By constructing a picture of society where racial harmony is the norm, the color-blind perspective functions to make white privilege invisible while removing from public discussion the need to maintain any social programs that are race-based.

... Starting with the deeply held belief that America is now a meritocracy, whites are able to imagine that the socio-economic success they enjoy relative to racial minorities is a function of individual hard work, determination, thrift and investments in education. The color-blind perspective removes from personal thought and public discussion any taint or suggestion of white supremacy or white guilt while legitimating the existing social, political and economic arrangements which privilege whites. This perspective insinuates that class and culture, and not institutional racism, are responsible for social inequality. Color-blindness allows whites to define themselves as politically progressive and racially tolerant as they proclaim their adherence to a belief system that does not see or judge individuals by the "color of their skin." This perspective ignores, as Ruth Frankenberg puts it, how whiteness is a "location of structural advantage societies structured in racial dominance."[3] Frankenberg uses the term "color and power evasiveness" rather than color-blindness to convey how the ability to ignore race by members of the dominant group reflects a position of power and privilege. Color-blindness hides white privilege behind a mask of assumed meritocracy while rendering invisible the institutional arrangements that perpetuate racial inequality. The veneer of equality implied in color-blindness allows whites to present their place in the racialized social structure as one that was earned.

Given this norm of color-blindness it was not surprising that respondents in this study believed that using race to promote group interests was a form of racism.

Joe, a student in his early twenties from a working class background, was quite adamant that the opportunity structure in the United States did not favor one racial group over another.

I mean, I think that the black person of our age has as much opportunity as me, maybe he didn't have the same guidance and that might hurt him. But I mean, he's got the same opportunities that I do to go to school, maybe even more, to get more money. I can't get any aid ... I think that blacks have the same opportunities as whites nowadays and I think it's old hat.

Not only does Joe believe that young blacks and whites have similar educational experiences and opportunity but it is his contention that blacks are more likely or able to receive money for higher education. The idea that race matters in any way, according to Joe, is anachronistic; it is "old hat" in a color-blind society to blame one's shortcomings on something as irrelevant as race.

Believing and acting as if America is now color-blind allows whites to imagine a society where institutional racism no longer exists and racial barriers to upward mobility have been removed. The use of group identity to challenge the existing racial order by making demands for the amelioration of racial inequities is viewed as racist because such claims violate the belief that we are a nation that recognizes the rights of individuals not rights demanded by groups. Sam, an upper middle class respondent in his 20's, draws on a pre- and post-civil rights framework to explain racial opportunity among his peers:

I guess I can understand in my parents' generation. My parents are older, my dad is almost 60 and my mother is in her mid 50's, ok? But the kids I'm going to school with, the minorities I'm going to school with, I don't think they should use racism as an excuse for not getting a job. Maybe their parents, sure, I mean they were discriminated against. But these kids have every opportunity that I do to do well.

In one generation, as Sam sees it, the color line has been erased. Like Sam's view that opportunity structure is open there is, according to Tara, a reason to celebrate the current state of race relations:

I mean, like you are not the only people that have been persecuted—I mean, yea, you have been, but so has every group. I mean if there's any time to be black in America it's now.

Seeing society as race-neutral serves to decouple past historical practices and social conditions from present day racial inequality as was the case for a number of respondents who pointed out that job discrimination had ended. Michelle was quite direct in her perception that the labor market is now free of discrimination stating that "I don't think people hire and fire because someone is black and white now." Ken also believed that discrimination in hiring did not occur since racial minorities now have legal recourse if discrimination occurs:

I think that pretty much we got past that point as far as jobs. I think people realize that you really can't discriminate that way because you will end up losing ... because you will have a lawsuit against you.

... The logic inherent in the color-blind approach is circular; since race no longer shapes life chances in a color-blind world there is no need to take race into account when discussing differences in outcomes between racial groups. This approach erases America's racial hierarchy by implying that social, economic and political power and mobility are equally shared among all racial groups. Ignoring the extent or ways in which race shapes life chances validates whites' social location in the existing racial hierarchy while legitimating the political and economic arrangements which perpetuate and reproduce racial inequality and privilege....

THE COST OF RACIALIZED PLEASURES

Being able to ignore or being oblivious to the ways in which almost all whites are privileged in a society cleaved on race has a number of implications. Whites derive pleasure in being told that the current system for allocating resources is fair and equitable. Creating and internalizing a color-blind view of race relations reflects how the dominant group is able to use the mass media, immigration stories of upward mobility, rags-to-riches narratives and achievement ideology to make white privilege invisible. Frankenberg argues that whiteness can be "displaced," as is the case with whiteness hiding behind the veil of color-blindness. It can also be made "normative" rather than specifically "racial," as is the case when being white is defined by white respondents as being no different than being black or Asian.[4] Lawrence Bobo and associates have advanced a theory of laissez-faire racism that draws on the color-blind perspective. As whites embrace the equality of opportunity narrative they suggest that

> laissez-faire racism encompasses an ideology that blames blacks themselves for their poorer relative economic standing, seeing it as a function of perceived cultural inferiority. The analysis of the bases of laissez-faire racism underscores two central components: contemporary stereotypes of blacks held by whites, and the denial of societal (structural) responsibility for the conditions in black communities.[5]

As many of my respondents make clear if the opportunity structure is open ("It doesn't matter what color you are"), there must be something inherently wrong with racial minorities or their culture that explains group level differences.

... [T]he form color-blindness takes as the nation's hegemonic political discourse is a variant of laissez-faire racism. Historian David Roediger contends that in order for the Irish to be absorbed into the white race in the mid-nineteenth century "the imperative to define themselves as whites came from the particular 'public and psychological wages' whiteness offered" these new immigrants.[6] There is still a "wage" to whiteness, that element of ascribed status whites automatically receive because of their membership in the dominant group. But within the framework of color-blindness the imperative has switched from whites overtly defining themselves or their interests as white, to one where they claim that color is irrelevant; being white is the same as being black, yellow, brown or red....

My interviews with whites around the country suggest that in this post-race era of color-blind ideology Ellison's keen observations about race relations need modification. The question now is what are we to make of a young white man from the suburbs who listens to hip-hop, wears baggy hip-hop pants, a baseball cap turned sideways, unlaced sneakers and a oversized shirt emblazoned with a famous NBA player who, far from shouting racial epithets, lists a number of racial minorities as his heroes? It is now possible to define oneself as not being racist because of the clothes you wear, the celebrities you like or the music you listen to while believing that blacks or Latinos are disproportionately poor or over-represented in low pay, dead end jobs because they are part of a debased, culturally deficient group. Having a narrative that smoothes over the cognitive dissonance and oft time schizophrenic dance that whites must do when they navigate race relations is an invaluable source of pleasure.

NOTES

1. The Gallup Organization, "Black/White Relations in the U.S.," *The Gallup Poll Monthly* (June 10, 1997): 1–5; David Shipler, *A Country of Strangers: Blacks and Whites in America* (New York: Vintage, 1998).

2. David Moore, "Americans' Most Important Sources of Information: Local News," *The Gallup Poll Monthly* (September 1995): 2–5; David Moore and Lydia Saad, "No Immediate Signs that Simpson Trial Intensified Racial Animosity," *The Gallup Poll Monthly* (October 1995): 2–5; Kaiser Foundation, *The Four Americas: Government and Social Policy through the Eyes of America's Multi-Racial and Multi-Ethnic Society* (Menlo Park, CA: Kaiser Family Foundation, 1995).

3. O. Ruth Frankenberg, "The Mirage of an Unmarked Whiteness," in *The Making and Unmaking of Whiteness*, eds. Birget Brander Rasmussen, Eric Klineberg, Irene J. Nexica, and Matt Wray (Durham: Duke University Press, 2001).

4. Ibid., 76.

5. Lawrence Bobo and James R. Kluegel, "Status, Ideology, and Dimensions of Whites' Racial Beliefs and Attitudes: Progress and Stagnation," in *Racial Attitudes in the 1990s: Continuity and Change*, eds. Steven A. Tuch and Jack K. Martin, 95 (Westport, CT: Praeger Publishers, 1997).

6. David Roediger, *The Wages of Whiteness: Race and the Making of the American Working Class*, 137 (New York: Verso Press, 1991).

KEY CONCEPTS

color-blind racism   prejudice   white privilege

DISCUSSION QUESTIONS

1. Summarize Gallagher's argument for why a color-blind attitude is still a privileged attitude. What does color-blindness *not* see when viewing race relations in America?

2. What is the problem with a generation of individuals who do not judge others by the "color of their skin"? Can the individualistic ideology of color-blindness coexist with a society of racist practices?

33

Barack Obama and the Politics of Race

The Myth of Postracism in America

MARTELL TEASLEY AND DAVID IKARD

Teasley and Ikard discuss the idea that President Barack Obama's election reflects a move toward a "post-racial society." Although many people want to believe this, they argue that a society where glaring inequalities of race persist can hardly be described as post-racial.

... For many citizens, including a significant segment of the African American population, Barack Obama's election does mean that the time has come to foreclose the discourse on race. An undeniable reality is that his presidency has engendered a new and indeed intoxicating feeling of optimism across race, class, and gender lines and pressed many of us to reassess, if not overhaul, our basic assumptions about the ways that "race matters" in the 21st century. Even though it is important not to underestimate the symbolic and real significance of Obama's historic presidency and the groundswell of interracial enthusiastic and goodwill that has accompanied it, it is equally as important not to overestimate it either. Using Obama's election as hard evidence that we have transcended race in the United States, many political proponents of postracial thinking are agitating for the end to all race- and ethnicity-centered social policy mechanisms aimed at reducing social inequities.

SOURCE: Teasley, Martell, and David Ikard. 2010. "Barack Obama and the Politics of Race: The Myth of Postracism in America." *Journal of Black Studies* 40: 411–425.

... In this essay, we demonstrate that despite all the advancements we have made to explode racial inequalities—advancements that have doubtless cleared the way for Barack Obama's historic rise to the presidency of the United States —we have a significant way yet to go and on multiple socioeconomic fronts before we can actualize true racial transcendence. Highlighting the salient disconnect between the Obama-inspired optimism among African Americans that a race-free society is imminent and the realities on the ground that reveal a decidedly bleaker social and economic outlook, we consider, at once, the pitfalls of postracial thinking as it pertains to African American agency and policy formation to end social inequities and the potential for Obama's "rhetoric of hope."

... THE DYNAMICS OF RACE AND POLITICS IN AMERICA

The notion of a postracial society has been with us for some time. It became a convenient tool of the political Right as a form of backlash to affirmative action policies enacted during the 1960s and the 1970s. It is a cause championed by those who benefit from its use as a form of social capital in maintaining the status quo of the American power structure (Marable, 2000) where African Americans and those of Hispanic/Latino decent own less than 5% of the wealth yet constitute at least 25% of the U.S. population; where the median net worth of African American families in the United States is $20,600 and that of Latino families is $18,600, compared to the $140,700 median net worth of Whites. African Americans, then, compose only 14.6% of the median net worth of Whites, and Latino families even less at 13.2% (Muhammad, 2008). Given these economic disparities, we are interested in the tangible analysis of race as a form of social capital, and we seek to assess substantive issues that are affected by postracial thinking and actions.

... THE MYTH OF A POSTRACIAL SOCIETY

We submit that many proponents of postracial thinking give race what logicians (Copi, 1978) refer to as *existential import*—the notion of attributing a tangible property to race that really does not exist—when it is politically convenient to do so or to intentionally obfuscate the realities of clear and present social inequity for oppressed populations in the United States.

... THE SYMBOLIC CAPITAL OF HOPE

Obama's famous rhetorical dexterity has given progressives as well as centrists reasons to believe he shares their values and outlook. (Hayes, 2008, p. 14)

... Obama's success is the outgrowth and culmination of a history of inspirational Black rhetoricians who have galvanized large interracial constituencies around the idea of hope and the possibility for racial reconciliation and equality. [Atwater 2009] cites the 1984 presidential campaign of Jessie Jackson as a key historical marker in the rise of the charismatic image and ethos of a unifying Black leader in American national politics. She asserts that his campaign simultaneously brought together national and international backing and transcended feeling of White supremacy through a unifying sense of universal possibility. Obama's "audacity of hope" rhetoric is for Atwater a revised and updated version of Jackson's "keep hope alive" slogan. Common rhetorically to both is their

> use of symbols to get Americans to care about ... [and] regain hope and faith in this country, and to believe that we [Blacks and Whites] are more alike than we are different with a common destiny and a core set of values. (p. 123)

However we contextualize Obama's legacy, the fact is his campaign and presidency have ushered in a new feeling of optimism in American, especially for African Americans. An April 2009 *CBS News/New York Times* poll bears this out, showing that the election of Barack Obama as president of the United States indicates that for the first time many Black Americans feel good about the overall state of race relations in the United States: "[59%] of African-Americans say race relations in the U.S. are good, compared with only 29% who thought so less than a year ago, before the election of Barack Obama" ("State of Race Relations," 2009, p. 1). Of Blacks, 61%, and 81% of Whites, agree that there has been real progress in diminishing racial discrimination in the United States since the 1960s. This compares to a December 1996 poll where only 37% of African Americans felt that real progress in race relations has been made since the 1960s. What is more, a December 2008 article in *The Economist* reports that 80% of Black Americans polled say that Obama's victory is "a dream come true," and 96% of them think it will improve race relations ("Search for the Promised Land," 2008).

The crucial issue becomes if this new and growing optimism will result in substitutive structural change. Though this is certainly a many-sided issue with an infinite number of potential outcomes, we focus on what we see as the best- and worst-case scenarios based on what we have witnessed thus far in Obama's presidency. To consider the best-case scenario first, African Americans are much more likely now than ever before in history to feel that they have agency because Barack Obama broke through not only a glass ceiling insofar as ascending to the highest office in the country but also a collective psychological racial barrier. Assuming that this African American optimism is sustainable beyond the euphoria of the historical election, it may result in a renewed energy to fight against social inequities based on race. It seems all but certain that it will encourage more African American participation in politics at the national and state levels. If people feel their perspectives and actions matter, there is a high probability that they will be more inclined to get involved in reshaping their communities for the better. For the youngest generation of African Americans, it may prove the biggest advantage of all because, unlike generations before them that hoped for, preached about, and agitated for these types of changes, they have at their disposal palpable evidence of what is possible. We can also find optimism in the fact that Obama has directly and repeatedly engaged the problem of social inequalities based on race, even if in strategic and lofty rhetoric that emphasizes nation over race, forward thinking over historical reckoning, interracial healing over group accountability. If we take into account the dynamics of racialized thinking that inform and complicate how Obama can talk about race and speak to social inequities based on race, we can view his fence-straddling rhetoric as a necessary, if regrettable, political ploy.

... But alas, there are significant, and perhaps even dangerous, consequences bound up in the, at times, uncritical optimism that abounds in Barack Obama's presidency and his near-hypnotic rhetoric of hope. Inspiring though it may be to Americans in general and African Americans in particular, Obama's rhetoric of hope has the potential to engender a false sense of hope, masking the realities of gross racial/ethnic disparities and inequality and worsening economic conditions, not only for many Black communities but also for the majority of Americans in general. The fact is that wealth disparities are increasing in the United States, particularly for the majority of African Americans and Hispanics/Latinos (Muhammad, 2008). To illustrate, the top 1% of income earners (the superrich) took home an average of $29.6 million in income in 2006, as compared to only $5.4 million in 1980 for this same group. Compare this to 1980, when "families in the bottom 90 percent averaged $30,446 in income, after adjusting for inflation, $72 more than the $30,374 comparable families earned in 2006" (Thompson, 2009, p. 25). In his essay "Race and Extreme Inequality" Muhammad (2008) explains that the small wealth gains of recent decades for Blacks and Latinos have all but evaporated with the subprime mortgage meltdown. When compared to Whites, Blacks are 3 times more likely to receive a subprime home loan and 4 times more likely to receive a subprime refinancing home loan. It is estimated that African Americans will lose between $71 million and $92 million during the snbprime financial meltdown; similarly, Hispanics will lose between $75 million and $92 million in the marketable worth of their homes from subprime loans.

We find it ironic that the media largely ignore the increasing economic and wealth inequality but are fascinated to the point of obsession with the less substantive properties of a postracial national dialogue. Void of critical analysis, and fixated on the possibility of a postracial America, the discourse on many pressing economic issues as they pertain to social inequality, class, and status is romanticized at best and vacuous at worst (Bobo & Charles, 2009). For example, the discourse on the national economic recession is framed as a discussion of how the challenges of American financial institutions will affect international markets and the plight of the middle class, but not the poor. Fiscal policies crafted by the executive branch of government and endorsed by Congress brazenly rewarded negligent investments with a financial bailout for lucrative banks and their executives at the expense of tax payers; this was succeeded by state legislators and municipalities cutting services and benefits to public universities and schools and health care programs for the poor. Banks and corporations cannot fail, but families and communities can. There is no discourse or discussion on the poor, who are disproportionately Black and Hispanic, and how they will fare in this era of economic downturn for the United States (Muhammad, 2008).

... Generally speaking, the social and economic success that Barack Obama has enjoyed before and after his rise to the presidency is atypical to the experiences of most African Americans. But the country's inherent fixation with race as an a priori in its national discourse and as an accompanying explanation for many of its social ills—mostly to the disadvantage of Black people—is something that even a master rhetorician like Obama may not be able to overcome, as his drop in presidential approval ratings indicates. As Julian Bond notes in an April 2009 forum on affirmative action, wealth, race, and ethnicity, "Changes in our society; not least in the election of our first African-American president, do not signal a shift in our [racial] temperature ... [or] mean the difficulty of the climb [to socioeconomic prosperity] has been erased for all others" (Curtis, 2009, p. 6). Moreover, a 2008 Pew Research poll shows that nearly half (45%) of African Americans born to middle-income parents during the post–civil rights era have descended into near poverty or poverty as adults (Younge, 2008).

But, the danger here is not just for African Americans: The symbolic capital that facilitates the development of a postracial euphoria, that beguiles America at the beginning of the Obama administration, has created a sort of mystique that somehow quells the clamor of national moral degradation and the impact of near financial implosion on the poor. It is not so much the ascendency of Barack Obama as the symbolic hope that he brings for all of America, renewing the country's "Camelot" experience, invoking the country's stance as a leader in the democratic world, and championing the country's subconscious psychic fixation with race as an inescapable, romantic national saga.

To be subjects of their own experience and destiny in America, Black people will need many structural changes in American society. These changes should include social policy enactments and monetary investments that reduce educational and health care disparities and outcomes, greater investment and revitalization in the nation's declining urban infrastructure where nearly 70% of African Americans reside, and grants for college and business entrepreneurship within Black communities (Muhammad, 2008). All of these suggested structural changes are similar to what FDR and the U.S. Congress enacted for White Americans at the end of World War II at the exclusion of Black Americans (Muhammad, 2008).

Although we acknowledge the significant milestone and meaning of the Obama presidency, we contend that in a capitalistic economy and society, freedom and self-determination in many ways constitute, at a minimum, economic viability and the accumulation of wealth. To this end, President Obama's rhetoric of hope cannot change the fact that African Americans compose 12% of the U.S. population but possess only approximately 3% of the country's wealth, that they have gained less than 2 percentage points in terms of wealth accumulation since post–Civil War Reconstruction (Davis & Bent-Goodley, 2004).

The truth of the matter is that African Americans would like nothing more than to end racism and racialized thinking in America. Many African Americans not only understand their victimization based on race in the past but also understand the more sophisticated methods of racial inequality that reduce their capacity to engage in self-determination today. The problem for African Americans in terms of racialized thinking is not that they are fixated on race as the central theme in the progression of their humanity and their participation in the American experience. Rather, it is, as W. E. B. Du Bois prophesized nearly a century ago, their "unforgivable [B]lackness" that America will not put to rest.

REFERENCES

Atwater, D. F. (2007). Senator Barack Obama: The rhetoric of hope and the American dream. *Journal of Black Studies*. 38. 121–129.

Bobo, L. D., & Charles, C. Z. (2009). Race in the American mind: From the Moynihan report to the Obama candidacy. *Annals of the American Academy of Political and Social Science*. 621(1). 243–259.

Copi, I. M. (1978). *Introduction to logic* (5th ed.). New York: Macmillan.

Curtis, K. (2009, April 16). Miller center debate: Should inequality be addressed by race-or-class-based affirmative action? *UVA Today*. Retrieved May 1, 2009, from http://www.virginia.edu/uvatoday/newsRelease.php?id=847l

Davis, K., & Bent-Goodley, T. B. (Eds.). (2004). *The Color of Social Policy*. Alexandria, VA: Council on Social Work Education.

Hayes, C. (2008, December 29). The pragmatist: It's a label Obama has embraced, but what does it mean? *The Nation*. pp. 13–16.

Kittles, R. (2008, January). *Race, Genetics, and Health*. Paper presented at the Society for Social Work Research conference, New Orleans, LA.

Marable, M. (2000). *How Capitalism Underdeveloped Black America: Problems in Race, Political Economy, and Society*. Boston: South End.

Muhammad, D. (2008, June 30). Race and extreme inequality. *The Nation*. p. 26.

Search for the promised land: What will Barack Obama's presidency mean for race relations? (2008, December 8). *The Economist*. Retrieved March 12, 2009, from http://www.economist.com/world/unitedstates/displayStory.cfm?story_id=12725114

The state of race relations. (2009, April 27). *CBS News/New York Times*. Retrieved May 7, 2009, from http://www.swamppolitics.com/news/politics/blog/2009/04/27/CBS%20poll%20on%20race%20relations.pdf

Thompson, G. (June 30, 2008). Meet the wealth gap. *The Nation*. pp. 18–27.

Younge, G. (2008, June 30). Beneath the radar. *The Nation*. p. 10.

KEY CONCEPTS

post-racial society   racial inequality   subprime mortgage

DISCUSSION QUESTIONS

1. What do people mean when they claim that the United States is post-racial? Is the meaning the same for White people as for African American people?
2. What realities of racial inequality do Teasley and Ikard identify? How well known are they to different groups in this society?

34

Mexican Americans and Immigrant Incorporation

EDWARD E. TELLES

Edward Telles reviews the complex and diverse characteristics of today's Mexican immigrants to the United States. In doing so, he questions the assumption that today's Mexican immigrants will have a similar experience as did European immigrants of the past.

The European American experience of incorporation is often described using the language and framework of "assimilation," wherein immigrants or their descendants eventually become an indistinguishable part of the dominant or mainstream society. However, an increasing number of sociologists argue that this may not always be true: today's immigrants are far less homogenous and encounter distinct circumstances and conditions when they arrive in the U.S. and as they become part of its society. For example, unlike the immigration of predominately low-skilled Europeans in the late 19th and early 20th centuries, today's immigrants are mostly from Latin America and Asia, they have varied skills and educational backgrounds, and many work in labor markets that offer fewer opportunities than before. The experience of today's immigrants with American society and culture, in other words, is more varied and uncertain than the old models can allow.

At the extreme, pundits like political scientist Samuel Huntington have argued that some new immigrants have not assimilated (or will not assimilate) and

SOURCE: Telles, Edward E. "Mexican Americans and Immigrant Incorporation."
Contexts 9: 28–33.

so they are a threat to American national unity. Similar, though usually more muted, claims about immigrant assimilation often involve cultural, economic and political worries about the new immigrants, which incidentally were similar to those raised during previous cycles of immigration. In any case, a careful examination of the evidence is important in order to design appropriate immigration and immigrant incorporation policies.

For examining the full range and complexity of the contemporary incorporation process, Mexican Americans, with their history, size, and internal diversity, are a very useful group. Their multiple generations since immigration, variation in their class backgrounds, the kinds of cities and neighborhoods they grew up in, and their skin color may reveal much about diverse patterns of immigrant incorporation in American society today. Unlike the study of most other non-European groups, the study of Mexican Americans allows analysts to examine the sociological outcomes of adults into the third and fourth generations since immigration.

SOME HISTORY

According to the U.S. Census Bureau, about 30 million people of Mexican origin currently live in the United States, and 13 million of them are immigrants. Mexicans comprise the largest group of immigrants in the U.S.—28 percent—so what happens to them and their descendants largely reflects what will happen to today's immigrants in general.

Moreover, Mexicans have been "coming to America" for over 150 years (before Americans came to them), and so there are several generations of U.S.-born Mexican Americans for us to study. (Ironically, analysts have mostly overlooked the fact that Mexican immigration is part of the old, or classic, period of immigration—seen as primarily European—as well as the new.) Each of these generations, successively more removed from the first-generation immigrant experience, informs our understanding of incorporation.

But first, we must start with approximately 100,000 Mexicans who instantly became Americans following the annexation of nearly half of Mexico's one-time territory. Since that year, Mexican immigration has been continuous, with a spike from 1910 through 1930. A second peak, beginning in 1980, continues today.

Mexico shares a 2,000-mile border with the United States. Until recently, Mexican immigration has been largely seasonal or cyclical and largely undocumented. The relative ease of entry and tight restrictions set by the U.S. government on immigrant visas for Mexicans have created a steady undocumented flow, which has increased in recent years. Demographers estimate that 7 million undocumented Mexican immigrants now live in the U.S.

The issue of race has also been important to the Mexican American experience throughout history. The U.S. based its conquest of the formerly Mexican territory (the current U.S. Southwest) on ideas of manifest destiny and the racial inferiority of the area's racially mixed inhabitants. Throughout the 19th and early 20th centuries, race-based reasoning was often used to segregate and limit Mexican American mobility. However, prior to the civil rights movement, Mexican American leaders strategically emphasized their Spanish roots and sought a white status for the group to diminish their racial stigma.

These leaders associated their belief in whiteness with the goal of middle-class assimilation, which they saw as possible for groups like southern and central Europeans, who were not considered fully white at the time. Indeed, historians like David Roediger show that European Americans were able to become white and thus fully included in American society through state benefits, such as home-ownership subsidies, that were largely denied to African Americans.

Mexican Americans didn't, however, succeed in positioning themselves on the "white track." Jim Crow-like segregation persisted against them until the 1960s, when a Chicano movement in response to discrimination in education and other spaces emerged among young Mexican Americans. The movement encouraged ethnic and racial pride by opposing continued discrimination and exclusion and drew on symbols of historic colonization.

Only a few Mexican Americans today can trace their ancestry to the U.S. Southwest prior to 1848, when it was part of Mexico, but this experience arguably has implications for the Mexican-origin population overall. This history of colonization and subsequent immigration, the persistence of racial stigmatization by American society, and the particular demographics involved in Mexican immigration and settlement make the Mexican American case unique and informative.

THE MEXICAN AMERICAN STUDY PROJECT, 1965 TO 2000

In 1993, my collaborator, Vilma Ortiz, and I stumbled upon several dusty boxes containing the questionnaires for a 1965 representative survey of Mexican Americans in Los Angeles and San Antonio. We believed that a follow up survey of these respondents and their children would provide a rare but much-needed understanding of the intergenerational incorporation experiences of the Mexican American population. Indeed, based upon this data set, we initiated a 35-year longitudinal study. In 2000, we set out to re-interview 684 of the surviving respondents and 758 of their children.

The original respondents were fairly evenly divided into three generations: immigrants (1st generation), the children of immigrants (2nd), and the grandchildren of immigrants (or later generations-since-immigration—the 3rd+). Their children, then, are of the 2nd, 3rd, and 4th+ generations. Using their responses from 2000, we examined change across these four generations regarding education, socioeconomic status, language, intermarriage, residential segregation, identity, and political participation.

We found that Mexican Americans experienced a diverse pattern of incorporation in the late 20th century. This included rapid assimilation on some dimensions, slower assimilation and even ethnic persistence on others, and persistent socioeconomic disadvantage across generations.

In terms of English language acquisition and development of strong American identities, these Mexican Americans generally exhibit rapid and complete assimilation by the second generation. They show slower rates of assimilation on language, religion, intermarriage, and residential integration, although patterns can also indicate substantial ethnic persistence. For example, 36 percent of the 4th generation continues to speak Spanish fluently (although only 11 percent can read Spanish), and 55 percent feel their ethnicity is very important to them (but, often also feel that "being American" is very important to them). Spanish fluency clearly erodes over each generation, but only slowly.

The results for education and socioeconomic status show far more incomplete assimilation. Schooling rapidly improves in the 2nd generation compared to the 1st but an educational gap with non-Hispanic whites remains in the 3rd and even by the 4th and 5th generation among Mexican Americans. (This stands in contrast to the European immigrants of the previous century who experienced full educational assimilation by the 3rd.) Although we see that conditions for Mexican Americans in 2000 have reportedly improved from their parents in 1965, the education and socioeconomic status gap with non-Hispanic white Americans remains large, regardless of how many generations they have been in the U.S. The 2000 U.S. Census showed that, among 35 to 54 year olds born in the U.S., only 74 percent of Mexican Americans had completed high school compared to 90 percent of non-Hispanic whites, 84 percent of blacks, and 95 percent of Asians.

… Educational assimilation remains elusive, but complete linguistic assimilation—or the loss of Spanish bilingualism—is nearly reached by the 5th generation.

Indeed, consistent with at least a dozen other studies, our evidence suggests that when the education of parents and other factors are similar across generational groups, educational attainment actually decreases in each subsequent generation.

THE CONTINUING IMPORTANCE OF RACE AND ETHNICITY

A high percentage of the Mexican Americans in our study claim a non-white racial identity. Even into the 3rd and 4th generations, the majority see themselves as non-white and believe they are stereotyped because of their ancestry. Nearly half report personal incidents of racial discrimination. Race continues to be important for them, and Mexican continues to be a race-like category in the popular imagination in much of the Southwest. In addition, the predominance and undocumented status of Mexican immigration coupled with large doses of anti-Mexican nativism may stigmatize all members of the group, whether immigrant or U.S.-born.

In many places, Mexican Americans are intermediate in the racial hierarchy, situated between whites and blacks (and newly arrived Mexican immigrants). Our survey did not directly examine the process through which race or racial stigma limits Mexican Americans. However, based on our in-depth interviews and other evidence, it seems that this occurs through both personal and institutional racial discrimination as well as through the internalization of a race-based stigma (which may affect life strategies and ambitions, especially during schooling). The geographical proximity of an underdeveloped and misunderstood Mexico and the persistent immigration of poorly educated (and often undocumented) Mexican workers may also reinforce the low status and the self-perceptions of Mexican Americans.

Low levels of education across generations also slows assimilation on other dimensions. Less-educated Mexican Americans of all generations earn less, are in less prestigious occupations, and are less likely to own their home than if they had more education. They are also more likely to live among, befriend, and marry other Mexican Americans; tend to have more children than their more-educated counterparts; are less likely to strongly identify as *American;* are less likely to vote; and are more tied to the Democratic party.

Finally, the large size and urban concentration of this population facilitates in-group interaction and limits exposure to out-group members. It also provides a large market for Spanish language media. Along with these, the continuous flow of immigrants from Mexico reinforces Spanish language fluency and use and provides incentives for later generation Mexican Americans to continue speaking Spanish. Also, the common use of Spanish language may raise nativist ire, which, in turn, may sharpen ethnic and racial identities for later generation Mexican Americans.

LESSONS FOR IMMIGRANT INCORPORATION

The Mexican American incorporation experience is not easy to sum up or generalize. But in many ways, that is precisely the point. The findings from the Mexican American Study Project demonstrate a range of outcomes and experiences. There are dimensions on which Mexican Americans assimilate as would be expected by the traditional (and most optimistic) theories. At the same time, there are other domains in which their experience is one of limited assimilation and even ethnic persistence. Particularly problematic is their experience in the educational realm, which leads to persistent socio-economic disadvantage across generations. Racial differences and stigmas can further contribute to these disadvantages, though the persistence of linguistic and other ethnic differences may be beneficial in other ways.

Perhaps because of immigration's centrality to the economy and social policies regarding immigrant incorporation, the heated immigration debates today are largely about whether or how long it will take the descendants of immigrants to assimilate in terms of schooling and the job market. In framing the debates about immigrant incorporation simply in these terms, we have neglected other dimensions of that process. The Mexican American case clearly demonstrates the multifaceted nature of the incorporation experience. Moreover, it has clear implications for how Americans—scholars and policy makers as well as the lay public—think about the incorporation of new generations of immigrants in their midst.

For example, there is a tendency to exaggerate the consistency of assimilation across dimensions. While examining the heterogeneous Mexican American population, we have shown that incorporation on particular dimensions may directly affect others and that the speed and direction of these dimensions may vary in unexpected ways.

To be certain, we have found that education affects nearly all other dimensions of assimilation. Moreover, we have also found that residential integration is a key intermediate variable where low education impedes one's ability to afford housing in an integrated middle class neighborhood, which in turn slows other dimensions such as intermarriage. A generation later, children who grew up in integrated neighborhoods and whose parents were intermarried are more likely to assimilate themselves. There may also be gradual assimilation on dimensions like retaining an ethnic language and increasing intermarriage, at the same time that there is rapid assimilation on learning English or no assimilation on educational attainment after the 2nd generation.

The study of Mexican Americans also points to the importance of looking at the diversity of the immigrant incorporation experience within groups. Previous findings mostly compare group averages or statistical distributions. We find, for example, that Mexican Americans in the second generation and beyond have lower educational levels and are more likely to end up with working class jobs than other groups. But, we also found a diversity of economic experiences among Mexican Americans, ranging from a few who move into the middle class and fall out of the ethnic community to others who are poor and are strongly rooted in the ethnic community, even into the 4th generation.

We often forget about the importance of history. This is understandable since many immigrant groups arrived at a specific time point so most group members experienced the same historical events. Most Italians that came to the United States, for example, arrived in the first fifteen years of the 20th century and experienced World War I as immigrants, World War II as 2nd generation ethnics, and as 3rd-generation Italian Americans fully integrated into the American mainstream by the 1970s.

For Mexican Americans, though, successive waves of immigrants have led to generations that experienced different historical events. We found that the experiences of incorporation for Mexican Americans depend largely on where they are inserted in history. The Mexican American Study Project disentangled generations-since-immigration from historical generations. By doing so, we found, for example, that the educational gap with whites has been narrowing for adults educated in the 1970s and 80s compared to those educated at mid-century. Spanish fluency has also diminished in recent decades for Mexican Americans of comparable generations-since-immigration. These are both indicators of group assimilation over historical time, though educational assimilation does not necessarily occur over generations-since-immigration.

Connected with this is the importance of examining multiple generations and at ages when they have completed their education and are well into their careers. Other empirical studies of incorporation have examined only the second generation that are in their 20s at the oldest, compared to their immigrant parents. This is largely due to the policy-related concerns of funders and researchers about how the children of the current wave of immigrants are faring. Our respondents, though, include the 3rd and 4th generation as well and are in their 30s, 40s and 50s, ages when they are more likely to have formed families and to have already availed themselves of the second chances that American society often provides, including the GED and occupational skills training. This gives us a fuller picture of incorporation.

Previous studies of incorporation have also generally over-looked local context. We also showed substantial variation in how Mexican Americans growing up in Los Angeles and San Antonio were incorporated. Overall, Mexican Americans in San Antonio had more ethnic lifestyles and behaviors, including retaining Spanish fluency into the third and fourth generation, but they were more politically conservative and identified as white to a greater extent than their Angelino counterparts. However, educational disadvantage was similar in the two urban areas. Variations in urban contexts are likely to affect how some immigrants or groups of immigrants and their descendants incorporate into society, especially as some areas place greater demographic or political pressures on assimilation. These factors may help account for differences in the incorporation of Mexican Americans compared to European Americans, whose ancestors arrived to New York and other east coast cities.

Finally, many previous studies of incorporation have emphasized a core to which immigrants and their descendants assimilate. But the case of Mexican Americans reminds us of the importance of a long-standing *Mexican American* core, which has arguably been a dominant model for assimilation for descendants of Mexican immigrants in many Southwest urban areas. This ethnic-based core represents models for Mexican American incorporation including acceptable occupations or class positions as well as cultural styles and models of political action.

Americans like to repeat the American narrative of immigrant success and assimilation, but that story doesn't describe the experience of many of today's immigrants. Even worse, to insist on the assimilation narrative as the story of all immigrants ignores the need for policies that address the specific needs and situations of different groups of immigrants. This neglect—born of a certain kind of historical optimism—comes at the peril of the lives of many Americans. But it also limits educational policies appropriate for the American economy, which increasingly requires an educated, employed, and integrated workforce and populace to maintain its international edge.

Perhaps the most basic and important lesson of the Mexican American incorporation experience, then, is the danger of trying to understand all immigrants with a single, one-size-fits-all model.

KEY CONCEPTS

assimilation   ethnic group   immigration   immigrant
                                                                       incorporation

DISCUSSION QUESTIONS

1. Why does Telles suggest that it is important to understand diversity *among*
    Mexican Americans if we are to fully understand their experience of incor-
    poration into American society?

2. In what ways do the experiences of today's Mexican immigrants differ from
    the experiences of earlier European immigrants? What does this suggest for
    the assimilation of Mexican immigrants into U.S. society and for social pol-
    icies to aid Mexican immigrants?

Applying Sociological Knowledge: An Exercise for Students

Consider how racial and ethnic groups are portrayed in the media. Take one type of television show (sitcom, sports broadcast, news, or drama) and watch two such shows for a specified period of time (say, one hour each). As you watch, systematically observe how different racial and ethnic groups (including Whites) are depicted. You should keep a tally of how many characters from different racial or ethnic groups are shown and how they are portrayed. Then ask yourself how images in the media shape racial stereotypes and group prejudice.

✳

PART XI

Gender

35

The Social Construction of Gender

MARGARET L. ANDERSEN

In this essay, Margaret Andersen outlines the meaning of the term "social construction of gender." She discusses the difference between the terms "sex" and "gender" and defines sexuality as it relates to both. After a brief discussion of the cultural basis of gender, the essay outlines the difference between a gender roles conceptualization of gender and the gendered institutions approach.

To understand what sociologists mean by the phrase *the social construction of gender*, watch people when they are with young children. "Oh, he's such a boy!" someone might say as he or she watches a 2-year-old child run around a room or shoot various kinds of play guns. "She's so sweet," someone might say while watching a little girl play with her toys. You can also see the social construction of gender by listening to children themselves or watching them play with each other. Boys are more likely to brag and insult other boys (often in joking ways) than are girls; when conflicts arise during children's play, girls are more likely than boys to take action to diffuse the conflict (McCloskey and Coleman, 1992; Miller, Danaber, and Forbes, 1986).

To see the social construction of gender, try to buy a gender-neutral present for a child—that is, one not specifically designed with either boys or girls in mind. You may be surprised how hard this is, since the aisles in toy stores are highly stereotyped by concepts of what boys and girls do and like. Even products such as diapers, kids' shampoos, and bicycles are gender stereotyped. Diapers for boys are packaged in blue boxes; girls' diapers are packaged in pink. Boys wear diapers with blue borders and little animals on them; girls wear diapers with pink borders with flowers. You can continue your observations by thinking about how we describe children's toys. Girls are said to play with dolls; boys play with action figures!

When sociologists refer to the social construction of gender, they are referring to the many different processes by which the expectations associated with being a boy (and later a man) or being a girl (later a woman) are passed on through society. This process pervades society, and it begins the minute a child is born. The exclamation "It's a boy!" or "It's a girl!" in the delivery room sets a course that from that moment on influences multiple facets of a person's life.

SOURCE: Margaret L. Andersen. 2003. *Thinking About Women: Sociological Perspectives on Sex and Gender.* Boston, MA: Allyn and Bacon, pp. 19–24. Reprinted with permission.

Indeed, with the modern technologies now used during pregnancy, the social construction of gender can begin even before one is born. Parents or grandparents may buy expected children gifts that reflect different images, depending on whether the child will be a boy or a girl. They may choose names that embed gendered meanings or talk about the expected child in ways that are based on different social stereotypes about how boys and girls behave and what they will become. All of these expectations—communicated through parents, peers, the media, schools, religious organizations, and numerous other facets of society—create a concept of what it means to be a "woman" or be a "man." They deeply influence who we become, what others think of us, and the opportunities and choices available to us. The idea of the social construction of gender sees society, not biological sex differences, as the basis for gender identity. To understand this fully, we first need to understand some of the basic concepts associated with the social construction of gender.

SEX, GENDER, AND SEXUALITY

The terms *sex, gender,* and *sexuality* have related, but distinct, meanings within the scholarship on women. Sex refers to the biological identity and is meant to signify the fact that one is either male or female. One's biological sex usually establishes a pattern of gendered expectations, although, … biological sex identity is not always the same as gender identity; nor is biological identity always as clear as this definition implies.

Gender is a social, not biological, concept, referring to the entire array of social patterns that we associate with women and men in society. Being "female" and "male" are biological facts; being a woman or a man is a social and cultural process—one that is constructed through the whole array of social, political, economic, and cultural experiences in a given society. Like race and class, gender is a social construct that establishes, in large measure, one's life chances and directs social relations with others. Sociologists typically distinguish sex and gender to emphasize the social and cultural basis of gender, although this distinction is not always so clear as one might imagine, since gender can even construct our concepts of biological sex identity.

Making this picture even more complex, sexuality refers to a whole constellation of sexual behaviors, identities, meaning systems, and institutional practices that constitute sexual experience within society. This is not so simple a concept as it might appear, since sexuality is neither fixed nor unidimensional in the social experience of diverse groups. Furthermore, sexuality is deeply linked to gender relations in society. Here, it is important to understand that sexuality, sex, and gender are intricately linked social and cultural processes that overlap in establishing women's and men's experiences in society.

Fundamental to each of these concepts is understanding the significance of culture. Sociologists and anthropologists define culture as "the set of definitions of reality held in common by people who share a distinctive way of life" (Kluckhohn, 1962:52). Culture is, in essence, a pattern of expectations about what are appropriate behaviors and beliefs for the members of the society; thus, culture provides prescriptions for social behavior. Culture tells us what we ought to do, what we ought to think, who we ought to be, and what we ought to expect of others....

The cultural basis of gender is apparent especially when we look at different cultural contexts. In most Western cultures, people think of *man* and *woman* as dichotomous categories—that is, separate and opposite, with no overlap between the two. Looking at gender from different cultural viewpoints challenges this assumption, however. Many cultures consider there to be three genders, or even more. Consider the Navaho Indians. In traditional Navaho society, the *berdaches* were those who were anatomically normal men but who were defined as a third gender and were considered to be intersexed. Berdaches married other men. The men they married were not themselves considered to be berdaches; they were defined as ordinary men. Nor were the berdaches or the men they married considered to be homosexuals, as they would be judged by contemporary Western culture....

Another good example for understanding the cultural basis of gender is the *hijras* of India (Nanda, 1998). Hijras are a religious community of men in India who are born as males, but they come to think of themselves as neither men nor women. Like berdaches, they are considered a third gender. Hijras dress as women and may marry other men; typically, they live within a communal subculture. An important thing to note is that hijras are not born so; they choose this way of life. As male adolescents, they have their penises and testicles cut off in an elaborate and prolonged cultural ritual—a rite of passage marking the transition to becoming a hijra....

These examples are good illustrations of the cultural basis of gender. Even within contemporary U.S. society, so-called "gender bending" shows how the dichotomous thinking that defines men and women as "either/or" can be transformed. Cross-dressers, transvestites, and transsexuals illustrate how fluid gender can be and, if one is willing to challenge social convention, how easily gender can be altered. The cultural expectations associated with gender, however, are strong, as one may witness by people's reactions to those who deviate from presumed gender roles....

In different ways and for a variety of reasons, all cultures use gender as a primary category of social relations. The differences we observe between men and women can be attributed largely to these cultural patterns.

THE INSTITUTIONAL BASIS OF GENDER

Understanding the cultural basis for gender requires putting gender into a sociological context. From a sociological perspective, gender is systematically structured in social institutions, meaning that it is deeply embedded in the social structure of society. Gender is created, not just within family or interpersonal relationships (although these are important sources of gender relations), but also within the structure of all major social institutions, including schools, religion, the economy, and the state (i.e., government and other organized systems of authority such as the police and the military). These institutions shape and mold the experiences of us all.

Sociologists define institutions as established patterns of behavior with a particular and recognized purpose; institutions include specific participants who share expectations and act in specific roles, with rights and duties attached to them. Institutions define reality for us insofar as they exist as objective entities in our experience....

Understanding gender in an institutional context means that gender is not just an attribute of individuals; instead, institutions themselves *are gendered*. To say that an institution is gendered means that the whole institution is patterned on specific gendered relationships. That is, gender is "present in the processes, practices, images and ideologies, and distribution of power in the various sectors of social life" (Acker, 1992:567). The concept of a gendered institution was introduced by Joan Acker, a feminist sociologist. Acker uses this concept to explain not just that gender expectations are passed to men and women within institutions, but that the institutions themselves are structured along gendered lines. Gendered institutions are the total pattern of gender relations—stereotypical expectations, interpersonal relationships, and men's and women's different placements in social, economic, and political hierarchies. This is what interests sociologists, and it is what they mean by the social structure of gender relations in society.

Conceptualizing gender in this way is somewhat different from the related concept of gender roles. Sociologists use the concept of social roles to refer to culturally prescribed expectations, duties, and rights that define the relationship between a person in a particular position and the other people with whom she or he interacts. For example, to be a mother is a specific social role with a definable set of expectations, rights, and duties. Persons occupy multiple roles in society; we can think of social roles as linking individuals to social structures. It is through social roles that cultural norms are patterned and learned. Gender roles are the expectations for behavior and attitudes that the culture defines as appropriate for women and men.

The concept of gender is broader than the concept of gender roles. *Gender* refers to the complex social, political, economic, and psychological relations between women and men in society. Gender is part of the social structure—in other words, it is institutionalized in society. *Gender roles* are the patterns through which gender relations are expressed, but our understanding of gender in society cannot be reduced to roles and learned expectations.

The distinction between gender as institutionalized and gender roles is perhaps most clear in thinking about analogous cases—specifically, race and class. Race relations in society are seldom, if ever, thought of in terms of "race roles." Likewise, class inequality is not discussed in terms of "class roles." Doing so would make race and class inequality seem like matters of interpersonal interaction. Although race, class, and gender inequalities are experienced within interpersonal interactions, limiting the analysis of race, class, or gender relations to this level of social interaction individualizes more complex systems of inequality; moreover, restricting the analysis of race, class, or gender to social roles hides the power relations that are embedded in race, class, and gender inequality (Lopata and Thorne, 1978).

Understanding the institutional basis of gender also underscores the interrelationships of gender, race, and class, since all three are part of the institutional framework of society. As a social category, gender intersects with class and race; thus, gender is manifested in different ways, depending on one's location in the race and class system. For example, African American women are more likely than White women to reject gender stereotypes for women, although they are more accepting than White women of stereotypical gender roles for children. Although this seems contradictory, it can be explained by understanding that African American women may reject the dominant culture's view while also hoping their children can attain some of the privileges of the dominant group (Dugger, 1988).

Institutional analyses of gender emphasize that gender, like race and class, is a part of the social experience of us all—not just of women. Gender is just as important in the formation of men's experiences as it is in women's (Messner, 1998). From a sociological perspective, class, race, and gender relations are systemically structured in social institutions, meaning that class, race, and gender relations shape the experiences of all. Sociologists do not see gender simply as a psychological attribute, although that is one dimension of gender relations in society. In addition to the psychological significance of gender, gender relations are part of the institutionalized patterns in society. Understanding gender, as well as class and race, is central to the study of any social institution or situation. Understanding gender in terms of social structure indicates that social change is not just a matter of individual will—that if we changed our minds, gender would disappear. Transformation of gender inequality requires change both in consciousness and in social institutions....

REFERENCES

Acker, Joan. 1992. "Gendered Institutions: From Sex Roles to Gendered Institutions" *Contemporary Sociology* 21 (September): 565–569.

Dugger, Karen. 1988. "The Social Location of Black and White Women's Attitudes" *Gender & Society* 2 (December): 425–448.

Kluckhohn, C. 1962. *Culture and Behavior*. New York: Free Press.

Lopata, Helene Z., and Barne Thome. 1978. "On the Term 'Sex Roles'" *Signs* 3 (Spring): 718–721.

McCloskey, Laura A., and Lerita M. Coleman. 1992. "Difference Without Dominance Children's Talk in Mixed- and Same-Sex Dyads" *Sex Roles* 27 (September): 241–258.

Messner, Michael A. 1998. "The Limits of 'The Male Sex Role' An Analysis of the Men's Liberation and Men's Rights Movements' Discourse." *Gender & Society* 12 (June): 255–276.

Miller, D., D. Danaber, and D. Forbes. 1986. "Sex-related Strategies for Coping with Interpersonal Conflict in Children Five and Seven" *Developmental Psychology* 22 543–548.

Nanda, Serena. 1998. *Neither Man Nor Woman: The Hijras of India.* Belmont, CA: Wadsworth.

KEY CONCEPTS

gender   gendered institution   gender socialization

DISCUSSION QUESTIONS

1. Walk through a baby store. Can you easily identify products for girls and for boys? Could you easily purchase gender-neutral clothing?

2. Consider an occupation that is traditionally men's work or traditionally women's work. What happens when a member of the other gender works in that field? What stereotypes and derogatory assumptions do we make about a woman working in a man's occupation or a man working in a woman's occupation?

36

Guyland

MICHAEL S. KIMMEL

Michael Kimmel explores "Guyland" as a social space where young men are caught between adolescence and adulthood. According to Kimmel, Guyland has particular social characteristics that stem from some of the social structural changes in gender relations that mark the current period of American society.

Welcome to Guyland.

SOURCE: Kimmel, Michael S., 2008. *Guyland: The Perilous World Where Boys Become Men.* New York: Harper.

Guyland is the world in which young men live. It is both a stage of life, a liminal undefined time span between adolescence and adulthood that can often stretch for a decade or more, and a place, or, rather, a bunch of places where guys gather to be guys with each other, unhassled by the demands of parents, girlfriends, jobs, kids, and the other nuisances of adult life. In this topsy-turvy, Peter-Pan mindset, young men shirk the responsibilities of adulthood and remain fixated on the trappings of boyhood, while the boys they still are struggle heroically to prove that they are real men despite all evidence to the contrary.

Males between 16 and 26 number well over 22 million—more than 15 percent of the total male population in the United States. The "guy" age bracket represents the front end of the single most desirable consumer market, according to advertisers. It's the group constantly targeted by major Hollywood studios, in part because this group sees the same shoot-em-up action film so many times on initial release. They're targeted in several of the most successful magazine launches in recent memory, magazines like *Men's Health, Maxim, FHM, Details,* and *Stuff.* Guys in this age bracket are the primary viewers of the countless sports channels on television. They consume the overwhelming majority of recorded music, video games, and computer technology, and they are the majority of first-time car buyers.

Yet aside from assiduous market research, Guyland is a *terra incognita*; it has never been adequately mapped. Many of us only know we've landed there when we feel distraught about our children, anxious that they have entered, or will be entering, a world that we barely know. We sense them moving away from us, developing allegiances and attitudes we neither understand nor support. Recently, a teacher at a middle school told me about his own 16-year-old son, Nick. "When we're together, he's excited, happy, curious, and so connected," he told me.

> "But when I drove him to school this morning, I watched an amazing transformation. In the car, Nick was speaking animatedly about something. As we arrived at his school, though, I saw him scan the playground for his friends. He got out of the car, still buoyant, with a bounce in his step. But as soon as he caught sight of his friends he instantly fell into that slouchy 'I don't give a shit' amble that teenagers get. I think I actually watched him become a 'guy'!"

Parents often feel we no longer know them—the young guys in our lives.

Just what are they doing in their rooms at all hours of the night? And what are they doing in college? And why are they so aimless and directionless when they graduate that they take dead-end jobs and move back home? When they come home for college vacations, we wonder just who is this new person who talks about ledge parties and power hours—and what happened to the motivated young man who left for college with such high hopes and a keen sense of purpose. And guys themselves often wonder where they left their dreams.

Every time we read about vicious gay-baiting and bullying in a high school, every time the nightly news depicts the grim horror of a school shooting, every time we hear about teen binge drinking, random sexual hookups, or a hazing death at a college fraternity, we feel that anxiety, that dread. And we ask ourselves, "Could that be my son?" Or, "Could that be my friend, or even my boyfriend?" Or, even "could that be *me*?"

Well, to be honest, probably not. Most guys are not predators, not criminals, and neither so consumed with adolescent rage nor so caught in the thrall of masculine entitlement that they are likely to end up with a rap sheet instead of a college transcript. But most guys know other guys who *are* chronic substance abusers, who *have* sexually assaulted their classmates. They swim in the same water, breathe the same air. Those appalling headlines are only the farthest extremes of a continuum of attitudes and behaviors that stretches back to embrace so many young men, and that so circumscribes their lives that even if they don't want to participate, they still must contend with it.

Guyland is not some esoteric planet inhabited only by alien creatures— despite how alien our teenage and 20-something sons might seem at times. It's the world of everyday "guys." Nor is it a state of arrested development, a case of prolonged adolescence among a cadre of slackers. It has become a stage of life, a "demographic," that is now pretty much the norm. Without fixed age boundaries, young men typically enter Guyland before they turn 16, and they begin to leave in their mid to late 20s. This period now has a definable shape and texture, a topography that can be mapped and explored. A kind of suspended animation between boyhood and manhood, Guyland lies between the dependency and lack of autonomy of boyhood and the sacrifice and responsibility of manhood. Wherever they are living, whatever they are doing, and whomever they are hooking up with, Guyland is a dramatically new stage of development with its own rules and limitations. It is a period of life that demands examination—and not just because of the appalling headlines that greet us on such a regular basis. As urgent as it may seem to explore and expose Guyland because of the egregious behaviors of the few, it may be more urgent to examine the ubiquity of Guyland in the lives of almost everyone else.

It's easy to observe "guys" virtually everywhere in America—in every high school and college campus in America, with their baseball caps on frontward or backward, their easy smiles or anxious darting eyes, huddled around tiny electronic gadgets or laptops, or relaxing in front of massive wide-screen hi-def TVs, in basements, dorms, and frat houses. But it would be a mistake to assume that each conforms fully to a regime of peer-influenced and enforced behaviors that I call the "Guy Code," or shares all traits and attitudes with everyone else. It's important to remember that individual guys are not the same as "Guyland."

In fact, my point is precisely the opposite. Though Guyland is pervasive—it is the air guys breathe, the water they drink—each guy cuts his own deal with it as he tries to navigate the passage from adolescence to adulthood without succumbing to the most soul-numbing, spirit-crushing elements that surround him every day.

Guys often feel they're entirely on their own as they navigate the murky shallows and the dangerous eddies that run in Guyland's swift current. They often stop talking to their parents, who "just don't get it." Other adults seem equally clueless. And they can't confide in one another lest they risk being exposed for the confused creatures they are. So they're left alone, confused, try-ing to come to terms with a world they themselves barely understand. They couch their insecurity in bravado and bluster, a fearless strut barely concealing a tremulous anxiety. They test themselves in fantasy worlds and in drinking con-tests, enduring humiliation and pain at the hands of others.

All the while, many do suspect that something's rotten in the state of Manhood. They struggle to conceal their own sense of fraudulence, and can smell it on others. But few can admit to it, lest all the emperors-to-be will be revealed as disrobed. They go along, in mime.

Just as one can support the troops but oppose the war, so too can one ap-preciate and support individual guys while engaging critically with the social and cultural world they inhabit. In fact, I believe that only by understanding this world can we truly be empathic to the guys in our lives. We need to enter this world, see the perilous field in which boys become men in our society because we desperately need to start a conversation about that world. We do boys a great disservice by turning away, excusing the excesses of Guyland as just "boys being boys"—because we fail to see just how powerful its influence really is. Only when we begin to engage in these conversations, with open eyes and open hearts—as parents to children, as friends, as guys themselves—can we both re-duce the risks and enable guys to navigate it more successfully. This book is an attempt to map that terrain in order to enable guys—and those who know them, care about them, love them—to steer a course with greater integrity and hon-esty, so they can be true not to some artificial code, but to themselves.

JUST WHO ARE THESE GUYS?

The guys who populate Guyland are mostly white, middle-class kids; they are college-bound, in college, or have recently graduated; they're unmarried. They live communally with other guys, in dorms, apartments, or fraternities. Or they live with their parents (even after college). Their jobs, if they have them, are modest, low-paying, low-prestige ones in the service sector or entry-level corpo-rate jobs that leave them with plenty of time to party. They're good kids, by and large. They blend into the crowd, drift with the tide, and often pass unnoticed through the lecture halls and multistory dorms of America's large college campuses.

Of course, there are many young people of this age group who are highly motivated, focused, with a clear vision and direction in their lives. Their stories of resilience and motivation will provide a telling rejoinder to many of the dom-inant patterns of Guyland. There are also just as many who immediately move back home after college, directionless, with a liberal arts BA that qualifies them for nothing more than a dead-end job making lattes or folding jeans. So while a few of them might jump right into a career or graduate school immediately after college, many more simply drift for a while, comforting themselves with the assurances that they have plenty of time to settle down later, after they've had their fun.

In some respects, Guyland can be defined by what guys do for fun. It's the "boyhood" side of the continuum they're so reluctant to leave. It's drinking, sex, and video games. It's watching sports, reading about sports, listening to sports on the radio. It's television—cartoons, reality shows, music videos, shoot-em-up movies, sports, and porn—pizza, and beer. It's all the behavior that makes the real grownups in their lives roll their eyes and wonder, "When will he *grow up*?!"

There are some parts of Guyland that are quite positive. The advancing age of marriage, for example, benefits both women and men, who have more time to explore career opportunities, not to mention establishing their identities, before committing to home and family. And much of what qualifies as fun in Guyland is relatively harmless. Guys grow out of a lot of the sophomoric humor—if not after their "sophomore" year, then at least by their mid-twenties.

Yet, there is a disturbing undercurrent to much of it as well. Teenage boys spend countless hours blowing up the galaxy, graphically splattering their computer screens in violent video games. College guys post pornography everywhere in their dorm rooms; indeed, pornographic pictures are among the most popular screen savers on male college students' computers. In fraternities and dorms on virtually every campus, plenty of guys are getting drunk almost every night, prowling for women with whom they can hook up, and chalking it all up to harmless fun. White suburban boys don do-rags and gangsta tattoos appropriating inner-city African-American styles to be cool. Homophobia is ubiquitous; indeed, "that's so gay" is probably the most frequently used put-down in middle schools, high schools, and college today. And sometimes gay-baiting takes an ugly turn and becomes gay-bashing.

All the while, these young people are listening to shock jocks on the radio, laughing at cable-rated T&A on the current generation's spin-offs of "The Man Show" and watching Spike TV, the "man's network," guffawing to sophomoric body-fluid humor of college circuit comedians who make Beavis and Butt-head sound quaint. They're laughing at clueless henpecked husbands on TV sitcoms; snorting derisively at guys who say the wrong thing on beer ads; snickering at duded-up metrosexuals prancing around major metropolitan centers drinking Cosmos and imported vodka. Unapologetically "politically incorrect" magazines, radio hosts, and television shows abound, filled with macho bluster and bikini-clad women bouncing on trampolines. And the soundtrack in these new boys' clubhouses, the sonic wallpaper in every dorm room and every shared apartment, is some of the angriest music ever made. Nearly four out of every five gangsta rap CDs are bought by suburban white guys. It is not just the "boys in the hood" who are a "menace to society." It's the boys in the "burbs."

Occasionally, the news from Guyland is shocking—and sometimes even criminal. There are guys who are drinking themselves into oblivion on campus on any given night of the week, organizing parties where they spike women's drinks with Rohypnol (the date rape drug), or just try to ply them with alcohol to make them more compliant—and then videotaping their conquests. These are the guys who are devising elaborately sadomasochistic hazing rituals for high-school athletic teams, collegiate fraternities, or military squads.

It is true, of course, that white guys do not have a monopoly on appalling behavior. There are plenty of young black and Latino boys who are equally desperate to prove their manhood, to test themselves before the watchful evaluative eyes of other guys. But only among white boys do the negative dynamics of Guyland seem to play themselves out so invisibly. Often, when there's news of young black boys behaving badly, the media takes on a "what can you expect?" attitude, failing to recognize that expecting such behavior from black men is just plain racism. But every time white boys hit the headlines, regardless of how frequently, there is an element of shock, a collective, "How could this happen? He came from such a good family!" Perhaps not identifying the parallel criminal behavior among white guys adds an additional cultural element to the equation: identification. Middle-class white families see the perpetrators as "our guys." We know them, we are them, they cannot be like that.

Though Guyland is not exclusively white, neither is it an equal-opportunity venture. Guyland rests on a bed of middle-class entitlement, a privileged sense that you are special, that the world is there for you to take. Upwardly mobile minorities feel the same tugs between claiming their rightful share of good times and delaying adult responsibilities that the more privileged white guys feel. But it often works itself out differently for them. Because of the needs and expectations of their families, they tend to opt for a more traditional trajectory. Indeed, many minority youths have begun to move into those slots designated for the ambitious and motivated, just at the moment that those slots are being abandoned by white guys having fun.

... And while the American college campus is Guyland Central, guys who don't go to college have ample opportunities—in the military, in police stations and firehouses, on every construction site and in every factory, in every neighborhood bar—for the intimately crude male bonding that characterizes Guyland's standard operating procedure. Sure, some working-class guys cannot afford to prolong their adolescence; their family needs them, and their grownup income, too badly. With no college degree to fall back on, and parents who are not financially able or willing to support a prolonged adolescence, they don't have the safety net that makes Guyland possible. But they find other ways, symbolic or real, at work or at play, to hold onto their glory days—or they become so resentful they seethe with jealous rage at the privileged few who seem able to delay responsibility indefinitely.

... The camaraderie of working-class guys long celebrated in American history and romanticized in Hollywood films—the playful bonding of the locker room, the sacrificial love of the foxhole, the courageous tenacity of the firehouse or police station—has a darker side. Homophobic harassment of the new guys, racial slurs, and seething sexism often lie alongside the casual banter of the band of brothers, and this is true in both the working-class bar and the university coffee house.

... Guyland revolves almost exclusively around other guys. It is a social space as well as a time zone—a pure, homosocial Eden, uncorrupted by the sober responsibilities of adulthood. The motto of Guyland is simple: "Bros Before Hos." (Long "o" in both Bro and Ho.) Just about every guy knows this—knows that his "brothers" are his real soul mates, his real life-partners. To them he swears allegiance and will take their secrets to his grave. And guys do not live in Guyland all the time. They take temporary vacations—when they are alone with their girlfriends or even a female friend, or when they are with their parents, teachers, or coaches.

GIRLS IN GUYLAND—BABES IN BOYLAND

What about girls? Guys love girls—all that homosociality might become suspect if they didn't! It's *women* they can't stand. Guyland is the more grownup version of the clubhouse on *The Little Rascals*—the "He-Man Woman Haters Club." Women demand responsibility and respectability, the antitheses of Guyland. Girls are fun and sexy, even friends, as long as they respect the centrality of guys' commitment to the band of brothers. And when girls are allowed in, they have to play by guy rules—or they don't get to play at all.

Girls contend daily with Guyland—the constant stream of pornographic humor in college dorms or libraries, or at countless work stations in offices across the country; the constant pressure to shape their bodies into idealized hyper-Barbies. Guyland sets the terms under which girls try to claim their own agency, develop their own senses of self. Guyland sets the terms of friendship, of sexual activity, of who is "in" and who is decidedly "out." Girls can even *be* guys—if they know something about sports (but not too much), enjoy casual banter about sex (but not too actively), and dress and act in ways that are pleasantly unthreatening to boys' fragile sense of masculinity.

… The entire landscape of Guyland is structured by the massive social and economic changes in the United States over the past several decades. As Susan Faludi documented in her book *Stiffed*, men who once found meaning and social value in their work are increasingly pushed into lower-wage service occupations; as the economy has shifted from a culture of production to a culture of consumption men experience their masculinity less as providers and protectors, and more as consumers, as "ornaments." Many men feel "downsized"—both economically and emotionally; they feel smaller, less essential, less like real men.

At the same time, women have entered every single arena once completely dominated by men. In the last three decades of the twentieth century, virtually every all-male college went coed, the military integrated as did police stations, and firehouses, and every single profession and occupation. Where once there were so many places where men could validate their masculinity, proving it in the eyes of other men, there are today fewer and fewer places where they aren't also competing with women.

It might seem ironic that Guyland encompasses an ever-expanding age spectrum, from mid-teens through the late twenties at the same time that the social space of Guyland is shrinking enormously. But young men are seeking what used to be so easy to find by pushing the age limits of their boyhoods as far into their twenties as they possibly can. That is why they are often so defensive: they've lost the casual ease of proving themselves to other guys that they once took for granted.

Yes, young men have always wanted to prove themselves, and that is nothing new. But today that desire has a distinct tone of desperation to it. In a world where their entitlement is eroding, where the racism and sexism that supported white male privilege for decades is taking hits left and right, where women are "everywhere they want to be," and affirmative action has provided at least some opportunities to minorities, the need for a "Band of Brothers" feels stronger than ever.

KEY CONCEPTS

Adolescence   hegemonic masculinity   homophobia
masculinity

DISCUSSION QUESTIONS

1. What is "Guyland" and what evidence does Kimmel use to analyze it as a unique social space?
2. What evidence of "Guyland" have you seen in your social environment, and how do you think it affects the men and the women in this environment?

37

Do Workplace Gender Transitions Make Gender Trouble?

KRISTEN SCHILT AND CATHERINE CONNELL

The research presented in this piece focuses on a group of transgender people and the challenges they face in the workplace. As biological males transition into women and biological females transition into men, the interactions with coworkers are strained. The interview data presented here offer a unique glimpse into how the "gender trouble" created by these transsexual transitions emerge and how it is negotiated and handled.

SOURCE: Schilt, Kristen, and Catherine Connell. 2007. "Do Workplace Gender Transitions Make Gender Trouble?" *Gender, Work and Organization*, 14: 596–618.

INTRODUCTION

Gendered expectations for workers are deeply embedded in workplace structures (Acker, 1990; Britton, 2004; Gherardi, 1995; Padavic and Reskin, 2002; Valian, 1999; Williams, 1995). Employers often bring their gender schemas about men and women's abilities to bear on hiring and promotion decisions, leading men and women to face very different relationships to employment and advancement (Acker, 1990; Britton, 2004; Valian, 1999; Williams, 1995). However, when an employer hires a man to do a 'man's job,' he or she typically does not expect this man to announce that he intends to become a woman and remain in the same job. Open workplace gender transitions—situations in which an employee undergoes a 'sex change' and remains in the same job—present an interesting challenge to this gendered division of labour. While the varied mechanisms that hold occupational sex segregation in place often are hidden, gender transitions can throw them into high relief. Becoming women at work, for example, can mean that transwomen lose high-powered positions they are seen as no longer suited for (Griggs, 1998). On the other hand, becoming men can make transmen more valued workers than they were as women (Schilt, 2006). Beyond illuminating deeply naturalized gendered workplace hierarchies, however, these open transitions also have the potential to make workplace 'gender trouble' (Butler, 1990), as transsexual/transgender people denaturalize the assumed connection between gender identity, genitals and chromosomal makeup when they 'cross over' at work.[1]

This article considers the impact of open gender transitions on binary conceptions of gender within the context of the workplace. Drawing on in-depth interviews, we illustrate how transsexual/transgender people and their co-workers socially negotiate gender identity during the transformative process of open workplace transitions. As gendered behavioural expectations for men and women can vary greatly depending on organizational cultures and occupational contexts (Britton, 2004; Connell, 1995; Salzinger, 2003), transmen and transwomen must develop a sense of how to facilitate same-gender and cross-gender interactions as new men or new women in their specific workplaces. In this renegotiation process, some of the interviewees in our study adopt what can be termed 'alternative' femininities and masculinities—gender identities that strive to combat gender and sexual inequality. However, regardless of their personal commitments to addressing sexism in the workplace, many transmen and transwomen are enlisted post-transition into workplace interactions that reproduce deeply held cultural beliefs about men and women's 'natural' abilities and interests. We argue that the strength of these enlistments, and the lack of viable alternative interactional scripts in the context of the workplace, limit the political possibilities of open workplace transitions. Rather than 'undoing gender' (Butler, 2004) in the workplace, then, individuals who cross over at work find themselves either anchored to their birth gender through challenges to the authenticity of their destination gender, or firmly repatriated into 'the other side' of the gender binary.[2]

Putting our focus on the context of the workplace, we examine how our interviewees negotiate cross-gender and same-gender workplace interactions after their open workplace transitions. Rather than causing gender trouble, transmen and transwomen report that their co-workers either hold them accountable to their birth gender, or repatriate them into the 'other side' of the gender binary. Even when our interviewees want to challenge this rigid binary thinking about gender, they can feel pressured to downplay their opposition because of their need to maintain steady employment. In the conclusion we argue that the potential of transgender/transsexual people to undo gender at work can be constrained by their real-life need to keep their jobs, as well as the slow-to-change gendered organizational cultures of most workplaces.

METHODS

This research was conducted with 28 transsexual/transgender people in Los Angeles, California and Austin, Texas between 2003 and 2005. Both cities have vibrant transgender communities, as well as recently adopted citywide employment protections for transgender workers. Seventeen interviewees came from the first author's study of the workplace experiences of transmen in southern California and 11 interviewees came from the second author's study of transwomen's workplace experiences in Austin, Texas. The interviewees were recruited from transgender activist groups, transgender support group meetings, transgender list-serves and personal contacts. All the interviews were conducted either in the interviewees' offices or homes, the interviewees' friends' homes, or at local cafés. The interviews ranged from two to four hours. All interviews were recorded, transcribed and later coded for analytic purposes.

In both studies, transsexual/transgender people's experiences at work both before and after their gender transition were examined via semi-structured interviews. In addition to their own personal sense of how their workplace experiences changed, the interviewees also were asked about the reactions of their co-workers and employers. This type of self-report data on interactional events presents a limited perspective, as we were not able to check the interviewees' perceptions with observational methods, such as participant observation. As workplace transitions are sensitive events, many interviewees also were uncomfortable with the idea of their co-workers being interviewed about their perceptions of the transition process. However, while observational data and co-worker interviews would have been a useful addition to this study, we do not feel that our reliance on self-report data is a fatal flaw of this research project. As Garfinkel (1967) argues, transgender/transsexual people can be considered 'practical methodologists' in regard to how gender 'works,' as they are forced to negotiate social identities as new men or new women. This negotiation process carries a deep awareness of gendered nuances in interactions, meaning that transmen and transwomen often clearly recognize differences in treatment 'before and after'—especially as changes in treatment often mean a gain of privilege (for transmen) or a loss of privilege (for transwomen).

DOING/PERFORMING GENDER AND THE
POSSIBILITIES OF GENDER TROUBLE

In 'Doing gender,' West and Zimmerman (1987) argue that while individuals have many different social identities that are constantly adopted, dropped, or made more salient in certain contexts, 'we are always women or men' (1987, p. 15). Emphasizing this omni-revelance of gender, West and Zimmerman make the case that individuals are always doing gender, as gender is a social process that is constantly negotiated, rather than something innate to men or women.

They posit that 'participants in interaction organize their various and manifold activities to reflect or express gender, and they are disposed to perceive the behavior of others in similar light' (1987, p. 4). As everyone is accountable to gendered expectations, gender inequality is socially produced and maintained in interactions, as men continually do dominance while women do deference. Yet, while doing gender, and thus doing inequality, is an active process, they argue, drawing on the ethnomethodological theory of Harold Garfinkel (1967), that this interactional work becomes obscured by incorrigible propositions about gender that position masculinity and femininity as binary opposites that occur as natural offshoots of biology (that is, chromosomes and genitals).

West and Zimmerman's conception of doing gender shares a great deal with Butler's theorization of gender performativity in *Gender Trouble* (1990). Butler draws on post-structuralist and psychoanalytic theories rather than sociological theories of symbolic interaction to frame her argument. Arguing against conceptions of gender as the result of a priori internal essences, she frames masculinity and femininity as citational—or, as Prosser paraphrases it, 'culturally intelligible stylized act[s]' (1998, p. 32) that are given meaning only through their constant repetition. Emphasizing the necessity of social norms to provide meaning to gender, Butler notes in *Undoing Gender*,

> One only determines 'one's own' sense of gender to the extent that
> social norms exist that support and enable that act of claiming gender for
> oneself. One is dependent on this 'outside' to lay claims to what is one's
> own. (Butler, 2004, p. 7)

With this emphasis on social norms, Butler, like West and Zimmerman, positions gender as an overarching system that restricts possibilities of gender expressions for men and women at the same time as it provides necessary structure for a 'livable life' (Butler, 2004, p. 8).

Seeing gender as a social product or a citation for which there is no original opens up the possibility of de-gendering the social world (Lorber, 2005), ungendering one's self (Bornstein, 1994), or undoing gender altogether (Butler, 2004). Envisioning how gender could be done differently in ways that alleviate inequality and remove restrictions on gender expressions requires a serious reconsideration of binary thinking on gender. As Gherardi notes:

> If we are to escape the gender trap, if we are to free ourselves of the
> idea that there exist two and only two types of individuals, if we are to

ensure that social differentiation is no longer based on sexual differentiation, we must destabilize all thought which dichotomizes (either male
or female) and hierarchies (male as one, as the norm, and female as the
other, the second sex). (1995, p. 4)

However, while some practical suggestions about how to de-emphasize the
social importance of gender have been offered (Lorber, 2005), challenging binary
thinking about gender often remains located in the realm of theoretical possibilities of the gender crossing of transgender people.

In theoretical conceptions of gender performativity and doing gender, what
can be referred to as the 'transgender subject,' a person who gender crosses in
some way, plays a central role. In 'Doing gender' (West and Zimmerman,
1987), how Agnes—a young woman who was born male—learns to *do* femininity without a biological claim to womanhood is used to illustrate the everyday
social production of gender.

Butler, in contrast, uses male-bodied individuals who perform femininity
in drag productions as a concrete example of gender performativity (Butler,
1990, 1993). As she notes, 'drag implicitly reveals the imitative structure of gender itself—as well as its contingency' (Butler, 1990, p. 130). Interestingly, Agnes
and Butler's drag performers can be seen as at odds with one another in their
theoretical deployment. While drag performers can be seen as purposefully playing with gender, or even purposefully challenging binary views on gender,
Agnes, like many individuals who can be classified as 'transsexuals', seeks 'very
pointedly to be nonperformative, to be constative, quite simply, to *be*' (Prosser,
1998, p. 35, emphasis in text). What these concepts share, however, is representing fixed, allegorical representations of transgender subjects—subjects who are
removed from actual, lived social contexts.

An empirical question that emerges from these theoretical conceptions is
whether transgender subjects can purposefully, or as an unintended consequence of their decision to gender cross, actually undo gender? Until recently
this question could be asked only rhetorically, as individuals who gender
crossed were 'encouraged' by medical and psychological personnel to 'pass' in
their destination gender (that is, not to reveal their gender transition) (Bolin,
1988; Bornstein, 1994; Stone, 1991). With the rise in transgender activism in
the last 15 years (Califia, 1997; Frye, 2000; Green, 2004; Whittle, 2002), people
who gender cross are increasingly opting to openly identify as transgender or
transsexual.

Open workplace transitions—the focus of this article—offer a new interactional context for examining the question of whether transgender people make
gender trouble for binary views on gender. In these transitions, individuals undergo gender changes while remaining in the same job. These transitions bring
gendered practices and interactions in the workplace—an organizational context
that is heavily gendered (Acker, 1990; Britton, 2004; Williams, 1995)—into
sharp focus, as transsexual/transgender people are actively violating the assumed
naturalness and permanency of gender (Garfinkel, 1967) under the gaze of
co-workers and supervisors.

These open workplace transitions are not necessarily motivated by a political desire for visibility. Rather, for many individuals, undergoing a gender transition is a way to make life more livable; staying in the same job during this process can provide welcome stability to a life that is in physical and emotional transition on many levels. Additionally, transgender/transsexual people can avoid losing any of their job history—a common sacrifice that accompanies the decision to find a new job where one's gender transition is unknown. Rather, co-workers can neutralize the transformative potential of open workplace transitions by encouraging transgender people to conform to rigid stereotypes of how men and women 'just are'.

This challenge theoretically could lead to an undoing of gender inequality in the workplace, as co-workers and employers could begin to rethink their stereotypes about gender, work performance, and 'natural' abilities. However, as we demonstrate in this article, even when transsexual/transgender people openly challenge what they view as sexist gender dynamics in the workplace, or attempt to introduce co-workers to a continuum of gender identities, the theoretical possibilities of undoing gender at work via open workplace transitions are not always born out. Rather, co-workers often neutralize the transformative potential of open workplace transitions, as they actively enlist transgender people to conforming to rigid stereotypes of how men and women 'just are'. By repatriating transgender and transsexual people into the gender binary, co-workers repair the potential trouble to their naturalized attitudes about gender—demonstrating, we argue, that in the gendered context of the workplace, the theoretical potential of gender crossing does not necessarily translate into an undoing of gender.

ANALYSIS

Workplaces are not gender-neutral locations filled with bodies, but rather complex sites in which gender expectations are embedded in workplace structures and reproduced in interactions (Acker, 1990; Britton, 2004; Williams, 1995). Gendered behavioural expectations for men and women vary greatly depending on organizational cultures and occupational contexts (Connell, 1995; Salzinger, 2003)....

In undertaking an open workplace transition, transgender people face the task of doing gender in their new social identity in a way that fits with both gendered workplace expectations and their personal gender ideologies.... Rather than challenging binary views on gender, we demonstrate that these renegotiations often push transmen and transwomen towards reproducing workplace gender hierarchies that privilege masculinity and devalue femininity, thereby reaffirming their co-workers' belief in the naturalness of the gender binary in the workplace.

Negotiating Cross-Gender Interactions

While doing gender reinforces socially constructed differences between men and women (West and Zimmerman, 1987), it also reproduces notions of a natural connection between men or between women. This presumption of a natural connection between people of the same gender can create specific behavioural expectations, such as demands for men in all-male groups to talk about sports and sex, or women in all-female groups to talk about romance and menstrual cycles. Even though transmen, for example, may not have felt that they fit into all women interactions at work prior to their transition, they still were often included in this female bonding based on their birth gender. Once transitions are officially announced, however, both transmen and transwomen can find themselves struggling—in what are now cross-gender relationships—to negotiate new gender boundaries and new interactional styles, and challenges to the authenticity of their destination gender.

New Gender Boundaries   After the public announcement of their gender transition in the workplace, both transmen and transwomen describe the erection of new cross-gender boundaries in workplace interactions. As transmen increasingly develop a masculine appearance, many find that they are less frequently included in 'girl talk' at work—generally conversations about appearance and dress, menstruation and romantic interests. For transmen who describe themselves as 'always already men' (Rubin, 2003) despite being born with a female body, these new boundaries are a relief. Illustrating this type of reaction, Aaron says:

> Even when I was living as a female, I never did get the way women interacted. And I was always on the outside of that, so I never really felt like one of them.

For Paul, who transitioned while working in one of the 'women's professions' (nursing and teaching), these new boundaries signaled a welcomed end to being held accountable to stereotypical feminine interactional expectations, such as noticing new hairstyles or offering compliments about hair and clothing—interactions he describes as not coming to him naturally.

Some transwomen also express relief about the cessation of gendered expectations to participate in stereotypically masculine interactions. Laura, who transitioned in a professional job, notes that her actual interactions with men changed little with her transition. When working as a man, he—at the time—had removed himself from 'guy talk'; generally conversations about cars, sports, and the sexual objectification of women, as participating in these types of interactions did not fit with his personal sense of being a woman, despite having a male body. Now that she has made her feminine identity public at work, she feels that men in her workplace have a new interpretative frame for understanding these boundaries. 'I think they understand a little more why I could have cared less who won the football game!'

Rather than challenging their ideas about the permanency of gender, interviewees felt that co-workers reincorporated their pre-transition interactions into an understanding of 'being transgender' and the innateness of gendered interests. In other words, Laura's lack of interest in football and Paul's lack of participation in the feminine niceties were re-evaluated as proof that transgender people are somehow trapped in the wrong body, a situation that is made right through a gender transition.

Not all interviewees felt a sense of relief at the creation or sudden acceptance of cross-gender boundaries. For some transmen who formerly identified as queer, bisexual or lesbian women, these new boundaries create a sense of sadness and exclusion. Describing this feeling, Elliott, who transitioned in a retail job, says:

> It's just like a little bit more of a wall there [with women] because I am not one of the girls anymore.... Like [women] have to get to know me better before they can be really relaxed with me.... I grew up surrounded by women and now to have women be kind of leery of me, it's a very strange thing.

Transmen who are saddened by their perceptions of a new distance between themselves and women in the workplace still try to be respectful of these boundaries. This acceptance, however, does little to challenge notions of the gender binary; rather, conceding a loss of participation in 'women's space' reifies divisions between men and women as natural. Yet, showing the importance of context in theorizing the potential of gender crossing to undo gender, transmen who seek to have masculine social identities at work have few other options, as most workplaces do not provide accepted interactional scripts for men who want to be just one of the girls. In order to keep social relationships smooth during the turmoil of an open workplace transition, then, transsexual/transgender people can hesitate to create additional challenges to gendered workplace expectations, as they desire to retain steady and comfortable employment.

In some cases, these new cross-gender boundaries can translate into workplace penalties. Agape, who transitioned in a high-tech company, remembers her boss worrying that taking oestrogen would adversely affect her programming abilities. She says:

> I think he just doesn't really have that high of an opinion of women. I think it's just he, he thinks fire and aggression is what gets things done.... And I guess he sees women as being more passive, and was worried my productivity would decrease.

Lana faced a similar situation. As a man, Lana co-owned a professional business, a company she describes as a 'real boys' club,' with three other men. While they began their business as close friends, the friendship did not survive the announcement that Lana—he at the time—intended to become a woman. Moving from being a hegemonically masculine man who did not outwardly acknowledge his inner feminine gender identity, Lana's transition disrupted the homosocial bonds of the company's power elite. After multiple expressions of their discomfort about both the transition and having a woman as a business partner, Lana was forced out of the company. Underscoring the gendered aspect of these drastic new boundaries, she recalls that during the negotiations to buy her shares of the company:

> The only thing I remember [my business partner] saying in the entire three days was, 'How you expect to run a company when all you're going to be thinking about is nail polish?'

As this comment suggests, Lana's partners locate their challenge to her in terms of gendered expectations that women cannot be serious business partners because they are too concerned with frivolities of appearance. Her transition does not undo gendered expectations, but rather is reincorporated into a workplace gender hierarchy that disadvantages and devalues women and femininity.

Interactional Styles   Open workplace gender transitions reveal the gender dynamics behind what are considered workplace-appropriate cross-gender interactions. Both transmen and transwomen recount their sudden realization that changing gender at work requires a renegotiation of once comfortable interactional styles. In some cases, transmen and transwomen make personal decisions about changing cross-gender interactions in an effort to meet their personal ideals of how men and women should act—such as the case of Preston, who transitioned from woman to man in a blue-collar job. When working as a woman, Preston—who publicly identified as a lesbian—describes frequently engaging in joking, sexualized banter with both men and women in her dyke-friendly workplace. However, after his transition, he suddenly felt uncomfortable engaging in similar conversations:

> I used to flirt a lot as a lesbian! It was easy for me to flirt … Since I have transitioned, a lot of the stuff that I could say as a dyke is so inappropriate! [laughs]. There is this one woman at work … she is just really straight. Very much. I used to tease her about … switching sides…. And if I say that now [as a man], it is just like so fucking inappropriate. There is no way for me to find a justification for that, even though that is the history of our relationship. It is the history of how we have interacted with one another…. [It] could be perceived wrong. Even though my motive for it hasn't changed, but it is still inappropriate.

Preston's sense of discomfort with this interactional style translates into adopting a policing role toward sexualized banter at work. His co-workers were surprised at his behavioural changes, as he used to engage in the same type of behaviour he now critiques. However, for him, this type of behavioural change was necessary, as he did not want to enact a form of masculinity that can be construed as sexist. As he gains cultural competency in the variety of ways men and women interact, he might feel more leeway to adopt different interactional styles, as some men in his workplace did engage in sexualized banter with women. At the onset of transition, however, many transmen err on the side of caution by policing their behaviour, as—even with legal protections for gender identity in the workplace—they can feel vulnerable as openly transgender employees.

Other changes to cross-gender interactional styles can be a result of implicit or explicit pressure from co-workers. Ellen, who transitioned from man to woman in a customer service job, describes implicit pressure to tone down stereotypically masculine styles of interaction:

> There is one thing that really drives me crazy—when I'm asked for my opinion on a subject [from men], I have to remember—'Do not express it as firmly as I actually believe'.

While she personally does not wish to change, she realizes that muting opinions and emotions is the predominant interactional style for women in her workplace. She continues:

> At work I tend not to trumpet my own horn very much, and the workplace environment demands that [women keep quiet]. I don't know if that's anything about me as transgender. I think that's just being a woman.

While this change to her interactional style does reproduce men doing dominance and women doing deference (West and Zimmerman, 1987), Ellen, like Preston, feels she needs to make these concessions in order to gain a feminine social identity at work and to keep friendly relationships with her co-workers.

Pressures to change cross-gender interactional styles also can be explicit. Several transmen describe women in their workplaces enlisting them into what can be termed 'gender rituals' (Goffman, 1977), stereotypical interactions that are typically played out between heterosexual men and women. After the announcement of their transition from woman to man, transmen recount women raising expectations that they will now, as men, do any requisite heavy lifting around the workplace, such as changing office water bottles, moving furniture, or carrying heavy boxes. Interestingly, this change in behavioural expectations occurs almost immediately after the transition announcement. The change was so rapid that many transmen were, at first, not sure how to make sense of these new expectations. Kelly, who transitioned in a semi-professional job, notes:

> Before [transition] no one ever asked me to do anything really and then [after], this one teacher, she's like, 'Can you hang this up? Can you move this for me?... Like if anything needed to be done in this room, it was me. Like she was just, 'Male—okay, you do it'. That took some adjusting. I thought she was picking on me for a while. And then I realized that she just, she just assumes that I'm gonna do all that stuff.

Ken describes a similar experience in his semi-professional workplace. While his co-workers were slow to adopt masculine pronouns with him, women did enlist him in performing masculine-coded duties in the workplace immediately after his transition announcement, such as carrying heavy items to the basement and unloading boxes.

Negotiating Same-Gender Interactions

In crossing over at work, transmen and transwomen must grapple with a new same-gender reference group. Fitting into same-gender groups is particularly salient in these types of open workplace transitions, as transmen and transwomen eventually must gain access to same-gender sanctuaries in the workplace: the men's room and the women's room respectively. People in these new same-gender reference groups, men for transmen and women for transwomen, might be expected to offer opposition to their transitioning colleague, as they do not have a biological claim to these gendered spaces. Yet, our interviewees generally report being included in same-gender spaces and interactions, and, in some cases, being taken on as gender apprentices. As with cross-gender interactions, these renegotiated interactions do not disrupt binary views on gender, but rather repatriate transmen and transwomen into us versus them enactments of masculinity and femininity.

Inclusion in Same-Gender Spaces   Attempting to gain access to same-gender spaces in the workplace, such as bathrooms and locker rooms, can create a great deal of anxiety for people transitioning at work. In transsexual/transgender autobiographies, bathroom horror stories abound, such as Kate Bornstein's (1994) description of being forced to use a women's room that was under construction and several floors away from her desk when she transitioned at IBM. However, while negotiating these spaces can take finesse, most transmen and transwomen in this study describe being included in same-gender spaces at work.

For transmen, this inclusion often came by direct invitation from men. Keith, who transitioned in a blue-collar job, waited to move into the men's locker room until he had been taking testosterone for several months and felt that he passed as a man. Elliott took a similar approach at his retail job. Eventually men in their workplaces began to ask Keith and Elliott when they were going to start using the locker room, signalling that they were open to this change. Once in the locker room, men also went out of their way to make Keith and Elliott feel comfortable by including them in locker room social interactions. Douglas and Trevor had similar experiences when they began to use the men's room in their social services jobs. Both of them worked predominantly with gay men who explicitly encouraged them to feel comfortable using the men's restroom. Douglas' co-workers even held a mock ceremony in which they presented him with his own key to the men's room.

Transwomen also describe their inclusion in same-gender workplace spaces positively. Carolina notes that she eats lunch with the 'girls' at work now, as she is included in this woman space. Agape describes her excitement at her new ability to interact with women in the bathroom, saying, 'It's so nice to be able chat in the bathroom—it's so nice! I never realized I was missing that because I never had it before!' However, this same-gender inclusion was not so readily available for all transwomen. Jackie recalls an initial coldness toward her on the part of other women at work. Eventually she found acceptance by befriending one socially influential woman. She says of the process:

> It's like a waterfall. It took one person to get through … and then the rest came forward. You know, at every workplace there are 'the girls', you know, the social clique…. As soon as [one woman] started being nice to me, the rest followed.

Interactional Styles   Open workplace transitions bring with them opportunities for engaging in same-gender interactions as new men or new women. In describing their relationships with other women in the workplace post-transition, transwomen express a new sense of freedom. When they were working as men, many transwomen had very stereotypically masculine workplace personas, as do many pre-transition transwomen (Griggs, 1998). Achieving these personas meant that they did not acknowledge their personal sense of relating more to women and 'women's interests' than to men and 'men's interests'. As women, however, they now are able to openly express interest in feminine things that they often denied when they were working as men. Describing this new freedom, Laura, who transitioned in a professional job, notes:

> I got a bigger field of friends in this building. And of course, they're all female, because we all have lots of talk about. You know, I have grandchildren that range from two and up, so you know, we can talk about kids, we can talk about babies, you know, just about anything any other woman would talk about is what I'm knowledgeable and like to talk about. I like to cook, I like to sew. So it makes it pretty easy.

Prior to her transition, she did not, as a man, attempt the same types of interactions with women, as she worried these interactions would be seen as inappropriate, or she would be labelled as an atypical or gay man.

Illustrating the greater leeway for women to admit interest in activities coded as masculine (Thorne, 1993), transmen do not recount having to hide their preference for masculinity—indeed, most transmen in this study describe themselves as embodying this preference in their personal appearance. Yet, many transmen recount men explicitly engaging them in 'guy talk' immediately after their announcement of their impending transition. Kelly, who transitioned in a semi-professional workplace where men were the minority, says:

> I definitely notice that the guys ... they will say stuff to me that I know they wouldn't have said before [when I was working as a woman].... And like one guy, he never talked to me before. I think he was uncomfortable [that I was a lesbian].... Recently we were talking and he was talking about his girlfriend and he's like, 'I go home and work it [have sex] for exercise'. And I know he would never have said that to me before.

Jake describes a similar enlisting into what can be described as masculine gender rituals. In his professional workplace, he recounts making few changes in his interactional style with men in his workplace. However, he notes:

> One of the funny things that happened that was gender specific was that a lot of my male colleagues, at least at first, started kind of like slapping me on the back [laughs]. But I think it was with more force than they probably slapped each other on the back.... And it was not that I had gained access to 'male privilege' but they were trying to affirm to me that they saw me as a male.... That they were going to try to be supportive and that was the way they were going to be supportive of me as a guy, or something of the sort [laughs]. Slapping me on the back.

As he remarks, he does not take this backslapping as a signal that these men have forgotten his birth gender. Rather, he interprets these actions as a kind of social validation performance his colleagues are acting out in an attempt to signal acceptance. The awkwardness of these backslaps illustrates his colleagues' own hyperawareness of trying to casually do gender man to man. As Jake actively cultivates a transman identity, he, like Kelly, is uncomfortable with this incorporation, as he perceives it as intended to gloss over his life history. Yet, while he is able to disrupt this incorporation momentarily by mentioning things from his life as a woman, such as when he was a Girl Scout, he notes that men in his workplace appear more comfortable trying to relate to him as 'just a guy' rather than a transman with a female history.

Gender Apprenticing   Rather than challenging the authenticity of transmen and transwomen's destination gender, some same-gender colleagues took their transitioning colleague on as what can be described as a gender apprentice. For transmen, this form of apprenticing typically came from heterosexual men who sought to socialize them into how to be a man. Colin, who transitioned in a professional workplace, remembers being stopped by the director of his office the first day he came to work in a tie. 'He's like, "Oh no, no, no. That's not a good tie. Come here!" And he showed me how to tie a Windsor knot.' In this situation, his older colleague adopts Colin as a younger protégé, teaching him 'masculine' knowledge—how to tie a Windsor knot—that typically is handed down from father to son.

While transwomen describe less frequent occurrences of gender apprenticing, they are, in contrast to transmen, typically appreciative of these apprenticing efforts of women they work with. This gender difference in reactions may be a result of the different reactions to gender crossing. In other words, as women, pre-transition transmen have more leeway for adopting masculine appearances and behaviour, which gives them more experience with masculinity when they transition. As men, on the other hand, pre-transition transwomen face severe social sanctions for expressing interest in feminine styles and behaviour. This difference, an adult version of the 'tomboy/sissy' dynamic (Thorne, 1993), means that transwomen have little experience with how to do femininity once they become women, and are appreciative of women's efforts to socialize them. Describing this reaction, Laura, who transitioned in a professional job, recounts how moved she was when a woman at her work took her shopping for make-up:

> I've had other women here help me with make-up.... There's a lady here who said, 'Oh, let me do it! I can show you simple things'. I said, 'Ok!' and we went shopping, that kind of thing.... I forgot how it was brought up—but somebody said, 'You know, Crystal loves to do makeup'. And I said, 'I'm gonna have to get a hold of her and see what her ideas are'. And that is how it began. And so—I paid for lunch and we ran out to Target or someplace and picked out a few things that would work. And she taught me how to put it on.

Laura initiates her own apprenticing by directly approaching her colleague for make-up advice. This apprenticing allows her to develop more confidence in her feminine appearance. While other transwomen describe being 'allowed' to engage in 'girl talk' about children, romance, and fashion, they do not recount such strong incorporation into their destination gender by women as transmen do from men.

DISCUSSION

In this article we consider the impact of open gender transitions on binary conceptions of gender in the context of the workplace…This article shows that in open workplace transitions, co-workers, rather than transmen and transwomen, can overdo and reinforce gender. This over-doing of gender typically occurs when co-workers attempt to demonstrate their acceptance of their transitioning colleague.

As everyone re-negotiates the meaning of gender and sexual difference made visible by open workplace transitions, binary thinking about gender is often upheld and the resulting gender hierarchies interwoven with heteronormativity and sexism are reproduced. Transmen and transwomen can be frustrated by the rigid gender expectations placed upon them by co-workers. For some transwomen, facing the devaluation of femininity in the workplace is detrimental to their careers, as they are rejected from powerful homosocial men's networks or classified as less able workers. The reactions transmen and transwomen describe to their gender transitions suggest that co-workers may face more anxieties about how to properly do gender in open workplace transitions than their transitioning colleague. Rather than causing gender trouble, however, these anxieties result in a reinforcement of binary views on gender through the reproduction of gendered hierarchies that disadvantage women and rigid adherence to the 'right' way to do gender.

Demonstrating how shared birth gender may impact on reactions to surgery, transwomen also face workplace resistance to their transition predominantly from heterosexual men at work. Transwomen have noted that biological men have a visceral reaction to the news of their 'sex change,' as they can vividly imagine the removal of their own penis (Griggs, 1998). Transwomen also are viewed as giving up male privilege, particularly if they were hegemonically masculine prior to transition. From this place of being the same, heterosexual men may have difficulty justifying why a man would want to move to a lower gender status, and remove his penis; the literal signifier of patriarchal dominance.

Transmen, on the other hand, are moving up the gender hierarchy. Women, then, may balk at what they perceive as gaining gender privilege. Exemplifying this attitude, several transmen noted that some women in their workplaces made comments that transmen were—now as men—going to benefit from the 'good ole boys' network'. Even with this challenge, however, the gender binary is upheld; transitioning individuals are simply returned to the 'reality' of their birth gender (women trying to be men, men acting like women) rather than being viewed as a third gender, or being repatriated into the 'other side' of the binary.

Looking at the behaviour of heterosexual men co-workers from the perspective of the social construction of masculinity, and the dovetailing of doing gender and doing heterosexuality, these incorporation responses in are not entirely surprising. While these men may not see their transman colleague as really a man, they recognize that in this particular context—the workplace—this identity is being socially validated. And, as their colleague begins to look like a man with hormone therapy, raising oppositions to their masculinity becomes problematic. Transmen noted that heterosexual men asked them questions about the physical aspects of their transition when it was first announced at work, such as 'Are you gonna grow a dick?' But as they began to look like men with the use of hormone therapy, they are enlisted into 'guy talk' about sex with women and working out. As homophobia governs men's interactions with one another (Connell, 1995; Messner, 1997), men appear to realize that showing a continued interest in the genitalia of someone who looks like a man can render them suspect to other co-workers. To avoid charges of homosexuality, it is easier for heterosexual men to 'forget' about the transition and try to relate to transmen on the level just being a guy.

These reactions show that in open workplace transitions, co-workers, transmen and transwomen all are renegotiating and managing gender and sexual difference. Yet, within this identity work, these data suggest there is little initial challenge in the workplace to naturalized attitudes about the immutability of gender, binary views about the complimentary nature of masculinity and femininity, or gendered workplace hierarchies. In other words, the mere introduction of a visibly transgender subject does not result in an undoing of gender or the creation of gender alternatives, such as a third gender category or a gender continuum (Bornstein, 1994; Garber, 1992).

In conclusion, we suggest that theoretical conceptions about the transformative potential of gender performances that are not in line with birth gender (Butler, 1990, 2004) should pay close attention to context, as well as the way in which these performances are socially interpreted. While intentional gender trouble performances can have political possibilities, such as in certain drag performances (Rupp and Taylor, 2004), they also can—as in the context of the workplace—be repatriated into a binary or dismissed as inauthentic.

NOTES

1. The term 'transgender' has become an umbrella term for a variety of gender identities (Green, 2004). 'Transsexuals', individuals who transition from one recognized gender category to another, can be placed under the transgender rubric. However, some individuals in this study prefer to identify as transsexual rather than transgender. Our use of 'transsexual/transgender' in this article is an effort to allow them to maintain their subjectivity, rather than choosing a label for them. We also employ the terms 'transwomen' and 'transmen,' as these terms—though not representative of specific identities—allow us a less cumbersome way to talk about gender differences between the collectives of our interviewees.

2. We use the term 'destination gender' to refer to social gender identity transsexual/transgender people seek to attain with their transition. In other words, the destination gender for transmen is male, while the destination gender for transwomen is female. The destination gender is in opposition to what we term 'birth gender', the gender transsexual/transgender people are assigned at birth.

REFERENCES

Acker, Joan. (1990). Hierarchies, jobs, bodies: A theory of gendered organizations. *Gender & Society*, 4, 2, 139–58.

Bolin, Anne. (1988). *In Search of Eve: Transsexual Rite of Passage*. South Hadley, MA: Bergin & Garvey.

Bornstein, Kate. (1994). *Gender Outlaw: On Men, Women, and the Rest of Us*. New York: Routledge.

Britton, Dana. (2004). *At Work in the Iron Cage: The Prison as Gendered Organization*. New York: New York University Press.

Buffer, Judith. (1990). *Gender Trouble: Feminism and the Subversion of Identity*. New York: Routledge.

Butler, Judith. (1993). *Bodies That Matter: The Discursive Limits of Sex*. New York: Routledge.

Butler, Judith. (2004). *Undoing Gender*. New York: Routledge.

Califia, Patrick. (1997). *Sex Changes: The Politics of Transgenderism*. San Francisco, CA: Cleis Press.

Connell, Robert. (1995). *Masculinities*. Berkeley, CA: University of California Press.

Frye, Phyllis Randolph. (2000). Facing discrimination, organizing for freedom: The transgender community. In D'Emilio, John, Turner, William and Vaid, Urvashi *Creating Change: Sexuality, Public Policy, and Civil Rights*, pp. 451–468. New York: St. Martin's Press.

Garber, Marjorie. (1992). *Vested Interests: Cross-dressing and Cultural Anxiety*. Routledge: New York.

Garfinkel, Harold. (1967). *Studies in Ethnomethodology*. Englewood Cliffs, NJ: Prentice-Hall.

Gherardi, Sylvia. (1995). *Gender, Symbolism and Organizational Cultures*. London: Sage.

Green, Jamison. (2004). *Becoming a Visible Man*. Nashvillle, TN: Vanderbilt University Press.

Griggs, Claudine. (1998). *S/he: Changing Sex and Changing Clothes*. New York: Berg.

Lorber, Judith. (2005). *Breaking the Bowls: Degendering and Feminist Change*. New York: W. W. Norton.

Messner, Michael. (1997). *The Politics of Masculinities: Men in Movements*. Thousand Oaks, CA: Sage.

Padavic, Irene and Reskin, Barbara. (2002.) *Women and Men at Work*, 2nd ed. Thousand Oaks, CA: Pine Forge Press.

Prosser, Jay. (1998). *Second Skins: The Body Narratives of Transsexuality*. New York: Columbia University Press.

Rubin, Henry. (2003). *Self Made Men: Identity and Embodiment among Transsexual Men*. Nashville, TN: Vanderbilt University Press.

Rupp, Leila and Taylor, Verta. (2004). *Drag Queens at the 801 Cabaret*. Chicago, IL: University of Chicago Press.

Salzinger, Leslie. (2003). *Genders in Production: Making Workers in Mexico's Global Factories*. Berkeley, CA: University of California Press.

Schilt, Kristen. (2006). Just one of the guys? How transmen make gender visible at work. *Gender & Society*, 20,4, 465–90.

Stone, Sandy. (1991). The empire strikes back: A post-transsexual manifesto. In Epstein, Julia and Straub, Kristina (eds.), *Body Guards: the Cultural Politics of Gender Ambiguity*, pp. 280–304. New York: Routledge.

Thorne, Barrie. (1993). *Gender Play: Girls and Boys in School*. New Brunswick, NJ: Rutgers University Press.

Valian, Virginia. (1999). *Why So Slow?: The Advancement of Women*. Cambridge, MA: MIT Press.

West, Candace and Zimmerman, Don. (1987). Doing gender. *Gender & Society*, 1,1, 125–51.

Whittle, Stephen. (2002). *Respect and Equality: Transsexual and Transgender Rights*. Portland, OR: Cavendish.

Williams, Christine. (1995). *Still a Man's World: Men Who Do 'Women's' Work*. Berkeley, CA: University of California Press.

KEY CONCEPTS

transsexual   doing gender   gender trouble

DISCUSSION QUESTIONS

1. How is the concept of "doing gender" represented in this article? What are some examples of someone who was female at birth acting in masculine ways? What are some examples of someone who was male at birth acting in feminine ways?

2. Are there other situations when "gender trouble" becomes a problem?

38

Trading on Heterosexuality

College Women's Gender Strategies and Homophobia

LAURA HAMILTON INDIANA UNIVERSITY

In this piece, Laura Hamilton explores women's homophobia on college campuses. The data presented are from a larger study at a Midwestern "party" school. The women participate in fraternity and off-campus parties in ways that are consistent with heterosexual expectations where women act in erotic and sexualized ways for the attention of men. Lesbians are therefore marginalized in this party atmosphere. This piece offers a unique look into the otherwise overlooked homophobia among women.

Scholars note that homophobia plays a central role in the construction of masculinities (Connell 1987; Corbett 2001; Kimmel 2001; Pascoe 2005). Indeed, as Corbett (2001) notes, the term *faggot* stands in for more than sexual insult: It connotes a failure to be fully masculine. "Real" men repudiate the feminine or that which they perceive to be weak, powerless, and inconsequential (Kimmel 2001). The hegemonic form of masculinity thus supports men's dominance over women and other men in subordinated positions because of race, class, or sexuality (Connell 1995). The literature on masculinities suggests that homophobia occurs when men try to perform hegemonic masculinity. By verbally or physically attacking men whom they perceive as not masculine, men may reassert their own manhood (Corbett 2001; Kimmel 2001; Pascoe 2005). When relying solely on this conceptualization, homophobia takes on gendered characteristics, underscoring a particular masculine manifestation of antihomosexual behaviors as quintessentially homophobic.

Past research seems to support the association of homophobia with men: For instance, studies often find that women have more positive attitudes toward homosexuality (Loftus 2001). Giddens (1992, 28) has even predicted that women will be the vanguard in creating a space for "the flourishing of homosexuality." Yet, it is possible that women's homophobia remains obscured when conceptualizing homophobia as a singular phenomenon. As Stein (2005) suggests, homophobia can take many forms and operate through multiple

SOURCE: Hamilton, Laura. 2007. "Trading on Heterosexuality: College Women's Gender Strategies and Homophobia." *Gender & Society* 21: 145–172.

mechanisms. Homophobia may also be central to the development of certain feminine selves but not in the same way as for masculine selves. Because women and men are in different positions with regard to power, women's homophobia may support gendered identities that are most successful in garnering men's approval (Rich 1980). Some women may distance themselves from others who do not perform the erotic selves that they perceive as valued by men. These women may exhibit homophobia to maintain the believability of their traditionally feminine identities.

In this article, I draw on ethnographic and interview data from a women's floor of a residence hall on a public university campus to suggest that heterosexual women may display homophobia against lesbians as they negotiate status in a gender-inegalitarian erotic market. First, I describe the Greek party scene on this campus, the erotic hierarchy linked to it, and lesbians' low ranking within this hierarchy. I then explain that women who were active partiers excluded lesbians from social interactions and spaces while critical partiers and nonpartiers were more inclusive. Finally, I describe how heterosexual women conceptualized the same-sex eroticism that they used to garner men's attention and the consequences that this had for lesbians. I conclude by discussing how gender inequality and heteronormitivity combine to create homophobia among women.

GENDER STRATEGIES: "TRADING ON" HETEROSEXUALITY

Scholars have used Swidler's (1986) concept of "strategies of action" to show how women create "gender strategies" that help them navigate inegalitarian gender conditions. A gender strategy is a course of action that attempts to solve a problem using the cultural conceptions of gender available to the individual (Handler 1995; Hochschild 1989). Gender strategies are thus both cognitive and behavioral. They are not, however, always reflexive. In interaction, decisions and actions often occur quickly and nonreflexively. Women may fall into well-established patterns of behavior that pull from available cultural definitions of femininity and masculinity. Consequently, they can engage in gender strategies without awareness of the gendered aspects of their actions (P. Y. Martin 2003).

Gender strategies involve the use of particular gender presentations over others. These presentations do not reflect preexisting internal qualities but become engrained in people's bodies through the constant repetition of particular movements, acts, and thoughts (Butler 1990; K. A. Martin 1998). Premised on gender difference, heterosexuality is one of the key mechanisms through which women and men learn to embody gender. Given women's subordinate position, much of what makes a woman traditionally feminine is her ability and desire to attract a man (Bartky 1990). Women learn to produce feminine bodies and to have desires for men that conform to heterosexual imperatives. Many of the roles from which they gain their identities—such as girlfriend, wife, and mother—further emphasize the centrality of heterosexuality to gender identity (Jackson 1996).

EROTIC MARKETS AND HETEROSEXUAL PRIVILEGE

A ubiquitous element of youth cultures, erotic markets are expanding to include larger segments of the population for longer periods of their lives. Erotic markets are public sexualized scenes in which individuals present erotic selves that are subject to the judgments and reactions of others (Collins 2004). These markets require a mass of individuals who share similar assumptions about the kinds of sexual activity that are open for negotiation and how to interpret the sexual activity that does occur.

Many erotic markets operate using heteronormative cultural logics. This does not mean that all people within these scenes are heterosexual or that all erotic behaviors in this scene occur between women and men; rather, the available cultural understandings in heterosexual erotic markets reflect heteronormative ideas about sexuality, what "sex" is, and for whom it is performed. Because heterosexuality presumes gender difference, these meanings also code "real" sex as that which is penetrative or initiated by men and position women as desired objects rather than desiring subjects (Armstrong 1995; Jackson 1996). As a result, same-sex eroticism between conventionally feminine women becomes a performance for men, one that inevitably ends in heterosexual sex (Jenefsky and Miller 1998).

Within erotic markets, hierarchical rankings sort individuals by both successful participation *and* perceived desirability to potential partners. These rankings often transfer into other social relationships, marking status even when individuals are outside of erotic markets. Rankings are determined, in part, through social activities that are "organized by flirtation and sexual carousing" (Collins 2004, 253). Individuals who are not skilled, interested, or successful at engaging in these activities face exclusion from this avenue to status and the social networks of those who are high status. They must also perform gender in ways that others recognize as legitimate and desirable. For women within heterosexual erotic markets, this means performing a conventionally feminine identity.

Heterosexual relations are often organized in ways that benefit men (Jackson 1996). Past research has documented the gender imbalance in power, resources, and status that operates in erotic markets on college campuses—particularly those in which Greek organizations are present (Armstrong, Hamilton, and Sweeney 2006; Boswell and Spade 1996; Handler 1995; Holland and Eisenhart 1990; Martin and Hummer 1989; Stombler and Martin 1995). In these situations, women can use heterosexual performances to access benefits through their relations with men (Schwalbe et al. 2000). Many women who identify as heterosexual are privileged in heterosexual erotic markets in ways that lesbians are not and invest in maintaining their privilege (Rich 1980). These investments may not be fully conscious—women's participation in the heterosexual erotic system can preclude the kind of social contact with lesbians that fosters acceptance.

SOCIAL DISTANCE: ASSESSING HOMOPHOBIA
AMONG WOMEN

Social distance is the degree of closeness that people are willing to tolerate in their interactions with a stigmatized group (Gentry 1987). Goffman's (1963) work on stigma suggests that people often avoid encounters with stigmatized individuals because of interactional ambiguities and a fear of contamination by association. Inserting social distance is one way to mitigate these perceived costs of engaging in social interaction with "different" individuals (Milner 2004).

Particularly among women, homophobia often appears as a form of social distance. Socialized into "niceness," women may not always participate in the direct, aggressive, and publicly visible behaviors that many equate with homophobia among men (Gilligan 1982; K. A. Martin 2003). Research on adolescents suggests that women often use exclusionary projects—such as the maintenance of social distance—to mark the difference between themselves and "others" (Eder 1985; Merten 1997). In college, lesbians pose unique interactional threats to heterosexual women if they fail to engage in the appropriate erotic activities or present traditionally gendered selves in heterosexual erotic markets. Heterosexual women may also feel that lesbians are sexualizing the previously "safe" (i.e., heterosexual) backstage area of the residence hall floor. By maintaining social distance from lesbians, many heterosexual women assuage their fears of status contamination and quell anxieties about their own sexuality.

METHOD AND DATA

Data for this study are from ethnographic observation, individual, and group interviews conducted at a large midwestern research university as part of a project on collegiate life. One goal of the project is to understand how dominant groups on campus maintain and reproduce environments in which they are privileged. For example, all 43 of the women in this study were white. In addition, most came from middle- to upper-class families, identified as heterosexual, and had traditionally feminine gender presentations. Only two identified as lesbian, six were from working-class families, one was born outside of the United States, and another was isolated for her noncompliance with norms of appearance. Therefore, most embodied a femininity that the prevailing erotic market of the campus rewarded—if they chose to participate.

Most of the data were collected as part of an ethnography conducted throughout the 2004-2005 academic year on a women's floor in a mixed-gender residence hall that was identified by students and staff as a "party dorm." The title does not refer to partying within the residence hall itself; instead, students are attracted to this residence hall because it offers the most direct route into the dominant party scene on campus. Students from all residence halls gather outside of this and other party dorms en route to parties, making a party dorm a good site to study the dominant party culture on campus. Roughly one-third of incoming students are housed in party dorms; these residence halls feed the greatest number of students into the Greek system (which includes about 20 percent of students). While students cannot choose to live in party dorms, they can request certain areas of the campus. Some selectivity does occur, as many students pick particular areas because of the party dorms within them. Yet, even party dorms include students who are at least initially less party oriented.

A research team including one faculty member, five graduate students, and three undergraduates conducted the ethnography; five team members identified as heterosexual women and one as a gay man. Our team occupied a room on the floor we were observing. During the first semester, at least one member of the research team was there three to four weekday afternoons and evenings and one to two weekend afternoons and evenings per week. In the second semester, I was there two weekday evenings and one weekend evening weekly. Members of the team took notes about each interaction after the observation periods were completed. Interviews with floor residents occurred throughout the academic year and lasted between 1.5 to 2.5 hours. After each interview, we took notes. I conducted the majority of interviews but also relied on data collected by others.

Researchers formed different types of relationships with women based on their age, position in the university, and shared interests and/or tastes. As I identify as white, upper–middle-class, and heterosexual and have a fairly traditional gender presentation, I was able to connect with most women on the floor. Yet, this did not hinder me in forming close relationships with the out lesbians on the floor or several of the working-class women. As women on the floor generally associated with those of similar status, they often did not realize that individual researchers also knew others on the floor. This allowed me to move among different social groups with ease. Researchers only brought up sensitive topics in interview settings. However, discussions about issues such as sexuality did occur spontaneously. Our relationships with respondents did not change perceptibly after completing interviews or observing these sensitive discussions, perhaps because we did not reveal our own political and social attitudes.

Of the 53 women in the hall, we interviewed 43. This article focuses on the 43 residents with whom we completed interviews. All of these women were first- or second-year students. As older students—particularly seniors—may age out of the party culture, this study is most representative of processes occurring in the early years of college. Although all women on the floor were part of the ethnography, interview data allowed me to confirm social distance to lesbians. During interviews, women referred to their actual contact with lesbians on campus, what—if anything—they did to maintain social distance, and their preferred level of contact. In no case did women present attitudes that did not match observed behaviors toward out lesbians on the hall. Based on observations, the 11 women who are not included in the article are representative of other women in the hall in terms of their orientation to partying and fall into levels of social distance in proportions similar to the rest of the hall.

I accepted the sexual orientation that women claimed across multiple data points—in interviews, surveys that we administered, and interactions with friends or the research team. The women who identified as heterosexual did not indicate otherwise across any of these settings in the course of an entire academic year. Recognizing that sexual identity may be concealed, is fluid, and may vary across multiple dimensions (i.e., political, social, sexual, etc.), it is entirely plausible that some of the heterosexual women in this study may privately see themselves as bisexual or lesbian or acknowledge this in different social contexts or during later periods of their lives. I am limited to the reported self-understandings of sexual identity that were in play during the ethnography. Regardless of their self-understandings in other aspects of their lives, the women who claimed public heterosexual identities could profit in keeping social distance from out lesbians. As I discuss later, many of them did simultaneously imitate same-sex erotic practices, but they generally did so *only with an audience of men.*

I also include data from a group interview with lesbian and bisexual women on campus conducted in spring 2004 to examine the impact of heterosexual women's same-sex eroticism on other women's experiences of social space on campus. This group interview was obtained though student organizations on campus and covered a variety of topics including sexuality, relationships, partying, and the Greek scene. Although they are marginalized, gay, lesbian, bisexual, and transgendered students do have resources geared toward recognizing their needs; for instance, they have access to alternative housing, support groups, discussion forums, rights-oriented organizations, and a few social venues. Yet, these resources and institutional policies against discrimination by sexual orientation do little to challenge the heterosexual social world on campus, and, for first-year students who are placed in "party dorms," knowledge of them may be limited.

THE EROTIC HIERARCHY OF THE GREEK PARTY SYSTEM

Although only one of the many social "games" on campus, the Greek party scene is the largest and most well known among students. Many arrive on campus anticipating participation in the drunken social world portrayed by MTV and other youth media as "college"; in fact, students often head off to party before they attend their first college class or unpack their possessions. Erotic interactions between men and women play a central part in this world.

The Greek Party Scene

The Greek party scene is a sexualized social arena that is temporally and spatially specific. It occurs in the evenings in fraternities and in popular bars known for their laxness in enforcing laws against underage drinking. All of these fraternities are effectively white organizations; the few Black and multicultural Greek organizations do not have on-campus houses that can accommodate large parties. Although fraternity houses host parties with varying themes, they all revolve around a predictable "party routine" in which women are expected to drink, flirt, and socialize (Armstrong et al. 2006). Bars often serve as secondary sites for those who party at fraternities. Thus, the party scene achieves a level of co-hesiveness. As one respondent put it, regardless of where you are, "it's the same party, exactly the same frat, the same people.' Fraternities have a monopoly on this scene because they provide "free" alcohol to underaged women who other-wise might not be able to obtain it. This resource, combined with little univer-sity policing and private ownership of communal spaces, allows them to dictate almost every aspect of the parties they hold (Armstrong et al. 2006).

For example, many fraternities operated a one-way transport system in front of the "party dorm" that we observed. Starting the week before school began, fraternity men waited in the latest sport utility vehicles to drive women to their parties. First-year women clustered in this area and had little control over their destinations. Fraternity men also dictated party themes, most pressuring women to arrive scantily clad. Women described attending parties such as "Golf Pro/ Tennis Ho," "Trophy Wife and James Bond Husband," "Playboy Mansion," and "CEO/Secretary Ho." In addition, fraternities screened admission into their parties. One evening I observed a fraternity member selecting what appeared to be the most attractive and scantily clad women to receive the first ride; some-times he even split up friendship groups. Women also reported that fraternity men rejected non-Greek men to create a favorable gender ratio. Finally, frater-nity men determined the flow of guests and alcohol in their houses. Several women described men luring them into private spaces to receive alcohol. One noted that "Every guy [asks] you wanna drink, you wanna, oh, come see this ... oh, let's close this, and closes the door, and I just get so annoyed."

Women's Erotic Status

The party scene privileged individuals who actively participated in the erotic market. Because fraternity men controlled important party resources, one had to attract their attention to be included in the party. A woman explained, "Well, I flirt with guys.... I just pretty much do that so we can go play flippy cup (a drinking game) or get free beer." The lesbians on the hall found this exchange to be intolerable. One described a party she attended as follows:

> I was uncomfortable ... in the sense that all of the girls kind of have to compete with each other to get the alcohol, and it just screams so much like prostitution to me. You know, even if they're not literally having sex with the guys, it's just like they're ... selling their flirtiness for beer or something, and that's just so not me.

She felt that fraternity men treated women who were unwilling to "trade on" their erotic interest as lower status and less deserving of alcohol. For this reason, she no longer attended Greek parties. The other lesbian on the hall never attempted to attend, stating, "I will never go. I don't want to go. It's not my scene at all."

Most heterosexual women who partied found men's erotic attention both important and rewarding. One woman noted that the best thing about "kissing guys" at parties was not physical pleasure but "know[ing] that a guy's attracted to you and is willing to kiss you. It's kinda ... like a game to play just to see." Women even felt that not receiving this attention could be damaging to one's self-esteem. A woman with a long-term boyfriend described the costs of not seeking men's approval: "I was like the little conservative, country bumpkin in my outfit. I was like, no, I'm not going to get any of the attention. They're not going to waste their time with me....You need to flirt; that's good for your confidence." Failing to signal interest in obtaining men's approval could also result in embarrassment. Another woman said that she was mortified when she unknowingly showed up at the "CEO/Secretary Ho" party dressed as an actual secretary wearing a long-sleeved blouse and a knee-length skirt. When she walked in the door, a fraternity member flashed her a sarcastic thumbs-up, telling her, "Nice outfit."

The importance that most women placed on men's erotic interest translated into a clear hierarchy among them. At the top of this hierarchy stood "the blonde." By definition, all "blonde" women were white, having tan skin and light-colored hair. They were also thin, trendy, and sociable. Women felt that men found all of these traits to be desirable. One woman explained that being "blonde" was when "all the guys are like, 'Oh my god they are so hot.'" The seemingly organic nature of the "blonde" appearance belayed the extensive bodily work that went into managing a "blonde" body. For example, navigating the line between "good" and "bad" tan (looking "orange," as the women put it) involved knowing how to tan and when to stop. Many women struggled to maintain slender physiques while engaged in a party lifestyle that involved drinking a lot of beer and eating late-night pizza. Money was also essential; women often used colored contacts, hair straighteners, and salon hair coloring to appear more "blonde."

"Blondeness" also implied erotic interest in and appeal to heterosexual men. Part of indicating their interest in men involved actively working to avoid signaling homosexuality. For example, a woman told me about having a rainbow-colored arm cast in junior high, noting that she would never get one now as people might think that she was a lesbian. These women often assumed that others who did not exhibit a high-status gender presentation were lesbians. During a discussion in a dorm room one evening, several of them recoiled with disgust at a picture of tennis star Serena Williams, noting that her extremely defined muscles made her look "mannish" and like a lesbian. Because sexual orientation is not necessarily visually apparent, they equated gender conformity with sexual conformity. Most heterosexual women believed that this method could detect lesbians, whom they assumed to be "boyish." Both out lesbians on the floor dressed "sportier" than other women (often in sweatpants or T-shirts and rarely in makeup—even at night). After the women came out as lesbians, others insisted that they already had guessed based on their appearance. As one noted,

"Definitely you can tell … there are people that have the stereotype…. They've got a way about them that they're probably gay."

Although heterosexual women generally did not believe that lesbians could be "hot," several did revaluate their ranking of lesbians based on this possibility. When I asked one woman how she would feel about having a "hot" lesbian roommate, she explained,

> If my roommate was a lesbian and she was more feminine, I think I would be more comfortable…. [If she was] like me—she looked girly—it wouldn't matter if she liked guys or girls. But if it was someone that was really boyish, I think it would be hard for me to feel comfortable.

As Gamson (1998) noted of talk show audiences, heterosexual women on the hall often found the idea of lesbians who conformed to gender norms less problematic than those who did not. Regardless of her actual availability to men, the "hot lesbian" would at least look available.

However, if she were unwilling to enter the party scene and "sell her flirtiness for beer," a hot lesbian—like any other woman—would find her access to erotic status severely limited. The lesbians on the floor were thus doubly disadvantaged; first, by their refusal to participate in the erotic market and then by their choice not to perform "blondeness."

MAPPING SOCIAL DISTANCE FROM LESBIANS

Women on the floor had varied relationships to the Greek party scene. Most were highly invested in this scene, but a number were critical or opted out of the party scene altogether. [W]omen also differed in their willingness to interact, establish relationships, and share personal space with lesbians. All of the women who were most involved in the party scene fell into the two outer rings of social distance, while those who invested less required less social distance from lesbians.

Active Partiers

I defined active partiers as those women who (a) reported attending a fraternity party at least once a week for the majority of the academic year and (b) generally expressed satisfaction with this scene. Thirty women met these criteria; 19 of them joined sororities. I spent hours talking to women in this group as they prepared hair, makeup, and outfits for "going out." For most of them, partying was one of the major activities of college life. One avid partier explained, "I guess the only things I feel like I do here are study and party. My life is split between those things." Many emphasized the thrill of dressing "all sexy" for these parties. As a woman noted of a Playboy mansion party, "It's an excuse for everyone to just like dress in the sluttiest little thing that they can pull off without looking like complete trash. It was just so fun because you have an excuse to just like let loose, and there were so many people there." Along with the erotic energy of this scene, they also took pleasure in drinking. One woman exclaimed, "I almost feel getting drunk is like—I'm so happy! I guess that's what we mostly do." These women also felt that partying was a ubiquitous part of campus life ("There's always a party going on here, you know?") in which almost every student was perceived to participate ("That's what practically everybody is doing on the weekends").

Among active partiers, there was a distinction between women who felt that homosexuality was "never okay" and those who felt that homosexuality was "okay for others but not in my space": For the six heterosexual women in the "never okay" category, religious beliefs were the guiding principle shaping their desired level of social distance from gays and lesbians. All of these women grew up in religious communities or rural towns. They saw homosexuality as a clear-cut moral issue; it was always wrong for both men and women. As two room-mates explained:

R1: We've been sheltered around diversity.... Everybody's a farmer, everybody's the same.

R2: Hardly [any] gays or anything, so we're not used to all this gay pride stuff, and it's like, What are they doing? Read the Bible.

None, however, were part of Christian groups on campus; in fact, these groups did not approve of participation in the sexualized and wild party scene. Instead, women in the "never okay" category used religious objections that reflected the cultural logics of their homogenous hometown communities rather than intense personal involvement with organized religion.

These women were frank in interviews and with peers about their beliefs, often saying homosexuality physically disgusted them. They struggled with what they felt to be an offensive new environment in which different values prevailed. For example, one night a frustrated woman told me and several others how tired she was of looking at "that" (the GLBT—Gay, Lesbian, Bisexual, and Transgendered—Rainbow week bulletin board just outside her door). She said vehemently, "I just want to take a big black marker and write "straighten up ... straighten up your future."

Others generally tolerated these women's verbal denouncements of homosexuality, labeling them as ignorant or provincial only during interviews. However, one of these women faced exclusion on the floor because she rejected her assigned lesbian roommate in such a negative fashion. This woman slandered her lesbian roommate as a "dyke" to others on the floor, engaged in loud verbal assaults on her, and made a show of changing her clothing elsewhere. Eventually the lesbian roommate chose to move out, and a friend with a similar conservative religious background moved in. Many floor residents ostracized this woman and her new roommate. She reported, "A lot of people don't say hi, don't smile, don't acknowledge us because they think I am this bad person."

Gamson (1998) notes that in talk shows, a similar process often occurs; audiences will isolate individuals with the most prejudiced views so as to define themselves as comparatively tolerant. Floor residents identified these women as "bigots" because they directly mistreated a lesbian, explicitly made it about her sexuality, cited religious morality, and made it political by hanging a "Vote for

Bush" poster on their door. As Eliasoph and Lichterman (2003) note, Americans generally avoid political discussion because they see it as too divisive. A resident elaborated, "It could maybe make her uncomfortable if her roommate was a lesbian … [but] she shouldn't go around blabbing it." Although this woman understood the desire not to share a room with a lesbian, she found the public and unsophisticated way in which the other woman handled it to be objectionable.

The 24 women in the homosexuality as "okay for others but not in my space" category displayed this sort of sophistication, walking the line between the competing values of openness to diversity and dissociation from low-status lesbians. They were aware of the need to respect the discourse of diversity acceptance promoted by university staff and officials through numerous and visible "Celebrate Diversity" decorations and activities. The increasing visibility of gay characters in television shows that many of them watched (including *The O.C.* and MTV's *The Real World*) also signaled that appearing gay friendly was hip and fashionable. These women responded in interviews as to "prove" their tolerance, often in comparison to others with less socially acceptable views. As one noted, "It just doesn't faze me. I think you can do whatever you want to do, and I never was brought up that gay is wrong—like shaking the Bible." Another woman explained, "I mean, if they want to be gay, that's great. I don't have anything against it. I would rather someone come out … than being scared, but I dunno because I'm not."

Awareness of cultural values for diversity, however, does not always translate into acceptance of marginalized groups. Researchers suggest that most whites now engage in "symbolic racism," framing their negative views toward other racial groups in ways that do not seem outwardly racist while continuing to engage in more indirect forms of discrimination (Schuman, Steeth, and Bobo 1988). Women in this category similarly avoided openly prejudiced statements about homosexuality or gays and lesbians as a group but still kept lesbians out of their social spheres. As one woman carefully noted of lesbians, "There's always going to be people that are different, but here I'm not friends with those people." Their lack of lesbian friends was not a consequence of the circumstances. The women in this group were generally aware of the two lesbians on the hall, mentioning their presence in interviews and interactions. One of the lesbians even noted that she was friendly with some of these women until she came out; then it was as if they were strangers.

Women in this group managed to avoid lesbians on the hall without appearing to contradict their "openness" to diversity by using the language of taste. Rather than highlighting lesbians' sexual preference as problematic, these women cited differences in interests, personal styles, or social chemistry. For example, one woman in this category also moved away from her lesbian roommate but maintained that it was mismatched personalities that led to her switch. As she explained during the interview, "I just don't like living with her because it's hard. We don't talk, and so I don't like that atmosphere." Even when talking with each other about lesbians on the hall, women in this group rarely said they disliked lesbians because of their sexual orientation. One woman told me about a conversation she had with another heterosexual woman who said that she would not want to move in with the lesbian (despite tension with her current roommate) because "she's bigger and she's weird." When I looked confused about this, the woman I was interviewing leaned in and whispered, "[She's] a lesbian."

The preference for social distance from lesbians was most apparent when I asked women to consider lesbians in their personal space. Almost invariably, they were concerned about being "checked out" by lesbian room- or floormates: "I'd be freaked out changing. I know I sound so close-minded, but truthfully I would be like scared. Like, is she watching me change or will she hit on me?" In a residence hall where private space is public space, lesbians introduce interactional ambiguity. Women were familiar with men eroticizing them—even on their floor and in their rooms—but lesbians added the possibility of an unfamiliar sexual gaze. One woman noted, "Having a lesbian on the floor has scared me. When I'm in the shower and I know she's next to me … I get nervous. 'Cause I never thought about a girl looking at me that way." In her opinion, even shared floor spaces were more comfortable when assumed to be heterosexual.

Critical Partiers and Nonpartiers

These 13 women shared an orientation to the party system that was different than for women in the outer two levels of social distance. They afforded the party scene less importance, choosing to define themselves through other avenues. I identified five of these women as critical partiers because, although they participated in the party scene, they consistently critiqued it. When I asked one critical partier where she partied the most, she said, "Whichever [place] I hate the least that week." The only critical partier or nonpartier to join the Greek system, she did so as an attempt to make friends but refused to be "fake and try to please people." She disliked the elitism of the party scene, where "people base stuff on money," judging others through their ownership of designer goods. Another woman maintained that she did not always have to party in the Greek scene, stating, "Tell me what I'm missing out on that I can't find with other people…. It's not worth it to me." She and her roommate, another critical partier, chose not to go through the process of visiting and eventually joining sororities, often referred to as "rush." They even posted what they called an "anti-rush" message on their door: "Yes, that's right. We quit. The two females who live in this room have been officially disqualified [by choice] from the rush process." This was a bold move in a context where the Greek system was highly valued.

I defined 11 women as nonpartiers because they chose to opt out of the Greek party system. Many of these women noted that they did not enjoy partying. One explained, "I'm not a big party person…. I'm not a big person on drinking, and I don't like being around people that are totally drunk, acting like idiots." Several of them reported being made fun of by other floor members. One woman, for instance, told me that a floormate chided, "You haven't drank and you're at college? Come on." For several who opted out, financial or personal issues led them to value school differently.

> Some of these girls don't even go to class. It's like they just live here. They stay up until 4:00 in the morning. [I want to ask,] "Do you guys go to class? Like what's your deal?... You're paying a lot of money for this.... If you want to be here, then why aren't you trying harder?"

Her contempt reflects the fact that partying is also a classed activity—one that not everyone can afford.

Critical partiers and nonpartiers fell into two groups: those who were willing to have lesbian friends and those who were willing to consider public lesbian identities for themselves. The 11 women in the homosexuality as "not my choice, but okay for my friends" category believed in the benefits of diversity in college. Many were curious about meeting new people and learning about their experiences. One woman, a friend of a lesbian on the floor, enthusiastically detailed the positives of having a lesbian roommate, exclaiming, "I don't care, I'd be inquisitive! I'd want to know about ... what they've dealt with and what their views are on gay pride like 'I wear the rainbow.'" Another woman talked about her interest in the gay and lesbian community in the town, sparked by her contact with a bisexual woman. She explained, "I'm really into everything different. Anything to have an experience is just so cool." For these women, contact with gays and lesbians was seen as a form of personal enrichment—"an experience" that could be consumed.

A few women in this group, however, felt that one's sexual orientation was not what made someone a desirable friend. They typically noted that a lesbian roommate would be the same as any other roommate—only a problem if they did not get along with each other. As one woman noted, she "loved" her lesbian friend because she was "real" and "down-to-earth." Another described her close lesbian friend, saying,

> She says what's on her mind and I love that. I just I get a kick out of half of the things she says, and the other half I just really appreciate that she's honest.... That's a good quality in anyone.

These two women did not think their friend's "gayness" made her fun and unique; they each enjoyed their lesbian friend for other aspects of her personality such as candor, wit, and sincerity.

While perhaps they were positive influences in the lives of individual lesbians, these women did not challenge the overall marginalization of lesbians. When they protested the exclusion of their lesbian friends, they did so in private or anonymously. For example, several of them privately expressed fierce hatred for the "bigots" on the floor. But, rather than speaking to the resident assistant or bringing up tolerance as a floor issue, they admitted to secretly writing things on the door of these two women. They realized that other people on the floor were not as accepting of lesbians as they were but did nothing to change this. As one noted, "It just sucks that not everyone can be open-minded."

The two women who felt that homosexuality was "okay for me, okay for my friends" had the smallest social distance from homosexuality. Both identified as lesbians and felt shunned because of their sexuality. One explained of her roommate, "She was really nice the day I met her, and then after I told her I was gay, she changed." Based on this reaction, the woman felt she needed to be careful about who she told. When I asked her if many people knew at first that she was a lesbian, she said, "No, take a look at the floor I was on, the building I was in. Of course not. I didn't want people to gang up on me." When she moved into a new residence hall, she decided that it was best to signal her sexuality only subtly by hanging her gay pride flag. She noted, "I learned my lesson from directly telling a roommate.... Honestly, I would have kept it a secret until I found out how she felt about it. If she was against it, I would have kept it secret." This woman was willing to hide her sexuality rather than face a negative reaction.

The other woman had a similar experience with her roommate. She and her roommate never actually talked about her sexuality nor did the roommate directly confront her about it. Instead, the roommate indirectly signaled her disgust for lesbianism. The woman who identified as a lesbian explained,

> I was watching this show on VH1. It showed a clip from Melissa Etheridge's wedding, and I remember she made this disgusted noise and commented on it, but it stuck out to me obviously—like I see how you feel about this.

A week later this roommate had "a full out conversation about how disgusting [lesbianism] was" with a woman who lived next door. This conversation occurred in the lesbian's room, but the women who were talking never acknowledged her presence. Even in her own private space, she felt dismissed and ignored.

Both lesbians also reported that they felt unwelcome on campus. When I asked one woman if she thought that students were accepting of homosexuality, she made a clear distinction between how she felt in Adams (a relatively small alternative residence hall she moved to midyear) versus other places on campus. She noted that in most places, "I just feel uncomfortable. Like here (Adams), I'm totally comfortable with everybody.... If they're Adams kids I know that they're accepting." Outside this pocket, however, the two lesbians often felt isolated. As the other woman explained, "I assumed that everyone was straight, just like the rest of the world, where everyone assumes that everyone is straight." In classes, she reported sensing intolerance that kept her from being more open.

> There have been a couple times that I've kind of come close to saying something about [my sexuality], and I hold myself back because I know. For example ... my English teacher is actually out and he's alluded to it a couple of times, talking about his partner. And the reactions that I've seen [my classmates] have to it, have kept me from [disclosing it].

For both women, college life involved constantly monitoring their surroundings, determining when they could be open, when they could not, and what spaces allowed them to be "lesbians."

APPROPRIATING LESBIAN EROTICISM

Although lesbians received clear messages that their sexuality was not welcome, same-sex eroticism of a certain kind thrived. Active partiers frequently engaged in same-sex sexual behaviors in the party scene. Their ability to do so without social stigma depended on maintaining social distance from those who identified as lesbians. Heterosexual women's appropriation of lesbian eroticism for their own use put lesbians in a difficult position. Woman-to-woman eroticism had its place on campus among those who identified as heterosexual, but out lesbians often encountered disgust or hostility.

Same-Sex Eroticism among "Straight Girls"

Only active partiers, those in the two outer rings of social distance from lesbians, participated in same-sex eroticism (4 out of 6 women in the "never okay" group; 17 out of 24 women in the "okay for others but not in my space" group). Same-sex eroticism included kissing (on the mouth, often involving tongues) and fondling (of breasts and buttocks), particularly while dancing; no heterosexual women reported oral or digital stimulation of the genitals. These women openly discussed such behaviors with researchers, talked about them with their friends, and posted pictures of themselves kissing women in their rooms and on the Internet. Heterosexual women who were more open about homosexuality did not either engage in the same behavior or advertise it in the same way.

As Jenefsky and Miller (1998) note, the performance of lesbianism for men may signal heterocentric eroticism. Women on the floor who engaged in this behavior claimed that they intended their same-sex kissing for an audience of heterosexual men. Several noted that they liked to get reactions from men. One described, "You get guys that you just like to see their expressions. It's just so funny to see them be like, 'Oh my god, I can't believe you just did that, that was awesome." Another woman explained, "Guys said, 'Do it, do it!' just screwing around…. [They] were like, 'These girls are going to kiss!' So you think you're cooler and guys think you're cooler." The value in the same-sex kiss, therefore, was in the attention that it could garner from men. Like a sexy outfit or new stilettos, heterosexual women could deploy same-sex eroticism as a statement of style to get attention amid a sea of scantily dressed young women. One resident even noted that unlike doing drugs, this way of getting attention did not cause bodily harm.

Heterosexual women were careful to claim that their kisses had little meaning behind them, noting that they were not involved and not "serious." They often contextualized their behaviors so that others (and perhaps themselves) would interpret them as heterosexual. As two roommates told me when I asked if they had ever seen two girls kissing,

R1: Well, sometimes we're drunk. (Both laughing)

R2: Like trashed.

R1: We have a wall of shame of pictures.

R2: Sometimes we get a little out of control and trashed, but it's not like we're going crazy on each other. Like, it's just to be funny. It's random kisses. It's not serious.

R1: Right (laughs). It's not like I want you or anything. Eww.

Women often attributed these kisses to alcohol. Among this crowd, however, intoxication was rarely an embarrassing state. Drunken pictures were most likely to make it into public view as they provided proof that one could party hard. Same-sex sexuality was just another way to mark oneself as edgy and spontaneous—"stepping outside of your box," as one woman called it.

Floor residents who employed woman-to-woman eroticism were careful to distinguish their behaviors from those whom they considered to be "real lesbians." As many felt that lesbians were identifiable through their unfeminine appearance, they seemed sure that those in their social networks were heterosexual even if sexual orientation was never a topic of conversation. As one respondent noted,

R: It's totally different if you're into it. Like lesbians or something. It's just your friend.

I: How can you tell like if somebody is really into it or not?

R: I don't know. I always just assumed everyone wasn't. Just 'cause it's people I knew. I've never seen real lesbians kiss.

All of these women agreed that you only kissed close friends whom you trusted to be heterosexual. One even described it as a "bonding" activity between her and another woman on the floor. When they saw other women kissing at parties, they usually applied the same assumptions.

These women felt that encountering lesbians making out in the heterosexual space of the party scene was unlikely. They understood that women achieved status and even basic inclusion in the party scene through their ability to attract men. In their eyes, most lesbians were incapable of doing so; lesbians were "boyish" and "weird" and therefore unlikely to be "hot" or "blonde." They assumed that lesbians simply could not succeed in passing as heterosexual women. This assumption allowed them to construct seeming boundaries between their same-sex erotic practices and those of who they deemed to be "real" or "actual" lesbians. The maintenance of these boundaries played a central role in their ability to maintain heterosexual identities and define their behaviors as hetero-, rather than homo-, erotic.

Reducing Lesbian Spaces

Heterosexual women's enactment of same-sex eroticism worked to further marginalize lesbians. Displays of eroticism between women perceived as undesirable to heterosexual men invited ridicule or worse. Because the heterosexual party scene encompassed all Greek houses, many off-campus houses, and all but a few bars in town, lesbians were effectively excluded (both by choice and by design) from most public erotic spaces in town. A lesbian in a focus group suggested that heterosexual women even encroached on the few lesbian- and gay-friendly party spots. Della's, the bar to which she referred, is a widely known gay bar.

> One night I was at Della's and waiting for my friends to meet me there. I'm sitting alone at this table, and a group of approximately 50 girls in matching T-shirts with sorority lettering across the front, came in, took over the dance floor, and were makin' out and givin' lap dances to each other.... I called [my friend] and I was talkin' to her about how just disgusted I was by it because it's making a mockery of us. These two girls overheard me 'cause I was being loud (laughter).... And I tried to explain to them that if I went to the straight bar with my girlfriend and stood next to her, let alone kissed her, that would not be okay. But that these little girls kissing and giggling is A-okay because it's implied that there's no pleasure there or that it's to please men rather than to please themselves.

This woman experienced the sorority women's presence in her space as invasive and their behavior as insulting. Acting as heterosexual "tourists," these sorority women consumed the experience of the "exotic other" but could safely leave it behind (Casey 2004). As most erotic spaces privileged their sexuality, they felt entitled enough to invade one of the few lesbian-identified spaces in pursuit of a thrill.

None of the women in the focus group felt that heterosexual women's use of same-sex eroticism would lead to claiming a lesbian identity. One explained, "There doesn't seem to be any ... authentic lesbian in between there." However, heterosexual women's enactment of same-sex eroticism in a gay bar suggests that their appropriation may not be only about garnering men's attention. It is possible that claiming a heterosexual identity allows them to enjoy experimentation with other women. On the floor, two roommates told me and another woman about a night when they danced together naked. They did this alone and were not recounting the story to get men's attention. Yet neither described this experience as a "lesbian" encounter, instead jokingly dismissing it as something to do when they were bored. They may have privately experienced this as a moment of questioning their sexuality; however, their ability to tell others without facing challenges to their heterosexual identity was dependent on the existence of out lesbians from whom they could differentiate themselves.

As Casey (2004) notes, heterosexual women's intrusion into gay and lesbian identified spaces can reduce lesbians' comfort, safety, and sense of inclusion. Women who claimed heterosexual identities may have experienced freedom from men's gaze and possibly played with same-sex desire while in the bar; however, as a result of their intrusion, lesbians lost the right to define the meaning of same-sex eroticism in their own space. By claiming same-sex eroticism as a heterosexual practice, heterosexual women made lesbian desire invisible and reconfigured it as a performance for men. Ironically, in the lesbian bar take-over, heterosexual women took up space with their bodies and their sexuality—

something that scholars find to be particularly difficult for women (K. A. Martin 1998; Tolman 2002). Yet they did so only at the cost of women who were more disenfranchised on campus than they were.

DISCUSSION

The literature on masculinities suggests that men's dominance over women encourages adherence to heteronormative ideals of manhood that support aggression against gays (Connell 1987; Corbett 2001; Pascoe 2005). These analyses present the flipside of that story; women's efforts to navigate inegalitarian gender contexts may fortify their efforts to meet heteronormative standards of femininity. Although disadvantaged relative to men, heterosexual women may raise their status among other women by distancing themselves from those who do not perform traditionally feminine identities. Lesbians, who often avoid signaling availability to men through behavior or appearance, thus encounter systematic social exclusion.

Past scholarship may have minimized homophobia among women because it does not look the same as among men. Men's homophobia often takes the form of physical or verbal violence against gay men. My analyses suggest that homophobia among women instead renders lesbians socially invisible. For example, when someone covertly dismantled the Rainbow Week bulletin board in the hall, no one, save the resident assistant, said anything. The unceremonious removal of the board and its subsequent replacement with healthy eating suggestions fittingly represented the situation of the lesbians on the floor. Most of the floor was so busy avoiding them, they were almost socially nonexistent.

The problem of lesbian visibility is deeply rooted in heteronormative cultural meanings that are fundamentally gendered. They reflect the idea that women's sexuality is a direct consequence of men's desire, socially transforming sex acts between women into erotic fodder for heterosexual men (Jenefsky and Miller 1998). When heterosexual women engage in same-sex eroticism for an audience of men or in lesbian-identified spaces, they make it difficult for lesbians to mark their erotic activities as nonheterosexual. In contrast, heterosexual men may feel that homosexuality is a persistent threat (Armstrong 1995). As a result of the fundamentally gendered nature of sexuality, heterosexual men often ward off accusations of homosexuality while lesbians have to struggle to make their sexuality visible.

Because of the invisibility of lesbian sexuality, lesbians often have to deliberately signal their unavailability to men through dress, group affiliation, and choice of social space (Armstrong 1995). This may mean both choosing not to participate in heterosexual erotic markets and creating a less feminine gender appearance. As my .data suggests, however, these are two key mechanisms through which women can gain status in gender inegalitarian conditions. Many lesbians face a dilemma: They can make their lesbian identity visible and face social invisibility or struggle with the invisibility of their sexual identity but benefit from social inclusion. Women's homophobia thus relies on heteronormative understandings of sexuality to keep lesbians marginalized.

My analyses suggest that homophobia among women (heterosexism) is tightly linked to gender inequality (sexism). When disempowered, women may rely on gender strategies that access compensatory benefits through their relationships with men (Schwalbe et al. 2000). These gender strategies require traditionally feminine gender presentations that become the primary form of embodied capital available to women in specific social contexts. First, women have to be in disadvantaged positions vis-à-vis men to need their "patronage" for achieving status. Social and structural inequalities that divide men and women into two different groups also naturalize the gender differences that they produce. Second, heteronormative cultural logics that assume the "otherness" of appropriate sexual partners must be in play. These logics privilege women who work to attract and please men, often through their gender performances. Although ultimately supporting their own subordination, women who benefit from these conditions can rely on homophobia to maintain the status quo.

My work also indicates that women's embodied capital is race and class specific. "Blonde" gender presentations are only possible for those who can produce long, straight blonde hair and "tan" skin. In addition, this appearance requires knowledge about styles and trends and the money necessary to buy and embody them. Not everyone is, therefore, capable of producing the kind of femininity that can bring benefits in the erotic market of the Greek party scene. Although my analyses do not detail how race or class statuses impact women's gender strategies, it is possible that women with reduced access to the rewards of heterosexual performance have more room for gender flexibility. However, some Black feminists such as Collins (1991) and Smith (1982) indicate that when heterosexuality is among the few privileges available to women, they may invest heavily in "maintaining 'straightness'" (Smith 1982, 171). For this reason, women marginalized because of their race, ethnicity, and/or class background may exercise strategies of social distancing from lesbians with more vehemence. This remains a topic for further examination.

REFERENCES

Armstrong, Elizabeth A. 1995. Traitors to the cause? Understanding the lesbian/gay "bisexuality debates." In *Bisexual politics: Theories, queries, & visions*, edited by N. Tucker. Binghamton, NY: Haworth.

Armstrong, Elizabeth A., Laura Hamilton, and Brian Sweeney. 2006. Sexual assault on campus: A multilevel, integrative approach to party rape. *Social Problems* 53:483–99.

Bartky, Sandra. 1990. *Femininity and domination*. New York: Routledge.

Boswell, A. Ayres, and Joan Z. Spade. 1996. Fraternities and collegiate rape culture: Why are some fraternities more dangerous places for women? *Gender & Society* 10: 133–47.

Butler, Judith. 1990. *Gender trouble: Feminism and the subversion of identity.* New York: Routledge.

Casey, Mark. 2004. De-dyking queer space(s): Heterosexual female visibility in gay and lesbian spaces. *Sexualities* 7:446–61.

Collins, Patricia Hill. 1991. *Black feminist thought: Knowledge, consciousness, and the politics of empowerment.* New York: Routledge.

Collins, Randall. 2004. *Interaction ritual chains.* Princeton, NJ: Princeton University Press.

Connell, R. W. 1987. *Gender and power: Society, the person, and sexual politics.* Palo Alto, CA: Stanford University Press.

———. 1995. *Masculinities.* Berkeley: University of California Press.

Corbett, Ken. 2001. Faggot = Loser. *Studies in Gender and Sexuality* 2:3–28.

Eder, Donna. 1985. The cycle of popularity: Interpersonal relations among female adolescents. *Sociology of Education* 58:154–65.

Eliasoph, Nina, and Paul Lichterman. 2003. Culture in interaction. *American Journal of Sociology* 108:735–94.

Gamson, Josh. 1998. *Freaks talk back: Tabloid talk shows and sexual nonconformity.* Chicago: University of Chicago Press.

Gentry, Cynthia S. 1987. Social distance regarding male and female homosexuals. *Journal of Social Psychology* 127:199–208.

Giddens, Anthony. 1992. *The transformation of intimacy: Sexuality, love, and eroticism in modern societies.* Cambridge, UK: Polity.

Gilligan, Carol. 1982. *In a different voice: Psychological theory and women's development.* Cambridge, MA: Harvard University Press.

Goffman, Erving. 1963. *Stigma: Notes on the management of spoiled identity.* Englewood Cliffs, NJ: Prentice Hall.

Handler, Lisa. 1995. In the fraternal sisterhood: Sororities as gender strategy. *Gender & Society* 9:236–55.

Hochschild, Arlie Russell. 1989. *The second shift.* New York: Avon Books.

Holland, Dorothy C., and Margaret C. Fisenhart. 1990. *Educated in romance: Women, achievement, and college culture.* Chicago: University of Chicago Press.

Jackson, Stevi. 1996. Heterosexuality and feminist theory. In *Theorising heterosexuality,* edited by D. Richardson. Buckingham, UK: Open University Press.

Jenefsky, Cindy, and Diane H. Miller. 1998. Phallic intrusion: Girl-girl sex in Penthouse. *Women's Studies International Forum* 21:375–85.

Kimmel, Michael S. 2001. Masculinity as homophobia: Fear, shame, and silence in the construction of gender identity. In *The Masculinities Reader,* edited by S. Whitehead and F. Barrett. Cambridge, UK: Polity.

Loftus, Jeni. 2001. America's liberalization in attitudes toward homosexuality, 1973 to 1998. *American Sociological Review* 66:762–82.

Martin, Karin A. 1998. Becoming a gendered body: Practices of preschools. *American Sociological Review* 63:494–511.

———. 2003. Giving birth like a girl. *Gender & Society* 17:54–72.

Martin, Patricia Y. 2003. "Said and done" versus "saying and doing": Gendering practices, practicing gender at work. *Gender & Society* 17:342–66.

Martin, Patricia Y., and Robert A. Hummer. 1989. Fraternities and rape on campus. *Gender & Society* 3:457–73.

Merten, Don E. 1997. The meaning of meanness: Popularity, competition, and conflict among junior high school girls. *Sociology of Education* 70:175–91.

Milner, Murray Jr. 2004. *Freaks, geeks, and cool kids: American teenagers, schools, and the culture of consumption.* New York: Routledge.

Pascoe, C. J. 2005. "Dude, you're a fag": Adolescent masculinity and the fag discourse. *Sexualities* 8:329–46.

Rich, Adrienne. 1980. Compulsory heterosexuality and lesbian existence. *Signs* 5: 631–60.

Schuman, Howard, Charlotte Steeth, and Lawrence Bobo. 1988. *Racial attitudes in America: Trends and interpretations.* Cambridge, MA: Harvard University Press.

Schwalbe, Michael, Sandra Godwin, Daphne Holden, Douglas Schrock, Shealy Thompson, and Michele Wolkomir. 2000. Generic processes in the reproduction of inequality: An interactionist analysis. *Social Forces* 79:419–52.

Smith, Barbara. 1982. Toward a Black feminist criticism. In *But some of us are brave*, edited by G. T. Hull, P. B. Scott, and B. Smith. Old Westbury, NY: Feminist Press.

Stein, Arlene. 2005. Make room for Daddy: Anxious masculinity and emergent homophobias in neopatriarchical politics. *Gender & Society* 19:601–20.

Stombler, Mindy, and Patricia Y. Martin. 1995. Bringing women in, keeping women down: Fraternity "little sister" organizations. *Journal of Contemporary Ethnography* 23: 150–84.

Swidler, Anne. 1986. Culture in action: Symbols and strategies. *American Sociological Review* 51:273–86.

Tolman, Deborah. 2002. *Dilemmas of desire: Teenage girls talk about sexuality.* Cambridge, MA: Harvard University Press.

KEY CONCEPTS

homophobia   heterosexual privilege   heterosexism

DISCUSSION QUESTIONS

1. How do women express their own homophobia, according to Hamilton? In what ways does the social atmosphere described in this article marginalize lesbian college students?

2. Have you witnessed some of these same behaviors and interactions in your college experience? What are some ways that social interactions in college can be less sexualized? Is this a reasonable solution?

Applying Sociological Knowledge: An Exercise for Students

Imagine that you wake up tomorrow morning and are a member of the other gender. Make a list of all the things about yourself that you think would have changed. After making your list, make a note of which characteristics are biological or physical, which are attitudinal, which involve behavior, and which are institutional. Then ask yourself how individual and institutional forms of gender are related.

✳

PART XII

Sexuality and Intimate Relationships

39

'Dude, You're a Fag': Adolescent Masculinity and the Fag Discourse

C.J. PASCOE

In this piece, Pascoe summarizes the results from her field research in a working-class, suburban high school in California. She examines the use of the word "fag" as an insult among and between adolescent males. Using gender and queer theory, the author presents evidence that there is a discourse that uses the word "fag" and negative homosexual stereotypes in interactions among high school boys. The term "fag" is not often used to refer to actual homosexual boys. Instead, heterosexual high school boys use the term to mock or tease other heterosexual boys. The "fag discourse," then, is a tool for establishing and highlighting the masculinity of the person using the language.

'There's a faggot over there! There's a faggot over there! Come look!' yelled Brian, a senior at River High School, to a group of 10-year-old boys. Following Brian, the 10-year-olds dashed down a hallway. At the end of the hallway Brian's friend, Dan, pursed his lips and began sashaying towards the 10-year-olds. He minced towards them, swinging his hips exaggeratedly and wildly waving his arms. To the boys Brian yelled, 'Look at the faggot! Watch out! He'll get you!' In response the 10-year-olds raced back down the hallway screaming in terror. (From author's fieldnotes)

The relationship between adolescent masculinity and sexuality is embedded in the specter of the faggot. Faggots represent a penetrated masculinity in which 'to be penetrated is to abdicate power' (Bersani, 1987: 212). Penetrated men symbolize a masculinity devoid of power, which, in its contradiction, threatens both psychic and social chaos. It is precisely this specter of penetrated masculinity that functions as a regulatory mechanism of gender for contemporary American adolescent boys.

Feminist scholars of masculinity have documented the centrality of homophobic insults to masculinity (Lehne, 1998; Kimmel, 2001) especially in school settings (Wood, 1984; Smith, 1998; Burn, 2000; Plummer, 2001; Kimmel, 2003). They argue that homophobic teasing often characterizes masculinity in

SOURCE: Pascoe, C. J. 2005. "'Dude, You're a Fag': Adolescent Masculinity and the Fag Discourse." *Sexualities* 8: 329–346.

adolescence and early adulthood, and that anti-gay slurs tend to primarily be directed at other gay boys.

This article both expands on and challenges these accounts of relationships between homophobia and masculinity. Homophobia is indeed a central mechanism in the making of contemporary American adolescent masculinity. This article both critiques and builds on this finding by (1) pointing to the limits of an argument that focuses centrally on homophobia, (2) demonstrating that the fag is not only an identity linked to homosexual boys but an identity that can temporarily adhere to heterosexual boys as well and (3) highlighting the racialized nature of the fag as a disciplinary mechanism.

'Homophobia' is too facile a term with which to describe the deployment of 'fag' as an epithet. By calling the use of the word 'fag' homophobia—and letting the argument stop with that point—previous research obscures the gendered nature of sexualized insults (Plummer, 2001). Invoking homophobia to describe the ways in which boys aggressively tease each other overlooks the powerful relationship between masculinity and this sort of insult. Instead, it seems incidental in this conventional line of argument that girls do not harass each other and are not harassed in this same manner. This framing naturalizes the relationship between masculinity and homophobia, thus obscuring the centrality of such harassment in the formation of a gendered identity for boys in a way that it is not for girls.

'Fag' is not necessarily a static identity attached to a particular (homosexual) boy. Fag talk and fag imitations serve as a discourse with which boys discipline themselves and each other through joking relationships. Any boy can temporarily become a fag in a given social space or interaction. This does not mean that those boys who identify as or are perceived to be homosexual are not subject to intense harassment. But becoming a fag has as much to do with failing at the masculine tasks of competence, heterosexual prowess and strength or in anyway revealing weakness or femininity, as it does with a sexual identity. This fluidity of the fag identity is what makes the specter of the fag such a powerful disciplinary mechanism. It is fluid enough that boys police most of their behaviors out of fear of having the fag identity permanently adhere and definitive enough so that boys recognize a fag behavior and strive to avoid it.

The fag discourse is racialized. It is invoked differently by and in relation to white boys' bodies than it is by and in relation to African-American boys' bodies. While certain behaviors put all boys at risk for becoming temporarily a fag, some behaviors can be enacted by African-American boys without putting them at risk of receiving the label. The racialized meanings of the fag discourse suggest that something more than simple homophobia is involved in these sorts of interactions. An analysis of boys' deployments of the specter of the fag should also extend to the ways in which gendered power works through racialized selves.

THEORETICAL FRAMING

The sociology of masculinity entails a 'critical study of men, their behaviors, practices, values and perspectives' (Whitehead and Barrett, 2001: 14). Recent studies of men emphasize the multiplicity of masculinity (Connell, 1995) detailing the ways in which different configurations of gender practice are promoted, challenged or reinforced in given social situations. . . .

Heeding Timothy Carrigan's admonition that an 'analysis of masculinity needs to be related as well to other currents in feminism' (Carrigan et al., 1987: 64), in this article I integrate queer theory's insights about the relationships between gender, sexuality, identities and power with the attention to men found in the literature on masculinities. Like the sociology of gender, queer theory destabilizes the assumed naturalness of the social order (Lemert, 1996). Queer theory is a 'conceptualization which sees sexual power as embedded in different levels of social life' and interrogates areas of the social world not usually seen as sexuality (Stein and Plummer, 1994). In this sense queer theory calls for sexuality to be looked at not only as a discrete arena of sexual practices and identities, but also as a constitutive element of social life (Warner, 1993; Epstein, 1996).

. . . This article does not seek to establish that there are homosexual boys and heterosexual boys and the homosexual ones are marginalized. Rather this article explores what happens to theories of gender if we look at a *discourse* of sexualized identities in addition to focusing on seemingly static identity categories inhabited by men. This is not to say that gender is reduced only to sexuality, indeed feminist scholars have demonstrated that gender is embedded in and constitutive of a multitude of social structures—the economy, places of work, families and schools. In the tradition of post-structural feminist theorists of race and gender who look at 'border cases' that explode taken-for-granted binaries of race and gender (Smith, 1994), queer theory is another tool which enables an integrated analysis of sexuality, gender and race.

As scholars of gender have demonstrated, gender is accomplished through day-to-day interactions (Fine, 1987; Hochschild, 1989; West and Zimmerman, 1991; Thorne, 1993). In this sense gender is the 'activity of managing situated conduct in light of normative conceptions of attitudes and activities appropriate for one's sex category' (West and Zimmerman, 1991: 127). Similarly, queer theorist Judith Butler argues that gender is accomplished interactionally through 'a set of repeated acts within a highly rigid regulatory frame that congeal over time to produce the appearance of substance, of a natural sort of being' (Butler, 1999: 43). Specifically she argues that gendered beings are created through processes of citation and repudiation of a 'constrictive outside' (Butler, 1993: 3) in which is contained all that is cast out of a socially recognizable gender category. The 'constitutive outside' is inhabited by abject identities, unrecognizably and unacceptably gendered selves. The interactional accomplishment of gender in a Butlerian model consists, in part, of the continual iteration and repudiation of this abject identity. Gender, in this sense, is 'constituted through the force of exclusion and abjection, on which produces a constitutive outside to the subject, an abjected outside, which is, after all, 'inside' the subject as its own founding repudiation' (Butler, 1993: 3). This repudiation creates and reaffirms a 'threatening specter' (Butler, 1993: 3) of failed, unrecognizable gender, the existence of which must be continually repudiated through interactional processes.

I argue that the 'fag' position is an 'abject' position and, as such, is a 'threatening specter' constituting contemporary American adolescent masculinity. The fag discourse is the interactional process through which boys name and repudiate this abjected identity. Rather than analyzing the fag as an identity for homosexual boys, I examine uses of the discourse that imply that any boy can become a fag, regardless of his actual desire or self-perceived sexual orientation. The threat of the abject position infuses the faggot with regulatory power. This article provides empirical data to illustrate Butler's approach to gender and indicates that it might be a useful addition to the sociological literature on masculinities through highlighting one of the ways in which a masculine gender identity is accomplished through interaction.

METHOD

Research Site

I conducted fieldwork at a suburban high school in north-central California which I call River High. River High is a working class, suburban 50-year-old high school located in a town called Riverton. With the exception of the median household income and racial diversity (both of which are elevated due to Riverton's location in California), the town mirrors national averages in the percentages of white collar workers, rates of college attendance, and marriages, and age composition (according to the 2000 census). It is a politically moderate to conservative, religious community. Most of the students' parents commute to surrounding cities for work.

On average Riverton is a middle-class community. However, students at River are likely to refer to the town as two communities: 'Old Riverton' and 'New Riverton'. A busy highway and railroad tracks bisect the town into these two sections. River High is literally on the 'wrong side of the tracks', in Old Riverton. Exiting the freeway, heading north to Old Riverton, one sees a mix of 1950s-era ranch-style homes, some with neatly trimmed lawns and tidy gardens, others with yards strewn with various car parts, lawn chairs and appliances. Old Riverton is visually bounded by smoke-puffing factories. On the other side of the freeway New Riverton is characterized by wide sidewalk-lined streets and new walled-in home developments. Instead of smokestacks, a forested mountain, home to a state park, rises majestically in the background. The teens from these homes attend Hillside High, River's rival.

River High is attended by 2000 students. River High's racial/ethnic breakdown roughly represents California at large: 50 percent white, 9 percent African-American, 28 percent Latino and 6 percent Asian (as compared to California's 46, 6, 32, and 11 percent respectively, according to census data and school records). The students at River High are primarily working class.

Research

I gathered data using the qualitative method of ethnographic research. I spent a year and a half conducting observations, formally interviewing 49 students at River High (36 boys and 13 girls), one male student from Hillside High, and conducting countless informal interviews with students, faculty and administrators. I concentrated on one school because I explore the richness rather than the breadth of data.

I recruited students for interviews by conducting presentations in a range of classes and hanging around at lunch, before school, after school at various events talking to different groups of students about my research, which I presented as 'writing a book about guys'. The interviews usually took place at school, unless the student had a car, in which case he or she met me at one of the local fast food restaurants where I treated them to a meal. Interviews lasted anywhere from half an hour to two hours.

The initial interviews I conducted helped me to map a gendered and sexualized geography of the school, from which I chose my observation sites. I observed a 'neutral' site—a senior government classroom, where sexualized meanings were subdued. I observed three sites that students marked as 'fag' sites—two drama classes and the Gay/Straight Alliance. I also observed two normatively 'masculine' sites—auto-shop and weightlifting. I took daily field notes focusing on how students, faculty and administrators negotiated, regulated and resisted particular meanings of gender and sexuality. I attended major school rituals such as Winter Ball, school rallies, plays, dances and lunches. I would also occasionally 'ride along' with Mr Johnson (Mr J.), the school's security guard, on his battery-powered golf cart to watch which, how and when students were disciplined. Observational data provided me with more insight to the interactional processes of masculinity than simple interviews yielded. If I had relied only on interview data I would have missed the interactional processes of masculinity which are central to the fag discourse.

Given the importance of appearance in high school, I gave some thought as to how I would present myself, deciding to both blend in and set myself apart from the students. In order to blend in I wore my standard graduate student gear—comfortable, baggy cargo pants, a black t-shirt or sweater and tennis shoes. To set myself apart I carried a messenger bag instead of a backpack, didn't wear makeup, and spoke slightly differently than the students by using some slang, but refraining from uttering the ubiquitous 'hecka' and 'hella'.

The boys were fascinated by the fact that a 30-something white 'girl' (their words) was interested in studying them. While at first many would make sexualized comments asking me about my dating life or saying that they were going to 'hit on' me, it seemed eventually they began to forget about me as a potential sexual/romantic partner. Part of this, I think, was related to my knowledge about 'guy' things. For instance, I lift weights on a regular basis and as a result the weightlifting coach introduced me as a 'weight-lifter from U.C. Berkeley, telling the students they should ask me for weight-lifting advice. Additionally, my taste in movies and television shows often coincided with theirs. I am an avid fan of the movies 'Jackass' and 'Fight Club', both of which contain high levels of violence and 'bathroom' humor. Finally, I garnered a lot of points among boys because I live off a dangerous street in a nearby city famous for drug deals, gang fights and frequent gun shots.

WHAT IS A FAG?

'Since you were little boys you've been told, "hey, don't be a little faggot,"' explained Darnell, an African-American football player, as we sat on a bench next to the athletic field. Indeed, both the boys and girls I interviewed told me that 'fag' was the worst epithet one guy could direct at another. Jeff, a slight white sophomore, explained to me that boys call each other fag because 'gay people aren't really liked over here and stuff.' Jeremy, a Latino Junior told me that this insult literally reduced a boy to nothing, 'To call someone gay or fag is like the lowest thing you can call someone. Because that's like saying that you're nothing.'

Most guys explained their or other's dislike of fags by claiming that homophobia is just part of what it means to be a guy. For instance Keith, a white soccer-playing senior, explained, 'I think guys are just homophobic.' However, it is not just homophobia, it is a *gendered* homophobia. Several students told me that these homophobic insults only applied to boys and not girls. For example, while Jake, a handsome white senior, told me that he didn't like gay people, he quickly added, 'Lesbians, okay that's *good*.' Similarly Cathy, a popular white cheerleader, told me 'Being a lesbian is accepted because guys think "oh that's cool."' Darnell, after telling me that boys were told not to be faggots, said of lesbians, 'They're [guys are] fine with girls. I think it's the guy part that they're like ewwww!' In this sense it is not strictly homophobia, but a gendered homophobia that constitutes adolescent masculinity in the culture of this school. However, it is clear, according to these comments, that lesbians are 'good' because of their place in heterosexual male fantasy, not necessarily because of some enlightened approach to same-sex relationships. It does however, indicate that using only the term homophobia to describe boys' repeated use of the word 'fag' might be a bit simplistic and misleading.

Additionally, girls at River High rarely deployed the word 'fag' and were never called 'fags'. I recorded girls uttering 'fag' only three times during my research. In one instance, Angela, a Latina cheerleader, teased Jeremy, a well-liked white senior involved in student government, for not ditching school with her, 'You wouldn't 'cause you're a faggot.' However, girls did not use this word as part of their regular lexicon. The sort of gendered homophobia that constitutes adolescent masculinity does not constitute adolescent femininity. Girls were not called dykes or lesbians in any sort of regular or systematic way. Students did tell me that 'slut' was the worst thing a girl could be called. However, my field notes indicate that the word 'slut' (or its synonym 'ho') appears one time for every eight times the word 'fag' appears. Even when it does occur, 'slut' is rarely deployed as a direct insult against another girl.

Highlighting the difference between the deployment of 'gay' and 'fag' as insults brings the gendered nature of this homophobia into focus. For boys and girls at River High 'gay' is a fairly common synonym for 'stupid'. While this word shares the sexual origins of 'fag', it does not *consistently* have the skew of gender-loaded meaning. Girls and boys often used 'gay' as an adjective referring to inanimate objects and male or female people, whereas they used 'fag' as a noun that denotes only un-masculine males. Students used 'gay' to describe anything from someone's clothes to a new school rule that the students did not like, as in the following encounter:

> In auto-shop Arnie pulled out a large older version black laptop computer and placed it on his desk. Behind him Nick said 'That's a gay laptop! It's five inches thick!'

A laptop can be gay, a movie can be gay or a group of people can be gay. Boys used 'gay' and 'fag' interchangeably when they refer to other boys, but 'fag' does not have the non-gendered attributes that 'gay' sometimes invokes.

While its meanings are not the same as 'gay', 'fag' does have multiple meanings which do not necessarily replace its connotations as a homophobic slur, but rather exist alongside. Some boys took pains to say that 'fag' is not about sexuality. Darnell told me 'It doesn't even have anything to do with being gay.' J.L., a white sophomore at Hillside High (River High's cross-town rival) asserted 'Fag, seriously, it has nothing to do with sexual preference at all. You could just be calling somebody an idiot you know?' I asked Ben, a quiet, white sophomore who wore heavy metal t-shirts to auto-shop each day, 'What kind of things do guys get called a fag for?' Ben answered 'Anything ... literally, anything. Like you were trying to turn a wrench the wrong way, "dude, you're a fag." Even if a piece of meat drops out of your sandwich, "you fag!"' Each time Ben said 'you fag' his voice deepened as if he were imitating a more masculine boy. While Ben might rightly *feel* like a guy could be called a fag for 'anything ... literally, anything', there are actually specific behaviors which, when enacted by most boys, can render him more vulnerable to a fag epithet. In this instance Ben's comment highlights the use of 'fag' as a generic insult for incompetence, which in the world of River High, is central to a masculine identity. A boy could get called a fag for exhibiting any sort of behavior defined as non-masculine (although not necessarily behaviors aligned with femininity) in the world of River High: being stupid, incompetent, dancing, caring too much about clothing, being too emotional or expressing interest (sexual or platonic) in other guys. However, given the extent of its deployment and the laundry list of behaviors that could get a boy in trouble, it is no wonder that Ben felt like a boy could be called 'fag' for 'anything'.

One-third (13) of the boys I interviewed told me that, while they may liberally insult each other with the term, they would not actually direct it at a homosexual peer. Jabes, a Filipino senior, told me

> I actually say it [fag] quite a lot, except for when I'm in the company of an actual homosexual person. Then I try not to say it at all. But when I'm just hanging out with my friends I'll be like, 'shut up, I don't want you hear you any more, you stupid fag'.

Similarly J.L. compared homosexuality to a disability, saying there is 'no way' he'd call an actually gay guy a fag because

> There's people who are the retarded people who nobody wants to associate with. I'll be so nice to those guys and I hate it when people make fun of them. It's like, 'bro do you realize that they can't help that?' And then there are gay people. They were born that way.

According to this group of boys, gay is a legitimate, if marginalized, social identity. If a man is gay, there may be a chance he could be considered masculine by other men (Connell, 1995). David, a handsome white senior dressed smartly in khaki pants and a white button-down shirt said, 'Being gay is just a lifestyle. It's someone you choose to sleep with. You can still throw around a football and be gay.' In other words there is a possibility, however slight, that a boy can be gay and masculine. To be a fag is, by definition, the opposite of masculine, whether or not the word is deployed with sexualized or non-sexualized meanings. In explaining this to me, Jamaal, an African-American junior, cited the explanation of popular rap artist, Eminem,

> Although I don't like Eminem, he had a good definition of it. It's like taking away your title. In an interview they were like, 'you're always capping on gays, but then you sing with Elton John.' He was like 'I don't mean gay as in gay'.

This is what Riki Wilchins calls the 'Eminem Exception. Eminem explains that he doesn't call people "faggot" because of their sexual orientation but because they're weak and unmanly' (Wilchins, 2003). This is precisely the way in which this group of boys at River High uses the term 'faggot'. While it is not necessarily acceptable to be gay, at least a man who is gay can do other things that render him acceptably masculine. A fag, by the very definition of the word, indicated by students' usages at River High, cannot be masculine. This distinction between 'fag' as an unmasculine and problematic identity and 'gay' as a possibly masculine, although marginalized, sexual identity is not limited to a teenage lexicon, but is reflected in both psychological discourses (Sedgwick, 1995) and gay and lesbian activism.

BECOMING A FAG

'The ubiquity of the word faggot speaks to the reach of its discrediting capacity' (Corbett, 2001: 4). It is almost as if boys cannot help but shout it out on a regular basis—in the hallway, in class, across campus as a greeting, or as a joke. In my fieldwork I was amazed by the way in which the word seemed to pop uncontrollably out of boys' mouths in all kinds of situations. To quote just one of many instances from my fieldnotes:

> Two boys walked out of the P.E. locker room and one yelled 'fucking faggot!' at no one in particular.

This spontaneous yelling out of a variation of fag seemingly apropos of nothing happened repeatedly among boys throughout the school.

The fag discourse is central to boys' joking relationships. Joking cements relationships between boys (Kehily and Nayak, 1997; Lyman, 1998) and helps to manage anxiety and discomfort (Freud, 1905). Boys invoked the specter of the fag in two ways: through humorous imitation and through lobbing the epithet at one another. Boys at River High imitated the fag by acting out an exaggerated 'femininity', and/or by pretending to sexually desire other boys. As indicated by the introductory vignette in which a predatory 'fag' threatens the little boys, boys at River High link these performative scenarios with a fag identity. They lobbed the fag epithet at each other in a verbal game of hot potato, each careful to deflect the insult quickly by hurling it toward someone else. These games and imitations make up a fag discourse which highlights the fag not as a static but rather as a fluid identity which boys constantly struggle to avoid.

In imitative performances the fag discourse functions as a constant reiteration of the fag's existence, affirming that the fag is out there; at any moment a boy can become a fag. At the same time these performances demonstrate that the boy who is invoking the fag is *not* a fag. By invoking it so often, boys remind themselves and each other that at any point they can become fags if they are not sufficiently masculine.

> Mr. McNally, disturbed by the noise outside of the classroom, turned to the open door saying 'We'll shut this unless anyone really wants to watch sweaty boys playing basketball.' Emir, a tall skinny boy, lisped 'I wanna watch the boys play!' The rest of the class cracked up at his imitation.

Through imitating a fag, boys assure others that they are not a fag by immediately becoming masculine again after the performance. They mock their own performed femininity and/or same-sex desire, assuring themselves and others that such an identity is one deserving of derisive laughter. The fag identity in this instance is fluid, detached from Emir's body. He can move in and out of this 'abject domain' while simultaneously affirming his position as a subject.

Boys also consistently tried to put another in the fag position by lobbing the fag epithet at one another.

> Going through the junk-filled car in the auto-shop parking lot, Jay poked his head out and asked 'Where are Craig and Brian?' Neil, responded with 'I think they're over there', pointing, then thrusting his hips and pulling his arms back and forth to indicate that Craig and Brian might be having sex. The boys in auto-shop laughed.

This sort of joke temporarily labels both Craig and Brian as faggots. Because the fag discourse is so familiar, the other boys immediately understand that Neil is indicating that Craig and Brian are having sex. However these are not necessarily identities that stick. Nobody actually thinks Craig and Brian are homosexuals. Rather the fag identity is a fluid one, certainly an identity that no boy wants, but one that a boy can escape, usually by engaging in some sort of discursive contest to turn another boy into a fag. However, fag becomes a hot potato that no boy wants to be left holding. In the following example, which occurred soon after the 'sex' joke, Brian lobs the fag epithet at someone else, deflecting it from himself:

> Brian initiated a round of a favorite game in auto-shop, the 'cock game'. Brian quietly, looking at Josh, said, 'Josh loves the cock,' then slightly louder, 'Josh loves the cock.' He continued saying this until he was yelling 'JOSH LOVES THE COCK!' The rest of the boys laughed

hysterically as Josh slinked away saying 'I have a bigger dick than all you mother fuckers!'

These two instances show how the fag can be mapped, momentarily, on to one boy's body and how he, in turn, can attach it to another boy, thus deflecting it from himself.

These examples demonstrate boys invoking the trope of the fag in a discursive struggle in which the boys indicate that they know what a fag is—and that they are not fags. This joking cements bonds between boys as they assure themselves and each other of their masculinity through repeated repudiations of a non-masculine position of the abject.

RACING THE FAG

The fag trope is not deployed consistently or identically across social groups at River High. Differences between white boys' and African-American boys' meaning making around clothes and dancing reveal ways in which the fag as the abject position is racialized.

Clean, oversized, carefully put together clothing is central to a hip-hop identity for African-American boys who identify with hip-hop culture. Richard Majors calls this presentation of self a 'cool pose' consisting of 'unique, expressive and conspicuous styles of demeanor, speech, gesture, clothing, hairstyle, walk, stance and handshake', developed by African-American men as a symbolic response to institutionalized racism (Majors, 2001: 211). Pants are usually several sizes too big, hanging low on a boy's waist, usually revealing a pair of boxers beneath. Shirts and sweaters are similarly oversized, often hanging down to a boy's knees. Tags are frequently left on baseball hats worn slightly askew and sit perched high on the head. Meticulously clean, unlaced athletic shoes with rolled up socks under the tongue complete a typical hip-hop outfit.

This amount of attention and care given to clothing for white boys not identified with hip-hop culture (that is, most of the white boys at River High) would certainly cast them into an abject, fag position. White boys are not supposed to appear to care about their clothes or appearance, because only fags care about how they look. Ben illustrates this:

> Ben walked in to the auto-shop classroom from the parking lot where he had been working on a particularly oily engine. Grease stains covered his jeans. He looked down at them, made a face and walked toward me with limp wrists, laughing and lisping in a in a high pitch sing-song voice 'I got my good panths all dirty!'

Ben draws on indicators of a fag identity, such as limp wrists, as do the boys in the introductory vignette to illustrate that a masculine person certainly would not care about having dirty clothes. In this sense, masculinity, for white boys, becomes the carefully crafted appearance of not caring about appearance, especially in terms of cleanliness.

However, African-American boys involved in hip-hop culture talk frequently about whether or not their clothes, specifically their shoes, are dirty:

> In drama class both Darnell and Marc compared their white Adidas basketball shoes. Darnell mocked Marc because black scuff marks covered his shoes, asking incredulously 'Yours are a week old and they're dirty—I've had mine for a month and they're not dirty!' Both laughed.

Monte, River High's star football player echoed this concern about dirty shoes when looking at the fancy red shoes he had lent to his cousin the week before, told me he was frustrated because after his cousin used them, the 'shoes are hella scuffed up'. Clothing, for these boys, does not indicate a fag position, but rather defines membership in a certain cultural and racial group (Perry, 2002).

Dancing is another arena that carries distinctly fag associated meanings for white boys and masculine meanings for African-American boys who participate in hip-hop culture. White boys often associate dancing with 'fags'. J.L. told me that guys think ''nSync's gay' because they can dance. 'nSync is an all white male singing group known for their dance moves. At dances white boys frequently held their female dates tightly, locking their hips together. The boys never danced with one another, unless engaged in a round of 'hot potato'. White boys often jokingly danced together in order to embarrass each other by making someone else into a fag:

> Lindy danced behind her date, Chris. Chris's friend, Matt, walked up and nudged Lindy aside, imitating her dance moves behind Chris. As Matt rubbed his hands up and down Chris's back, Chris turned around and jumped back startled to see Matt there instead of Lindy. Matt cracked up as Chris turned red.

However dancing does not carry this sort of sexualized gender meaning for all boys at River High. For African-American boys, dancing demonstrates membership in a cultural community (Best, 2000). African-American boys frequently danced together in single sex groups, teaching each other the latest dance moves, showing off a particularly difficult move or making each other laugh with humorous dance moves. Students recognized K.J. as the most talented dancer at the school. K.J. is a sophomore of African-American and Filipino descent who participated in the hip-hop culture of River High. He continually wore the latest hip-hop fashions. K.J. was extremely popular. Girls hollered his name as they walked down the hall and thrust urgently written love notes folded in complicated designs into his hands as he sauntered to class. For the past two years K.J. won first place in the talent show for dancing. When he danced at assemblies the room reverberated with screamed chants of 'Go K.J.! Go K.J.! Go K J.!' Because dancing for African-American boys places them within a tradition of masculinity, they are not at risk of becoming a fag for this particular gendered practice. Nobody called K.J. a fag. In fact in several of my interviews, boys of multiple racial/ethnic backgrounds spoke admiringly of K.J.'s dancing abilities.

IMPLICATIONS

These findings confirm previous studies of masculinity and sexuality that position homophobia as central to contemporary definitions of adolescent masculinity. These data extend previous research by unpacking multi-layered meanings that boys deploy through their uses of homophobic language and joking rituals. By attending to these meanings I reframe the discussion as one of a fag discourse, rather than simply labeling this sort of behavior as homophobia. The fag is an 'abject' position, a position outside of masculinity that actually constitutes masculinity. Thus, masculinity, in part, becomes the daily interactional work of repudiating the 'threatening specter' of the fag.

The fag extends beyond a static sexual identity attached to a gay boy. Few boys are permanently identified as fags; most move in and out of fag positions. Looking at 'fag' as a discourse rather than a static identity reveals that the term can be invested with different meanings in different social spaces. 'Fag' may be used as a weapon with which to temporarily assert one's masculinity by denying it to others. Thus 'fag' becomes a symbol around which contests of masculinity take place.

The fag epithet, when hurled at other boys, may or may not have explicit sexual meanings, but it always has gendered meanings. When a boy calls another boy a fag, it means he is not a man, not necessarily that he is a homosexual. The boys in this study know that they are not supposed to call homosexual boys 'fags' because that is mean. This, then, has been the limited success of the mainstream gay rights movement. The message absorbed by some of these teenage boys is that 'gay men can be masculine, just like you.' Instead of challenging gender inequality, this particular discourse of gay rights has reinscribed it. Thus we need to begin to think about how gay men may be in a unique position to challenge gendered as well as sexual norms.

This study indicates that researchers who look at the intersection of sexuality and masculinity need to attend to the ways in which racialized identities may affect how 'fag' is deployed and what it means in various social situations.... It is important to look at when, where and with what meaning 'the fag' is deployed in order to get at how masculinity is defined, contested, and invested in among adolescent boys.

REFERENCES

Almaguer, Tomas. (1991). 'Chicano Men: A Cartography of Homosexual Identity and Behavior'. *Differences* 3: 75–100.

Bersani, Leo. (1987). 'Is the Rectum a Grave?' *October* 43: 197–222.

Best, Amy. (2000). *From Night: Youth, Schools and Popular Culture*. New York: Routledge.

Burn, Shawn M. (2000). 'Heterosexuals' Use of "Fag" and "Queer" to Deride One Another: A Contributor to Heterosexism and Stigma'. *Journal of Homosexuality* 40: 1–11.

Butler, Judith. (1993). *Bodies that Matter*. New York: Routledge.

Butler, Judith. (1999). *Gender Trouble*. New York: Routledge.

Carrigan, Tim, Connell, Bob and Lee, John. (1987). 'Toward a New Sociology of Masculinity', in Harry Brod (ed.), *The Making of Masculinities: The New Men's Studies*, pp. 188–202. Boston, MA: Allen & Unwin.

Connell, R.W. (1995). *Masculinities*. Berkeley: University of California Press.

Corbett, Ken. (2001). 'Faggot—Loser'. *Studies in Gender and Sexuality* 2: 3–28.

Epstein, Steven. (1996). 'A Queer Encounter', in Steven Seidman (ed.). *Queer Theory/Sociology*, pp. 188–202. Cambridge, MA: Blackwell.

Fine, Gary. (1987). *With the Boys: Little League Baseball and Preadolescent Culture*. Chicago, IL: University of Chicago Press.

Freud, Sigmund. (1905). *The Basic Writings of Sigmund Freud*, (translated and edited by A.A. Brill). New York: The Modern Library.

Hochschild, Arlie. (1989). *The Second Shift*. New York: Avon.

Julien, Isaac and Mercer, Kobena. (1991). 'True Confessions: A Discourse on Images of Black Male Sexuality', in Essex Hemphill (ed.). *Brother to Brother: New Writings by Black Gay Men*, pp. 167–73. Boston, MA: Alyson Publications.

Kehily, Mary Jane and Nayak, Anoop. (1997). 'Lads and Laughter: Humour and the Production of Heterosexual Masculinities', *Gender and Education* 9: 69–87.

Kimmel, Michael. (2001). 'Masculinity as Homophobia: Fear, Shame, and Silence in the. Construction of Gender Identity', in Stephen Whitehead and Frank Barrett (eds). *The Masculinities Reader*, pp. 266–187. Cambridge: Polity.

Kimmel, Michael. (2003). 'Adolescent Masculinity, Homophobia, and Violence: Random School Shootings, 1982–2001', *American Behavioral Scientist* 46: 1439–58.

King, D. L.. (2004). *Double Lives on the Down Low*. New York: Broadway Books.

Lehne, Gregory. (1998). 'Homophobia among Men: Supporting and Defining the Male Role', in Michael Kimmel and Michael Messner (eds). *Men's Lives*, pp. 237–249. Boston, MA: Allyn and Bacon.

Lemert, Charles. (1996). 'Series Editor's Preface', in Steven Seidman (ed.), *Queer Theory/Sociology*. Cambridge, MA: Blackwell.

Lyman, Peter. (1998). 'The Fraternal Bond as a Joking Relationship: A Case Study of the Role of Sexist Jokes in Male Group Bonding', in Michael Kimmel and Michael Messner (eds), *Men's lives*, pp. 171–93. Boston, MA: Allyn and Bacon.

Majors, Richard. (2001). 'Cool Pose: Black Masculinity and Sports', in Stephen Whitehead and Frank Barrett (eds.), *The Masculinities Reader*, pp. 208–17. Cambridge: Polity.

Perry, Pamela. (2002). *Shades of White: White Kids and Racial Identities in High School*. Durham, NC: Duke University Press.

Plummer, David C. (2001). 'The Quest for Modern Manhood: Masculine Stereotypes, Peer Culture and the Social Significance of Homophobia', *Journal of Adolescence* 24: 15–23.

Riggs, Marlon. (1991). 'Black Macho Revisited: Reflections of a SNAP! Queen', in Essex Hemphill (ed.), *Brother to Brother: New Writings by Black Gay Men*, pp. 153–260. Boston, MA: Alyson Publications.

Sedgwick, Eve K.. (1995). "Gosh, Boy George, You Must be Awfully Secure in Your Masculinity!" in Maurice Berger, Brian Wallis and Simon Watson (eds.), *Constructing Masculinity*, pp. 11–20. New York: Routledge.

Smith, George W. (1998). 'The Ideology of "Fag": The School Experience of Gay Students', *The Sociological quarterly* 39: 309–35.

Smith, Valerie. (1994). 'Split Affinities: The Case of Interracial Rape', in Anne Herrmann and Abigail Stewart (eds.). *Theorizing Feminism*, pp. 155–70. Boulder, CO: Westview Press.

Stein, Arlene and Plummer, Ken. (1994). "'I Can't Even Think Straight": "Queer" Theory and the Missing Sexual Revolution in Sociology', *Sociological Theory* 12: 178 ff.

Thorne, Barrie. (1993). *Gender Play: Boys and Girls in School*. New Brunswick, NJ: Rutgers University Press.

Warner, Michael. (1993). 'Introduction', in Michael Warner (ed.), *Fear of a Queer Planet: Queer Politics and Social theory*, pp. vii–xxxi. Minneapolis: University of Minnesota Press.

West, Candace and Zimmerman, Don. (1991). 'Doing Gender', in Judith Lorber (ed.), *The Social Construction of Gender*, pp. 102–21. Newbury Park: Sage.

Whitehead, Stephen and Barrett, Frank. (2001). 'The Sociology of Masculinity', in Stephen Whitehead and Frank Barrett (eds.), *The Masculinities Reader*, pp. 472–6. Cambridge: Polity.

Wilchins, Riki. (2003). 'Do You Believe in Fairies?' *The Advocate*, 4 February.

Willis, Paul. (1981). *Learning to Labor: How Working Class Kids Get Working Class Jobs*. New York: Columbia University Press.

Wood, Julian. (1984). 'Groping Toward Sexism: Boy's Sex Talk', in Angela McRobbie and Mica Nava (eds.), *Gender and Generation*. London: Macmillan Publishers.

KEY CONCEPTS

queer theory   masculinity   adolescence

DISCUSSION QUESTIONS

1. Pascoe explains that the term "fag" is rarely, if ever, directed at someone who is actually a homosexual. Why do you think this is? How does the term "fag" get used among heterosexuals, but not as a label for homosexuals?

2. Pascoe states that very rarely did she overhear girls using the "fag" language? Can you explain why? Is this consistent with your own high-school experiences?

40

Gendered Sexuality in Young Adulthood

Double Binds and Flawed Options

LAURA HAMILTON AND ELIZABETH A. ARMSTRONG

This research examines the "hooking up" culture at college. While recent research has explored the lack of committed dating relationships among college heterosexuals, this study looks more specifically at how gender and class expectations create a double standard for women. For privileged women, they are expected to delay marriage and family in order to establish their own financial and career success. This expectation leads to more casual sexual experiences. These experiences, however, contradict the traditional stereotype of women as chaste and desiring relationships rather than a "hook-up." For less privileged women, they are expected to participate in the college culture of "hooking up;" yet they typically look for future marriage and family desires.

As traditional dating has declined on college campuses, hookups—casual sexual encounters often initiated at alcohol-fueled, dance-oriented social events—have become a primary form of intimate heterosexual interaction (England, Shafer, and Fogarty 2007; Paul, McManus, and Hayes 2000). Hookups have attracted attention among social scientists and journalists (Bogle 2008; Glenn and Marquardt 2001; Stepp 2007). To date, however, limitations of both data and theory have obscured the implications for women and the gender system. Most studies examine only the quality of hookups at one point during college and rely, if implicitly, on an individualist, gender-only approach. In contrast, we follow a group of women as they move through college—assessing all of their sexual experiences. We use an interactionist approach and attend to how both gender and class shape college sexuality. Our analyses offer a new interpretation of this important issue, contribute to gender theory, and demonstrate how to conduct an interactionist, intersectional analysis of young adult sexuality.

SOURCE: Hamilton, Laura, and Elizabeth A. Armstrong. 2009. "Gendered Sexuality in Young Adulthood: Double Binds and Flawed Options." *Gender & Society* 23: 589–616.

GENDER THEORY AND COLLEGE SEXUALITY

Research on Hooking Up

Paul, McManus, and Hayes (2000) and Glenn and Marquardt (2001) were the first to draw attention to the hookup as a distinct social form. As Glenn and Marquardt (2001, 13) explain, most students agree that "a hook up is anything 'ranging from kissing to having sex,' and that it takes place outside the context of commitment." Others have similarly found that *hooking up* refers to a broad range of sexual activity and that this ambiguity is part of the appeal of the term (Bogle 2008). Hookups differ from dates in that individuals typically do not plan to do something together prior to sexual activity. Rather, two people hanging out at a party, bar, or place of residence will begin talking, flirting, and/or dancing. Typically, they have been drinking. At some point, they move to a more private location, where sexual activity occurs (England, Shafer, and Fogarty 2007). While strangers sometimes hook up, more often hookups occur among those who know each other at least slightly (Manning, Giordano, and Longmore 2006).

England has surveyed more than 14,000 students from 19 universities and colleges about their hookup, dating, and relationship experiences. Her Online College Social Life Survey (OCSLS) asks students to report on their recent hookups using "whatever definition of a hookup you and your friends use." Seventy-two percent of both men and women participating in the OCSLS reported at least one hookup by their senior year in college. Of these, roughly 40 percent engaged in three or fewer hookups, 40 percent between four and nine hookups, and 20 percent 10 or more hookups. Only about one-third engaged in intercourse in their most recent hookups, although—among the 80 percent of students who had intercourse by the end of college—67 percent had done so outside of a relationship.

Ongoing sexual relationships without commitment were common and were labeled "repeat," "regular," or "continuing" hookups and sometimes "friends with benefits" (Armstrong, England, and Fogarty 2009; Bogle 2008; Glenn and Marquardt 2001). Ongoing hookups sometimes became committed relationships and vice versa; generally, the distinction revolved around the level of exclusivity and a willingness to refer to each other as "girlfriend/boyfriend" (Armstrong, England, and Fogarty 2009). Thus, hooking up does not imply interest in a relationship, but it does not preclude such interest. Relationships are also common among students. By their senior year, 69 percent of heterosexual students had been in a college relationship of at least six months.

To date, however, scholars have paid more attention to women's experiences with hooking up than relationships and focused primarily on ways that hookups may be less enjoyable for women than for men. Glenn and Marquardt (2001, 20) indicate that "hooking up is an activity that women sometimes find rewarding but more often find confusing, hurtful, and awkward." Others similarly suggest that more women than men find hooking up to be a negative experience (Bogle 2008, 173; Owen et al. 2008) and focus on ways that hookups may be harmful to women (Eshbaugh and Gute 2008; Grello, Welsh, and Harper 2006).

This work assumes distinct and durable gender differences at the individual level. Authors draw, if implicitly, from evolutionary psychology, socialization, and psychoanalytic approaches to gender—depicting women as more relationally oriented and men as more sexually adventurous (see Wharton 2005 for a review). For example, despite only asking about hookup experiences, Bogle (2008, 173) describes a "battle of the sexes" in which women want hookups to "evolve into some semblance of a relationship," while men prefer to "hook up with no strings attached" (also see Glenn and Marquardt 2001; Stepp 2007).

The battle of the sexes view implies that if women could simply extract commitment from men rather than participating in hookups, gender inequalities in college sexuality would be alleviated. Yet this research—which often fails to examine relationships—ignores the possibility that women might be the losers in both hookups and relationships. Research suggests that young heterosexual women often suffer the most damage from those with whom they are most intimate: Physical battery, emotional abuse, sexual assault, and stalking occur at high rates in youthful heterosexual relationships (Campbell et al. 2007; Dunn 1999). This suggests that gender inequality in college sexuality is systemic, existing across social forms.

Current research also tends to see hooking up as solely about gender, without fully considering the significance of other dimensions of inequality. Some scholars highlight the importance of the college environment and traditional college students' position in the life course (Bogle 2008; Glenn and Marquardt 2001). However, college is treated primarily as a context for individual sexual behavior rather than as a key location for class reproduction. Analyzing the role of social class in sex and relationships may help to illuminate the appeal of hookups for both college women and men.

Gender Beliefs and Social Interaction

Contemporary gender theory provides us with resources to think about gender inequality in college sexuality differently. Gender scholars have developed and refined the notion of gender as a social structure reproduced at multiple levels of society: Gender is embedded not only in individual selves but also in interaction and organizational arrangements (Connell 1987; Glenn 1999; Risman 2004). This paper focuses on the interactional level, attending to the power of public gender beliefs in organizing college sexual and romantic relations.

Drawing on Sewell's (1992) theory of structure, Ridgeway and Correll (2004, 511) define gender beliefs as the "cultural rules or instructions for enacting the social structure of difference and inequality that we understand to be gender." By believing in gender differences, individuals "see" them in interaction and hold others accountable to this perception. Thus, even if individuals do not internalize gender beliefs, they must still confront them (Ridgeway 2009).

Ridgeway and coauthors (Ridgeway 2000; Ridgeway and Correll 2004) assert that interaction is particularly important to the reproduction of gender inequality because of how frequently men and women interact. They focus on the workplace but suggest that gendered interaction in private life may be intensifying in importance as beliefs about gender difference in workplace competency diminish (Correll, Benard, and Paik 2007; Ridgeway 2000; Ridgeway and Correll 2004). We extend their insights to sexual interaction, as it is in sexuality and reproduction that men and women are believed to be most different. The significance of gender beliefs in sexual interaction may be magnified earlier in the life course, given the amount of time spent in interaction with peers and the greater malleability of selves (Eder, Evans, and Parker 1995). Consequently, the university provides an ideal site for this investigation.

The notion that men and women have distinct sexual interests and needs generates a powerful set of public gender beliefs about women's sexuality. A belief about what women should not do underlies a *sexual double standard*: While men are expected to desire and pursue sexual opportunities regardless of context, women are expected to avoid casual sex—having sex only when in relationships and in love (Crawford and Popp 2003; Risman and Schwartz 2002). Much research on the sexuality of young men focuses on male endorsement of this belief and its consequences (e.g., Bogle 2008; Kimmel 2008; Martin and Hummer 1989). There is an accompanying and equally powerful belief that normal women should always want love, romance, relationships, and marriage—what we refer to as the *relational imperative* (also see Holland and Eisenhart 1990; Martin 1996; Simon, Eder, and Evans 1992). We argue that these twin beliefs are implicated in the (re)production of gender inequality in college sexuality and are at the heart of women's sexual dilemmas with both hookups and relationships.

An Intersectional Approach

Gender theory has also moved toward an intersectional approach (Collins 1990; Glenn 1999). Most of this work focuses on the lived experiences of marginalized individuals who are situated at the intersection of several systems of oppression (McCall 2005). More recently, scholars have begun to theorize the ways in which systems of inequality are themselves linked (Beisel and Kay 2004; Glenn 1999; McCall 2005). Beisel and Kay (2004) apply Sewell's (1992) theory of structure to intersectionality, arguing that structures intersect when they share resources or guidelines for action (of which gender beliefs would be one example). Using a similar logic, we argue that gender and class intersect in the sexual arena, as these structures both rely on beliefs about how and with whom individuals should be intimate.

Like gender, class structures beliefs about appropriate sexual and romantic conduct. Privileged young Americans, both men and women, are now expected to defer family formation until the mid-twenties or even early-thirties to focus on education and career investment—what we call the *self-development imperative* (Arnett 2004; Rosenfeld 2007). This imperative makes committed relationships less feasible as the sole contexts for premarital sexuality. Like marriage, relationships can be "greedy," siphoning time and energy away from self-development (Gerstel and Sarkisian 2006; Glenn and Marquardt 2001). In contrast, hookups offer sexual pleasure without derailing investment in human capital and are increasingly viewed as part of life-stage appropriate sexual experimentation. Self-protection—both physical and emotional—is central to this logic, suggesting the rise of a strategic approach to sex and relationships (Brooks 2002; Illouz 2005). This approach is reflected in the development of erotic market-places offering short-term sexual partners, particularly on college campuses (Collins 2004).

In this case, gender and class behavioral rules are in conflict. Gender beliefs suggest that young women should avoid nonromantic sex and, if possible, be in a committed relationship. Class beliefs suggest that women should delay relation-ships while pursuing educational goals. Hookups are often less threatening to self-development projects, offering sexual activity in a way that better meshes with the demands of college. We see this as a case wherein structures intersect, but in a contradictory way (Friedland and Alford 1991; Martin 2004; Sewell 1992). This structural contradiction has experiential consequences: Privileged women find themselves caught between contradictory expectations, while less privileged women confront a foreign sexual culture when they enter college.

METHOD

The strength of our research strategy lies in its depth: We conducted a longitu-dinal ethnographic and interview study of a group of women who started college in 2004 at a university in the Midwest, collecting data about their entire sexual and romantic careers. Like McCall (2005), we see an "intercategorical" approach to intersectionality as ideal; however, space and data limitations prevent us from theorizing structural intersection along all axes of inequality and analyzing the experiences of all of the various possible locations in relation to these structures. However, the richness of our data allows us to reveal taken-for-granted gender and class beliefs organizing the college sexual arena. While the data are at the individual level, our goal is to illustrate how the intersection of gender and class as structures creates dilemmas for college women.

Ethnography and Longitudinal Interviews

A research team of nine, including the authors, occupied a room on an all-female floor in a mixed-gender dormitory....

Fifty-three 18- to 20-year-old unmarried women (51 freshmen, two sopho-mores) lived on the floor for at least part of the year. No one opted out of the ethnographic study. All but two identified as heterosexual. All participants were white, a result of low racial diversity on campus overall and racial segregation in campus housing. Sixty-eight percent came from middle-, upper-middle-, or upper-class backgrounds; 32 percent came from working- or lower-middle-class backgrounds. Forty-five percent were from out of state; all of these women were from upper-middle-class or upper-class families. Thirty-six percent, mostly wealth-ier women, joined sororities in their first year.

The residence hall in which they lived was identified by students and staff as one of several "party dorms." The term refers to the presumed social orientation of the modal resident, not to partying within the dorm itself. Students reported that they requested these dormitories if they were interested in drinking, hooking up, and joining the Greek system. This orientation places them in the thick of American youth culture. Few identified as feminist, and all presented a traditionally feminine appearance (e.g., not one woman had hair shorter than chin length). Most planned to marry and have children.

We observed throughout the academic year, interacting with participants as they did with each other—watching television, eating meals, helping them dress for parties, sitting in as they studied, and attending floor meetings. We let the women guide our conversations, which often turned to "boys," relationships, and hooking up. We also refrained from revealing our own predispositions, to the extent that women openly engaged in homophobic and racist behaviors in front of us. Our approach made it difficult for women to determine what we were studying, which behaviors might be interesting to us, and in which ways we might be judgmental. Consequently, we believe they were less likely to either underreport or exaggerate sexual behavior, minimizing the effects of social desirability.

We conducted interviews with 41 of the 53 women on the floor during their first year, 37 the following year, 35 when they were juniors (two were seniors), and 43 when they were seniors (one had graduated, and one was a fifth-year senior). Forty-six (87 percent) women were interviewed producing 156 interviews. Interviews ranged from 45 minutes to two and a half hours and covered partying, sexuality, relationships, friendships, classes, employment, religion, and relationships with parents. This holistic approach enabled us to see how sexual and romantic interactions intersected with the rest of the women's lives. In collecting data over time, we saw women move back and forth among hookups and relationships—expressing dissatisfaction with both.

Data Analysis, Presentation, and Overview

… All interviews are followed by a number indicating the participant and wave of the interview (e.g., 37-3).…

… Our goal is not to generalize from the experiences of our participants but rather to bring an interactional and intersectional approach to college sexuality. However, it is useful to offer a brief overview of participant sexual and romantic careers. Thirty-three of 44 women (75 percent) from whom we collected complete trajectories reported at least one hookup by their senior year. All but one (95 percent) reported at least one college relationship, and 32 (72 percent) reported relationships of six months or longer. Living in a party dorm may have encouraged hooking up, and the women we studied may have been particularly sought after as girlfriends. Yet rates of participation in hookups and relationships are consistent with the OCSLS data. Thirty-three women (75 percent) cycled between both over the course of college. Ten participated in relationships only, and one had no sexual or romantic involvements. Relationships typically involved sexual intercourse, while sexual activity in hookups ranged from kissing to intercourse. All but four (91 percent) had intercourse before college graduation—a rate that is higher than in the OCSLS.

THE POWER OF GENDER BELIEFS

A battle of the sexes approach suggests that women have internalized a relational orientation but are unable to establish relationships because hooking up—which men prefer—has come to dominate college sexual culture. Rather than accepting stated individual-level preferences at face value, we focus on the interactional contexts in which preferences are formed and expressed. We show that gender beliefs about what women should and should not do posed problems for our participants in both hookups and relationships.

The "Slut" Stigma

Women did not find hookups to be unproblematic. They complained about a pervasive sexual double standard. As one explained, "Guys can have sex with all the girls and *it makes them more of a man*, but if a girl does then all of a sudden she's a ho, and she's not as quality of a person" (10-1, emphasis added). Another complained, "Guys, they can go around and have sex with a number of girls and they're not called anything" (6-1). Women noted that it was "easy to get a reputation" (11-1) from "hooking up with a bunch of different guys" (8-1) or "being wild and drinking too much" (14-3). Their experiences of being judged were often painful; one woman told us about being called a "slut" two years after the incident because it was so humiliating (42-3).

Fear of stigma constrained women's sexual behavior and perhaps even shape their preferences. For example, several indicated that they probably would "make out with more guys" but did not because "I don't want to be a slut" (27-2). Others wanted to have intercourse on hookups but instead waited until they had boyfriends. A couple hid their sexual activity until the liaison was "official." One said, "I would not spend the night there [at the fraternity] because that does not look good, but now everyone knows we're boyfriend/girlfriend, so it's like my home now" (15-1). Another woman, who initially seemed to have a deep aversion to hooking up, explained, "I would rather be a virgin for as much as I can than go out and do God knows who." She later revealed a fear of social stigma, noting that when women engage in nonromantic sex, they "get a bad reputation. I know that I wouldn't want that reputation" (11-1). Her comments highlight the feedback between social judgment and internalized preference.

Gender beliefs were also at the root of women's other chief complaint about hookups—the disrespect of women in the hookup scene. The notion that hooking up is okay for men but not for women was embedded in the organization of the Greek system, where most parties occurred: Sorority rules prohibited hosting parties or overnight male visitors, reflecting notions about proper feminine behavior. In contrast, fraternities collected social fees to pay for alcohol and viewed hosting parties as a central activity. This disparity gave fraternity men almost complete control over the most desirable parties on campus—particularly for the underage crowd (Boswell and Spade 1996; Martin and Hummer 1989).

Women reported that fraternity men dictated party transportation, the admittance of guests, party themes such as "CEO and secretary ho," the flow of alcohol, and the movement of guests within the party (Armstrong, Hamilton, and Sweeney 2006). Women often indicated that they engaged in strategies such as "travel[ing] in hordes" (21-1) and not "tak[ing] a drink if I don't know where it came from" (15-1) to feel safer at fraternity parties. Even when open to hooking up, women were not comfortable doing so if they sensed that men were trying to undermine their control of sexual activity (e.g., by pushing them to drink too heavily, barring their exit from private rooms, or refusing them rides home). Women typically opted not to return to party venues they perceived as unsafe. As one noted, "I wouldn't go to [that house] because I heard they do bad things to girls" (14-1). Even those interested in the erotic competition of party scenes tired of it as they realized that the game was rigged.

The sexual double standard also justified the negative treatment of women in the party scene—regardless of whether they chose to hook up. Women explained that men at parties showed a lack of respect for their feelings or interests—treating them solely as "sex objects" (32-1). This disregard extended to hookups. One told us, "The guy gets off and then it's done and that's all he cares about" (12-4). Another complained of her efforts to get a recent hookup to call: "That wasn't me implying I wanted a relationship—that was me implying I wanted respect" (42-2). In her view, casual sex did not mean forgoing all interactional niceties. A third explained, "If you're talking to a boy, you're either going to get into this huge relationship or you are nothing to them" (24-3). This either-or situation often frustrated women who wanted men to treat them well regardless of the level of commitment.

The Relationship Imperative

Women also encountered problematic gender beliefs about men's and women's different levels of interest in relationships. As one noted, women fight the "dumb girl idea"—the notion "that every girl wants a boy to sweep her off her feet and fall in love" (42-2). The expectation that women should want to be in relationships was so pervasive that many found it necessary to justify their single status to us. For example, when asked if she had a boyfriend, one woman with no shortage of admirers apologetically explained, "I know this sounds really pathetic and you probably think I am lying, but there are so many other things going on right now that it's really not something high up on my list…. I know that's such a lame-ass excuse, but it's true" (9-3). Another noted that already having a boyfriend was the only "actual, legitimate excuse" to reject men who expressed interest in a relationship (34-3).

Certainly, many women wanted relationships and sought them out. However, women's interest in relationships varied, and almost all experienced periods during which they wanted to be single. Nonetheless, women reported pressure to be in relationships all the time. We found that women, rather than struggling to get into relationships, had to work to avoid them.

The relational imperative was supported by the belief that women's relational opportunities were scarce and should not be wasted. Women described themselves as "lucky" to find a man willing to commit, as "there's not many guys like that in college" (15-1). This belief persisted despite the fact that most women were in relationships most of the time....

Gender beliefs may also limit women's control over the terms of interaction within relationships. If women are made to feel lucky to have boyfriends, men are placed in a position of power, as presumably women should be grateful when they commit. Women's reports suggest that men attempted to use this power to regulate their participation in college life. One noted, "When I got here my first semester freshman year, I wanted to go out to the parties ... and he got pissed off about it.... He's like, 'Why do you need to do that? Why can't you just stay with me?'" (4-2). Boyfriends sometimes tried to limit the time women spent with their friends and the activities in which they participated.

Women also became jealous; however, rather than trying to control their boyfriends, they often tried to change themselves. One noted that she would "do anything to make this relationship work."... Other women changed the way they dressed, their friends, and where they went in the attempt to keep boyfriends.

INTERSECTIONALITY: CONTRADICTIONS BETWEEN CLASS AND GENDER

Existing research about college sexuality focuses almost exclusively on its gendered nature. We contend that sexuality is shaped simultaneously by multiple intersecting structures. In this section, we examine the sexual and romantic implications of class beliefs about how ambitious young people should conduct themselves during college. Although all of our participants contended with class beliefs that contradicted those of gender, experiences of this structural intersection varied by class location. More privileged women struggled to meet gender and class guidelines for sexual behavior, introducing a difficult set of double binds. Because these class beliefs reflected a privileged path to adulthood, less privileged women found them foreign to their own sexual and romantic logics.

More Privileged Women and the Experience of Double Binds

The Self-Development Imperative and the Relational Double Bind The four-year university is a classed structural location. One of the primary reasons to attend college is to preserve or enhance economic position. The university culture is thus characterized by the self-development imperative, or the notion that individual achievement and personal growth are paramount. There are also accompanying rules for sex and relationships: Students are expected to postpone marriage and parenthood until after completing an education and establishing a career.

For more privileged women, personal expectations and those of the university culture meshed. Even those who enjoyed relationships experienced phases in college where they preferred to remain single. Almost all privileged women (94 percent) told us at one point that they did not want a boyfriend. One noted, "All my friends here … they're like, 'I don't want to deal with [a boyfriend] right now. I want to be on my own'" (37-1). Another eloquently remarked, "I've always looked at college as the only time in your life when you should be a hundred percent selfish.… I have the rest of my life to devote to a husband or kids or my job … but right now, it's my time" (21-2).

The notion that independence is critical during college reflected class beliefs about the appropriate role for romance that opposed those of gender. During college, relational commitments were supposed to take a backseat to self-development. As an upper-middle-class woman noted, "College is the only time that you don't have obligations to anyone but yourself.… I want to get settled down and figure out what I'm doing with my life before [I] dedicate myself to something or someone else" (14-4). Another emphasized the value of investment in human capital: "I've always been someone who wants to have my own money, have my own career so that, you know, 50 percent of marriages fail.… If I want to maintain the lifestyle that I've grown up with … I have to work. I just don't see myself being someone who marries young and lives off of some boy's money" (42-4). To become self-supporting, many privileged women indicated they needed to postpone marriage. One told us, "I don't want to think about that [marriage]. I want to get secure in a city and in a job.… I'm not in any hurry at all. As long as I'm married by 30, I'm good" (13-4). Even those who wanted to be supported by husbands did not expect to find them in college, instead setting their sights on the more accomplished men they expected to meet in urban centers after college.

More privileged women often found committed relationships to be greedy—demanding of time and energy. As one stated, "When it comes to a serious relationship, it's a lot for me to give into that. [What do you feel like you are giving up?] Like my everything.… There's just a lot involved in it" (35-3). These women feared that they would be devoured by relationships and sometimes struggled to keep their self-development projects going when they did get involved. As an upper-class woman told us, "It's hard to have a boyfriend and be really excited about it and still not let it consume you" (42-2). This situation was exacerbated by the gender beliefs discussed earlier, as women experienced pressure to fully devote themselves to relationships.…

With marriage far in the future, more privileged women often worried about college relationships getting too serious too fast. All planned to marry—ideally to men with greater earnings—but were clear about the importance of temporary independence.…

For more privileged women, contradictory cultural rules created what we call *the relational double bind*. The relational imperative pushed them to participate in committed relationships; however, relationships did not mesh well with the demands of college, as they inhibited classed self-development strategies. Privileged women struggled to be both "good girls" who limited their sexual activity to relationships and "good students" who did not allow relational commitments to derail their educational and career development.

The Appeal of Hookups and the Sexual Double Bind   In contrast, hookups fit well with the self-development imperative of college. They allowed women to be sexual without the demands of relationships. For example, one upper-class woman described hooking up as "fun and nonthreatening." She noted, "So many of us girls, we complain that these guys just want to hook up all the time. I'm going, these guys that I'm attracted to … get kind of serious." She saw her last hookup as ideal because "we were physical, and that was it. I never wanted it to go anywhere" (34-2). Many privileged women understood, if implicitly, that hooking up was a delay tactic, allowing sex without participation in serious relationships.

As a sexual solution for the demands of college, hooking up became incorporated into notions of what the college experience should be. When asked which kinds of people hook up the most, one woman noted, "All…. The people who came to college to have a good time and party" (14-1). With the help of media, alcohol, and spring break industries, hooking up was so institutionalized that many took it for granted. One upper-middle-class woman said, "It just happens. It's natural" (15-1). They told us that learning about sexuality was something they were supposed to be doing in college. Another described, "I'm glad that I've had my one-night stands and my being in love and having sex…. Now I know what it's supposed to feel like when I'm with someone that I want to be with. I feel bad for some of my friends…. They're still virgins" (29-1).

High rates of hooking up suggest genuine interest in the activity rather than simply accommodation to men's interests. Particularly early in college, privileged women actively sought hookups. One noted, "You see a lot of people who are like, 'I just want to hook up with someone tonight.'… It's always the girls that try to get the guys" (41-1). Data from the OCSLS also suggest that college women like hooking up almost as much as men and are not always searching for something more. Nearly as many women as men (85 percent and 89 percent, respectively) report enjoying the sexual activity of their last hookup "very much" or "somewhat," and less than half of women report interest in a relationship with their most recent hookup.

In private, several privileged women even used the classed logic of hooking up to challenge stereotyped portrayals of gender differences in sexuality. As one noted, "There are girls that want things as much as guys do. There are girls that want things more, and they're like, 'Oh it's been a while [since I had sex].' The girls are no more innocent than the guys…. People think girls are jealous of relationships, but they're like, 'What? I want to be single'" (34-1). When asked about the notion that guys want sex and girls want relationships another responded, "I think that is the absolute epitome of bullshit. I know so many girls who honestly go out on a Friday night and they're like, 'I hope I get some ass tonight.' They don't wanna have a boyfriend! They just wanna hook up with someone. And I know boys who want relationships. I think it goes both ways" (42-2). These women drew on gender-neutral understandings of sexuality characteristic of university culture to contradict the notion of women's sexuality as inevitably and naturally relational.

Hookups enabled more privileged women to conduct themselves in accordance with class expectations, but as we demonstrated earlier, the enforcement of gender beliefs placed them at risk of sanction. This conflict gets to the heart of a *sexual double bind*: While hookups protected privileged women from relationships that could derail their ambitions, the double standard gave men greater control over the terms of hooking up, justified the disrespectful treatment of women, supported sexual stigma, and produced feelings of shame.

Less Privileged Women and the Experience of Foreign Sexual Culture

Women's comfort with delaying commitment and participating in the hookup culture was shaped by class location. College culture reflects the beliefs of the more privileged classes. Less privileged women arrived at college with their own orientation to sex and romance, characterized by a faster transition into adulthood. They often attempted to build both relationships and career at the same time. As a result, the third of the participants from less privileged backgrounds often experienced the hookup culture as foreign in ways that made it difficult to persist at the university.

Less privileged women had less exposure to the notion that the college years should be set aside solely for educational and career development. Many did not see serious relationships as incompatible with college life. Four were married or engaged before graduating—a step that others would not take until later. One reminisced, "I thought I'd get married in college.... When I was still in high school, I figured by my senior year, I'd be engaged or married or something.... I wanted to have kids before I was 25" (25-4). Another spoke of her plans to marry her high school sweetheart: "I'll be 21 and I know he's the one I want to spend the rest of my life with.... Really, I don't want to date anybody else" (6-1).

Plans to move into adult roles relatively quickly made less privileged women outsiders among their more privileged peers. One working-class woman saw her friendships dissolve as she revealed her desire to marry and have children in the near future. As one of her former friends described,

> She would always talk about how she couldn't wait to get married and have babies.... It was just like, Whoa. I'm 18.... Slow down, you know? Then she just crazy dropped out of school and wouldn't contact any of us.... The way I see it is that she's from a really small town, and that's what everyone in her town does ... get married and have babies. That's all she ever wanted to do maybe?... I don't know if she was homesick or didn't fit in. (24-4)

This account glosses over the extent to which the working-class woman was pushed out of the university—ostracized by her peers for not acclimating to the self-development imperative and, as noted below, to the campus sexual climate. In fact, 40 percent of less privileged women left the university, compared to 5 percent of more-privileged women. In all cases, mismatch between the sexual culture of women's hometowns and that of college was a factor in the decision to leave.

Most of the less privileged women found the hookup culture to be not only foreign but hostile. As the working-class woman described above told us,

> I tried so hard to fit in with what everybody else was doing here.... I think one morning I just woke up and realized that this isn't me at all; I don't like the way I am right now.... I didn't feel like I was growing up. I felt like I was actually getting younger the way I was trying to act. Growing up to me isn't going out and getting smashed and sleeping around.... That to me is immature. (28-1)

She emphasized the value of "growing up" in college. Without the desire to postpone adulthood, less privileged women often could not understand the appeal of hooking up. As a lower-middle-class woman noted, "Who would be interested in just meeting somebody and then doing something that night? And then never talking to them again?... I'm supposed to do this; I'm supposed to get drunk every weekend. I'm supposed to go to parties every weekend ... and I'm supposed to enjoy it like everyone else. But it just doesn't appeal to me" (5-1). She reveals the extent to which hooking up was a normalized part of college life: For those who were not interested in this, college life could be experienced as mystifying, uncomfortable, and alienating.

The self-development imperative was a resource women could use in resisting the gendered pull of relationships. Less privileged women did not have as much access to this resource and were invested in settling down. Thus, they found it hard to resist the pull back home of local boyfriends, who—unlike the college men they had met—seemed interested in marrying and having children soon. One woman noted after transferring to a branch campus, "I think if I hadn't been connected with [my fiancé], I think I would have been more strongly connected to [the college town], and I think I probably would have stayed" (2-4). Another described her hometown boyfriend: "He'll be like, 'I want to see you. Come home.'... The stress he was putting me under and me being here my first year. I could not take it" (7-2). The following year, she moved back home. A third explained about her husband, "He wants me at home.... He wants to have control over me and ... to feel like he's the dominant one in the relationship.... The fact that I'm going to school and he knows I'm smart and he knows that I'm capable of doing anything that I want ... it scares him" (6-4). While she eventually ended this relationship, it cost her an additional semester of school.

Women were also pulled back home by the slut stigma, as people there—perhaps out of frustration or jealousy—judged college women for any association with campus sexual culture. For instance, one woman became distraught when a virulent sexual rumor about her circulated around her hometown, especially when it reached her parents. Going home was a way of putting sexual rumors to rest and reaffirming ties that were strained by leaving.

Thus, less-privileged women were often caught between two sexual cultures. Staying at the university meant abandoning a familiar logic and adopting a privileged one—investing in human capital while delaying the transition to adulthood. As one explained, attending college led her to revise her "whole plan": "Now I'm like, I don't even need to be getting married yet [or] have kids.... All of [my brother's] friends, 17- to 20-year-old girls, have their ... babies, and I'm like, Oh my God.... Now I'll be able to do something else for a couple years before I settle down ... before I worry about kids"(25-3). These changes in agenda required them to end relationships with men whose life plans diverged from theirs. For some, this also meant cutting ties with hometown friends. One resolute woman, whose friends back home had turned on her, noted, "I'm just sick of it. There's nothing there for me anymore. There's absolutely nothing there" (22-4).

DISCUSSION

The Strengths of an Interactional Approach

Public gender beliefs are a key source of gender inequality in college heterosexual interaction. They undergird a sexual double standard and a relational imperative that justify the disrespect of women who hook up and the disempowerment of women in relationships—reinforcing male dominance across social forms. Most of the women we studied cycled back and forth between hookups and relationships, in part because they found both to be problematic. These findings indicate that an individualist, battle of the sexes explanation not only is inadequate but may contribute to gender inequality by naturalizing problematic notions of gender difference.

We are not, however, claiming that gender differences in stated preferences do not exist. Analysis of the OCSLS finds a small but significant difference between men and women in preferences for relationships as compared to hookups: After the most recent hookup, 47 percent of women compared to 37 percent of men expressed some interest in a relationship. These differences in preferences are consistent with a multilevel perspective that views the internalization of gender as an aspect of gender structure (Risman 2004). As we have shown, the pressure to internalize gender-appropriate preferences is considerable, and the line between personal preferences and the desire to avoid social stigma is fuzzy. However, we believe that widely shared beliefs about gender difference contribute more to gender inequality in college heterosexuality than the substantively small differences in actual preferences.

The Strengths of an Intersectional Approach

An intersectional approach sheds light on the ambivalent and contradictory nature of many college women's sexual desires. Class beliefs associated with the appropriate timing of marriage clash with resilient gender beliefs—creating difficult double binds for the more privileged women who strive to meet both. In the case of the relational double bind, relationships fit with gender beliefs but pose problems for the classed self-development imperative. As for the sexual double bind, hookups provide sexual activity with little cost to career development, but a double standard penalizes women for participating. Less privileged women face an even more complex situation: Much of the appeal of hookups derives from their utility as a delay strategy. Women who do not believe that it is desirable to delay marriage may experience the hookup culture as puzzling and immature.

An intersectional approach also suggests that the way young heterosexuals make decisions about sexuality and relationships underlies the reproduction of social class. These choices are part of women's efforts to, as one privileged participant so eloquently put it, "maintain the lifestyle that I've grown up with." Our participants were not well versed in research demonstrating that college-educated women benefit from their own human capital investments, are more likely to marry than less educated women, and are more likely to have a similarly well-credentialed spouse (DiPrete and Buchmann 2006). Nonetheless, most were aware that completing college and delaying marriage until the mid-to-late twenties made economic sense. Nearly all took access to marriage for granted, instead focusing their attention on when and whom they would marry.

Theoretical Contributions and Directions for Future Research

In this article, we link multilevel and intersectional approaches, using Sewell's (1992) theory of structure as a bridge. We focus on the intersection of gender and class beliefs about sexuality on the interactional level. Our approach suggests that gender intersects with a variety of other structures at all levels—the individual, interactional, and organizational. This opens up a wide range of analytical possibilities. Scholars might look at intersections occurring at other levels of structure or with structures in addition to social class. For example, the reproduction of racial categories depends on rules limiting sexual and romantic contact across racial boundaries. A next step would be to investigate how the intersection of race, class, and gender structures shapes sexual experiences in college.

REFERENCES

Armstrong, Elizabeth A., Paula England, and Alison C. K. Fogarty. (2009). Orgasm in college hookups and relationships. In *Families as they really are*, edited by B. Risman. New York: Norton.

Armstrong, Elizabeth A., Laura Hamilton, and Brian Sweeney. (2006). Sexual assault on campus: A multilevel, integrative approach to party rape. *Social Problems* 53:483–99.

Arnett, Jeffrey Jensen. (2004). *Emerging adulthood: The winding road from the late teens through the twenties*. New York: Oxford.

Beisel, Nicola, and Tamara Kay. (2004). Abortion, race, and gender in nineteenth century America. *American Sociological Review* 69:498–518.

Bogle, Kathleen A. (2008). *Hooking up: Sex, dating, and relationships on campus.* New York: New York University Press.

Boswell, A. Ayres, and Joan Z. Spade. (1996). Fraternities and collegiate rape culture: Why are some fraternities more dangerous places for women? *Gender & Society* 10:133–47.

Brooks, David. (2002). Making it: Love and success at America's finest universities. *The Weekly Standard*, December 23.

Campbell, Jacquelyn C., Nancy Glass, Phyllis W. Sharps, Kathryn Laughon, and Tina Bloom. (2007). Intimate partner homicide. *Trauma, Violence, & Abuse* 8:246–69.

Collins, Patricia Hill. (1990). *Black feminist thought: Knowledge, consciousness, and the politics of empowerment.* Boston: Unwin Hyman.

Collins, Randall. (2004). *Interaction ritual chains.* Princeton, NJ: Princeton University Press.

Connell, R. W. (1987). *Gender and power: Society, the person, and sexual politics.* Stanford, CA: Stanford University Press.

Correll, Shelley J., Stephen Benard, and In Paik. (2007). Getting a job: Is there a motherhood penalty? *American Journal of Sociology* 112:1297–1338.

Crawford, Mary, and Danielle Popp. (2003). Sexual double standards: A review and methodological critique of two decades of research. *Journal of Sex Research* 40:13–26.

DiPrete, Thomas A., and Claudia Buchmann. (2006). Gender-specific trends in the value of education and the emerging gender gap in college completion. *Demography* 43:1–24.

Dunn, Jennifer L. (1999). What love has to do with it: The cultural construction of emotion and sorority women's responses to forcible interaction. *Social Problems* 46:440–59.

Eder, Donna, Catherine Colleen Evans, and Stephen parker. (1995). *School talk: Gender and adolescent culture.* New Brunswick, NJ: Rutgers University press.

England, Paula, Emily Fitzgibbons Shafer, and Alison C. K. Fogarty. (2007). Hooking up and forming romantic relationships on today's college campuses. In *The gendered society reader,* edited by M. Kimmel. New York: Oxford University Press.

Eshbaugh, Elaine M., and Gary Gute. (2008). Hookups and sexual regret among college women. *Journal of Social Psychology* 148:77–89.

Friedland, Roger, and Robert R. Alford. (1991). Bringing society back in: Symbols, practices, and institutional contradictions. In *The new institutionalism in organizational analysis,* edited by W. W. Powell and P. J. DiMaggio, 232–63. Chicago: University of Chicago Press.

Gerstel, Naomi, and Natalia Sarkisian. (2006). Marriage: The good, the bad, and the greedy. *Contexts* 5:16–21.

Glenn, Evelyn Nakano. (1999). The soda construction and institutionalization of gender and race: An integrative framework. In *Revisioning gender,* edited by M. M. Ferree, J. Lorber, and B. B. Hess, 3–43. Thousand Oaks, CA: Sage.

Glenn, Norval, and Elizabeth Marquardt. (2001). *Hooking up, hanging out, and hoping for Mr. Right: College women on mating and dating today.* New York: Institute for American Values.

Grello, Catherine M., Deborah P. Welsh, and Melinda M. Harper. (2006). No strings attached: The nature of casual sex in college students. *Journal of Sex Research* 43:255–67.

Holland, Dorothy C., and Margaret A. Eisenhart. (1990). *Educated in romance: Women, achievement, and college culture.* Chicago: University of Chicago Press.

Illouz, Eva. (2005). *Cold intimacies: The making of emotional capitalism.* Cambridge, UK: Polity.

Kimmel, Michael. (2008). *Guyland: The perilous world where boys become men.* New York: HarperCollins.

Manning, Wendy D., Peggy C. Giordano, and Monica A. Longmore. (2006). Hooking up: The relationship contexts of "nonrelationship" sex. *Journal of Adolescent Research* 21:459–83.

Martin, Karin. (1996). *Puberty, sexuality, and the self: Boys and girls at adolescence.* New York: Routledge.

Martin, Patricia Yancey. (2004). Gender as a social institution. *Social Forces* 82:1249–73.

Martin, Patricia Yancey, and Robert A. Hummer. (1989). Fraternities and rape on campus. *Gender & Society* 3:457–73.

McCall, Leslie. (2005). The complexity of intersectionality. *Signs: Journal of Women in Culture and Society* 30:1771–1800.

Owen, Jesse J., Galena K. Rhoades, Scott M. Stanley, and Frank D. Fincham. (2008). "Hooking up" among college students: Demographic and psychosocial correlates. *Archives of Sexual Behavior*, http://www.springerlink.com&content/44j645v7v38013u4/fulltext.html.

Paul, Elizabeth L., Brian McManus, and Allison Hayes. (2000). "Hookups": Characteristics and correlates of college students' spontaneous and anonymous sexual experiences. *Journal of Sex Research* 37:76–88.

Ridgeway, Cecilia L. (2000). Limiting inequality through interaction: The end(s) of gender. *Contemporary Sociology* 29:110–20.

Ridgeway, Cecilia L. (2009). Framed before we know it: How gender shapes social relations. *Gender & Society* 23:145–60.

Ridgeway, Cecilia L., and Shelley J. Correll. (2004). Unpacking the gender system: A theoretical perspective on gender beliefs and social relations. *Gender & Society* 18:510–31.

Risman, Barbara, and Pepper Schwartz. (2002). After the sexual revolution: Gender politics in teen dating. *Contexts* 1:16–24.

Risman, Barbara J. (2004). Gender as a social structure: Theory wrestling with activism. *Gender & Society* 18:429–50.

Rosenfeld, Michael J. (2007). *The age of independence: Interracial unions, same-sex unions and the changing American family.* Cambridge, MA: Harvard University press.

Sewell, William H. (1992). A theory of structure: Duality, agency, and transformation. *American Journal of Sociology* 98:1–29.

Simon, Robin W., Donna Eder, and Cathy Evans. (1992). The development of feeling norms underlying romantic love among adolescent females. *Social Psychology Quarterly* 55:29–46.

Stepp, Laura Sessions. (2007). *Unhooked: How young women pursue sex, delay love, and lose at both.* New York: Riverhead.

Wharton, Amy S. (2005). *The sociology of gender: An introduction to theory and research.* Maiden, MA: Blackwell.

KEY CONCEPTS

gender theory   hooking up   class privilege

DISCUSSION QUESTIONS

1. According to the research presented by Hamilton and Armstrong, how do class and gender intersect to create competing expectations for sexual activity in college? What do the authors mean by the "double standard" for women of privilege and women of less privilege?

2. Are the results presented in this article consistent with your own experiences in college? Are dating relationships less common then casual hook-ups? Do you see a disadvantage for women in this type of social atmosphere?

41

Hetero-Romantic Love and Heterosexiness in Children's G-Rated Films

KARIN A. MARTIN AND EMILY KAZYAK

Martin and Kazyak explore the way heterosexual relationships are presented in children's films. Specifically, they look at how films use visual cues, music, and character interaction to make heterosexuality normal and expected. While G-rated films are considered acceptable for children and are expected to be free of sexual scenes and sexual innuendo, this research shows instead that children's films are laden with sexual images that reinforce heterosexuality as the dominant, normal, expected, accepted, and only form of interaction between characters.

SOURCE: Martin, Karin A., and Emily Kazyak. 2009. "Hetero-Romantic Love and Heterosexiness in Children's G-Rated Films." *Gender & Society* 23: 315–336.

Multiple ethnographic studies suggest that by elementary school, children understand the normativity of heterosexuality That is, by elementary school, children have a heteronormative understanding of the world (Best 1983; Renold 2002, 2005; Thorne 1993). Yet we know little about what children bring with them to the peer cultures these ethnographers describe and how these understandings develop before elementary school. Martin (2009) finds that mothers' conversations with young children normalize heterosexuality, but children's social worlds are larger than the mother-child dyad. Research on adolescence suggests that alongside parents and peers, the media are important in shaping cultural understandings of sexuality (Kim et al. 2007; Ward 1995, 2003). This article provides a beginning step toward understanding the role of the media in the development of children's heteronormativity. We ask, How are heteronormativity and heterosexuality constructed in children's top-selling G-rated movies between 1990 and 2005?

HETERONORMATIVITY

Heteronormativity includes the multiple, often mundane ways through which heterosexuality overwhelmingly structures and "pervasively and insidiously" orders "everyday existence" (Jackson 2006, 108; Kitzinger 2005). Heteronormativity structures social life so that heterosexuality is always assumed, expected, ordinary, and privileged. Its pervasiveness makes it difficult for people to imagine other ways of life. In part, the assumption and expectation of heterosexuality is linked to its status as natural and biologically necessary for procreation (Lancaster 2003). Anything else is relegated to the nonnormative, unusual, and unexpected and is, thus, in need of explanation. Specifically, within heteronormativity, homosexuality becomes the "other" against which heterosexuality defines itself (Johnson 2005; Rubin 1984).

But not just any kind of heterosexuality is privileged. Heteronormativity regulates those within its boundaries as it marginalizes those outside of it. According to Jackson (2006), heteronormativity works to define more than normative sexuality, insofar as it also defines normative ways of life in general. Heteronormativity holds people accountable to reproductive procreative sexuality and traditional gendered domestic arrangements of sexual relationships, and it is linked to particular patterns of consumerism and consumption (Ingraham 1999). In other words, while heteronormativity regulates people's sexualities, bodies, and sexual relationships (for both those nonheterosexuals on the "outside" and heterosexuals on the "inside"), it regulates nonsexual aspects of life as well.

Heteronormativity also privileges a particular type of heterosexual. Among those aspects desired in heterosexuals, Rubin (1984) includes being married, monogamous, and procreative. We might also include that heterosexuality is most sanctioned when it is intraracial and that other inequalities, like race and class, intersect and help construct what Rubin calls "the inner charmed circle" in a multitude of complicated ways (e.g., Whose married sex is most sanctioned?

Whose reproductive sex is most normal?). Heteronormativity also rests on gender asymmetry, as heterosexuality depends on a particular type of normatively gendered women and men (Jackson 2006). In this article, we examine how children's movies construct heterosexuality to better understand what information is available in media that might contribute to children's heteronormative social worlds.

CHILDREN, MEDIA, AND MOVIES

The media are an important avenue of children's sexual socialization because young children are immersed in media-rich worlds. Thirty percent of children under three years old and 43 percent of four- to six-year-olds have a television in their bedrooms, and one-quarter of children under six years old have a VCR/DVD player in their bedrooms (Rideout, Vandewater, and Wartella 2003). Since the deregulation of television in the 1980s, there has been more and more content produced on television for children. Children's programming produced for television, however, must still meet educational regulations. Films produced with young children as a significant intended portion of the audience are under no such obligations. However, to attract young children (and their parents) to films, filmmakers must get their movies a G-rating. Film producers are interested in doing this because the marketing advantages that accompany a successful children's film are enormous (Thomas 2007). The Motion Picture Association of America rates a film G for "General Audience" if the film "contains nothing in theme, language, nudity, sex, violence or other matters that, in the view of the Rating Board, would offend parents whose younger children view the motion picture.... No nudity, sex scenes or drug use are present in the motion picture" (Motion Picture Association of America 2009). Thus, a G-rating signals that these films expect young children in their audience.

We examine the top-selling G-rated movies to challenge the idea that these movies are without (much) sexual content and the notion that young children are therefore not exposed to matters relating to sexuality. As theorists of heteronormativity suggest, heterosexuality is pervasive, and we want to examine how it makes its way into films that are by definition devoid of sexuality. If heteronormativity structures social life well beyond the sexual arena, then it is likely at work even in films that announce themselves as free of sexuality.

We look at movies themselves rather than children's reception of them because of the difficulty of research with young children generally, especially around issues of sexuality (Martin, Luke, and Verduzco-Baker 2007) and around media (Thomas 2007). Parents, human subjects review boards, and schools all serve as barriers to research with children on these topics. Given that we know little about how heteronormativity is constructed for children, examining the content of these films seems a logical first step before asking what children take from them. Although we will not be able to say whether or which accounts of heteronormativity children take away with them after watching these movies, current research about children's relationships to such movies indicates that children are engaged with these media and the stories they tell. Enormous numbers of children watch Disney and other G-rated children's movies. In a 2006 survey of more than 600 American mothers of three- to six-year-olds, only 1 percent reported that their child had not seen any of the films we analyze here; half had seen 13 or more (Martin, Luke, and Verduzco-Baker 2007).

Many children also watch these movies repeatedly (Mares 1998). The advent of videos made it possible for children to watch and rewatch movies at home. In fact, preschool children enjoy watching videos/DVDs repeatedly, and this has implications for the way they comprehend their messages. Crawley et al. (1999) discovered that children comprehended more from repeated viewing. Repeated viewing may also mean that jokes or innuendo intended for adults in these films may become more visible and curious, if not more intelligible, to young children. Further work by Schmitt, Anderson, and Collins (1999) also suggests that young children's attention is most focused and content best understood when watching media that includes animation, child characters, nonhuman characters, animals, frequent movement, and purposeful action (as opposed to live action; adults, especially adult men; and characters who only converse without much action). These are prominent features of most of the G-rated films we analyze here, suggesting that they are certainly vehicles for children's attention and comprehension.

We also know young children are engaged by many such films as the plots and toys marketed from them are used in many creative ways in children's fantasy and play. Not only do movies make social worlds visible on screen, but the mass marketing surrounding these movies invites young people to inhabit those worlds (Giroux 1996). These media not only offer what is normal but also actively ensure that children understand it and compel them to consume it (Schor 2004). Researchers have demonstrated the depth of children's engagement with such media and how they adapt it for their own uses ... [T]here is evidence that children certainly incorporate such media into their learning and play.

Some scholarship has begun to look at what kinds of narratives, accounts, and images are available in children's movies, and especially in Disney movies. Most useful for our purposes is the research on gender (Thompson and Zerbinos 1995; Witt 2000) and on gender and race stereotypes in young children's media (Giroux 1996; Hurley 2005; Mo and Shen 2000; Pewewardy 1996; Witt 2000). Most of this research indicates that there are fewer portrayals of women and of nonwhites and that those portrayals often rely on stereotypes. Analyses of the stereotypes and discourses of race and gender sometimes embed some discussion of sexuality within them. A smattering of research on race examines how some racial/ethnic groups are portrayed as exoticized and more sexualized than white women (Lacroix 2004). Research that examines gender construction in the media sometimes links heterosexuality and romantic love to femininity and discusses the importance of finding a man/prince for the heroines (Junn 1997; Thompson and Zerbinos 1995). But heterosexuality is a given in such analyses. The existing research does not fully analyze how heterosexuality is constructed in these films.

In a different vein, media scholars have offered queer readings of some children's and especially Disney films (Byrne and McQuillan 1999; Griffin 2000). Employing a poststructuralist lens that privileges the radically indeterminate meaning of texts, Byrne and McQuillan (1999) highlight how certain characters and story lines in Disney movies can be read as queer. They discuss the many queer or ambiguous characters populating these films, such as Quasimodo and the gargoyles in *The Hunchback of Notre Dame*. They describe the character Mulan as a "transvestite bonanza," representing "Disney's most sustained creation of lesbian chic" (1999, 143). Moreover, they highlight the queerness of certain story lines in Disney movies. For instance, they argue that homosocial desire and bonds between men structure many of the films, and they explicate the queerness of the portrayal of monstrous desire, a desire that threatens the family unit, in *Beauty and the Beast*. These readings do not argue that particular characters or plots are gay or lesbian per se; rather, they emphasize their queer potential. Similarly, Griffin (2000) aims to queer Disney by analyzing how gay and lesbian viewers might understand these films with gay sensibilities. He highlights how Disney characters who do not fit into their societies echo the feeling of many gays and lesbians. He also argues that many characters (especially villains) lend themselves to queer readings because of how they overperform their gender roles. Villainesses often look like drag queens, such as Ursula in *The Little Mermaid*, a character modeled after the transvestite star Divine. These analyses rest on the desire to destabilize the meanings of characters and story lines in movies to open them up and discover their queer potential. This scholarship, however, presumes a sophisticated and knowledgeable reader of culture. It does not consider children as the audience or address whether such readings are possible for young children. It overlooks, for example, that while there are transvestite characters like Mulan, the Mulan toys marketed to children were feminine, long-haired, non-sword-wielding ones (Nguyen 1998), perhaps making such readings less sustainable for children even if they are possible. Again, we will need research on what children take away from such media to address these issues.

OUR RESEARCH

In this article, we do not aim to do a queer reading of these films as such readings have already been done. Instead, we analyze *how* heterosexuality is constructed in children's G-rated films. We ask not how characters might be read as queer but what accounts these films offer of heterosexuality and how such accounts serve heteronormativity. Unpacking the construction of heterosexuality in these films is a first step toward understanding what social-sexual information is available to the children who watch them.

SAMPLE AND METHOD

The data for this study come from all the G-rated movies released (or rereleased) between 1990 and 2005 that grossed more than $100 million in the United States (see Table 1). Using this sample of widely viewed films overcomes the

T A B L E 1   Sample: $1000 Million GRated Movies, 1900–2005

Movie	Year	Produced By	Hetero-Romantic Story Line			Hetrosexuality	
			Any Reference	Major Plot	Minor Plot	Sexiness	Ogling of Women's Bodies
Chicken Little	2005	Disney	x		x	x	
The Polar Express	2004	Castle Rock					
Finding Nemo	2003	Disney/Pixar	x				
The Santa Clause 2	2002	Disney	x	x			
Monsters, Inc.	2001	Disney/Pixar	x		x		x
The Princess Diaries	2001	Disney	x	x		x	x
Chicken Run	2000	Dreamworks	x		x		x
Tarzan	1999	Disney	x		x		
Toy Story 2	1999	Disney/Pixar	x			x	x
A Bug's Life	1998	Disney/Pixar	x		x		x
Mulan	1998	Disney	x		x		x
The Rugrats Movie	1998	Nickelodeon	x				
101 Dalmatians	1996	Disney	x		x		
The Hunchback of Notre Dame	1996	Disney	x	x		x	x
Toy Story	1995	Disney/Pixar	x			x	
Pocahontas	1995/2005	Disney	x	x			
The Lion King	1994/2002	Disney	x	x			
Aladdin	1994	Disney	x	x		x	x
Beauty and the Beast	1991/2002	Disney	x	x		x	x
The Little Mermaid	1989/1997	Disney	x	x		x	x

limitations of previous analyses of children's, and especially Disney, movies, which often focus on a few particular examples. Here we have tried to examine all the most viewed films within this genre and time period. The films in our sample were extremely successful and widely viewed, as evidenced by their sales numbers in theaters. Home videos/DVDs sales and rentals of these films are also very high (Arnold 2005), including direct-to-video/DVD sequels of many of these films, for example, *Lion King 1.5, Ariel's Beginning,* and *Beauty and the Beast's Enchanted Christmas.* While the audience for these films is broader than children, children are certainly centrally intended as part of the audience. G is the rating given to films that contain nothing that "would offend parents whose younger children view the motion picture" according to the Motion Picture Association of America (2009). Sixteen (80 percent) of these films are animated, and 17 are produced by Disney, a major producer of children's consumption and socialization (Giroux 1997).

After collecting this sample, the first author screened all the films and then trained three research assistants to extract any story lines, images, scenes, songs, or dialogue that depicted anything about sexuality, including depictions of bodies, kissing, jokes, romance, weddings, dating, love, where babies come from, and pregnancy. The research assistants then wrote descriptions of the scenes in which they found material related to sexuality. They described the visuals of the scenes in as vivid detail as possible and transcribed the dialogue verbatim. Two research assistants watched each film and extracted the relevant material. The first author reconciled the minimal differences between what each research assistant included by rescreening the films herself and adding or correcting material.

RESULTS AND DISCUSSION

We describe two ways that heterosexuality is constructed in these films. The primary account of heterosexuality in these films is one of hetero-romantic love and its exceptional, magical, transformative power. Secondarily, there are some depictions of heterosexuality outside of this model. Outside of hetero-romantic love, heterosexuality is constructed as men gazing desirously at women's bodies. This construction rests on gendered and racialized bodies and is portrayed as less serious and less powerful than hetero-romantic love.

Magical, Exceptional, Transformative Hetero-Romantic Love

Hetero-romantic love is the account of heterosexuality that is most developed in these films. Only two films have barely detectable or no hetero-romantic references (see Table 1). In eight of these films hetero-romance is a major plot line, and in another seven films it is a secondary story line. Those films not made by Disney have much less hetero-romantic content than those made by Disney.

Films where we coded hetero-romantic love as a major plot line are those in which the hetero-romantic story line is central to the overall narrative of the film. In *The Little Mermaid,* for instance, the entire narrative revolves around the romance between Ariel, a mermaid, and Eric, a human. The same is true of movies like *Beauty and the Beast, Aladdin,* and *Santa Claus 2*. There would be no movie without the hetero-romantic story line for these films. In others, the hetero-romantic story line is secondary. For example, in *Chicken Run* the romance develops between Ginger and Rocky as they help organize the chicken revolt—the heart of the movie—although the movie ends with them coupled, enjoying their freedom in a pasture. While removing the hetero-romantic story line would still leave other stories in place in such films, the romance nonetheless exists. In other movies, like *Toy Story,* references are made to hetero-romance but are not developed into a story line. For instance, this film suggests romantic interest between Woody and Little Bo Peep, but their romance is not woven throughout the film.

While our focus is on the construction of heterosexuality, we recognize that other stories exist in these films. For instance, there are stories about parent-child relationships (e.g., Chicken Little wants his father to be proud of him; Nemo struggles against his overprotective father). Stories about workers, working conditions, and collective revolt also appear, for instance, in *Monsters, Inc.* (whose characters, working for the city's power company that relies on scaring children to generate electricity, successfully stop an evil corporate plan to kidnap children and eventually change their policy to making children laugh) and *Chicken Run* (whose main character, Ginger, successfully organizes all of her fellow chickens to escape their farm after learning of the farmers' plan to begin turning them into chicken pies). Though certainly there is much analysis that could be done around such stories, we do not do so here. Rather, we turn our attention to the hetero-romantic story lines and the work they do in constructing heterosexuality.

Theorists of heteronormativity suggest that the power of heteronormativity is that heterosexuality is assumed, mundane, ordinary, and expected. In contrast, we find that in these films, while it is certainly assumed, heterosexuality is very often not ordinary or mundane. Rather, romantic heterosexual relationships are portrayed as a special, distinct, exceptional form of relationship, different from all others. Characters frequently defy parents, their culture, or their very selves to embrace a hetero-romantic love that is transformative, powerful, and (literally) magical. At the same time, these accounts are sometimes held in tension with or constructed by understandings of the naturalness of heterosexuality.

These films repeatedly mark relationships between cross-gender lead characters as special and magical by utilizing imagery of love and romance. Characters in love are surrounded by music, flowers, candles, magic, fire, ballrooms, fancy dresses, dim lights, dancing, and elaborate dinners. Fireflies, butterflies, sunsets, wind, and the beauty and power of nature often provide the setting for—and a link to the naturalness of—hetero-romantic love. For example, in *Beauty and the Beast,* the main characters fall in love frolicking in the snow; Aladdin and Jasmine fall in love as they fly through a starlit sky in *Aladdin*; Ariel falls in love as she discovers the beauty of earth in *The Little Mermaid*; Santa and his eventual bride ride in a sleigh on a sparkling snowy night with snow lightly falling over only their heads in *Santa Claus 2*; and *Pocahontas* is full of allusion to water, wind, and trees as a backdrop to the characters falling in love. The characters often say little in these scenes. Instead, the scenes are overlaid with music and song that tells the viewer more abstractly what the characters are feeling. These scenes depicting hetero-romantic love are also paced more slowly with longer shots and with slower and soaring music.

These films also construct the specialness of hetero-romantic love by holding in tension the assertion that hetero-romantic relationships are simultaneously magical and natural. In fact, their naturalness and their connection to "chemistry" and the body further produce their exceptionalness. According to Johnson (2005), love and heterosexuality become interwoven as people articulate the idea that being in love is overpowering and that chemistry or a spark forms the basis for romantic love. These formulations include ideas about reproductive instincts and biology, and they work to naturalize heterosexuality. We see similar constructions at work in these G-rated movies where the natural becomes the magical. These films show that, in the words of Mrs. Pots from *Beauty and the Beast,* if "there's a spark there," then all that needs to be done is to "let nature take its course." However, this adage is usually not spoken. Rather, the portrayal of romantic love as occurring through chemistry or a spark is depicted by two characters gazing into each other's eyes and sometimes stroking each other's faces. The viewer usually sees the two characters up close and in profile as serious and soaring music plays as this romantic chemistry is not explained with words but must be felt and understood via the gazing eye contact between the characters. Disney further marks the falling in love and the triumphs of hetero-romantic love by wrapping the characters in magical swirls of sparks, leaves, or fireworks as they stare into each other's eyes. The music accompanying such scenes is momentous and triumphant.

We asked whether all sorts of relationships might be magical, special, and exceptional in similar ways, as it is possible that many types of relationships have these qualities in these imaginative fantasies where anything is possible. However, we found that romantic heterosexual relationships in G-rated movies are set apart from other types of relationships. This serves to further define them as special and exceptional. All other love relationships are portrayed without the imagery described above. The pacing of friendship scenes is also faster and choppier, and the music is quicker and bouncy. Nor do friendships and familial relationships start with a "spark."

Parent–child relationships are portrayed as restrictive, tedious, and protective. The child is usually escaping these relationships for the exciting adolescent or adult world. Friendships are also set aside as different from romantic love. There are many close friendships and buddies in these stories, and none are portrayed with the imagery of romantic love. Cross-gender friends are often literally smaller and a different species or object in the animated films, thus making them off limits for romance. For example, Mulan's friend is Mushoo, a small, red dragon; Pocahontas is friends with many small animals (a raccoon; a hummingbird); Ariel is looked after by Sebastian (a crab) and Flounder (a fish); and Belle is befriended by a range of small household items (teapot, candlestick, broom). Same–sex friendships or buddies are unusual for girls and women unless the friends are maternal (e.g., Willow in *Pocahontas,* Mrs. Pots in *Beauty and the Beast*). The lead male characters, however, often have comical buddies (e.g., Timon in *The Lion King,* Abu in *Aladdin,* the gargoyles in *The Hunchback of Notre Dame,* Mike in *Monsters, Inc.*). These friendships are often portrayed as funny, silly, gross, and fun but certainly not as serious, special, powerful, important, or natural. For example, in *The Lion King,* Timon (a meerkat), Pumba (a boar), and Simba (a lion) all live a carefree life together in the jungle as the best of friends, but Simba quickly deserts them for Nala, a female lion, once he is an adolescent. Throughout the film, Timon and Pumba provide comic relief from the serious business of the lions falling in (heterosexual) love and saving the kingdom. Thus, the construction of friendships and family relationships reveal that hetero-romantic relationships in contrast are serious, important, and natural.

Furthermore, while friendships provide comic relief and friends and family are portrayed as providing comfort or advice to lead characters, these relationships are not portrayed as transformative, powerful, or magical. Hetero-romantic love is exceptional in these films because it is constructed as incredibly powerful and transformative. Throughout many of these films with a primary plot about hetero-romantic love, such love is depicted as rebellious, magical, defiant, and with a power to transform the world. This is quite different from our understanding of heterosexuality as normative, ordinary, and expected. The hetero-romantic relationships in these films are extraordinary. Falling in heterosexual love can break a spell (*Beauty and the Beast*) or cause one to give up her identity (*The Little Mermaid*). It can save Santa Claus and Christmas (*Santa Claus 2*). It can lead children (e.g., Ariel, Jasmine, Pocahontas, Belle) to disobey their parents and defy the social rules of their culture (e.g., Jasmine, Pocahontas). It can stop a war that is imminent (*Pocahontas*) or change an age-old law (*Aladdin*).

Hetero-romantic love is constructed as being in a realm of freedom and choice, a realm where chemistry can flourish and love can be sparked and discovered. Thus, romantic love is so exceptional it is positioned "outside of the control of any social or political force" (Johnson 2005, 37). This construction appears in G-rated movies and intertwines race and heteronormativity as characters who are nonwhite critique arranged marriages as backward and old-fashioned and celebrate a woman's ability to choose her own husband. For example, in *Aladdin,* Jasmine protests the law that dictates that she must marry a prince and says, "The law is wrong. ... I hate being forced into this ... if I do marry, I want it to be for love." Later, Aladdin agrees with her that being forced to be married by her father is "awful." Pocahontas faces a similar dilemma, as her father insists that she marry Kocoum. When she disagrees and asks him, "Why can't I choose?" he says, "You are the daughter of the chief ... it is your time to take your place among our people." While arranged marriages are portrayed as something outdated, these characters "choose" whom they will love, thus simultaneously securing hetero-romantic love's naturalness and extraordinariness and its position beyond the prescriptions of any social-political context. In fact, their love changes these prescriptions in both of these examples. Jasmine and Aladdin's love overturns the age-old law that the princess must marry a prince when she is of age, and Pocahontas's love for John Smith ends the war between her tribe and colonizers. This transformative power of hetero-romantic love is echoed throughout these films.

Finally, we observe that hetero-romantic love is not sexually embodied in these films except through kissing. The power of hetero-romantic love is often delivered through a heterosexual kiss. A lot of heterosexual kissing happens in G-rated films. *Princess Diaries,* with its live-action teenage characters, contains the most explicit kissing, as the main character daydreams that a boy kisses her passionately, open-mouthed as she falls back against the lockers smiling and giggling. Most animated kisses are with closed mouths (or the viewer cannot fully see the mouths) and of shorter duration, but they are often even more powerful. Throughout these films, but especially in the animated ones, a heterosexual kiss signifies heterosexual love and in doing so is powerful.

Heterosexiness and the Heterosexual Gaze:
Heterosexuality Outside of Love

Thus far, we have described how heterosexuality is constructed through depictions of hetero-romantic love relationships in these films. There is also heterosexuality depicted outside of romantic relationships, though this heterosexuality is quite different and more ordinary. As such, it is depicted not as earnest or transformative but as frivolous, entertaining, and crude. This nonromantic heterosexuality is constructed through the different portrayals of women's and men's bodies, the heterosexiness of the feminine characters, and the heterosexual gaze of the masculine ones.

Heteronormativity requires particular kinds of bodies and interactions between those bodies. Thus, as heterosexuality is constructed in these films, gendered bodies are portrayed quite differently, and we see much more of some bodies than others. Women throughout the animated features in our sample are drawn with cleavage, bare stomachs, and bare legs. Women of color are more likely to be drawn as young women with breasts and hips and white women as delicate girls (Lacroix 2004). Men are occasionally depicted without their shirts, such as in *Tarzan*; or without much of a shirt, as in *Aladdin*; and in one scene in *Mulan*, it is implied that men have been swimming naked. However, having part of the body exposed is more common among the lead women characters and among the women who make up the background of the scenes.

Women's nudity is also often marked as significant through comment or reaction. Women are often "almost caught" naked by men. For example, Mia of the *Princess Diaries* has her dressing area torn down by jealous girls, almost revealing her naked to a group of male photographers. Mulan bathes in a lake when she thinks she is alone, but when male soldiers come to swim, Mushoo refers to her breasts, saying, "There are a couple of things they're bound to notice," and she sneaks away. Similarly, Quasimodo accidentally stumbles into Esmeralda's dressing area, and she quickly covers up with a robe and hunches over so as not to expose herself. She ties up her robe as Quasimodo apologizes again and again and hides his eyes. However, as he exits, he glances back toward her with a smile signifying for the viewer his love for her. A glimpse of her body has made her even more lovable and desirable.

Men's bodies are treated quite differently in these films. Male bodies, to the extent they are commented on at all, are the site of jokes. Men's crotches, genitals, and backsides are funny. For example, in *Hunchback of Notre Dame*, a cork from a bottle of champagne flies between a man's legs and knocks him over and the man yells in pain; later in that movie, during a fight, someone says, "That's hitting a little below the belt," and the woman says, "No this is!" and aims to strike him in the groin but is deflected by a sword. A boy in *Princess Diaries* is doubled over in pain as a baseball hits him in the groin. This scene is played as funny and the result of another character extracting her vengeance. *The Rugrats Movie* is full of jokes and images of boys' bare bottoms and penises. There are also references in other films to "a limp noodle" (*Mulan*) and "a shrinky winky" (*101 Dalmatians*). Mushoo in *Mulan* also jokes about male nudity, saying, "I hate biting naked butts." Women's genitals are never mentioned or invoked in any way. Their bodies are not the sites of jokes. Rather, women's bodies become important in the construction of heteronormative sexuality through their "sexiness" at which men gaze.

Much of the sexuality that these gendered bodies engage in has little to do with heterosexual sex narrowly defined as intercourse or even behaviors that might lead to it, but rather with cultural signs of a gendered sexuality for women. These signs are found in subplots, musical numbers, humorous scenes, and scenes depicting women's bodies, rather than in the main story lines of hetero-romantic true love. Such scenes contain sexual innuendo based in gesture, movement, tone of voice, and expression. Importantly, in all cases, sexiness is depicted as something women possess and use for getting men's attention. Sexiness is more often an attribute of female characters of color (e.g., Esmeralda, Jasmine, Ursula) (Hurley 2005) and is implicitly heterosexual given that the films construct the intended spectator of this sexiness as male (Mulvey 1975).

There are a few examples of white women depicted as "sexy," although these are more delimited and do not involve the main white women/girl characters. In *Princess Diaries*, a group of teenage friends are shown doing many of the same things as the animated women in *Aladdin*. They dance, shake their hips, make faces with curled and puckered lips and squinting eyes, play with their hair, and slap their hips. In *Beauty and the Beast*, a man is hit on the head for talking to a large-breasted woman with cleavage and much lipstick who moves and speaks in a sexy, flirtatious manner. *Toy Story 2* has a group of singing, dancing, nearly all-white Barbies who are ogled by the masculine toys. These scenes make it clear that women move and adorn their bodies and contort their faces for men.

While the women are being sexy, the (usually white) men are performing a different role as these films construct heterosexuality. As evident from some of the examples above, there is much explicit heterosexual gazing at or ogling of women's bodies in these films. Sometimes such gazing establishes that a woman is worth the pursuit of men and the fight for her that will develop the plot of the film, as in *Beauty and the Beast*. In an early scene in this film, when Belle walks out of a bookshop, three men who had been peering through the window turn around as if to pretend that they had not been staring. The man in the middle is then held up by the other two so that he can stare at Belle's backside as she walks away. All three men stare and then start to sing of her beauty. In other films, sexualized gazing is not so tightly attached to beauty but to the performance of heterosexual masculinity. In one instance in *Chicken Run*, the chickens are "exercising," and Rocky (a chicken) stares at Ginger's (a chicken) backside. She catches him, and he smiles, slyly. When the main characters refrain from overt ogling and sexual commentary, the "sidekicks" provide humor through this practice.

The objectifying gaze at women's bodies is often translated into objectifying, sexist language. Girl/women characters are called doll face, chicks, cuties, baby doll, angel face, sweet cheeks, bodacious, succulent little garden snail, tender oozing blossom, temptress snake, and tramp; and the boys/men say things like "I'll give you a tune up any time" and "give her some slack and reel her in." The desiring gazes, the commentary, and the depictions of them (large eyes, staring, open mouths, sound effects, and anxiousness) are constructed as competitive and conquering or frivolous, in stark contrast to the exceptional, magical, powerful hetero-romantic love described above. These depictions of heterosexual interactions have the effect of normalizing men's objectification of women's bodies and the heterosexual desire it signifies.

CONCLUSION

Despite the assumption that children's media are free of sexual content, our analyses suggest that these media depict a rich and pervasive heterosexual landscape. We have illustrated two main ways that G-rated films construct heterosexuality. First, heterosexuality is constructed through depictions of hetero-romantic love as exceptional, powerful, transformative, and magical. Second, heterosexuality is also constructed through depictions of interactions between gendered bodies in which the sexiness of feminine characters is subjected to the gaze of masculine characters. These accounts of heterosexuality extend our understandings of heteronormativity.

First, the finding that heterosexuality is constructed through heterosexiness points to the ways that heteronormativity intersects with gender, race, and class in its constructions. While heterosexuality is normalized and expected, it takes different forms for different sorts of bodies, and this is especially true for heterosexuality outside of romantic relationships. Second, the finding that hetero-romantic love is depicted as exceptional, powerful, and transformative runs counter to current theoretical understandings of heteronormativity's scaffolding being the ordinary, expected, everydayness of heterosexuality These films show heterosexuality to be just the opposite. Heterosexuality achieves a taken-for-granted status in these films not because it is ordinary but because hetero-romance is depicted as powerful. This finding in no way negates previous understandings of heteronormativity but rather extends another theoretical tenet—that is, that heterosexuality and its normativity are pervasive. Heterosexual exceptionalism extends the pervasiveness of heterosexuality and may serve as a means of inviting investment in it. Furthermore, heterosexuality is glorified here in mass culture but is also ordinary and assumed in everyday life. Thus, its encompassing pervasiveness lends it its power. Both ordinary and exceptional constructions of heterosexuality work to normalize its status because it becomes difficult to imagine anything other than this form of social relationship or anyone outside of these bonds.

Finally, we want to again emphasize that we cannot know what understandings and interpretations children might take away from these films or how they make sense of them alongside all the other social and cultural information they acquire. Others have shown that queer readings of such films are possible for adults (Griffin 2000). Children may have their own queer readings of such films. Without future work with children directly, we cannot know. However, these films are widely viewed by many very young children who are engaged with media rich worlds. It is likely that these accounts of heterosexuality make it into their understanding of the world in some way, albeit likely with layers of misunderstanding, reinterpretation, and integration with other information. Regardless, these films provide powerful portraits of a multifaceted and pervasive heterosexuality that likely facilitates the reproduction of heteronormativity.

REFERENCES

Arnold, Thomas K. (2005). Kids' DVDs are in a growth spurt. *USA Today*. http://www.usatoday.com/life/movies/news/2005-04-04-kids-dvds_x.htm.

Best, Raphaela. (1983). We've All Got Scars: What Boys and Girls Learn in Elementary School. Bloomington, IN: Indiana University Press.

Byrne, Eleanor, and Martin McQuillan. (1999). *Deconstructing Disney*. London: Pluto Press.

Crawley, Alisha M., D. R. Anderson, A. Wilder, M. Williams, and A. Santomero. (1999). Effects of repeated exposures to a single episode of the television program *Blue's Clues* on the viewing behaviors and comprehension of preschool children. *Journal of Educational Psychology* 91:630–37.

Giroux, Henry A. (1996). Animating youth: The Disneyfication of children's culture. In *Fugitive cultures: Race, violence, and youth*. New York: Routledge.

———. (1997). Are Disney movies good for your kids? In *Kinderculture: The corporate construction of childhood*, edited by Shirley R. Steinberg and Joe L. Kincheloe. Boulder, CO: Westview.

Griffin, Sean. (2000). *Tinker Belles and evil queens: The Walt Disney Company from the inside out*. New York: New York University Press.

Hurley, Dorothy L. (2005). Seeing white: Children of color and the Disney fairy tale princess. *Journal of Negro Education* 74:221–32.

Ingraham, Chrys. (1999). *White weddings: Romancing heterosexuality in popular culture*. New York: Routledge.

Jackson, Stevi. (2006). Gender, sexuality and heterosexuality: The complexity (and limits) of heteronormativity. *Feminist Theory* 7:105–21.

Johnson, Paul. (2005). *Love, heterosexuality, and society.* London: Routledge.

Junn, Ellen N. (1997). Media portrayals of love, marriage & sexuality for child audiences. Paper presented at the Biennial Meeting, Society for Research in Child Development, Washington, DC.

Kim, J. L., C. L. Sorsoll, K. Collins, and B. A. Zylbergold. (2007). From sex to sexuality: Exposing the heterosexual script on primetime network television. *Journal of Sex Research* 44:145.

Kitzinger, Celia. (2005). Heteronormativity in action: Reproducing the heterosexual nuclear family in after-hours medical calls. *Social Problems* 52:477–98.

Lacroix, Celeste. (2004). Images of animated others: The orientalization of Disney's cartoon heroines from *The Little Mermaid* to *The Hunchback of Notre Dame*. *Popular Communication* 2:213–29.

Lancaster, Roger. (2003). *The trouble with nature.* Berkeley: University of California Press.

Mares, M. L. (1998). Children's use of VCRs. *Annals of the American Academy of Political and Social Science* 557:120–31.

Martin, Karin A. (2009). Normalizing heterosexuality: Mothers' assumptions, talk, and strategies with young children. *American Sociological Review* 74:190–207.

Martin, Karin A., Katherine Luke, and Lynn Verduzco-Baker. (2007). The sexual socialization of young children: Setting the agenda for research. In *Advances in group processes*, vol. 6, *Social psychology of gender*, edited by Shelly Correll. Oxford, UK: Elsevier Science.

Mo, W., and W. Shen. (2000). A mean wink at authenticity: Chinese images in Disney's Mulan. *New Advocate* 13:129–42.

Motion Picture Association of America, "Film Ratings," Motion Picture Association of America, http://www.mpaa.org/FilmRatings.asp

Mulvey, Laura. (1975). Visual pleasure and narrative cinema. *Screen* 16:6–18.

Nguyen, Mimi. (1998). A feminist fantasia, almost. *San Jose Mercury News*, July 5.

Pewewardy, Cornel. (1996). The Pocahontas paradox: A cautionary tale for educators. *Journal of Navajo Education* 14:20–25.

Renold, Emma. (2002). Presumed innocence: (Hetero)Sexual, heterosexist and homophobic harassment among primary school girls and boys. *Childhood* 9:415–34.

———. (2005). *Girls, boys, and junior sexualities: Exploring children's gender and sexual relations in the primary school.* London: Routledge Falmer.

Rideout, V., E. A. Vandewater, and E. A. Wartella. (2003). *Zero to six: Electronic media in the lives of infants, toddlers, and preschoolers.* Washington, DC: Henry J. Kaiser Family Foundation.

Rubin, Gayle. (1984). Thinking sex: Notes for a radical theory of the politics of sexuality. In *Pleasure and danger*, edited by Carol Vance. Boston: Routledge.

Schmitt, K. L., D. R. Anderson, and P. A. Collins. (1999). Form and content: Looking at visual features of television. *Developmental Psychology* 35:1156–67.

Schor, Juliet B. (2004). *Born to buy: The commercialized child and the new consumer culture.* New York: Scribner.

Thomas, Susan Gregory. (2007). *Buy, buy, baby: How consumer culture manipulates parents.* New York: Houghton Mifflin.

Thompson, T. L., and B. Zerbinos. (1995). Gender roles in animated cartoons: Has the picture changed in 20 years? *Sex Roles: A Journal of Research* 32:651–73.

Thorne, Barrie. (1993). *Gender play: Girls and boys in school*. New Brunswick, NJ: Rutgers University Press.

Ward, L. Monique. (1995). Talking about sex: Common themes about sexuality in the prime-time television programs children and adolescents view most. *Journal of Youth and Adolescence* 24:595–615.

———. (2003). Understanding the role of entertainment media in the sexual socialization of American youth: A review of empirical research. *Developmental Review* 23:347–88.

Witt, Susan D. (2000). The influence of television on children's gender role socialization. *Childhood Education* 76:322–24.

KEY CONCEPTS

adolescence   sexuality   heteronormativity

DISCUSSION QUESTIONS

1. What do the authors mean by "heterosexiness" in films? Are gendered behaviors in the children's films consistent with an expectation of heterosexuality?

2. What about children's television programming? Are the images presented on television the same as those presented in films? Do you see the same patterns in films for young adults?

Applying Sociological Knowledge:
An Exercise for Students

Sexuality is generally thought to be a private matter, and yet public social norms very much shape and regulate sexuality. For a period of one week, observe and keep track of every comment you hear that seems to enforce certain sexual scripts. What have you heard? What assumptions does everyday talk make about heterosexuality? About gays and lesbians? How does everyday public talk shape social norms about sexuality?

✳

PART XIII

Social Institutions

A. Family

42

The Unfinished Revolution

How a New Generation Is Reshaping Family, Work, and Gender in America

KATHLEEN GERSON

Gerson's research on young, adult women and men asks how their values have been shaped by the changes in women's and men's roles at this time in history. She also asks how these young adults have been influenced by the structures of their own families. She concludes with a commentary on the resistance of social institutions to social change.

Whether they are judged as liberating or disastrous, the closing decades of the twentieth century witnessed revolutionary shifts in the ways new generations grow to adulthood. The march of mothers into the workplace, combined with the rise of alternatives to lifelong marriage, created a patchwork of domestic arrangements that bears little resemblance to the 1950s Ozzie and Harriet world of American nostalgia. By 2000, 60 percent of all married couples had two earners, while only 26 percent depended solely on a husband's income, down from 51 percent in 1970. In fact, in 2006, two-paycheck couples were more numerous than male-breadwinner households had been in 1970. During this same period, single-parent homes, overwhelmingly headed by women, claimed a growing proportion of American households.[1] To put this in perspective, not all female-headed households consist of a mother only, since many parents cohabit but do not marry. Nevertheless, in 2007, 33 percent of non-Hispanic white children and 60 percent of black children lived with one parent (up from 10 percent and 41 percent in 1970).[2] As today's young women and

SOURCE: Gerson, Kathleen. 2010. *The Unfinished Revolution: How a New Generation is Reshaping Family, Work, and Gender in America.* New York: Oxford University Press.

men have reached adulthood, two-income and single-parent homes outnumber married couples with sole (male) breadwinners by a substantial margin.

Equally significant, members of this new generation lived in families far more likely to change shape over time. While families have always faced predictable turning points as children are born, grow up, and leave home, today's young adults were reared in households where volatile changes occurred when parents altered their ties to each other or to the wider world of work. These young women and men grew up in a period when divorce rates were increasing, and a rising proportion of children were born into homes anchored either by a single mother or cohabiting but unmarried parents. Lifelong marriage, once the only socially acceptable option for bearing and rearing children, became one of several alternatives that now include staying single, breaking up, or remarrying.

This generation also came of age just as women's entry into the paid labor force began to challenge the once ascendant pattern of home-centered motherhood. In 1975, only 34 percent of mothers with children under the age of three held a paid job, but this number rose to 61 percent by 2000. This peak subsided slightly, with 57 percent of such mothers at work in 2004, but even this figure represents an enormous shift from earlier patterns. More telling, among mothers with children under eighteen, a full 71 percent are now employed.[3]

In fact, the recent ebbs and flows among working mothers with young children point to the competing pushes and pulls women continue to confront in balancing the needs of children and the demands of jobs. Even as women have strengthened their commitment to paid work, they have had to cope with unforeseen work-family conflicts. Growing up in this period, children observed women's massive shift from home to work, but they also watched their mothers move back and forth between full-time work, part-time work, and no job at all.

Finally, the rising uncertainty in men's economic fortunes has also reverberated in their children's lives. During the closing decades of the twentieth century, the "family wage," which once made it possible for most men (though certainly not all) to support nonworking wives, became a quaint relic of an earlier time. Whether at the factory or the office, a growing number of men faced unpredictable prospects as secure, well-paid careers offering the promise of upward mobility became an increasingly endangered species. Fathers who expected to be sole breadwinners found they needed their wives' earnings to survive. Like a life raft in choppy seas, second incomes helped keep a growing number of families afloat and allowed some fathers to change jobs if they hit a sudden dead end on a once promising career path. As more fathers could not live up to the "good provider" ethic, however, many left their families or were dismissed by mothers who saw little reason to care for a man who could not keep himself afloat. The changes in men's lives and economic fortunes provide another reason why many members of this generation experienced unpredictable ups and downs.

Coming of age in an era of more fluid marriages, less stable work careers, and profound shifts in mothers' ties to the workplace shaped the experiences of a new generation. Compared to their parents or grandparents, they are more likely to have lived in a home containing either one parent or a cohabiting but unmarried couple and to have seen married parents break up or single parents remarry. They are more likely to have watched a stay-at-home mother join the workplace or an employed mother pull back from work when the balancing act got too difficult. And they are more likely to have seen their financial stability rise or fall as a household's composition changed or parents encountered unexpected shifts in their job situations.

These intertwined changes in intimate relationships, work trajectories, and gender arrangements have created new patterns of living, working, and family-building that amount to no less than a social revolution. Yet this revolution also faces great resistance from institutions rooted in earlier eras. On the job, workers continue to experience enormous pressures to give uninterrupted full-time, and often overtime, commitment not just to move up but even stay in place. In the home, privatized caretaking leaves parents, especially mothers, coping with seemingly endless demands and unattainable standards. And the entrenched conflicts between work and family life place mounting strains on adult partnerships. The tensions between changing lives and resistant institutions have created dilemmas for everyone.

In all of these ways, the children of the gender revolution grew to adulthood amid unprecedented, unpredictable, and uneven changes. They now must build their lives in an irrevocably but uncertainly altered world.

THE VOICES OF A NEW GENERATION

What are the consequences of this widespread, but partial, social revolution? Where some see a generation shortchanged by working mothers and fragmenting households, others see one that can draw on more diverse and egalitarian models of family life. Where some see a resurgence of tradition, especially among those young women who want to leave the workplace, others see a deepening decline of commitment in the rising number of young adults living on their own. Whether judged to be worrisome or welcome, these contradictory views point to the continuing puzzles of the family and gender revolution. Has the rise of two-earner and single-parent households left children feeling neglected and insecure, or has it given them hope for the possibility of more diverse and flexible relationships? Will the young women and men reared in these changing circumstances turn back toward older patterns or seek new ways of building their families and integrating family and work?

… Poised between the dependency of childhood and the irrevocable investments of later life, young adulthood is a crucial phase in the human life course that represents both a time of individual transition and a potential engine for social change. Old enough to look back over the full sweep of their childhoods and forward to their own futures, today's young adults are uniquely positioned to help us see beneath the surface of popular debate to deeper truths. Their childhood experiences can tell us how family, work, and gender arrangements shape life chances, and their young adult strategies can, in turn, reveal how people use their experiences to craft new life paths and redefine the contours of change.

Regardless of their own family experiences, today's young women and men have grown up in revolutionary times. For better or worse, they have inherited new options and questions about women's and men's proper places. Now making the transition to adulthood, they have no well-worn paths to follow. Marriage no longer offers the promise of permanence, nor is it the only option for bearing and rearing children, but there is no clear route to building and maintaining an intimate bond. Most women no longer assume they can or will want to stay home with young children, but there is no clear model for how children should now be raised. Most men can no longer assume they can or will want to support a family on their own, but there is no clear path to manhood. Work and family shifts have created an ambiguous mix of new options *and* new insecurities, with growing conflicts between work and parenting, autonomy and commitment, time and money. Amid these social conflicts and contradictions, young women and men must search for new answers and develop innovative responses.

THE LIVES OF YOUNG WOMEN AND MEN

Each generation's experiences are both a judgment about the past and a statement about the future. To understand the sources of these outlooks and actions, we need to examine what C. Wright Mills argued is the core focus of "the sociological imagination"—the intersection of biography, history, and social structure.[4] This approach calls on us to investigate how specific social and historical contexts give shape to the transhistorical links between social arrangements and human lives, paying special attention to how societies and individuals develop. Such an approach is especially needed when social shifts erode earlier ways of life, reveal the tenuous nature of certainties once taken for granted, and create new social conditions and possibilities.

Following in this tradition, I examine the lives of a strategically situated group to ask and answer broad questions. How, why, and under what conditions does large-scale social change take place? What are its limits, and what shapes its trajectories? How do social arrangements affect individual lives, and how, in turn, does the cumulative influence of individual responses give unexpected shape to the course of change?

Using this pivotal generation as a window on change, I interviewed 120 young women and men between the ages of eighteen and thirty-two. As a whole, they lived through the full range of changes taking place in family life. Most lived in some form of nontraditional home before reaching eighteen. Forty percent had some experience growing up with a single parent, and another 7 percent saw their parents separate or divorce after they left home. About a third had two parents who held full-time jobs for a significant portion of their childhood, while 27 percent grew up in homes where fathers were consistent primary breadwinners and mothers worked intermittently or not at all. Yet even many of these traditional households underwent significant shifts as parents changed their work situation or marriages faced a crisis.

With an average age of twenty-four at the time of the interview, they are evenly divided between women and men, and about 5 percent (also evenly divided between women and men) openly identified as either lesbian or gay. Randomly chosen from a broad range of city and suburban neighborhoods dispersed widely throughout the New York metropolitan area, the group includes people from a broad range of racial, ethnic, and class backgrounds who were reared in all regions of the country, including the South, West, and Midwest as well as throughout the East.

About 46 percent had a middle-class or upper-middle-class background, while another 38 percent described a working-class upbringing and 16 percent lived in or on the edge of poverty (including 10 percent whose families received public assistance during some portion of their childhood). The group contained a similar level of racial and ethnic diversity. In all, 55 percent identified as non-Hispanic white, 22 percent as African-American, 17 percent as Latino or Latina, and 6 percent as Asian. As a group, they reflect the demographic contours of young adults throughout metropolitan America.

Everyone participated in a lengthy, in-depth life history interview in which they described their experiences growing up, reflected on the significance of these experiences, and considered their hopes and plans for the future. Focusing on processes of stability and change, the interview sought to uncover critical turning points in the lives of families and individuals, to discover the social contexts and events triggering these changes, and to explore how people imparted meaning and adopted coping strategies in response. Their life stories provide a surprising view on the social revolution this generation has inherited and whose future course it will shape.

THE VIEW FROM BELOW

What have young women and men concluded about their experiences in changing families? In contrast to the popular claim that this generation feels neglected by working mothers, unsettled by parental breakups, and wary of equality, they express strong support for working mothers and much greater concern with the quality of the relationship between parents than whether parents stayed together or separated. Almost four out of five of those who had work-committed mothers believe this was the best option, while half of those whose mothers did not have sustained work lives wish they had. On the controversial matters of divorce and single parenthood, a slight majority of those who lived in a single-parent home wish their biological parents had stayed together, but almost half believe it was better, if not ideal, for their parents to separate than to live in a conflict-ridden or silently unhappy home. Even more surprising, while a majority of children from intact homes think this was best, two out of five feel their parents might have been better off splitting up.

[This is] a generation more focused on *how well* parents met the challenges of providing economic and emotional support than on *what form* their families took.

They care about how their families unfolded, not what they looked like at any one point in time. Their narratives show that family life is a film, not a snapshot. Families are not a stable set of relationships frozen in time but a dynamic process that changes daily, monthly, and yearly as children grow. In fact, all families experience change, and even the happiest ones must adapt to changing contingencies—both in their midst and in the wider world—if they are to remain happy. No outcome is guaranteed. Stable, supportive families can become insecure and riven with conflict, while unstable families can develop supportive patterns and bonds.

Young women and men recount *family pathways* that moved in different directions as some homes became more supportive and others less so. These pathways undermine the usefulness of conceiving of families as types. Not only do many contemporary families change their form as time passes, but even those retaining a stable outward form can change in subtle but important ways as interpersonal dynamics shift.

By changing the focus from family types to family pathways, we can transcend the seemingly intractable debate pitting "traditional" homes against other family forms. The lives of these young women and men call into question a number of strongly held beliefs about the primacy of family structure and the supremacy of one household type. Their experiences point instead to the importance of processes of family change, the ways that social contexts shape a family's trajectory, and people's active efforts to cope with and draw meaning from their changing circumstances.

What explains why some family pathways remain stable or improve, while others stay mired in difficulty or take a downward course? *Gender flexibility* in breadwinning and caretaking provides a key to answering this question. In the place of fixed, rigid behavioral strategies and mental categories demarcating separate spheres for women and men, gender flexibility involves more equal sharing and more fluid boundaries for organizing and apportioning emotional, social, and economic care. Flexible strategies can take different forms, including sharing, taking turns, and expanding beyond narrowly defined roles, in addition to more straightforward definitions of equality, but they all transgress the once rigidly drawn boundaries between women as caretakers and men as breadwinners.

In a world where men may not be able or willing to support wives and children and women may need and want to pursue sustained work ties, parents (and other caretakers) could only overcome such family crises as the loss of a father's income or the decline of a mother's morale by letting go of rigid gender boundaries. As families faced a father's departure, a mother's frustration at staying home, or the loss of a parent's job, the ability of parents and other caretakers to respond flexibly to new family needs helped parents create more financially stable and emotionally supportive homes. Flexible approaches to earning and caring helped families adapt, while inflexible outlooks on women's and men's proper places left them ill prepared to cope with new economic and social realities. Although it may not be welcomed by those who prefer a clearer gender order, gender flexibility in earning and caring provided the most effective way for families to transcend the economic challenges and marital conundrums that imperiled their children's well-being.

FACING THE FUTURE

What, then, do young women and men hope and plan to do in their own lives? My interviews subvert the conventional wisdom here as well, whether it stresses the rise of "opt-out" mothers or the decline of commitment. Most of my interviewees hope to create lasting, egalitarian partnerships, but they are also doubtful about their chances of reaching this goal. Whether or not their parents stayed together, more than nine out of ten hope to rear children in the context of a satisfying lifelong bond. Far from rejecting the value of commitment, almost everyone wants to create a lasting marriage or marriage-like relationship.

Their affirmation of the value of commitment does not, however, reflect a desire for a relationship based on clear, fixed separate spheres for mothers and fathers. Instead, most want to create a flexible, egalitarian partnership with considerable room for personal autonomy. Whether reared by homemaker-breadwinning, dual-earner, or single parents, most women *and* men want a committed bond where they share both paid work and family caretaking. Three-fourths of those reared in dual-earner homes want their spouses to share breadwinning and caretaking, but so do more than two-thirds of those from traditional homes and close to nine-tenths of those with single parents. Four-fifths of the women want egalitarian relationships, but so do over two-thirds of the men.

Yet young women and men also fear it may not be possible to forge an enduring, egalitarian relationship or integrate committed careers with devoted parenting. Skeptical about whether they can find the right partner and worried about balancing family and work amid mounting job demands and a lack of caretaking supports, they are developing second-best fallback strategies as insurance against their worst-case fears. In contrast to their ideals, women's and men's fallback strategies diverge sharply.

Hoping to avoid being trapped in an unhappy marriage or deserted by an unfaithful spouse, most women see work as essential to their survival. If a supportive partner cannot be found, they prefer self-reliance over economic dependence within a traditional marriage. Most men, however, worry more about the costs equal sharing might exact on their careers. If time-greedy workplaces make it difficult to strike an equal balance between work and parenting, men prefer a neotraditional arrangement that allows them to put work first and rely on a partner for the lion's share of caregiving. As they prepare to settle for second best, women and men both emphasize the importance of work as a central source of personal identity and financial survival, but this stance leads them to pursue different strategies. Reversing the argument that women are returning to tradition, men are more likely to want to count on a partner at home. Women, on the other hand, are more likely to see paid work as essential to providing for themselves and their children in a world where they may not be able to count on a man.

The rise of self-reliant women, who stress emotional and economic autonomy, and neotraditional men, who grant women's choice to work but also want to maintain their position as the breadwinning specialist, portends a new work–family divide. But this division does not reflect the highest aspirations of most women or men. The debate about whether a new generation is rejecting commitment or embracing tradition does not capture the full story, because it does not distinguish between *ideals* and *fallback positions*. Young adults overwhelmingly hope to forge a lasting marriage or marriage-like relationship, to create a flexible and egalitarian bond with their intimate partner, and to blend home and work in their own lives. When it comes to their aspirations, women and men share many hopes and dreams. But fears that time-demanding workplaces, unreliable partners, and a dearth of caretaking supports will place these ideals out of reach propel them down different paths.

Drawing a distinction between ideals and enacted strategies resolves the ambiguity about the shape and direction of generational change. One-dimensional images—whether they depict resurgent traditionalism or family decline—cannot capture the complex, ambiguous experiences of today's young women and men. New generations *neither* wish to turn back to earlier gender patterns *nor* to create a brave new world of disconnected individuals. Most prefer instead to build a life that balances autonomy and commitment in the context of satisfying work and an egalitarian partnership.

Yet changing lives are colliding with resistant institutions, leaving new generations facing alternatives that are far less appealing. While institutional shifts such as the erosion of single-earner paychecks, the fragility of modern marriage, and the expanding options and pressures for women to work have made gender flexibility both desirable and necessary, demanding workplaces and privatized child rearing make work–family integration and egalitarian commitment difficult to achieve. Young women and men must reshape family, work, and gender amid an unfinished revolution. Whether they are able to create the world they want or will have to fall back on less desirable options remains an open question. Their struggles point to the social roots of these conflicts. They also make it clear that nothing less than the restructuring of work and caretaking will allow new generations to achieve the ideals they seek and provide the supports their own children will need.

NOTES

1. U.S. Census Bureau. 2007. "Single-Parent Households Showed Little Variation since 1994." Washington, DC.

2. Roberts, Sam. 2008a. "Most Children Still Live in Two-Parent Homes, Census Bureau Reports." *The New York Times*, February 21; Roberts, Sam. 2008b. "Two-Parent Black Families Showing Gains." *The New York Times*, December 17.

3. U.S. Census Bureau. 2006. "Current Population Survey Annual Social and Economic Supplement: Families and Living Arrangements: 2005." Washington, DC.

4. Mills, C. Wright. 1959. *The Sociological Imagination*. New York: Oxford University Press.

KEY CONCEPTS

divorce rate	family wage	gender roles
egalitarian	gendered revolution	marriage rate

DISCUSSION QUESTIONS

1. What are the specific social structural changes that Gerson argues are shaping the family and work expectations of current generations? How have these expectations been shaped by what the younger generation has observed in their parents' generation?

2. What does Gerson mean by "resistant institutions," and why are such institutions producing challenges for young men and women as they forge their families and work relationships?

43

Gay Marriage

STEVEN SEIDMAN

The debate about extending the right to marry to gays and lesbians is summarized in this article. Seidman outlines the key issues presented against gay marriage, including religious and historical arguments. Most often, those who are critical of gay marriage argue that it weakens the institution of marriage even further. Seidman then goes on to offer a rebuttal argument. While he recognizes the religious value placed on heterosexual unions, he argues that America is not a solely Christian country, that the dynamics of family structure have changed, and that gender roles within families have changed. Why, then, can't the gender of those who are married change?

M arriage is not just about love and intimacy between two adults. It is an institution. Marriage is recognized by the state, and those that marry get specific rights and benefits such as the right to spousal support, to bring wrongful

SOURCE: Steven Seidman. 2003. *The Social Construction of Sexuality.* New York: Norton, pp. 123–133.

death suits, to be listed as social security beneficiaries, to get legally recognized divorces, and to claim full parental rights. In America, only a man and woman can marry. The question raised by the gay marriage debate is whether the gender of a spouse should matter in determining who can marry. And, what role should the state have in regulating marriage and intimate choices?

Gay marriage was not an issue in the 1950s and early 1960s when gays and lesbians began their struggle for tolerance. It initially surfaced in public debate in the 1970s.... [A] more confident and assertive gay movement redefined homosexuality: it was no longer a stigma but was viewed as natural and good. The gay movement approached being gay as a positive basis of identity, community, and a fulfilling lifestyle.

By the mid-1980s, the gay movement sought full, across-the-board legal and social equality. Gays wanted to be equal citizens. They pursued equality at work, in schools, in the military, and in the eyes of the law. They also demanded to be treated respectfully in their families, by mass media, and elsewhere.

Gay marriage became a key issue in the 1990s as part of the pursuit of social equality. Without the right to marry, gays are second-class citizens. Also, many gays and lesbians were turning their attention to intimate relationships. The generation that came out in the late 1960s and 1970s was now middle-aged; like many straight Americans in their thirties, forties, and fifties, their attention often turned to creating families. Many gays and lesbians were in or sought long-term intimate relationships. It was inevitable that they would turn their political focus toward marriage. Furthermore, AIDS pushed the issue of the legal status of their intimate relationships into the heart of their lives. The AIDS crisis forced many gays and lesbians to address concerns such as health coverage of partners, hospital visitation rights, inheritance, and residency rights. AIDS compelled many gays and lesbians to invest enormous amounts of time, energy, and resources into caring for their partners, which made the legal status of their relationships an urgent concern.

Despite unfriendly social conditions, gays have always formed long-term relationships. For example, scholars have uncovered a long and complicated history of gay relationships in nineteenth century America. Sometimes women passed as men to form straight-seeming relationships; sometimes men or women lived together as housemates but were really lovers; sometimes individuals would marry but still carry on romantic, sometimes life-long, same-sex intimate relationships. By the 1990s, many gay men and lesbians were in long-term relationships; the lives of these couples were emotionally, socially, and financially intertwined in ways that were similar to straight marriages. Indeed, through either adoption or artificial insemination, many gay and lesbian couples were adding children to their families.

Gay marriage has today become a front-line issue for gays and lesbians—and not only in the United States. Gay marriage has become a worldwide issue. Many societies have already enacted laws that permit gays to marry (for example, Denmark, Sweden, Norway, and the Netherlands). In Canada, France, and Germany, laws have been passed that offer some type of state recognition to gay relationships.

Gay marriage might not have gained political traction in the United States were it not for broader changes in the American family. We are all aware that today families come in many sizes and shapes. Although the nuclear family, with a breadwinner husband and a stay-at-home wife, might still be an ideal for many Americans, it describes only a minority of actual households. The reality is a dizzying variety of families—cohabiting couples, one-parent households, combined families, marriages without children, lifelong partners who do not live together, and so on. The variety of intimate arrangements, the reality that almost half of all marriages end in divorce, the uncoupling of motherhood from marriage, and the lessening of the stigma attached to being single have all diminished the cultural authority of marriage. Still, make no mistake; marriage is not just one choice among others. The state supports marriage with a cluster of rights and benefits that no other intimate relationship is given. In addition, the right to marry continues to serve as a symbol of first-class citizenship. Many gays and lesbians want to marry, or at least want the right to marry, in order to become equal, respected citizens.

Changes in the social organization of intimacy have also contributed to raising doubts about the reasonableness of excluding gays from marriage. For most of our parents and grandparents, the organization of marriage was more or less fixed. Men were the breadwinners; their lives were focused on making a living. Women were wives and mothers; their lives were centered on domestic tasks. Within the household, men were expected to do "masculine" domestic activities (mow the lawn, take out the garbage, discipline the children), whereas women were responsible for "feminine" tasks such as cooking, cleaning, and child care. Gender shaped the very texture of intimacy. Men were supposed to initiate and direct sex; women were supposed to go along without showing too much interest or pleasure. Men made the big decisions (where to live or how to spend money); women arranged the social affairs of the couple. Today, many women work, have careers, and pursue interests outside the household; men are expected to perform household and child care tasks. We can perhaps reasonably speak of a somewhat new ideal of intimacy: marriage as a relationship between equals in which decisions are openly discussed and household roles are negotiated. In short, gender is becoming less important in organizing marriage. The gender of those that marry would then seem to matter less.

The battle over gay marriage is being fought on two fronts. A war is being waged in the legal courts and in the court of public opinion between mostly gay and lesbian advocates of gay marriage and straight critics. Public opinion is still on the side of critics, but this is beginning to change. There is also debate within the gay movement. Not all gays and lesbians think that pursuing marital rights is the right goal for a movement that once aspired to ideals of sexual liberation. Gay critics view marriage as inimical to a culture championing sexual variation.

Straight critics of gay marriage make several key points. Heterosexual marriage, they say, has deep roots in history. As far back as we know, marriage has always been between a man and a woman. There is something seemingly natural and right about heterosexual marriage. Moreover, heterosexual marriage is a cornerstone of the Judeo-Christian tradition. As some critics quip, God made Adam and Eve, not Adam and Steve. America has a secular government and was founded on secular principles, but a majority of Americans identify themselves as Christian.

Marriage is also a cornerstone of American society. It provides a stable, positive, moral environment essential for shaping good American citizens. Critics ask, will young people acquire clear gender identities without heterosexual marriage?

Boys look to their fathers to learn what it means to be a man; wives provide daughters with a clear notion of what it means to be a woman. Gay marriage would create gender confusion.

Critics also raise another issue: marriage has already been weakened by the ease and frequency of divorce, by the increase in rates of illegitimacy and single-parent families, and by families in which both parents work. Legalizing gay marriage would further undermine the stability and strength of this institution. Even more ominously, permitting gays to marry would open the door for all sorts of people to demand the right to marry—polygamists, children, friends, kin.

These objections express real anxieties on the part of Americans about the fragility of marriage. Such concerns should not quickly be dismissed. Marriage has been, and still is, according to virtually all social researchers, a cherished ideal for most Americans. Marriage is bundled with a number of hopes shared by many Americans—for a home, a family, and a sense of community. Instead of bringing additional change to this already weakened institution, critics say, we should find ways to strengthen it.

Advocates of gay marriage have offered forceful rebuttals to these criticisms. No one denies, they say, that heterosexual marriage has deep roots in history, though recent historical scholarship suggests that same-sex intimacies, including marriage, were not as exceptional as once believed. Setting aside past realities, though, the question that must be addressed is this: should the past or tradition always serve as a guide for the present? Consider that racism and sexism also have deep roots in history and social customs. Most of us agree that tradition should not be followed blindly, but examined in light of contemporary thinking and values. After all, imperfect people shaped past practices and traditions. Possibly the historical prejudice against homosexuality is similar to prejudices against non-whites and women. The battleground for debating the issue of gay marriage should be the present, not the past.

What about the Judeo-Christian disapproval of homosexuality? With all due respect, gay marriage advocates hold that, like other traditions, religious practices too have been made by ordinary individuals who often shared the prejudices of their time. Christian belief and tradition has changed many times in history. The question for Christians is whether gay marriage can be understood as consistent with the spirit of Christianity. Both pro and con arguments have been made in this regard; I'll leave this debate to the faithful. There is, in any event, a more compelling reason to be cautious about religious objections to gay marriage. America is not a Christian nation. Although Judeo-Christian traditions have shaped America, so too have non-Christian and secular traditions. America is, above all, a secular nation. Our government does not officially recognize or promote any particular religion and does not enact legislation, including laws regulating marriage, that need be aligned to any specific faith.

So, appeals to the past, to religious and secular traditions alone, should not exclude the right of gays to marry. But opponents have advanced additional arguments. Some argue that gays and lesbians are psychologically and morally unable to form stable marriages. This argument cannot be taken seriously because it relies on stereotypes that have been exposed and dismissed by social scientific research. The argument that if you extend marital rights to gays and lesbians then all sorts of people and relationships will clamor for the same rights is also not persuasive, since gays are not challenging the institution of marriage as a consensual relationship between two adults. Gays and lesbians simply want the same right of access as heterosexuals.

What are the positive arguments for gay marriage? The key issue, say advocates, is the meaning of marriage today. Too often, critics of gay marriage assume that marriage has always had the same meaning and social organization. This is not the case. Marriage has changed considerably, even in the short history of the United States. Consider that not too long ago marriage was possible only between adults of the same race. Antimiscegenation laws, initially enacted in Maryland in 1661 and not declared unconstitutional by the Supreme Court until 1967, forbid marriage between whites and non-whites, including blacks, Asians, and Native Americans. Or, consider that throughout most of American history, marriage was rigidly organized around gender roles. Until the twentieth century, women were the legal property of men; a wife could not own her own property and did not even have the right to her own wages. Marriage was thought to be fundamentally for the purpose of having children. Today, we recognize spouses as equal before the law, and gender roles are not enforced by the state. Many couples marry and remain childless by choice.

Marriage today has various meanings. For some, it is about creating a family; for others, it is about social and financial security. For many of us, marriage is fundamentally about love and forging an intimate life with another person. In this companionate ideal, individuals look to marriage to find a deep emotional, social, even spiritual union. In principle, gender plays less of a role in organizing marriage. Men and women share domestic duties; they attempt to negotiate a life together as equals. Spouses want to be respected and fulfilled as individuals. To be sure, social scientists have documented that gender still plays a considerable role in organizing intimacy. For example, women continue to do the lion's share of domestic tasks, including child care. But, today, individuals can demand that their unique wants and desires, regardless of their gender, be considered in the social organization of intimate relationships.

If marriage is about equal individuals forging an intimate, loving, mutually committed relationship, then the gender of the partners should be irrelevant. What should matter is whether the partners agree to marry, and whether they are caring, committed, respectful, responsible, and willing to communicate openly their respective needs and wants. Permitting gays to marry will not change the institution of marriage but strengthen an ideal of this institution as a relationship of loving intimacy between equals.

I think that some people oppose gay marriage because they are threatened by this egalitarian ideal of marriage. This ideal, after all, devalues the role of gender. Some men may fear a loss of status and power, and some women may fear economic insecurity and the loss or devaluation of their chief identity as full-time wives and mothers. For women and men who are deeply invested in gender roles and identities or in marriages that are fundamentally about having children, gay marriage may be viewed as a threat not because it challenges heterosexual privilege, but because it challenges a very specific and narrow idea of gender and the family....

So, for both moral and practical political reasons, I think it is important to defend the right of gays and lesbians to marry. Although I do not expect an end to state support of marriage anytime soon, I do anticipate the distribution of some marital rights to other intimate arrangements—those relationships that, in terms of their intimate ties, look a lot like marriage, that is, relationships involving longstanding emotional, sexual, social, and economic interdependence between two unrelated adults. These intimate unions merit state recognition. In fact, this is already happening. Many of the rights and benefits of marriage are now claimed by "domestic partnerships"—which are recognized in many cities, states, businesses, unions, and colleges—by "civil unions" in Vermont, and by common-law marriages, cohabitation, and single-parent households. In the short run, the best way to promote intimate diversity in the United States is to expand the range of intimate relationships that are recognized by the state.

KEY CONCEPTS

egalitarian   miscegenation laws

gender roles   sexual politics

DISCUSSION QUESTIONS

1. How does Seidman counter the argument that marriage is based on a long history of heterosexual unions? What points does he make to oppose this?

2. What traditional forms of marriage are no longer accepted? How has marriage changed over the years?

44

The Myth of the Missing Black Father

ROBERTA L. COLES AND CHARLES GREEN

Roberta Coles and Charles Green explore the common belief that Black men are largely absent as fathers. They challenge this idea through a careful review of

SOURCE: Coles, Roberta L., and Charles Green, eds., 2010. *The Myth of the Missing Black Father*. New York: Columbia University Press.

research on Black fathers and Black families, and they also identify the historical, economic, and demographic trends that are affecting Black family life.

The black male. A demographic. A sociological construct. A media caricature. A crime statistic. Aside from rage or lust, he is seldom seen as an emotionally embodied person. Rarely a father. Indeed, if one judged by popular and academic coverage, one might think the term "black fatherhood" an oxymoron. In their parenting role, African American men are viewed as verbs but not nouns; that is, it is frequently assumed that Black men *father* children but seldom *are* fathers. Instead, as the law professor Dorothy Roberts (1998) suggests in her article "The Absent Black Father," black men have become the symbol of *fatherlessness*. Consequently, they are rarely depicted as deeply embedded within and essential to their families of procreation. This stereotype is so pervasive that when black men are seen parenting, as Mark Anthony Neal (2005) has personally observed in his memoir, they are virtually offered a Nobel Prize.

But this stereotype did not arise from thin air. In 2000, only 16 percent of African American households were married couples with children, the lowest of all racial groups in America. On the other hand, 19 percent of Black households were female-headed with children, the highest of all racial groups. From the perspective of children's living arrangements, ... over 50 percent of African American children lived in mother-only households in 2004, again the highest of all racial groups. Although African American teens experienced the largest decline in births of all racial groups in the 1990s, still in 2000, 68 percent of all births to African American women were nonmarital, suggesting the pattern of single-mother parenting may be sustained for some time into the future (Martin et al. 2003). This statistic could easily lead observers to assume that the fathers are absent.

While it would be remiss to argue that there are not many absent black fathers, absence is only one slice of the fatherhood pie and a smaller slice than is normally thought. The problem with "absence," as is fairly well established now, is that it's an ill–defined pejorative concept usually denoting nonresidence with the child, and it is sometimes *assumed* in cases where there is no legal marriage to the mother. More importantly, absence connotes invisibility and noninvolvement, which further investigation has proven to be exaggerated (as will be discussed below). Furthermore, statistics on children's living arrangements also indicate that nearly 41 percent of black children live with their fathers, either in a married or cohabiting couple household or with a single dad.

These African American family-structure trends are reflections of large-scale societal trends—historical, economic, and demographic—that have affected all American families over the past centuries. Transformations of the American society from an agricultural to an industrial economy and, more recently, from an industrial to a service economy entailed adjustments in the timing of marriage, family structure, and the dynamics of family life. The transition from an industrial to a service economy has been accompanied by a movement of jobs out of cities; a decline in real wages for men; increased labor-force participation for women; a decline in fertility; postponement of marriage; and increases in divorce, nonmarital births, and single-parent and non-family households.

These historical transformations of American society also led to changes in the expected and idealized roles of family members. According to Lamb (1986), during the agricultural era, fathers were expected to be the "moral teachers"; during industrialization, breadwinners and sex-role models; and during the service economy, nurturers. It is doubtful that these idealized roles were as discrete as implied. In fact, LaRossa's (1997) history of the first half of the 1900s reveals that public calls for nurturing, involved fathers existed before the modern era. It is likely that many men had trouble fulfilling these idealized roles despite the legal buttress of patriarchy, but it was surely difficult for African American men to fulfill these roles in the context of slavery, segregation, and, even today, more modern forms of discrimination. A comparison of the socioeconomic status of black and white fathers illustrates some of the disadvantages black fathers must surmount to fulfill fathering expectations. According to Hernandez and Brandon (2002), in 1999 only 33.4 percent of black fathers had attained at least a college education, compared to 68.5 percent of white fathers. In 1998, 25.5 percent of black fathers were un- or underemployed, while 17.4 percent of white fathers fell into that category. Nearly 23 percent of black fathers' income was half of the poverty threshold, while 15 percent of white fathers had incomes that low.

The historical transformations were experienced across racial groups but not to the same extent. The family forms of all racial groups in America have become more diverse, or at least recognition of the diversity of family structure has increased, but the proportions of family types vary across racial groups. Because African American employment was more highly concentrated in blue-collar jobs, recent economic restructuring had harsher implications for black communities and families (Nelson 2004). The higher and more concentrated poverty levels and greater income and wealth inequality—both among African Americans and between African Americans and whites—expose African American men, directly and indirectly, to continued lower life expectancy, higher mortality, and, hence, a skewed gender ratio that leaves black women outnumbering black men by the age of eighteen.

All of these societal and family-level trends affect black men's propensity to parent and their styles of parenting in ways we have yet to fully articulate. For instance, Americans in general have responded to these trends by postponing marriage by two to four years over the last few decades, but that trend is quite pronounced among African Americans, to the point that it is estimated that whereas 93 percent of whites born from 1960 through 1964 will eventually marry, only 64 percent of blacks born in the same period ever will (Goldstein and Kenney 2001). Consequently, in 1970 married-couple families accounted for about 68 percent of all black families, but in 2000, after several decades of deindustrialization, only 46 percent were married couples. The downstream effect of marriage decline is that the majority of black children no longer live in married-couple homes.

Certainly, the skewed gender ratio mentioned earlier contributes to this declining marriage trend, but the role of other factors is under debate. Wilson (1987) and others have suggested that black men's underemployment, along with black women's higher educational attainment in relation to black men (and smaller wage gap than between white men and women, according to Roberts 1994), may decrease both men's and women's desire to marry and may hinder some black men's efforts to be involved fathers (Marsiglio and Cohan 2000). However, other research (Lerman 1989; Ellwood and Crane 1990) has found that even college-educated and employed black men have exhibited declines in marriage, and yet additional research points to attitudinal factors (South 1993; Tucker and Mitchell-Kernan 1995; Crissey 2005), with black men desiring marriage less than white and Latino men.

Other parenting trends may also be affected by black men's unique status. Their higher mortality rate and lower life expectancy may affect the timeline of parenting, increasing pressure to reproduce earlier. If married or cohabiting, black women's higher employment rate may increase the amount of time black men spend with their children (Fagan 1998). Higher poverty and collective values also pull extended family members into the mix, diffusing parenting responsibilities, which may lead to more protective or more neglectful styles of parenting.

Because of these society-wide and race-specific changes in family formation and gender roles, academia and popular culture have exhibited an increasing fascination with the diversifying definitions of masculinity and the roles men play in families, particularly as fathers.

… In conjunction with this increased amount of research and, in fact, frequently fueling the research, has been a proliferation of public and private programs and grants aimed at creating "responsible fatherhood." While many of the programs have been successful in educating men on how to be qualitatively better fathers, many have aimed primarily either at encouraging fathers to marry the mothers of their children or at securing child support. Marriage and child support are important aspects of family commitment, but marriage is no guarantee of attentive fathering, and garnished child support alone, particularly if it goes to the state and not to the mother and child, is hardly better parenting. Within this policy focus, African American men are most frequently attended to under the rubric of "fragile families" (Hobson 2002; Gavanas 2004; Mincy, Garfinkel, and Nepomnyaschy 2005). Although this classification may be intended to bring attention to structural supports that many families lack, once again it promulgates the idea that black men cannot be strong fathers.

Given the increased focus on fatherhood in scholarly and popular venues, what do we really know about black men and parenting?

So let's start … with what we know about nonresident or so-called absent fathers. Studies on this ilk of fathers indicate that generally a large portion of nonresident fathers are literally absent from their children's lives or, if in contact, their involvement decreases substantially over time. A number of memoirs by black men and women, sons and daughters of literally absent fathers, attest to the painful experience that this can be for the offspring—both sons and daughter—of these physically or emotionally missing father.

Although … anguished experiences are too common, they remain only one part, though often the more visible part, of the larger fatherhood picture. An increasing number of quantitative and qualitative studies find that of men who become fathers through nonmarital births, black men are least likely (when compared to white and Hispanic fathers) to marry or cohabit with the mother (Mott 1994; Lerman and Sorensen 2000). But they were found to have the highest rates (estimates range from 20 percent to over 50 percent) of visitation or provision of some caretaking or in-kind support (more than formal child support). For instance, Carlson and McLanahan's (2002) figures indicated that only 37 percent of black nonmarital fathers were cohabiting with the child (compared to 66 percent of white fathers and 59 percent of Hispanic), but of those who weren't cohabiting, 44 percent of unmarried black fathers were visiting the child, compared to only 17 percent of white and 26 percent of Hispanic fathers. These studies also suggested that black nonresident fathers tend to maintain their level of involvement over time longer than do white and Hispanic nonresident fathers (Danziger and Radin 1990; Taylor et al. 1990; Seltzer 1991; Stier and Tienda 1993; Wattenberg 1993; Coley and Chase-Lansdale 1999).

Sometimes social, fictive, or "other" fathers step in for or supplement nonresident biological fathers. Little research has been conducted on social fathers, but it is known they come in a wide variety: relatives, such as grandfathers and uncles; friends, romantic partners and new husbands of the mother, cohabiting or not; and community figures, such as teachers, coaches, or community-center staff. Although virtually impossible to capture clearly in census data, it is known that a high proportion of black men act as social fathers of one sort or another, yet few studies exist on this group of dads. Lora Bex Lempert's 1999 study of black grandmothers as primary parents found that many families rely on grandfathers, other male extended family members, or community members to fill the father's shoes, but unfortunately her study did not explore the experience of these men.

… McLanahan and Sandefur (1994) found that, compared to those who live in single-parent homes, black male teens who lived with stepfathers were significantly less likely to drop out of school and black teen females were significantly less likely to become teen mothers. The authors speculated that the income, supervision, and, role models that stepfathers provide may help compensate for communities with few resources and social control. Although they are often pictured as childless men, these social fathers may also be some other child's biological father, sometimes a nonresident father himself. Consequently, it is not easy and is certainly misleading to discuss fathers as if they come in discreet, nonoverlapping categories of biological or social.

A smaller amount of research has been conducted on black fathers in two-parent families, which are more likely to also be middle-class families. Allen (1981), looking at wives' reports, found black wives reported a higher level of father involvement in childrearing than did white wives. McAdoo (1988) and Bowman (1993) also concluded that black fathers are more involved than white fathers in childrearing. However, Roopnarine and Alimeduzzaman (1993), and Hossain and Roopnarine (1994) find no or insignificant racial differences in the level and quality of married fathers' involvement. Across races, fathers in married-couple families were about equally involved with their children, which in all cases was less than mothers.

In terms of parenting style, studies of black two-parent families have found that African American parenting styles tend to be more authoritarian, with an emphasis on obedience and control or monitoring, than those of white parents. This style difference is frequently explained by lower income and neighborhood rather than by race itself (Garcia-Coll 1990; Hofferth 2003). Bright and Williams (1996) conducted a small qualitative study of seven low- to middle-income black fathers in two-parent families in an urban area. They found these fathers worked collaboratively with their wives to nurture their children and that chief among their concerns were rearing children with high self-esteem, protecting their family members in unsafe environments, securing quality education, and having a close relationship with their children. Marsiglio (1991) also found black fathers to talk more and have positive engagement with their older children.

Finally, and ironically, most *absent* in the literature on black fatherhood have been those fathers who are most *present*: black, single full-time fathers. About 6 percent of black households are male-headed, with no spouse present; about half of those contain children under eighteen years old. These men also may be biological or adoptive fathers, but little is known about them.

… In sum, research on black fathers has been limited in quantity and has narrowly focused on nonmarital, nonresident fathers and only secondarily on dads in married-couple households. This oversight is not merely intentional, for black men are only about 6 percent of the U.S. population and obviously a smaller percent are fathers. They are not easy to access, particularly by an academy that remains predominantly white.

… We do not intend to decide which set of dads is better (whether by race or by type of father within races). We are not interested in the good dad-bad dad typology. We make the assumption that good fathering is best for children, but we also assume good fathering can take many forms and styles.

REFERENCES

Allen, W. 1981. "Mom, Dads, and Boys: Race and Sex Differences in the Socialization of Male Children." In *Black Men*, ed. L. Gary, 99–114. Beverly Hills, Calif.: Sage.

Bowman, P. 1993. "The Impact of Economic Marginality on African-American Husbands and Fathers." In *Family Ethnicity*, ed. H. McAdoo, 120–137. Newbury park, Calif.: Sage.

Bright, J. A., and C. Williams. 1996. "Child-rearing and Education in Urban Environments: Black Fathers' Perspectives." *Urban Education* 31(3): 245–60.

Carlson, M. J. and S. S. McLanahan. 2002. "Fragile Families, Father Involvement, and Public Policy" In *Handbook of Father Involvement: Multidisciplinary Perspectives*, ed. Catherine Tamis-LeMonda and Natasha Cabrera, 461–88. Mahwah, N.J.: Lawrence Erlbaum.

Coles, R. L., and P. L. Chase-Lansdale. 1999. "Stability and Change in Paternal Involvement Among Urban African American Fathers." *Journal of Family Psychology* 13(3): 1–20.

Crissey, S. R. 2005. "Race/Ethnic Differences in the Marital Expectations of Adolescents: The Role of Romantic Relationships." *Journal of Marriage and Family* 67: 697–709.

Danziger, S. and N. Radin. 1990. "Absent Does Not Equal Uninvolved: Predictors of Fathering in Teen Mother Families." *Journal of Marriage and the Family* 52(3): 636–42.

Elwood, D. T., and J. Crane. 1990. "Family Change Among Black Americans: What Do We Know?" *The Journal of Economic Perspectives* 4(4): 65–84.

Fagan, J. 1998. "Correlates of Low-Income African American and Puerto Rican Fathers' Involvement with Their Children." *Journal of Black Psychology* 24(3): 351–67.

Garcia-Coll, C. 1990. "Developmental Outcome of Minority Infants: A Process-Oriented Look Into our Beginnings. *Child Development* 61: 270–89.

Gavanas, A. 2004. *Fatherhood Politics in the United States: Masculinity, Sexuality, Race, and Marriage.* Chicago: University of Illinois Press.

Goldstein, J. R. and C. T. Kenney. 2001. "Marriage Delayed or Marriage Forgone? New Cohort Forecasts of First Marriage for U.S. Women," *American Sociological Review* 66(4): 506–19.

Hernandez, D. J., and P. D. Brandon. 2002. "Who Are the Fathers of Today?" In *Handbook of Father Involvement,* ed. C. S. Tamis-LeMonda and N. Cabrera, 33–62. Mahwah, N.J.: Lawrence Erlbaum.

Hobson, B. 2002. *Making Men Into Fathers.* Cambridge: Cambridge University Press.

Hofferth, S. 2003. "Race/Ethnic Differences in Father Involvement in Two-Parent Families: Culture, Context, or Economy?" *Journal of Family Issues* 24(2): 185–216.

Hossain, Z. and Roopnarine, J. 1994. "African-American Fathers' Involvement with Infants: Relationship to Their Functioning Style, Support, Education, and Income." *Infant Behavior and Development* 17: 175–84.

Lamb, M. E. 1986. "The Changing Role of Fathers." In *The Father's Role: Applied Perspectives,* ed. M. E. Lamb, 3–27. New York: Wiley.

LaRossa, R. 1997. *The Modernization of Fatherhood: A Social and Political History.* Chicago: University of Chicago Press.

Lempert, L. B. 1999. "Other Fathers: An Alternative Perspective on African American Community Caring." In *The Black Family: Essays and Studies,* ed. R. Staples, 189–201. Belmont, Calif.: Wadsworth.

Lerman, R. L. 1989. "Employment Opportunities of Young Men and Family Formation." *American Economic Review* (May): 62–66.

Lerman, R. and E. Sorensen. 2000. "Father Involvement with Their Nonmarital Children: Patterns, Determinants, and Effects on Their Earnings." *Marriage and Family Review* 29(2/3): 137–58.

Marsiglio, W. 1991. "Paternal Engagement Activities with Minor Children." *Journal of Marriage and the Family* 53: 973–86.

Marsiglio, W., and M. Cohan. 2000. "Contextualizing Father Involvement and Paternal Influence: Sociological and Qualitative Themes." In *Fatherhood: Research, Interventions, and Policies,* ed. H. E. Peters, G. W Peterson, S. K. Steinmetz, and R. D. Day, 75–95. New York: Haworth.

Martin, J. A., B. E. Hamilton, P. D. Sutton, S. J. Ventura, F. Menacker, and M. L. Munson. 2003. "Births: Final Data for 2002." *National Vital Statistics Reports* 52(10). Washington, D.C.: Government Printing Office.

McAdoo, J. L. 1988. "The Roles of Black Fathers in the Socialization of Black Children." In *Black Families*, ed. H. P. McAdoo, 257–69. Newbury Park, Calif.: Sage.

McLanahan, S., and G. Sandefur. 1994. *Growing Up with a Single Parent: What Hurts, What Helps*. Cambridge, Mass.: Harvard University Press.

Mincy, R., I. Garfinkel, and L. Nepomnyaschy. 2005. "In-Hospital Paternity Establishment and Father Involvement in Fragile Families." *Journal of Marriage and Family* 67: 611–26.

Mott, E. L. 1994. "Sons, Daughters, and Fathers' Absence: Differentials in Father-Leaving Probabilities and in Home Environments." *Journal of Family Issues* 5: 97–128.

Nelson, T. J. 2004. "Low-Income Fathers." *Annual Review of Sociology* 30: 427–51.

Pitts, L., Jr. 1999. *Becoming Dad: Black Men and the Journey to Fatherhood*. Atlanta: Longstreet.

Roberts, D. 1998. "The Absent Black Father." In *Lost Fathers: The Politics of Fatherlessness in America*, 144–61. New York: St. Martin's Press.

Roberts, S. 1994. "Black Women Graduates Outpace Male Counterparts." *New York Times*. October 31.

Roopnarine, J. L., and M. Ahmeduzzaman. 1993. "Puerto Rican Fathers' Involvement with Their Preschool-Age Children." *Hispanic Journal of Behavioral Sciences* 15(1): 96–107.

Seltzer, J. A. 1991. "Relationships Between Fathers and Children Who Live Apart: The Father's Role After Separation." *Journal of Marriage and the Family* 53: 79–101.

South, S. J. 1993. "Racial and Ethnic Differences in the Desire to Marry." *Journal of Marriage and Family* 55(2): 357–70.

Stier, H. and M. Tienda. 1993. "Are Men Marginal to the Family? Insights from Chicago's Inner City." In *Men, Work, and Family*, ed. J. C. Hood, 23–44. Newbury Park, Calif.: Sage.

Taylor, R., L. Chatters, M. B. Tucker, and E. Lewis. 1990. "Developments in Research on Black Families: A Decade Review." *Journal of Marriage and the Family* 52: 993–1014.

Tucker, M. B., and Mitchell-Kernan, C. 1995. "Trends in African American Family Formation: A Theoretical and Statistical Overview." In *The Decline in Marriage Among African American: Causes, Consequences, and Policy Implications*, ed. M. B. Tucker and C. Mitchell-Kernan, 3–26. New York: Russell Sage Foundation.

Wattenberg, E. 1993. "Paternity Actions and Young Fathers." In *Young Unwed Fathers: Changing Roles and Emerging Policies*, ed. R. Lerman and T. Ooms, 213–34. Philadelphia: Temple University Press.

Wilson, W. J. 1987. *The Truly Disadvantaged: The Inner City, the Underclass, and Public Policy*. Chicago: University of Chicago Press.

KEY CONCEPTS

fatherhood	industrialization	nuclear family
fictive fatherhood	myth of the absent father	

DISCUSSION QUESTIONS

1. What are the myths about Black fathers that Coles and Green identify in their article and how are they challenged by sociological research?

2. What are the social and historical trends that have shaped the experiences of Black men as fathers? How are these both similar to and different from the experiences of fathers in other racial or ethnic groups?

B. Religion

45

The Protestant Ethic and the Spirit of Capitalism

MAX WEBER

Max Weber's classic analysis of the Protestant ethic and the spirit of capitalism shows how cultural belief systems, such as a religious ethic, can support the development of specific economic institutions. His multidimensional analysis shows how capitalism became morally defined as something more than pursuing monetary interests and, instead, has been culturally defined as a moral calling because of its consistency with Protestant values.

The impulse to acquisition, pursuit of gain, of money, of the greatest possible amount of money, has in itself nothing to do with capitalism. This impulse exists and has existed among waiters, physicians, coachmen, artists, prostitutes, dishonest officials, soldiers, nobles, crusaders, gamblers, and beggars. One may say that it has been common to all sorts and conditions of men at all times and in all countries of the earth, wherever the objective possibility of it is or has been given. It should be taught in the kindergarten of cultural history that this naïve idea of capitalism must be given up once and for all. Unlimited greed for gain is not in the least identical with capitalism, and is still less its spirit. Capitalism may even be identical with the restraint, or at least a rational tempering, of this irrational impulse. But capitalism is identical with the pursuit of profit, and forever renewed profit, by means of continuous, rational, capitalistic enterprise....

SOURCE: Weber, Max. *The Protestant Ethic and the Spirit of Capitalism*, translated by Talcott Parsons, 17–27, 44–83, 157–83. New York: Scribner, 1958. Reprinted with permission.

If any inner relationship between certain expressions of the old Protestant spirit and modern capitalistic culture is to be found, we must attempt to find it, for better or worse, not in its alleged more or less materialistic or at least anti-ascetic joy of living, but in its purely religious characteristics....

In the title of this study is used the somewhat pretentious phrase, the *spirit* of capitalism. What is to be understood by it? The attempt to give anything like a definition of it brings out certain difficulties which are in the very nature of this type of investigation.

If any object can be found to which this term can be applied with any understandable meaning, it can only be an historical individual, i.e. a complex of elements associated in historical reality which we unite into a conceptual whole from the standpoint of their cultural significance....

"Remember, that *time* is money. He that can earn ten shillings a day by his labour, and goes abroad, or sits idle, one half of that day, though he spends but sixpence during his diversion or idleness, ought not to reckon *that* the only expense; he has really spent, or rather thrown away, five shillings besides.

"Remember, that *credit* is money. If a man lets his money lie in my hands after it is due, he gives me the interest, or so much as I can make of it during that time. This amounts to a considerable sum where a man has good and large credit, and makes good use of it....

"The most trifling actions that affect a man's credit are to be regarded. The sound of your hammer at five in the morning, or eight at night, heard by a creditor, makes him easy six months longer; but if he sees you at a billiard-table, or hears your voice at a tavern, when you should be at work, he sends for his money the next day; demands it, before he can receive it, in a lump." ...

Truly what is here preached is not simply a means of making one's way in the world, but a peculiar ethic. The infraction of its rules is treated not as foolishness but as forgetfulness of duty. That is the essence of the matter. It is not mere business astuteness, that sort of thing is common enough, it is an ethos. *This* is the quality which interests us.

When Jacob Fugger, in speaking to a business associate who had retired and who wanted to persuade him to do the same, since he had made enough money and should let others have a chance, rejected that as pusillanimity and answered that "he (Fugger) thought otherwise, he wanted to make money as long as he could," the spirit of his statement is evidently quite different from that of Franklin.[1] What in the former case was an expression of commercial daring and a personal inclination morally neutral, in the latter takes on the character of an ethically coloured maxim for the conduct of life. The concept spirit of capitalism is here used in this specific sense, it is the spirit of modern capitalism. For that we are here dealing only with Western European and American capitalism is obvious from the way in which the problem was stated. Capitalism existed in China, India, Babylon, in the classic world, and in the Middle Ages. But in all these cases, as we shall see, this particular ethos was lacking....

And in truth this peculiar idea, so familiar to us today, but in reality so little a matter of course, of one's duty in a calling, is what is most characteristic of the social ethic of capitalistic culture, and is in a sense the fundamental basis of it. It is an obligation which the individual is supposed to feel and does feel towards the content of his professional activity, no matter in what it consists, in particular no matter whether it appears on the surface as a utilization of his personal powers, or only of his material possessions (as capital).

Rationalism is an historical concept which covers a whole world of different things. It will be our task to find out whose intellectual child the particular concrete form of rational thought was, from which the idea of a calling and the devotion to labour in the calling has grown, which is, as we have seen, so irrational from the standpoint of purely eudæmonistic self-interest, but which has been and still is one of the most characteristic elements of our capitalistic culture. We are here particularly interested in the origin of precisely the irrational element which lies in this, as in every conception of a calling....

... Like the meaning of the word, the idea is new, a product of the Reformation. This may be assumed as generally known. It is true that certain suggestions of the positive valuation of routine activity in the world, which is contained in this conception of the calling, had already existed in the Middle Ages, and even in late Hellenistic antiquity. We shall speak of that later. But at least one thing was unquestionably new: the valuation of the fulfillment of duty in worldly affairs as the highest form which the moral activity of the individual could assume. This it was which inevitably gave every-day worldly activity a religious significance, and which first created the conception of a calling in this sense.... Late Scholasticism, is, from a capitalistic viewpoint, definitely backward. Especially, of course, the doctrine of the sterility of money which Anthony of Florence had already refuted.

... For, above all, the consequences of the conception of the calling in the religious sense for worldly conduct were susceptible to quite different interpretations. The effect of the Reformation as such was only that, as compared with the Catholic attitude, the moral emphasis on and the religious sanction of, organized worldly labour in a calling was mightily increased....

The real moral objection is to relaxation in the security of possession, the enjoyment of wealth with the consequence of idleness and the temptations of the flesh, above all of distraction from the pursuit of a righteous life. In fact, it is only because possession involves this danger of relaxation that it is objectionable at all. For the saints' everlasting rest is in the next world; on earth man must, to be certain of his state of grace, "do the works of him who sent him, as long as it is yet day." Not leisure and enjoyment, but only activity serves to increase the glory of God, according to the definite manifestations of His will.

Waste of time is thus the first and in principle the deadliest of sins. The span of human life is infinitely short and precious to make sure of one's own election. Loss of time through sociability, idle talk, luxury, even more sleep than is necessary for health, six to at most eight hours, is worthy of absolute moral condemnation. It does not yet hold, with Franklin, that time is money, but the proposition is true in a certain spiritual sense. It is infinitely valuable because every hour lost is lost to labour for the glory of God. Thus inactive contemplation is also valueless, or even directly reprehensible if it is at the expense of one's daily work....

It is true that the usefulness of a calling, and thus its favour in the sight of God, is measured primarily in moral terms, and thus in terms of the importance of the goods produced in it for the community. But a further, and, above all, in practice the most important, criterion is found in private profitableness. For if that God, whose hand the Puritan sees in all the occurrences of life, shows one of His elect a chance of profit, he must do it with a purpose. Hence the faithful Christian must follow the call by taking advantage of the opportunity. "If God shows you a way in which you may lawfully get more than in another way (without wrong to your soul or to any other), if you refuse this, and choose the less gainful way, you cross one of the ends of your calling, and you refuse to be God's steward, and to accept His gifts and use them for Him when He requireth it: you may labour to be rich for God, though not for the flesh and sin."

Wealth is thus bad ethically only in so far as it is a temptation to idleness and sinful enjoyment of life, and its acquisition is bad only when it is with the purpose of later living merrily and without care. But as a performance of duty in a calling it is not only morally permissible, but actually enjoined....

Let us now try to clarify the points in which the Puritan idea of the calling and the premium it placed upon ascetic conduct was bound directly to influence the development of a capitalistic way of life. As we have seen, this asceticism turned with all its force against one thing: the spontaneous enjoyment of life and all it had to offer....

On the side of the production of private wealth, asceticism condemned both dishonesty and impulsive avarice. What was condemned as covetousness, Mammomsm, etc., was the pursuit of riches for their own sake. For wealth in itself was a temptation. But here asceticism was the power "which ever seeks the good but ever creates evil"; what was evil in its sense was possession and its temptations. For, in conformity with the Old Testament and in analogy to the ethical valuation of good works, asceticism looked upon the pursuit of wealth as an end in itself as highly reprehensible; but the attainment of it as a fruit of labour in a calling was a sign of God's blessing. And even more important: the religious valuation of restless, continuous, systematic work in a worldly calling, as the highest means to asceticism, and at the same time the surest and most evident proof of rebirth and genuine faith, must have been the most powerful conceivable lever for the expansion of that attitude toward life which we have here called the spirit of capitalism.

When the limitation of consumption is combined with this release of acquisitive activity, the inevitable practical result is obvious: accumulation of capital through ascetic compulsion to save. The restraints which were imposed upon the consumption of wealth naturally served to increase it by making possible the productive investment of capital....

One of the fundamental elements of the spirit of modern capitalism, and not only of that but of all modern culture: rational conduct on the basis of the idea of the calling, was born—that is what this discussion has sought to demonstrate—from the spirit of Christian asceticism....

The Puritan wanted to work in a calling; we are forced to do so. For when asceticism was carried out of monastic cells into everyday life, and began to dominate worldly morality, it did its part in building the tremendous cosmos of the modern economic order. This order is now bound to the technical and economic conditions of machine production which today determine the lives of all the individuals who are born into this mechanism, not only those directly concerned with economic acquisition, with irresistible force....

Since asceticism undertook to remodel the world and to work out its ideals in the world, material goods have gained an increasing and finally an inexorable power over the lives of men as at no previous period in history. Today the spirit of religious asceticism—whether finally, who knows? has escaped from the cage. But victorious capitalism, since it rests on mechanical foundations, needs its support no longer. The rosy blush of its laughing heir, the Enlightenment, seems also to be irretrievably fading, and the idea of duty in one's calling prowls about in our lives like the ghost of dead religious beliefs. Where the fulfillment of the calling cannot directly be related to the highest spiritual and cultural values, or when, on the other hand, it need not be felt simply as economic compulsion, the individual generally abandons the attempt to justify it at all. In the field of its highest development, in the United States, the pursuit of wealth, stripped of its religious and ethical meaning, tends to become associated with purely mundane passions, which often actually give it the character of sport.

No one knows who will live in this cage in the future, or whether at the end of this tremendous development entirely new prophets will arise, or there will be a great rebirth of old ideas and ideals, or, if neither, mechanized petrification, embellished with a sort of convulsive self-importance. For of the last stage of this cultural development, it might well be truly said: "Specialists without spirit, sensualists without heart; this nullity imagines that it has attained a level of civilization never before achieved." ...

The modern man is in general, even with the best will, unable to give religious ideas a significance for culture and national character which they deserve. But it is, of course, not my aim to substitute for a one-sided materialistic an equally one-sided spiritualistic causal interpretation of culture and of history. Each is equally possible, but each, if it does not serve as the preparation, but as the conclusion of an investigation, accomplishes equally little in the interest of historical truth.

NOTE

1. The quotations are attributed to Benjamin Franklin.

KEY CONCEPTS

capitalism   Protestant ethic

DISCUSSION QUESTIONS

1. Weber is known for developing a multidimensional view of human society. What role does he see the Protestant ethic playing in the development of capitalism?

2. Weber's analysis sees Western capitalists as not pursuing money just for the sake of money, but because of the moral calling invoked by the Protestant ethic. Given the place of consumerism in contemporary society, how do you think Weber might modify his argument were he writing now? In other words, are there still remnants of the Protestant ethic in our beliefs about stratification? If so, how do they fit with contemporary capitalist values?

46

America and the Challenges of Religious Diversity

ROBERT WUTHNOW

Robert Wuthnow points out that Americans generally think of the United States as a Christian nation, but that idea is being challenged by the increasing diversity of religious practice in the United States. His article, based on an extensive survey of Americans' religious beliefs and practices, examines the challenges that religious diversity poses for national identity.

Questions about racial and ethnic differences and questions about the impact of immigration have attracted extraordinary interest in recent years. Questions about religion and its cultural effects are just as important. They involve beliefs and convictions, assumptions about good and evil, individual and group identities, and concerns about how to live together. These questions were not resolved during the nation's formative era. And they certainly have not faded away. The growing presence of American Muslims, Hindus, Buddhists, and other new immigrant groups makes these questions more pressing than ever. The United States has a strong tradition respecting the rights of diverse religious

SOURCE: Robert Wuthnow. 2005. *America and the Challenges of Religious Diversity.* Princeton, NJ: Princeton University Press.

communities. But American culture is also a product of its distinctive Christian heritage. This heritage exists in tension with the nation's religious and cultural diversity.

The tension between America's Christian heritage and its religious and cultural diversity became evident in the days following the September 11, 2001, attacks on the Pentagon and World Trade Center. In a speech to Congress, President Bush declared to Muslims, "We respect your faith." Yet Bush had also said that only Christians have a place in heaven. How did he reconcile these views? What did he mean by "respect"?

An apparent inconsistency in political rhetoric like this would hardly merit attention were it not for the fact that it points to something much deeper. American identity is an odd mixture of religious particularism and cultural pluralism. Although it is not an established religion, Christianity is the nation's majority religion, and its leaders and followers have often claimed it had special, if not unique, access to divine truth. Yet the reality of religious pluralism, including beliefs and practices different from those embraced by Christianity, has also had a profound impact on American culture. These strands of our national identity are not just contradictory or conflicting impulses. They are inextricably bound together in ways that feed our collective imagination and evoke questions about who we are.

Through a large number of in-depth interviews, data from a new national survey, and published materials about the past and present, I examine the terms in which the relationship between America's Christian heritage and its growing religious diversity is being debated. I emphasize the perceptions of ordinary Americans as well as those of community leaders and the languages in which these perceptions are framed. I argue that interpretations of religious diversity have been, and continue to be, a profound aspect of our national identity.

It has become popular among social observers to argue that American religion is so thoroughly composed of private beliefs and idiosyncratic practices that belief and practice ultimately do not matter. People pick and choose in whatever way helps them to get ahead (or, at least, to get along). Their beliefs are so shallow that inconsistencies make no difference. Some observers also argue that Americans can hold fundamentally incommensurate beliefs in their personal lives, but live amicably in public. This is a recent litany in the literature on pluralism. Let religious subgroups believe whatever they want to, the argument goes, but count on laws and norms of civic decorum to maintain social order. In this view, religion and civic life function without mutual influence. Pluralism is culturally uncomplicated.

The evidence I present here suggests that these views are wrong. I show that pluralism and religious practices are intertwined. How people think about pluralism is influenced by their religious convictions. And religious convictions are influenced by their experiences with pluralism. This means that cultural interpretations of religious questions matter. They matter, not so much as formal expressions of what theologians or religious organizations teach, but in the way that Michael Polanyi described the *tacit* knowledge in which all human behavior is inscribed. Tacit knowledge matters because we prefer to live in a world, even if it is a world of our own construction, that makes sense, rather than in a world without sense. Understood this way, it makes a difference how people think about questions of God, death, salvation, heaven, good and evil, other religions, and the teachings of their own tradition. It certainly matters to the many Muslims, Hindus, Buddhists, and practitioners of other non-Western religions who now make up a growing minority of the U.S. population. It also matters to Americans who claim to be Christians or Jews, or who are self-styled spiritual shoppers. They may sometimes deny that it does. But when they confront religious diversity, and when they think about what it means to be religious or spiritual in a diverse society, they articulate tacit assumptions about what it means to be human and what it means to be an American.

Religious identities matter to the collective life of society as well as to the personal lives of individuals. Religious identities are among the ways in which cultural assumptions about what is right and good, or better and best, are organized. Americans believe they are a special people with a distinctive mission to fulfill in the world. This belief is associated historically with our understanding of religion. To say that we are a Christian nation has been a normative statement as well as a descriptive one. Christian values and practices occupied a special place in our thinking. To say that some people were Christian implied that others were not. Our moral universe included assumptions about Jews, Muslims, Hindus, Buddhists, and practitioners of Native American religions. They, too, had duties to fulfill, roles in the cultural drama to perform. Religious diversity was inscribed in the moral order.

Another popular approach to religion among social scientists is to deal with it as if it were purely an expression of something else, such as class, race, gender, and region, or to explain its trends and patterns with reference to demography, organizations, leadership styles, and theories about rational choice. These reductionistic approaches give social scientists an excuse to avoid the *content* of religion. What people believe, or say they believe, and the language in which they make sense of their beliefs and practices are somehow, in the view of these scholars, too marginal, too normative, or too difficult to measure for any self-respecting social scientist to tackle. This is the point at which narrow definitions of disciplinary boundaries get in the way of knowledge.

I choose to emphasize what people think and the cultural idioms in which they express their thoughts. This is how people make sense of their beliefs and practices. It is how they negotiate meaning when faced with multiple religious teachings and traditions. If people were guided only by demography or social position, there would be no need to know what they say or think. But there is a well-established tradition in the social sciences (counting Max Weber and George Herbert Mead among those who observe it) that says that the meaning-making activity of humans is crucial to our understanding of society. Making sense of religious diversity is one of these meaning-making activities.

Still, listening to what people say would be of little value if their views merely echoed the writings of theologians and social philosophers. If ordinary people were guided by these writings, one would want to spend the time one has for scholarly reflection understanding these tomes and writing commentaries about them. Worthy as that may be, it does not provide much of a picture of the society in which we actually live.

When rank-and-file Americans talk about religious diversity, they disclose an implicit cultural text composed of narrative fragments from personal experience, from conversations with friends and neighbors, from the media, from books and magazines, and in many instances from ruminations about questions raised in Sunday school, a high school youth group, a course in comparative religions, or a visit to another country. It is possible to identify themes and variations in this subterranean text. Some people find ways to embrace religious diversity as fully as possible. Others assert loyalty to the tradition in which they were raised (or are presently involved), but acknowledge the validity of other traditions. A substantial number of Americans adamantly reject the truth of religions other than their own. In each of these orientations, people articulate a bricolage of ideas that both reflects and subverts public images of cultural diversity. Patterns of avoidance minimizing considered engagement among religious traditions are evident. And these avoidances illuminate the behavior of religious organizations and their leaders.

I do not argue that the present encounter with religious diversity is entirely new or without precedent. My argument is rather that America and American Christianity have always existed in a world of religious differences and with some awareness of these differences. I further argue, however, that this awareness is probably greater among rank-and-file Americans now than in the past because of mass communications, immigration, and our nation's role in the global economy. …

Whereas historical treatments of religious diversity in America have typically emphasized the divisions *within* Christianity (especially those separating Protestants and Catholics, Protestant denominations, and the various sects), my emphasis here is on the encounter between American Christians and other major religious traditions, such as Islam, Buddhism, and Hinduism. I understand that the tensions between Protestants and Catholics, or even between rival branches of Presbyterianism, were sometimes as fierce as anything evident currently between Christians and non-Christians. I am nevertheless interested in the fact that Christians have always had to formulate arguments about people who were clearly outside the Christian tradition by virtue of belonging to other major religious traditions. I am interested in how these arguments played into our national identity historically and how they are being revised at present.

My aim is not to encourage readers to conclude that religions are interchangeable. Nor do I believe the best way to live in a pluralistic society is to combine bits and pieces of several religions. I do insist that the growing religious diversity of our society poses a *significant cultural challenge*. The fact that Muslims, Hindus, and Buddhists are now a significant presence in the United States raises fundamental questions about our historic identity as a Christian nation. This new reality requires us to rethink our national identity and to face difficult choices about how pluralistic we are willing to be. It requires people of all religions, as well as scholars and community leaders, to take notice. If a person's best friend in elementary school belonged to a different religion, and if this person takes religion seriously, he or she will surely think about his or her faith differently than would have been the case if everyone in school belonged to the same religion. If one's neighbors and coworkers hold beliefs vastly different from one's own, this too will evoke a response. We can try to understand and become more aware of these influences, and thus make more informed choices about how we respond, rather than letting circumstances dictate our responses.

... News coverage from around the world includes images of religious leaders, adherents, and their places of worship. The nation's expansive economic and military activities render these images more newsworthy than they would have been in the past. Apart from media, exposure to the world's religions comes increasingly through first-hand encounters. During the last third of the twentieth century, approximately twenty-two million immigrants came to the United States. Like the surge of immigration that occurred between 1890 and 1920, most of these immigrants came from countries in which Christians are the dominant religion. Yet, in contrast to that earlier period, the recent immigration included millions of people from countries in which Christians are only a small minority. Thus, in little more than a generation, the United States has witnessed an unprecedented increase in the diversity of major religious traditions represented among its population. More Americans belong to religions outside of the Christian tradition than ever before. The new immigrants include large numbers of Muslims, Hindus, Buddhists, and followers of other traditions and spiritual practices. Their presence greatly increases the likelihood of personal interaction across these religious lines.

Recent immigrants and their descendants generally do not live isolated from other Americans in homogeneous enclaves. They frequently work in middle-class occupations and live in the same neighborhoods as other Americans do. Their mosques, temples, and meditation centers are often located in close proximity to churches and synagogues. The typical American, therefore, can more readily encounter people of other religions as neighbors, friends, and coworkers.

Diversity is always challenging, whether it is manifest in language differences or in modes of dress, eating, and socializing. Seeing people with different habits and lifestyles makes it harder to practice our own unreflectively. When religion is involved, these challenges are multiplied. Religious differences are instantiated [represented] in dress, food, holidays, and family rituals; they also reflect historic teachings and deeply held patterns of belief and practice. These beliefs and practices may be personal and private, but they cannot easily be divorced from questions about truth and morality. Believing that one's faith is correct and behaving in ways that reflect this belief may well be different in the presence of diversity than in its absence.

How have we responded to the religious diversity that increasingly characterizes our neighborhoods, schools, and places of work? Has it sunk into our awareness that the temple or mosque down the street is not just another church? Does it matter that our coworkers have radically different ideas of the sacred than we do? Or do we perceive these ideas as so different from our own? Are our views of America affected by having neighbors whose beliefs and lifestyles may run counter to our own? Does it bother us to read about hate crimes directed at Muslims or Hindus?

Historic interpretations of Christian teachings encourage Christians to practice the acceptance and love exhibited by Mother Teresa. Stories about Jesus' willingness to violate social boundaries separating Jews and Gentiles exemplify how Christianity may encourage openness to racial, ethnic, and cultural diversity.

Yet Christianity has also taught that only by accepting Jesus as their savior can believers overcome sinfulness and gain divine redemption. According to some interpretations of this teaching, the followers of other religions must convert to Christianity if they are to know God.

Throughout America's history, our sense of who we are has been profoundly influenced by our religious beliefs and practices. Christianity's claim to be the unique representative of divine truth has been one of these influences. We have thought of ourselves as a chosen people, a city on a hill, and a new Israel. We have considered ourselves defenders of the faith, a God-fearing people, and a Christian nation. At present, we remain one of the most religiously committed of all nations, at least if religious commitment is measured in numbers professing belief in God and attending services at houses of worship. Our identity is still marked by this fact. Many Americans take for granted that we are a Christian society, even if they implicitly make a place in this notion for Jews and unbelievers. Others take pride in our national accomplishments, our democratic traditions, and our extensive voluntary associations, assuming that these reflect Christian values.

If our understanding of what it means to be American reflects our religious heritage, our collective identity is also influenced by how we think about religious diversity. Until recently, we were able to think of ourselves as a Christian civilization, divided by the historic cleavages separating Protestants from Catholics and, among Protestants, Methodists from Baptists, Presbyterians from Episcopalians, Congregationalists from Quakers, and so on. We were a diverse nation because of the national origins from which the various denominational groups had come and because of racial, ethnic, and regional divisions in which religious disunity was embedded. We took pride in this diversity. It seemed like a mark of distinction.

We clearly do have a long history of religious diversity. This history has affected our laws, encouraging us to avoid governmental intrusion in religious affairs that might lead to an establishment of one tradition in favor of others. And it has taught us a kind of civic decorum that discourages blatant expressions of racist, ethnocentric, and nativist ideas. Yet it will not do, now in the face of new diversity, simply to rewrite our nation's history as a story of diversity and pluralism.

The reality of large numbers of Americans who are Muslims, Buddhists, Zoroastrians, Sikhs, Hindus, and followers of other non-Western religions poses a new challenge to American self-understandings. When Christian leaders and their followers think about it, they will have more trouble knowing what exactly to think about their neighbors who belong to these other religions than they ever did simply thinking about the differences between Methodists and Baptists or Protestants and Catholics. That is, *if* they stop to think about it.

But the truth is, we know very little at this point about how ordinary Americans are responding to religious diversity. And, for that matter, we know little more about how religious leaders are dealing with diversity. We do know, for example, that religious leaders occasionally form interfaith alliances that include representatives of the world's major religious traditions, and we know that other

leaders are sometimes quoted in newspapers as saying that the followers of a particular religion other than their own are condemned to hell. Such headlines,
however, seldom tell us much about how things are going in local communities
or what people really believe and think....

... How well are we managing to face the new challenges of religious and
cultural diversity? Are we merely managing in the sense of making do, muddling
our way by avoiding the issues whenever possible and responding superficially
whenever we must? Or are we managing better than that? Are we taking advantage of the opportunities that diversity provides and moving toward a more mature pluralism than we have known in the past?

These are, in my view, among the most serious questions we currently face
as a nation. In our public discourse about religion, we seem to be a society of
schizophrenics. On the one hand, we say casually that we are tolerant and have
respect for people whose religious traditions happen to be different from our
own. On the other hand, we continue to speak as if our nation is (or should
be) a Christian nation, founded on Christian principles, and characterized by
public references to the trappings of this tradition. That kind of schizophrenia
encourages behavior that no well-meaning people would want if they stopped
to think about it for very long. It allows the most open-minded among us to
get by without taking religion very seriously at all. It permits religious hate
crimes to occur without much public attention or outcry. The members of
new minority religions experience little in the way of genuine understanding.
The churchgoing majority seldom hear anything to shake up their comforting
convictions. The situation is rife with misunderstanding and, as such, holds little
to prevent outbreaks of religious conflict and bigotry. It is little wonder that
many Americans retreat into their private worlds whenever spirituality is mentioned. It is just easier to do that than to confront the hard questions about religious truth and our national identity.

KEY CONCEPTS

pluralism   religiosity   secular

DISCUSSION QUESTIONS

1. How has immigration changed the religious composition of the U.S.
   population?

2. How does Wuthnow explain the seeming contradiction between a national
   identity that respects religious diversity, while at the same time being centered in a Christian tradition?

3. What are some of the new questions that Wuthnow identifies as stemming
   from increased religious diversity in the United States?

47

Abandoned, Pursued, or Safely Stowed?

TIM CLYDESDALE

Tim Clydesdale examines the religious identification of college students, with an eye toward explaining how religious identity either does or does not affect the critical thinking of college students. He introduces the concept of "identity lock-boxes" to explain how young people position themselves in the context of American culture.

There is certainly no shortage of passionate opinions about what happens to the religious commitments of college students. Some religious conservatives allege that liberal professors undermine student faith, that college administrators permit conditions that foster sexual promiscuity and alcohol abuse, and that the result is a deliberate weakening of students' religious commitments. College professors and administrators respond that college is a time when students explore new possibilities and reflect critically on their new *adult* lives, and that any change in religious commitment is a result of these adults' own choices and individual learning processes. Left unstated, of course, is the opinion of many professors that traditional religious faiths are incompatible with liberal education, and the opinion of many religious conservatives that professors lead morally vacuous lives. This longstanding and deeply-rooted difference of opinion has undoubtedly helped to fuel the two-decade-plus expansion of religious college and university enrollments; that is, expansion at educational institutions that combine faith development with a liberal arts education.

But popular recognition of religion's influence in America, especially after the 2004 election, has given rise to a new interpretation. Several observers, who previously ignored religion, now argue that the vast majority of American college students "report high levels of spiritual interest and involvement," that over half affirm "reducing pain and suffering in the world" as a life goal, that "nearly half" of American college students "are on a quest" to identify a spiritual purpose for their lives, and that spiritual traditions provide resources which can inspire students' educational efforts.[1] By framing religion as "spirituality," this interpretation grants religious life legitimacy as an (optional) component in

SOURCE: Social Science Research Council. 2006. www.ssrc.org

college student "wellness," and provides market-savvy colleges with a rationale for expanding support of religious life on their campuses.

There is just one problem with this view, which is the same problem that the longer-standing views have: woefully inadequate evidence. For all the fears of religious conservatives, and all the claims of students' critical thinking by professors and college administrators, there is precious little evidence that college students either abandon their faith commitments *or* develop intellectual curiosity. And the evidence offered in support of claims about widespread pursuit of spiritual purpose or social justice among college students is as compelling as a survey about world peace completed by beauty pageant contestants. The real issue is not how many college students check off "an interest in spirituality," but how many actualize that interest in their everyday priorities. Social survey results need to be checked against grounded and contextualized understandings of college student lives—particularly if one wishes to draw conclusions about the lived culture of American college students. Such grounded and contextualized understandings are what this author has undertaken and summarizes here.[2]

What in-depth, longitudinal interviews and field research with college freshmen reveal is that most freshmen are thoroughly consumed with the everyday matters of navigating relationships, managing gratifications, handling finances, and earning diplomas—and that they stow their (often vague) religious and spiritual identities in an *identity lockbox* well before entering college. This lockbox protects religious identities, along with political, racial, gender, and civic identities, from tampering that might affect their holders' future entry into the American cultural mainstream. If religious identities were to shift to a religious or anti-religious extreme, for example, they could ruin a teen's mainstream standing and future trajectory. The same holds true for political, racial, gender, and civic identities. "Wrong" choices in any of these areas could put freshmen seriously out of step with mainstream culture and endanger their odds of attaining the privatized, consumer happiness that American youth have long been socialized to seek.

Not all college students make use of the lockbox to store religious identities, and these exceptions deserve close attention. But most college students do so because they view religion not unlike vegetables—as something that is "good for you" and part of adult life, but not as something all that relevant to their current stage as college students. As one freshman put it, "I feel like God dropped me off at college and said, 'I'll be back to pick you up in four years'." Note that this student, like many of his peers, planned to be picked up when he graduated—in the same place and by the same driver. It is not that his religious identity was unimportant (quite the contrary), only that he did not see its relevancy to his college education and campus experience. The same holds true of students' political, civic, racial, and gender identities. These identities, as undeveloped as they often are, play a critical role in guiding youth into the cultural mainstream of the United States. College students, of course, are not one-dimensional. There are those who peek inside their identity lockboxes, with varying frequency, and who consider some or all of its contents. Those who do this in a sustained manner, however, are proportionately few, and qualify as one of three exceptional types described below.

RELIGIOUS INVOLVEMENT VS. RELIGIOUS IDENTIFICATION

There is, to be sure, well-documented research on college students' decline in attendance at religious services and in other forms of involvement in organized religious life. This is not in dispute. But a decline in religious involvement is not equivalent to a decline in religious identification, and needs to be understood carefully. Freshmen whose religious involvements declined offer various reasons for their reduced involvement: a few choose to behave in keeping with the non-religious identities they had established previously, though as high school students they attended religious services to please their parents. Others visit a few religious services "out at college," do not find a service they "like," but still attend "every time" they are "back home." And many continue to attend worship services —just "a little less often" because "it can be really hard to get up that early on Sundays." (A national survey of college freshmen, in fact, found 57 percent reported attending religious services "frequently" or "occasionally" at the *end* of their first year of college.) What freshmen do *not* say, however, is that they have gained a critical perspective on religion because of attending college, and thus have ceased to identify themselves as a religious person. Religious identifications are not questioned during the freshman year, not because they are held to in widespread piety, but because doing so would require freshmen to give attention to these identities, and few have any interest in doing so.

Asking students the summer after their freshman year to describe their spiritual and religious beliefs brought forth nearly identical answers to those they gave as high school seniors. Post-freshman year interviewees still did not relate to the terms "spiritual" or "spirituality," and they still struggled to define such terms. "Being spiritual" meant "having morals" or "being religious," and fewer than half offered even that definition. As others have well-documented, the vast majority of American teens are *not* spiritual seekers, and the few interviewees who identified as spiritual did so within established religious traditions (e.g., "I pray the rosary and meditate every night"). Likewise, freshmen's religious identifications had not shifted in the slightest. It was as if these rising sophomores peeked inside their identity lockboxes, dusted off their religious identifications, and reported, "Yeah, I'm still religious." This was quite striking. Why would freshmen choose to preserve what were often vague religious identifications, and which often diminished as aspects of their regular activities? The answer lies in understanding the powerful effects of popular American moral culture on mainstream American teens.

RELIGION & POPULAR AMERICAN MORAL CULTURE

The American mainstream can be defined in many ways; it is defined here as including American households earning $25,000 or more a year, but excluding independently wealthy households. Members of mainstream households have a toehold (or better) in the "American dream," and they have been fully socialized into American culture. Culture, to use a computer analogy, is humanity's operating system. Without it, there would be no language, no communication, no knowledge, and no meaning. And like a computer operating system, culture gets installed with certain "default" settings that, unless overridden, determine how humans view their world and structure their everyday behavior. In the United States, the current default settings install a popular American moral culture that: celebrates personal effort and individual achievement, demonstrates patriotism, believes in God and a spiritual afterlife, values loyalty to family, friends, and coworkers, expects personal moral freedom, distrusts large organizations and bureaucracies, and conveys that happiness is found primarily in personal relationships and individual consumption. Unless these default settings are altered, typically to install more specific religious or nonreligious sub-cultural settings, this constellation of beliefs and practices is characteristic of most Americans.

Thus, one national study reports that most American teens consider religion to be "a very nice thing," and despite their specific religious tradition, essentially adhere to a faith in "divinely underwritten personal happiness and interpersonal niceness."[3] There are, to be sure, nonreligious teens who have no need for the divine and theistic elements of popular American moral culture: national surveys estimate 12–18 percent of American teens consider themselves *nonreligious*. And there are also, to be sure, strongly religious teens who subscribe to elaborate religious doctrines and particular moral codes: national surveys estimate 25–35 percent of American teens are *strongly religious*. The majority of American teens, however, about 55 percent, comprise the *semireligious* middle ground.[4] These teens believe in God and identify with a religious tradition, but their practical creed is essentially a combination of Benjamin Franklin's "God helps those who help themselves" and the "Golden Rule." In other words, their semireligious identities provide divine reinforcement for pursuing individual achievement and, as one interviewee put it, for "trying to be a nice person."

Semireligious identities therefore serve a specific purpose: they underwrite a popular American moral culture that has been inculcated since birth. Semireligious identities reinforce the mainstream cultural script that graduating from college leads to a good job, which leads to marriage, which leads to children, comfortable housing, and a good standard of living. To question these religious identities is thus to question the whole of the mainstream cultural script and the popular American moral culture that created it, and these college students see no benefit in doing so. Besides, given the myriad of personal relationships to navigate, gratifications to manage, money to earn and spend, and credentials to complete—there are more pressing daily matters to which college students must attend. Semireligious identities are therefore stowed in college students' identity lockboxes, often alongside political, racial, gender, and civic identities, and all are left undisturbed.

It is not just semireligious college freshmen who stow identities, however. So do many strongly religious and nonreligious freshmen. Many strongly religious freshmen do so because they have become proficient *compartmentalizers*. That is, they stow religious identities when in educational settings, and stow educational identities when in religious settings, and readily switch one with the other. And most nonreligious freshmen stow identities because they not only lack interest in religion, they also lack interest in their *non*-religion. Issues of religion, philosophy, ethics, or meaning are of no concern to them. These nonreligious, religious and strongly religious freshmen use identity lockboxes for the same reasons as semireligious teens: they too subscribe to popular American moral culture (with a few minor adjustments), and they too possess more than enough everyday concerns to occupy their attentions.

Hence, the vast majority of college freshmen approach their education not as intellectual explorers but as *practical credentialists*; they focus on degree completion (and on grades if they seek high-status credentials), and view the rest of their education as little more than a necessary nuisance. Popular American moral culture is dubious of large organizations and bureaucracies, and especially of higher education, and college students are both products and proponents of this moral culture.

RELIGIOUS & EDUCATIONAL EXCEPTIONS

There *are* exceptions to the above pattern. There are teens who enter college seeking to understand their own lives more fully and the wider world more thoughtfully. They take advantage of educational opportunities because they enjoy learning for its own sake, they pursue creative opportunities because these express deeper realities, or they serve needy communities because they desire a more just society. In short, they refuse to stow critical identities in identity lockboxes. Who are they? Some are the *future intelligentsia*—that is, the next generation of professors and allied professionals like psychologists, deans, journalists, and guidance counselors. Some are *religious skeptics* and atheists—that is, a subset of nonreligious teens who consider religion to be the chief obstacle to achieving social justice and equity. And some are *religious emissaries*—that is, a subset of strongly religious teens who refuse to stow or compartmentalize faith but are driven to understand it and engage it with the world. The existence and inclusion of this last category here may be surprising to some, but as scholars of contemporary American religion demonstrate repeatedly, religious communities thrive in American pluralism because they engage it thoughtfully, not retreat from it.[5] Even members of the most conservative religious communities know they possess the option to pursue any religion, or none—thus religious communities put much effort into attracting and keeping adherents, including intellectual appeals, and their teen emissaries become quite conversant in these matters.

What all of these teen exceptions share is a critical perspective on popular American moral culture. That perspective may be rooted in their possession of inquisitive and self-reflective minds, and in parental and educational nurturance of the same. It may be rooted in their alienation from "mindless" theism and in their relationships with like-minded mentors. Or it may be rooted in their personal and deep identification with a religious community that decries the superficiality of American culture. (Those in the first category, in fact, frequently qualify for one of the latter two categories.) Whatever its cause, these teens grow up doubting core elements of popular American moral culture and thus reflecting upon their deeper identities and broader perspectives on the world regularly. When they enter college, this does not change. They become highly-desirable students, because they genuinely engage with class materials and because they demonstrate intellectual curiosity, creative engagement, social awareness, or all three.

Some professors will point to these intellectually-engaged students as evidence of the value of liberal education. But these students' patterns of engagement pre-date their arrival at college, and while they take advantage of educational, creative and service opportunities during their college years, college is not the cause of their engagement. Further confounding some professors' perceptions is the temporary nature of the intellectual curiosity, creative engagement, or social awareness that many students demonstrate in class. Sometimes this temporariness has more genuine roots—for example, the marketing student who becomes enraptured with her opera performance course, but who subsequently pushes aside that interest to concentrate on her "more realistic" educational goals. And sometimes this temporariness is more Machiavellian—that is, a pose that grade-obsessed practical credentialists strike because they know intellectual curiosity, creative engagement, and social awareness are precisely what their professors want to see.

The actual proportion of American teens who possess both genuine and sustained intellectual curiosity, creative engagement, or social awareness is quite small. About 1–2 percent of American teens are atheists or religious skeptics (i.e., nonreligious teens with an active and sustained interest in their non-religiosity), another 10–15 percent are religious emissaries (i.e., non-compartmentalizing, religiously-driven teens), and perhaps one percent more are future intelligentsia (i.e., intellectually-engaged teens not already included above, often in semireligious transition), giving a total estimate of 12–18 percent. Their representation on college campuses, moreover, is uneven. These exceptional teens often enroll in more selective colleges, with religious skeptics gravitating to nonreligious colleges, and religious emissaries to religious colleges. None of this should undermine the importance of professors' efforts to encourage intellectual, creative, or social engagement among students, only professors' self-aggrandizing assumptions that liberal education is the cause of such engagement.

IMPLICATIONS

The enemy of developing critical thinking, creative engagement, and social awareness among college students is therefore not students' possession of religious identities—it is their widespread use of identity lockboxes. So, too, the enemy of a thoughtful and lasting religiosity among college students is not their pursuit of college education, but their widespread use of identity lockboxes. Thus, what hinders college students' development is neither religion nor liberal education, but the use of these lockboxes. College educators need to understand that religion, and devout religion in particular, can indeed be an ally in the cause of critical thinking and social awareness. Correspondingly, religious leaders need to understand that college education, and a liberal education in particular, can aid in the development of a thoughtful and meaningful religious identity.

College students are not, however, likely to end their use of identity lockboxes anytime soon. The power of college students' desire to keep within the American cultural mainstream is not likely to diminish, and may even enlarge, as America's new economic realities make entering the cultural mainstream even more difficult. College students know that companies are quick to reorganize, to relocate to less costly areas or nations, and to release even diligent and long-term employees. They know that downward mobility is a real possibility, and that better odds of attaining economic security come with a college diploma. Thus, college students are not, save for the exceptions above, going to risk using college as a time for developing intellectual curiosity, reflecting on identities—religious or otherwise, or understanding their interdependence with communities large and small. Doing so could move students outside the cultural mainstream and jeopardize their long-term futures—a risk too great for most college students to take.

Pleading with freshmen to swim against these economic and cultural currents is not the solution. Colleges and religious communities already do this extensively, and have likely seen as much gain as they will from such appeals. Freshman interviewees were quite aware of these appeals, and had long developed immunity to them. There is anecdotal evidence that once the daily life management project of freshman year is mastered, a window of opportunity opens to engage sophomores and juniors more deeply in both religious and nonreligious pursuits. But the established, everyday patterns of college student lives, combined with the narrowing of social circles during the freshman year, makes this a narrow window indeed.

Rather than "curse the cards" American culture has dealt, college educators and religious leaders should play the hand they have. Because higher education does not possess the cultural authority in America that it does in other societies, its educators need to become public intellectuals. College educators need to realize that the same cultural pluralism that challenges religious truth claims also challenges scholarly truth claims. It is not enough for college educators to be members of scholarly disciplines, where knowledge claims are accepted when they meet the criteria of that discipline. College educators must also earn the right to be heard by larger publics because they speak plainly, marshal evidence, evaluate dispassionately, and lead their audience to logical conclusions. And the first place where that right must be earned is with their most immediate audience: college students. In the same way, religious leaders must earn the right to be heard by larger publics by the same methods. College students should be respected as the individual arbiters of truth that they are, and encouraged to see that the skills of critical thinking, creative expression, and social engagement are as useful for faith as they are for learning.

Stowed identities benefit no one—not educators, not religious communities, and certainly not college freshmen. There exists no quick fix, either. Broad cultural and economic forces will ensure that the vast majority of American college freshmen continue to follow mainstream scripts that offer no guarantee of success or satisfaction. Still, it is better to know the real problem confronting freshmen development than to continually waste energy and resources addressing the wrong issues.

ENDNOTES

1. Alexander W. Astin & Helen S. Astin, "The Spiritual Life of College Students: A National Study of College Students' Search for Meaning and Purpose," (Los Angeles: Higher Education Research Institute at University of California, 2004); see also Alexander W. Astin, "Why Spirituality Deserves a Central Place in Liberal Education," *HERI Spirituality Newsletter* (Los Angeles: Higher Education Research Institute at University of California, Spring 2004).

2. The author conducted 125 in-depth interviews and did a year of field research during his research for *The First Year Out: Understanding American Teens after High School*, (Chicago: University of Chicago Press, forthcoming 2007). That project's panel interviews with teens were particularly insightful (i.e., two-wave interviews conducted just prior to a young person's high school graduation, then repeated 15 months later). The author has also interviewed 40 additional college students to date as part of his current "Life and Vocation of American Youth Project;" this project will eventually include more than 200 college students of varied religious backgrounds.

3. Christian Smith with Melinda Lundquist Denton, *Soul Searching: The Religious and Spiritual Lives of American Teenagers*, (New York: Oxford University Press, 2005), 124, 171.

4. See Smith 2005; Astin and Astin 2004; Christian Smith, Robert Faris, and Mark Regnerus, "Mapping American Adolescent Religious Participation," *Journal for the Scientific Study of Religion*, 41(4): 597–612 (December 2002).

5. Robert Wuthnow, *America and the Challenges of Religious Diversity* (Princeton, NJ: Princeton University Press, 2005); Christian Smith, *American Evangelicalism: Embattled and Thriving* (Chicago: University of Chicago Press, 1988).

KEY CONCEPTS

identity lockbox   religiosity   spirituality

DISCUSSION QUESTIONS

1. What evidence of college students' religious involvement and religious identification does Clydesdale present?

2. What does Clydesdale mean by an "identity lockbox," and how is this concept important to understanding the religious attitudes and identities of college students?

C. Education

48

From the Achievement Gap to the Education Debt: Understanding Achievement in U.S. Schools

GLORIA LADSON-BILLINGS

This article is a speech given in 2006 by the then president of the American Educational Research Association, Gloria Ladson-Billings. In her presidential address, she uses the analogy of the national debt and the national deficit to explain what has happened in American education. She argues that instead of focusing so much attention on the achievement gap between minority disadvantaged students and White privileged students, educational research and policy should focus on the education debt. The problem with the American system of education, according to Ladson-Billings, is that we are accumulating more debt in that all students are suffering from a poor system. The article calls for action, policy, and research that will help reverse this trend.

The questions that plague me about education research are not new ones. I am concerned about the meaning of our work for the larger public—for real students, teachers, administrators, parents, policymakers, and communities in real school settings. I know these are not new concerns; they have been raised by others, people like the late Kenneth B. Clark, who, in the 1950s, was one of the first social scientists to bring research to the public in a meaningful way. His

SOURCE: Ladson-Billings, Gloria. 2006. "From the Achievement Gap to the Education Dept: Understanding Achievement in U.S. Schools." *Educational Researcher* 35: 3–12.

work with his wife and colleague Mamie formed the basis for the landmark *Brown v. Board of Education* (1954) case that reversed legal segregation in public schools and other public accommodations. However, in his classic volume *Dark Ghetto: Dilemmas of Social Power,* first published in 1965, Clark took social scientists to task for their failure to fully engage and understand the plight of the poor:

> To my knowledge, there is at present nothing in the vast literature of social science treatises and textbooks and nothing in the practical and field training of graduate students in social science to prepare them for the realities and complexities of this type of involvement in a real, dynamic, turbulent, and at times seemingly chaotic community. And what is more, nothing anywhere in the training of social scientists, teachers, or social workers now prepares them to understand, to cope with, or to change the normal chaos of ghetto communities. These are grave lacks which must be remedied soon if these disciplines are to become *relevant* [emphasis added] to the stability and survival of our society. (p. xxix)

Clark's concern remains some 40 years later. However, the paradox is that education research has devoted a significant amount of its enterprise toward the investigation of poor, African American, Latina/o, American Indian, and Asian immigrant students, who represent an increasing number of the students in major metropolitan school districts. We seem to study them but rarely provide the kind of remedies that help them to solve their problems.

To be fair, education researchers must have the freedom to pursue basic research, just as their colleagues in other social sciences do. They must be able to ask questions and pursue inquiries "just because." However, because education is an applied field, a field that local states manage and declare must be available to the entire public, *most* of the questions that education researchers ask need to address the significant question that challenge and confound the public: Why don't children learn to read? What accounts for the high levels of school dropout among urban students? How can we explain the declining performance in mathematics and science at the same time that science and mathematics knowledge is exploding? Why do factors like race and class continue to be strong predictors of achievement when gender disparities have shrunk?

THE PREVALENCE OF THE ACHIEVEMENT GAP

One of the most common phrases in today's education literature is "the achievement gap." The term produces more than 11 million citations on Google. "Achievement gap," much like certain popular culture music stars, has become a crossover hit. It has made its way into common parlance and everyday usage. The term is invoked by people on both ends of the political spectrum, and few argue over its meaning or its import. According to the National Governors' Association, the achievement gap is "a matter of race and class. Across the U.S., a gap in academic achievement persists between minority and disadvantaged students and their white counterparts." It further states: "This is one of the most pressing education-policy challenges that states currently face" (2005). The story of the achievement gap is a familiar one. The numbers speak for themselves. In the 2005 National Assessment of Educational Progress results, the gap between Black and Latina/o fourth graders and their White counterparts in reading scaled scores was more than 26 points. In fourth-grade mathematics, the gap was more than 20 points (Education Commission of the States, 2005). In eighth-grade reading, the gap was more than 23 points, and in eighth-grade mathematics the gap was more than 26 points. We can also see that these gaps persist over time (Education Commission of the States).

Even when we compare African Americans and Latina/os with incomes comparabale to those of Whites, there is still an achievement gap as measured by standardized testing (National Center for Education Statistics, 2001). While I have focused primarily on showing this gap by means of standardized test scores, it also exists when we compare dropout rates and relative numbers of students who take advanced placement examinations; enroll in honors, advanced placement, and "gifted" classes; and are admitted to colleges and graduate and professional programs.

Scholars have offered a variety of explanations for the existence of the gap. In the 1960s, scholars identified cultural deficit theories to suggest that children of color were victims of pathological lifestyles that hindered their ability to benefit from schooling (Hess & Shipman, 1965; Bereiter & Engleman, 1966; Deutsch, 1963). The 1966 Coleman Report, *Equality of Educational Opportunity* (Coleman et al.), touted the importance of placing students in racially integrated classrooms. Some scholars took that report to further endorse the cultural deficit theories and to suggest that there was not much that could be done by schools to improve the achievement of African American children. But Coleman et al. were subtler than that. They argued that, more than material resources alone, a combination of factors was heavily correlated with academic achievement. Their work indicated that the composition of a school (who attends it), the students' sense of control of the environments and their futures, the teachers' verbal skills, and their students' family background all contribute to student achievement. Unfortunately, it was the last factor—family background—that became the primary point of interest for many school and social policies.

But I want to use this opportunity to call into question the wisdom of focusing on the achievement gap as a way of explaining and understanding the persistent inequality that exists (and has always existed) in our nation's schools. I want to argue that this all-out focus on the "Achievement Gap" moves us toward short-term solutions that are unlikely to address the long-term underlying problem.

NATIONAL DEBT VERSUS NATIONAL DEFICIT

Most people hear or read news of the economy every day and rarely give it a second thought. We hear that the Federal Reserve Bank is raising interest rates, or that the unemployment numbers look good. Our ears may perk up when we hear the latest gasoline prices or that we can get a good rate on a mortgage refinance loan. But busy professionals rarely have time to delve deeply into all things economic. Two economic terms—"national deficit" and "national debt"—seem to befuddle us. A deficit is the amount by which a government's, company's, or individual's spending exceeds income over a particular period of time. Thus, for each budget cycle, the government must determine whether it has a balanced budget, a budget surplus, or a deficit. The debt, however is the sum of all previously incurred annual federal deficits. Since the deficits are financed by government borrowing, national debt is equal to all government debt.

Most fiscal conservatives warn against deficit budgets and urge the government to decrease spending to balance the budget. Fiscal liberals do not necessarily embrace deficits but would rather see the budget balanced by increasing tax revenues from those most able to pay. The debt is a sum that has been accumulating since 1791, when the U.S. Treasury recorded it as $75,463,476.52 (Gordon, 1998).

But the debt has not merely been going up. Between 1823 and 1835 the debt steadily decreased, from a high of almost $91 million to a low of $ 33,733.05. The nation's debt hit the $1 billion mark in 1863 and the $1 trillion mark in 1981. Today, the national debt sits at more than $8 trillion. This level of debt means that the United States pays about $132,844,701,219.88 in interest each year. This makes our debt interest the third-largest expenditure in the federal budget after defense and combined entitlement programs such as Social Security and Medicare (Christensen, 2004).

Even in those years when the United States has had a balanced budget, that is, no deficits, the national debt continued to grow. It may have grown at a slower rate, but it did continue to grow. President Clinton bragged about presenting a balanced budget—one without deficits—and not growing the debt (King, J., 2000). However, the debt was already at a frighteningly high level, and his budget policies failed to make a dent in the debt.

THE DEBT AND EDUCATION DISPARITY

What does a discussion about national deficits and national debt have to do with education, education research, and continued education disparities? It is here where I began to see some metaphorical concurrences between our national fiscal situation and our education situation. I am arguing that our focus on the achievement gap is akin to a focus on the budget deficit, but what is actually happening to African American and Latina/o students is really more like the national debt. We do not have an achievement gap; we have an education debt.

... I have taken a somewhat different tack on this notion of the education debt. The yearly fluctuations in the achievement gap give us a short-range picture of how students perform on a particular set of achievement measures. Looking at the gap from year to year is a misleading exercise. Lee's (2002) look at the trend lines shows us that there was a narrowing of the gap in the 1980s both between Black and White students and between the Latina/o and White students, and a subsequent expansion of those gaps in the 1990s. The expansion of the disparities occurred even though the income differences narrowed during the 1990s. We do not have good answers as to why the gap narrows or widens. Some research suggests that even the combination of socioeconomic and family conditions, youth culture and student behaviors, and schooling conditions and practices do not fully explain changes in the achievement gap (Lee).

However, when we begin looking at the construction and compilation of what I have termed the education debt, we can better understand why an achievement gap is a logical outcome. I am arguing that the historical, economic, sociopolitical, and moral decisions and policies that characterize our society have created an education debt. So, at this point, I want to briefly describe each of those aspects of the debt.

THE HISTORICAL DEBT

Scholars in the history of education…, have documented the legacy of educational inequities in the United States. These inequities initially were formed around race, class, and gender. Gradually, some of the inequities began to recede, but clearly they persist in the realm of race. In the case of African Americans, education was initially forbidden during the period of enslavement. After emancipation we saw the development of freedmen's schools whose purpose was the maintenance of a servant class. During the long period of legal apartheid, African Americans attended schools where they received cast-off textbooks and materials from White schools. In the South, the need for farm labor meant that the typical school year for rural Black students was about 4 months long. Indeed, Black students in the South did not experience universal secondary schooling until 1968 (Anderson, 2002). Why, then, would we not expect there to be an achievement gap?

The history of American Indian education is equally egregious. It began with mission schools to convert and use Indian labor to further the cause of the church. Later, boarding schools were developed as General George Pratt asserted the need "to kill the Indian in order to save the man." This strategy of deliberate and forced assimilation created a group of people, according to Pulitzer Prize writer N. Scott Momaday, who belonged nowhere (Lesiak, 1991). The assimilated Indian could not fit comfortably into reservation life or the stratified mainstream. No predominately White colleges welcomed the few Indians who successfully completed the early boarding schools. Only historically Black colleges, such as Hampton Institute, opened their doors to them. There, the Indians studied vocational and trade curricula.

Latina/o students also experienced huge disparities in their education. In Ferg-Cadima's report *Black, White, and Brown: Latino School Desegregation Efforts in the Pre– and Post–Brown. v. Board of Education Era* (2004), we discover the longstanding practice of denial experienced by Latina/os dating back to 1848. Historic desegregation cases such as *Mendez v. Westminster* (1946) and the Lemon Grove Incident detail the ways that Brown children were (and continue to be) excluded from equitable and high-quality education.

It is important to point out that the historical debt was not merely imposed by ignorant masses that were xenophobic and virulently racist. The major leaders of the nation endorsed ideas about the inferiority of Black, Latina/o, and Native peoples. Thomas Jefferson (1816), who advocated for the education of the American citizen, simultaneously decried the notion that Blacks were capable of education. George Washington, while deeply conflicted about slavery, maintained a substantial number of slaves on his Mount Vernon Plantation and gave no thought to educating enslaved children.

A brief perusal of some of the history of public schooling in the United States documents the way that we have accumulated an education debt overtime. In 1827 Massachusetts passed a law making all grades of public school open to all pupils free of charge. At about the same time, most Southern states already had laws forbidding the teaching of enslaved Africans to read. By 1837, when Horace Mann had become head of the newly formed Massachusetts State Board of Education, Edmund Dwight, a wealthy Boston industrialist, felt that the state board was crucial to factory owners and offered to supplement the state salary with his own money. What is omitted from this history is that the major raw material of those textile factories, which drove the economy of the East, was cotton—the crop that depended primarily on the labor of enslaved Africans (Farrow, Lang, & Frank, 2005). Thus one of the ironies of the historical debt is that while African Americans were enslaved and prohibited from schooling, the product of their labor was used to profit Northern industrialists who already had the benefits of education.

This pattern of debt affected other groups as well. In 1864 the U.S. Congress made it illegal for Native Americans to be taught in their native languages. After the Civil War, African Americans worked with Republicans to rewrite state constitutions to guarantee free public education for all students. Unfortunately, their efforts benefited White children more than Black children. The landmark *Plessy v. Ferguson* (1896) decision meant that the segregation that the South had been practicing was officially recognized as legal by the federal government.

Although the historical debt is a heavy one, it is important not to overlook the ways that communities of color always have worked to educate themselves. Between 1865 and 1877, African Americans mobilized to bring public education to the South for the first time. Carter G. Woodson (1933/1972) was a primary critic of the kind of education that African Americans received, and he challenged African Americans to develop schools and curricula that met the unique needs of a population only a few generations out of chattel slavery.

THE ECONOMIC DEBT

As is often true social research, the numbers present a startling picture of reality. The economics of the education debt are sobering. The funding disparities that currently exist between schools serving White students and those serving students of color are not recent phenomena. Separate schooling allows for differential funding. In present-day dollars, the funding disparities between urban schools and their suburban counterparts present a telling story about the value we place on the education of different groups of students.

The Chicago public schools spend about $8,482 annually per pupil, while nearby Highland Park spends $17,291 per pupil. The Chicago public schools have an 87% Black and Latina/o population, while Highland Park has a 90% White population. Per pupil expenditures in Philadephia are $9,299 per pupil for the city's 79% Black and Latina/o population, while across City Line Avenue in Lower Merion, the per pupil expenditure is $17,261 for a 91% White population. The New York City public schools spend $11,627 per pupil for a student population that is 72% Black and Latina/o, while suburban Manhasset spends $22,311 for a student population that is 91% White (figures from Kozol, 2005).

One of the earliest things one learns in statistics is that correlation does not prove causation, but we must ask ourselves why the funding inequities map so neatly and regularly onto the racial and ethnic realities of our schools. Even if we cannot prove that schools are poorly funded *because* Black and Latina/o students attend them, we can demonstrate that the amount of funding rises with the rise in White students. This pattern of inequitable funding has occurred over centuries. For many of these populations, schooling was nonexistent during the early history of the nation; and, clearly, Whites were not prepared to invest their fiscal resources in these strange "others."

Another important part of the economic component of the education debt is the earning ratios related to years of schooling. The empirical data suggest that more schooling is associated with higher earnings; that is, high school graduates earn more money than high school dropouts, and college graduates earn more than high school graduates.

THE SOCIOPOLITICAL DEBT

The sociopolitical debt reflects the degree to which communities of color are excluded form the civic process. Black, Latina/o, and Native communities had little or no access to the franchise, so they had no true legislative representation. According to the Civil Rights Division of the U.S. Department of Justice, African Americans and other persons of color were substantially disenfranchised in many Southern states despite the enactment of the Fifteenth Amendment in 1870 (U.S. Department of Justice, Civil Rights Division, 2006).

The Voting Rights Act of 1965 is touted as the most successful piece of civil rights legislation ever adopted by the U.S. Congress (Grofman, Handley, & Niemi). This act represents a proactive attempt to eradicate the sociopolitical debt that had been accumulating since the founding of the nation.

Table 1 shows the sharp contrasts between voter registration rates before the Voting Rights Act of 1965 and after it. The dramatic changes in voter registration are a result of Congress's bold action. In upholding the constitutionality of the act, the Supreme Court ruled as follows:

> Congress has found that case-by-case litigation was inadequate to combat wide-spread and persistent discrimination in voting, because of the inordinate amount of time and energy required to overcome the

obstructionist tactics invariably encountered in these lawsuits. After enduring nearly a century of systematic resistance to the Fifteenth Amendment, Congress might well decide to shift the advantage of time and inertia from the perpetrators of the evil to its victims. (*South Carolina v. Katzenbach*, 1966; U.S. Department of Justice, Civil Rights Division, 2006)

It is hard to imagine such a similarly drastic action on behalf of African American, Latina/o, and Native American children in schools. For example, imagine that an examination of the achievement performance of children of color provoked an immediate reassignment of the nation's best teachers to the schools serving the most needy students. Imagine that those same students were guaranteed places in state and regional colleges and universities. Imagine that within one generation we lift those students out of poverty.

The closest example that we have of such a dramatic policy move is that of affirmative action. Rather that wait for students of color to meet predetermined standards, the society decided to recognize that historically denied groups should be given a preference in admission to schools and colleges. Ultimately, the major beneficiaries of this policy were White women. However, Bowen and Bok (1999) found that in the case of African Americans this proactive policy helped create what we now know as the Black middle class.

As a result of the sociopolitical component of the education debt, families of color have regularly been excluded from the decision-making mechanisms that should ensure that their children receive quality education. The parent–teacher organizations, school site councils, and other possibilities for democratic participation have not been available for many of these families. However, for a brief moment in 1968, Black parents in the Ocean Hill–Brownsville section of New York exercised community control over the public schools (Podair, 2003). African American, Latina/o, Native American, and Asian American parents have often advocated for improvements in schooling, but their advocacy often has been muted and marginalized. This quest for control of schools was powerfully captured in the voice of an African American mother during the fight for school desegregation in Boston. She declared: "When we fight about schools, we're fighting for our lives" (Hampton, 1986).

Indeed, a major aspect of the modern civil rights movement was the quest for quality schooling. From the activism of Benjamin Rushing in 1849 to the struggles of parents in rural South Carolina in 1999, families of color have been fighting for quality education for their children (Ladson-Billings, 2004). Their more limited access to lawyers and legislators has kept them from accumulating the kinds of political capital that their White, middle-class counterparts have.

THE MORAL DEBT

A final component of the education debt is what I term the "moral debt." I find this concept difficult to explain because social science rarely talks in these terms.

TABLE 1   **Black and White Voter Registration Rates (%) in Selected U.S. States, 1965 and 1988**

State	March 1965			November 1988		
	Black	White	Gap	Black	White	Gap
Alabama	19.3	69.2	49.9	68.4	75.0	6.6
Georgia	27.4	62.6	35.2	56.8	63.9	7.1
Louisiana	31.6	80.5	48.9	77.1	75.1	−2.0
Mississippi	6.7	69.9	63.2	74.2	80.5	6.3
North Carolina	46.8	96.8	50.0	58.2	65.6	7.4
South Carolina	37.3	75.7	38.4	56.7	61.8	5.1
Virginia	38.3	61.1	22.8	63.8	68.5	4.7

Note: From the website of the U.S. Department of Justice, Civil Rights Division, Voting Rights Section (*http://www.usdoj.gov/crt/voting/intro/intro_c.html*), "introduction to Federal Voting Rights Laws."

What I did find in the literature was the concept of "moral panics" (Cohen, 1972; Goode & Ben-Yehuda, 1994a, 1994b; Hall, Critcher, Jefferson, Clarke, & Roberts, 1978) that was popularized in British sociology. People in moral panics attempt to describe other people, groups of individuals, or events that become defined as threats throughout a society. However, in such a panic the magnitude of the supposed threat overshadows the real threat posed.

... [A] moral debt reflects the disparity between what we know is right and what we actually do. Saint Thomas Aquinas saw the moral debt as what human beings owe to each other in the giving of, or failure to give, honor to another when honor is due. This honor comes as a result of people's excellence or because of what they have done for another. We have no trouble recognizing that we have a moral debt to Rosa Parks, Martin Luther King, Cesar Chavez, Elie Wiesel, or Mahatma Gandhi. But how do we recognize the moral debt that we owe to entire groups of people? How do we calculate such a debt?

... What is it that we might owe to citizens who historically have been excluded from social benefits and opportunities? Randall Robinson (2000) states:

> No nation can enslave a race of people for hundreds of years, set them free bedraggled and penniless, pit them, without assistance in a hostile environment, against privileged victimizers, and then reasonably expect the gap between the heits of the two groups to narrow. Lines, begun parallel and left alone, can never touch up. (p. 74)

Robinson's sentiments were not unlike those of President Lyndon B. Johnson, who stated in a 1965 address at Howard University: "You cannot take a man who has been in chains for 300 years, remove the chains, take him to the starting line and tell him to run the race, and think that you are being fair" (Miller, 2005).

Taken together, the historic, economic, sociopolitical, and moral debt that we have amassed toward Black, Brown, Yellow, and Red children seems insurmountable, and attempts at addressing it seem futile. Indeed, it appears like a task for Sisyphus. But as legal scholar Derrick Bell (1994) indicated, just because something is impossible does not mean it is not worth doing.

WHY WE MUST ADDRESS THE DEBT

On the face of it, we must address it because it is the equitable and just thing to do. As Americans we pride ourselves on maintaining those ideal qualities as hallmarks of our democracy. That represents the highest motivation for paying this debt. But we do not always work from our highest motivations.

Most of us live in the world of the pragmatic and practical. So we must address the education debt because it has implications for the kinds of lives we can live and the kind of education the society can expect for most of its children. I want to suggest that there are three primary reasons for addressing the debt— (a) the impact the debt has on present education progress, (b) the value of understanding the debt in relation to past education research findings, and (c) the potential for forging a better educational future.

The Impact of the Debt on Present Education Progress

As I was attempting to make sense of the deficit/debt metaphor, educational economist Doug Harris (personal communication, November 19, 2005) reminded me that when nations operate with a large debt, some part of their current budget goes to service that debt. I mentioned earlier that interest payments on our national debt represent the third largest expenditure of our national budget. In the case of education, each effort we make toward improving education is counterbalanced by the ongoing and mounting debt that we have accumulated. That debt service manifests itself in the distrust and suspicion about what schools can and will do in communities serving the poor and children of color. Bryk and Schneider (2002) identified "relational trust" as a key component in school reform. I argue that the magnitude of the education debt erodes that trust and represents a portion of the debt service that teachers and administrators pay each year against what they might rightfully invest in helping students advance academically.

The Value of Understanding the Debt in Relation to Past Research Findings

The second reason that we must address the debt is somewhat selfish from an education research perspective. Much of our scholarly effort has gone into looking at educational inequality and how we might mitigate it. Despite how hard we try, there are two interventions that have never received full and sustained hypothesis testing—school desegregation and funding equity. Orfield and Lee (2006) point out that not only has school segregation persisted, but it has been transformed by the changing demographics of the nation. They also point out that "there has not been a serious discussion of the costs of segregation or the advantages of integration for our most segregated population, white students" (p. 5). So, although we may have recently celebrated the 50th anniversary of the *Brown* decision, we can point to little evidence that we really gave *Brown* a chance. According to Frankenberg, Lee, and Orfield (2003) and Orfield and Lee (2004), America's public schools are more than a decade into a process of resegregation. Almost three-fourths of Black and Latina/o students attend schools that are predominately non-White. More than 2 million Black and Latina/o students— a quarter of the Black students in the Northeast and Midwest—attend what the researchers call apartheid schools. The four most segregated states for Black students are New York, Michigan, Illinois, and California.

The funding equity problem, ... also has been intractable. In its report entitled *The Funding Gap 2005,* the Education Trust tells us that "in 27 of the 49 states studied, the highest-poverty school districts receive fewer resources than the lowest-poverty districts.... Even more states shortchange their highest minority districts. In 30 states, high minority districts receive less money for each child than low minority districts" (p. 2). If we are unwilling to desegregate our schools *and* unwilling to fund them equitably, we find ourselves not only backing away from the promise of the *Brown* decision but literally refusing even to take *Plessy* seriously. At least a serious consideration of *Plessy* would make us look at funding inequities.

In one of the most graphic examples of funding inequity, new teacher Sara Sentilles (2005) described the southern California school where she was teaching:

> At Garvey Elementary School, I taught over thirty second graders in a so-called temporary building. Most of these "temporary" buildings have been on campuses in Compton for years. The one I taught in was old. Because the wooden beams across the ceiling were being eaten by termites, a fine layer of wood dust covered the students' desks every morning. Maggots crawled in a cracked and collapsing area of the floor near my desk. One day after school I went to sit in my chair, and it was completely covered in maggots. I was nearly sick. Mice raced behind cupboards and bookcases. I trapped six in terrible traps called "glue lounges" given to me by the custodians. The blue metal window coverings on the outsides of the windows were shut permanently, blocking all sunlight. Someone had lost the tool needed to open them, and no one could find another.... (p. 72)

Rothstein and Wilder (2005) move beyond the documentation of the inequalities and inadequacies to their *consequences*. In the language that I am using in this discussion, they move from focusing on the gap to tallying the debt. Although they focus on Black–White disparities, they are clear that similar disparities exist between Latina/os and Whites and Native Americans and Whites. Contrary to conventional wisdom, Rothstein and Wilder argue that addressing the achievement gap is not the most important inequality to attend to. Rather, they contend that inequalities in health, early childhood experiences, out-of-school experiences, and economic security are also contributory and cumulative and make it near-impossible for us to reify the achievement gap as *the* source and cause of social inequality.

The Potential for Forging a Better Educational Future

Finally, we need to address what implications this mounting debt has for our future. In one scenario, we might determine that our debt is so high that the only thing we can do is declare bankruptcy. Perhaps, like our airline industry, we could use the protection of the bankruptcy laws to reorganize and design more streamlined, more efficient schooling options. Or perhaps we could be like developing nations that owe huge sums to the IMF and apply for 100% debt relief. But what would such a catastrophic collapse of our education system look like? Where could we go to begin from the ground up to build the kind of education system that would aggressively address the debt? Might we find a setting where a catastrophic occurrence, perhaps a natural disaster—a hurricane—has completely obliterated the schools? Of course, it would need to be a place where the schools weren't very good to begin with. It would have to be a place where our Institutional Review Board and human subject concerns would not keep us from proposing aggressive and cutting-edge research. It would have to be a place where people were so desperate for the expertise of education researchers that we could conduct multiple projects using multiple approaches. It would be a place so hungry for solutions that it would not matter if some projects were quantitative and others were qualitative. It would not matter if some were large-scale and some were small-scale. It would not matter if some paradigms were psychological, some were social, some were economic, and some were cultural. The only thing that would matter in an environment like this would be that education researchers were bringing their expertise to bear on education problems that spoke to pressing concerns of the public. I wonder where we might find such a place?

REFERENCES

Anderson, J. D. (2002, February 28). *Historical perspectives on Black academic achievement.* Paper presented for the Visiting Minority Scholars Series Lecture. Wisconsin Center for Educational Research, University of Wisconsin, Madison.

Apple, M., & Butas, K. (Eds.). (2006). *The subaltern speak: Curriculum, power and education struggles.* New York: Routledge.

Bell, D. (1994). *Confronting authority: Reflections of an ardent protester.* Boston: Beacon Press.

Bereiter, C., & Engleman, S. (1966). *Teaching disadvantaged children in preschool.* Englewood Cliffs, NJ: Prentice Hall.

Bowen, W., & Bok, D. (1999). *The shape of the river.* Princeton, NJ: Princeton University Press.

Brown v. Board of Education 347 U.S. 483(1954).

Bryk, A., & Schneider, S. (2002). *Trust in schools: A core resource for improvement.* New York: Russell Sage Foundation.

Christensen, J. R. (Ed.). (2004). *The national debt: A primer.* Haupauge, NY: Nova Science Publishers.

Clark, K. B. (1965). *Dark ghetto: Dilemmas of social power.* Hanover, NH: Wesleyan University Press.

Cohen, S. (1972). *Folk devils and moral panics: The creation of mods and rockers.* London: McGibbon and Kee.

Coleman, J., Campbell, E., Hobson, C., McPartland, J., Mood, A., Weinfeld, F. D., et al. (1966). *Equality of educational opportunity.* Washington, DC: Department of Health, Education and Welfare.

Deutsch, M. (1963). The disadvantaged child and the learning process. In A. H. Passow (Ed.), *Education in depressed areas* (pp. 163–179). New York: New York Bureau of Publications, Teachers College, Columbia University.

Education Commission of the States. (2005). *The nation's report card.* Retrieved January 2, 2006, from *http://nces.ed.gov/nationsreportcard*

Education Trust. (2005). *The funding gap 2005.* Washington, DC: Author.

Farrow, A., Lang, J., & Frank, J. (2005). *Complicity: How the North promoted, prolonged and profited from slavery.* New York: Ballantine Books.

Ferg-Cadima, J. (2004, May). *Black, White, and Brown: Latino school desegregation efforts in the pre– and post–Brown v. Board of Education era.* Washington, DC: Mexican-American Legal Defense and Education Fund.

Frankenberg, E., Lee, C., & Orfield, G. (2003, January). *A multiracial society with segregated schools: Are we losing the dream?* Cambridge, MA: The Civil Rights Project, Harvard University.

Goode, E., & Ben-Yehuda, N. (1994a). Moral panics: Culture, politics, and social construction. *Annual Review of Sociology, 20,* 149–171.

Goode, E., & Ben-Yehuda, N. (1994b). *Moral panics: The social construction of deviance.* Oxford: Blackwell.

Gordon, J. S. (1998). *Hamilton's blessing: The extraordinary life and times of our national debt.* New York: Penguin Books.

Grofman, B., Handley, L., & Niemi, R.G. (1992). *Minority representation and the quest for voting equality.* New York: Cambridge University Press.

Hall, S., Critcher, C., Jefferson, T., Clarke, J., & Roberts, B. (1978). *Policing the crisis: Mugging, the state, and law and order.* London: Macmillan.

Hampton, H. (Director). (1986). *Eyes on the prize* [Television video series]. Blackside Productions (Producer). New York: Public Broadcasting Service.

Hess, R. D., & Shipman, V. C. (1965). Early experience and socialization of cognitive modes in children. *Child Development, 36,* 869–886.

Jefferson, T. (1816, July 21). *Letter to William Plumer. The Thomas Jefferson Paper Series. 1. General correspondence, 1651–1827.* Retrieved September 11, 2006, from *http://rs6. loc.gov/cgi-bin/ampage*

King, J. (2000, May 1). *Clinton announces record payment on national debt.* Retrieved February 7, 2006, from *http://archives.cnn.com/2000/ALLPOLITICS/stories/05/01/ Clinton.debt*

Kozol, J. (2005). *The shame of the nation: The restoration of apartheid schooling in America.* New York: Crown Publishing.

Ladson-Billings, G. (2004). Landing on the wrong note: The price we paid for *Brown. Educational Researcher,* 33(7), 3–13.

Lee, J. (2002). Racial and achievement gap trends: Reversing the progress toward equity. *Educational Researcher,* 31(1), 3–12.

Lesiak, C. (Director). (1992). *In the White man's image* [Television broadcast]. New York: Public Broadcasting Corporation.

Mendez v. Westminster 64F. Supp. 544 (1946).

Miller, J. (2005, September 22). New Orleans unmasks apartheid American style [Electronic version]. *Black Commentator, 151.* Retrieved September 11, 2006, from *http:// www.blackcommentator.com/151/151_miller_new_orleans.html*

National Center for Education Statistics. (2001). *Education achievement and Black–White inequality.* Washington, DC: Department of Education.

National Governors' Association. (2005). *Closing the achievement gap.* Retrieved October 27, 2005, from *http://www.subnet.nga.org/educlear/achievment/*

Orfield. G., & Lee, C. (2004, January). *Brown at 50: King's dream or Plessy's nightmare?* Cambridge, MA: The Civil Rights Project, Harvard University.

Orfield. G., & Lee, C. (2006, January). *Racial transformation and the changing nature of segregation.* Cambridge, MA: The Civil Rights Project, Harvard University.

Plessy v. Ferguson 163 U.S. 537 (1896).

Robinson, R. (2000). *The debt: What America owes to Blacks.* New York: Dutton Books.

Rothstein, R., & Wilder, T. (2005, October 24). *The many dimensions of racial inequality.* Paper presented at the Social Costs of Inadequate Education Symposium, Teachers College, Columbia University, New York.

Sentilles, S. (2005). *Taught by America: A story of struggle and hope in Compton.* Boston: Beacon Press.

South Carolina v. Katzenbach 383 U.S. 301, 327–328 (1966).

U.S. Department of Justice, Civil Rights Division. (2006, September 7). *Introduction to federal voting rights laws.* Retrieved September 11, 2006, from *http://www.usdoj.gov/ crt/voting/intro/intro.htm*

Woodson, C. G. (1972). *The mis-education of the Negro.* Trenton, NJ: Africa World Press. (Original work published 1933)

KEY CONCEPTS

achievement gap   education debt

DISCUSSION QUESTIONS

1. What did this article tell us about the achievement gap in education? What has typically been the pattern between disadvantaged students and privileged students with regard to school performance?

2. Why is the achievement gap bad for everyone? In what ways does inequality in education hurt America as a whole?

49

Historic Reversals, Accelerating Resegregation, and the Need for New Integration Strategies

GARY ORFIELD AND CHUNGMEI LEE

Gary Orfield and Chungmei Lee present important data on how the nation's schools are undergoing increasing racial and class segregation. They argue that this reflects an abandonment of social policies and laws that, in the aftermath of the Brown v. Board of Education *decision in 1954 and the civil rights movement, had led to a decline in racial segregation.*

American schools, resegregating gradually for almost two decades, are now experiencing accelerating isolation, and this will doubtless be intensified by the recent decision of the U.S. Supreme Court. This June, the Supreme Court handed down its first major decision on school desegregation in 12 years in the Louisville and Seattle cases.[1] A majority of a divided Court told the nation both that the goal of integrated schools remained of compelling importance but that most of the means now used voluntarily by school districts are unconstitutional. As a result, most voluntary desegregation actions by school districts must now be changed or abandoned. As educational leaders and citizens across the country try to learn what they can do, and decide what they will do, we need to know how the nation's schools are changing, what the underlying trends are in the segregation of American students, and what the options are they might consider.

SOURCE: Civil Rights Project/*Proyecto Derechos Civiles*, University of California, Los Angeles August 2007.

The trends are those of increasing isolation and profound inequality. The consequences become larger each year because of the growing number and percentage of nonwhite and impoverished students and the dramatic relationships between educational attainment and economic success in a globalized economy. Almost nine-tenths of American students were counted as white in the early 1960s, but the number of white students fell 20 percent from 1968 to 2005, as the baby boom gave away to the baby bust for white families, while the number of blacks increased 33 percent and the number of Latinos soared 380 percent amid surging immigration of a young population with high birth rates. The country's rapidly growing population of Latino and black students is more segregated than they have been since the 1960s, and we are going backward faster in the areas where integration was most far-reaching.

Compared to the civil rights era, we have a far larger population of "minority" children and a major decline in the number of white students. Latino students, who are the least successful in higher education attainment, have become the largest minority population. We are in the last decade of a white majority in American public schools, and there are already minorities of white students in our two largest regions, the South and the West. When today's children become adults, we will be a multiracial society with no majority group, where all groups will have to learn to live and work successfully together. School desegregation has been the only major policy directly addressing this need, and that effort has now been radically constrained.

The schools are not only becoming less white but also have a rising proportion of poor children. The percentage of school children poor enough to receive subsidized lunches has grown dramatically. This is not because white middle class students have produced a surge in private school enrollment; private schools serve a smaller share of students than a half century ago and are less white. The reality is that the next generation is much less white because of the aging and small family sizes of white families, and the trend is deeply affected by immigration from Latin American and Asia. Huge numbers of children are growing up in families with very limited resources, and face an economy with deepening inequality of income distribution, where only those with higher education are securely in the middle class. It is a simple statement of fact to say that the country's future depends on finding ways to prepare groups of students who have traditionally fared badly in American schools to perform at much higher levels and to prepare all young Americans to live and work in a society vastly more diverse than ever in our past. Some of our largest states will face a decline in average educational levels in the near future as the racial transformation proceeds if the educational success of nonwhite students does not improve substantially.

From the "excellence" reforms of the Reagan era and the Goals 2000 project of the Clinton Administration to the No Child Left Behind Act of 2001, we have been trying to focus pressure and resources on making the achievement of minority children in segregated schools equal. The record to date justifies deep skepticism. On average, segregated minority schools are inferior in terms of the quality of their teachers, the character of the curriculum, the level of competition, average test scores, and graduation rates. This does not mean that desegregation solves all problems or that it always works, or that segregated schools do not perform well in rare circumstances, but it does mean that desegregation normally connects minority students with schools which have many potential advantages over segregated ghetto and barrio schools, especially if the children are not segregated at the classroom level.

Desegregation is often treated as if it were something that occurred after the *Brown* decision in the 1950s. In fact, serious desegregation of the black South only came after Congress and the Johnson Administration acted powerfully under the 1964 Civil Rights Act; serious desegregation of the cities only occurred in the 1970s and was limited outside the South. Though the Supreme Court recognized the rights of Latinos to desegregation remedies in 1973, there was little enforcement as the Latino numbers multiplied rapidly and their segregation intensified.

Resegregation, which took hold in the early 1990s after three Supreme Court decisions from 1991 to 1995 limiting desegregation orders, is continuing to grow in all parts of the country for both African Americans and Latinos and is accelerating the most rapidly in the only region that had been highly desegregated—the South. The children in United States schools are much poorer than they were decades ago and more separated in highly unequal schools. Black and Latino segregation is usually double segregation, both from whites and from middle class students. For blacks, more than a third of a century of progress in racial integration has been lost—though the seventeen states which had segregation laws are still far less segregated than in the 1950s when state laws enforced apartheid in the schools and the massive resistance of Southern political leaders delayed the impact of *Brown* for a decade. For Latinos, whose segregation in many areas is now far more severe than when it was first measured nearly four decades ago, there never was progress outside of a few areas and things have been getting steadily worse since the 1960's on a national scale. Too often Latino students face triple segregation by race, class, and language. Many of these segregated black and Latino schools have now been sanctioned for not meeting the requirements of No Child Left Behind, and segregated high poverty schools account for most of the "dropout factories" at the center of the nation's dropout crisis.

One would assume that a nation which now has more than 43 percent nonwhite students, but where judicial decisions are dissolving desegregation orders and fostering increasing racial and economic isolation, must have discovered some way to make segregated schools equal since the future of the country will depend on the education of its surging nonwhite enrollment which already accounts for more than two students of every five. You would suppose that it must have identified some way to prepare students in segregated schools to live and work effectively in multiracial neighborhoods and workplaces since experience in many racially and ethnically divided societies show that deep social cleavages, especially subordination of the new majority, could threaten society and its basic institutions. Those assumptions would be wrong. The basic judicial policies are to terminate existing court orders, to forbid most race-conscious desegregation efforts without court orders, and to reject the claim that there is a right to equal resources for the segregated schools. Not only do the federal courts not require either integration or equalization of segregated schools, but this means that they forbid state and local officials to implement most policies that have proven effective in desegregating schools. State and local politics will determine what, if anything, happens in terms of equalizing resources between segregated schools and privileged schools.

The basic educational policy model in the post-civil rights generation assumes that we can equalize schools without dealing with segregation through testing and accountability. It is nearly a quarter century since the country responded to the Reagan Administration's 1983 report, "A Nation at Risk," warning of dangerous shortcomings in American schools and demanding that "excellence" policies replace the "equity" policies of the 1960s. Since then almost every state has adopted the recommendations for the more demanding tests and accountability and more required science and math classes the report recommended. Congress and the last three Presidents have established national goals for upgrading and equalizing education. The best evidence indicates that these efforts have failed, both the Goals 2000 promise of equalizing education for nonwhite students by 2000 and the NCLB promise of closing the achievement gap with mandated minimum yearly gains so that everyone would be proficient by 2013. In fact, the previous progress in narrowing racial achievement gaps from the 1960s well into the 1980s has ended, and most studies find that there has been no impact from NCLB on the racial achievement gap. These reforms have been dramatically less effective in that respect than the reforms of the 1960s and '70s, including desegregation and anti-poverty programs. On some measures the racial achievement gaps reached their low point around the same time as the peak of black-white desegregation in the late 1980s.

Although the U.S. has some of the best public schools in the world, it also has too many far weaker than those found in other advanced countries. Most of these are segregated schools which cannot get and hold highly qualified teachers and administrators, do not offer good preparation for college, and often fail to graduate even half of their students. Although we have tried many reforms, often in confusing succession, public debate has largely ignored the fact that racial and ethnic separation continues to be strikingly related to these inequalities. As the U.S. enters its last years in which it will have a majority of white students, it is betting its future on segregation. The data coming out of the No Child Left Behind tests and the state accountability systems show clear relationships between segregation and educational outcomes, but this fact is rarely mentioned by policy makers.

The fact of resegregation does not mean that desegregation failed and was rejected by Americans who experienced it. Of course the demographic changes made full desegregation with whites more difficult, but the major factor, particularly in the South, was that we stopped trying. Five of the last seven Presidents actively opposed urban desegregation, and the last significant federal aid for desegregation was repealed 26 years ago in 1981.

The country risks becoming a nation where most of the new nonwhite majority of young people will be attending separate and inferior schools, and educators will be forbidden to take any direct action likely to bring down the color line. The experience in districts which have already been forbidden to carry out voluntary programs suggests that segregation may rapidly intensify. Obviously educators still face many choices that will be related to the intensity and degree of this resegregation, but there is no simple alternative.

One of the deepest ironies of this period is that never before has there been more evidence about the inequalities inherent in segregated education, the potential benefits for both nonwhite and white students, and the ways in which those benefits could be maximized.

Nearly 40 years after the assassination of Dr. Martin Luther King, Jr., we have now lost almost all the progress made in the decades after his death in desegregating our schools. It was very hard won progress that produced many successes and enabled millions of children, particularly in the South, to grow up in more integrated schools. Though it was often imperfectly implemented and sometimes poorly designed, school integration was, on average, a successful policy, linked to a period of social mobility and declining gaps in achievement and school completion and improved attitudes and understanding among the races. The experience under No Child Left Behind and similar high stakes testing and accountability policies that ignore segregation has been deeply disappointing, and the evidence from those tests show the continuing inequality of segregated schools even after many years of fierce pressure and sanctions on those schools and students.

It is time to think very seriously about the central proposition of the *Brown* decision, that segregated education is "inherently unequal," and think about how we can begin to regain the ground that has been lost. The pioneers whose decades of investigations and communication about the conditions of racial inequality helped make the civil rights revolution possible a half century ago should not be honored merely by naming schools and streets or even holidays after them but should be remembered as a model of the work that must be done, as many times as necessary, for as long as it takes, to return to the promise of truly equal justice under law in our schools, to insist that we have the kind of schools that can build and sustain a successful profoundly multiracial society.

NOTE

1. *Parents Involved In Community Schools V. Seattle School District No. 1 Et Al.* June 28, 2007.

KEY CONCEPTS

Brown v. Board of Education	de facto segregation	resegregation
de jure segregation	desegregation	

DISCUSSION QUESTIONS

1. What trends in school enrollment do Orfield and Lee identify, and how are these related to race? To social class and poverty?

2. Why have attempts at educational reform failed, according to Orfield and Lee, and what policy changes are suggested by their research on the re-segregation of the schools?

50

More Than "A is for Alligator"

How to Ensure Early Childhood Systems Help Break the Cycle of Poverty

LISA GUERNSEY

Guernsey begins this article with a description of two different types of pre-K classrooms for four-year-olds. The better approach to teaching young children and preparing them for school involves interaction and a variety of teaching tools in the classroom. She then outlines the Head Start program and other new initiatives to help children prepare for school. Her summary points to the need for more highly qualified teachers, more money for salaries and materials, and the link between school and child care. Educational research highlights the preschool years as important for establishing good students. This article makes that point clearly and argues that preschool should not be simply daycare for our youngest learners.

Early childhood programs have become Exhibit A in conventional accounts of how to eradicate inequality and poverty. Advocates for early childhood programs, most notably Head Start, routinely argue that such programs help children enter school ready to learn, increasing their likelihood of academic success and reducing the chances that they remain poor in adulthood.

But today's classroom realities make this difficult to realize. If disadvantaged kids are going to achieve in school and life, classrooms must be more than play spaces staffed with babysitters. Nor should children be subjected to sit-in-your-seats,

SOURCE: Guernsey, Lisa. 2010. "More Than 'A' is for Alligator: How to Ensure Early Childhood Systems Help Break the Cycle of Poverty." *Pathways* (Spring): 28–32.

miniaturized versions of school. Instead, the programs they attend must be high quality and developmentally appropriate—maddeningly difficult characteristics to define, let alone achieve.

Let me set the stage by comparing two hypothetical classrooms for four-year-olds.

Enter Classroom One. The teacher starts with the standard "circle time" in which the children gather in a circle on the rug. She reads the children a picture book about alligators, then dismisses them to tables where they receive photo-copied sheets showing an alligator next to the letter A. While the kids select their crayons, she asks them to repeat after her: "A, ah, alligator. A, ah, alligator." They answer back and begin coloring as if on autopilot: scribble, grab a new crayon, scribble, repeat.

Now enter Classroom Two. The teacher reads a book about alligators, takes a brief moment to point to the word "alligator," and notes that it starts with "A." She then asks the children what they know about alligators. One child mentions their sharp teeth, and the teacher probes, "Why do you think they have such sharp teeth?"

One child answers, "To eat!"

"Ah," the teacher says with a twinkle in her eye, "What do they eat, any-way? Spaghetti?"

"No!" the kids scream back.

After introducing the word "predator," the teacher passes around photographs of alligators and their prey. She asks the kids to stand up and stretch their arms out, raising one high and one low, then snapping them together. The kids giggle as they pretend to chomp one another. Later, they measure whether an alligator is big enough to cover their circle-time rug. As they unravel a piece of string cut to an alligator's average length, the children exclaim, "Alligators are huge!"

Even without the benefit of decades of developmental research, the reader can spot the advantages of Classroom Two. In this classroom, the teacher is able to move beyond the simple didactic lesson that "A" stands for alligator. By engag-ing the students in fun, developmentally appropriate activities and discussions, she is able to get the students not just thinking about the letter "A," but also about such abstract concepts as size, about the meaning of the word "predator," and maybe even a bit about the concept of ecosystem. But this lesson is not just better on its face. Reading research, for example, shows that children will have a much easier time learning to read and, more importantly, *comprehending* what they read, when they already have a base of vocabulary and content knowledge to lean on. It's pretty hard for an elementary school student to understand a passage about preda-tors if he has never even heard the word "predator" before and doesn't know what one is. Years of cognitive science show the importance of giving a child early and repeated interactions with words and concepts, enabling them to become part of a child's long-term memory so that the brain can easily call upon those memories when introduced to something new.

Sadly, Classroom Two is not the norm in today's early childhood programs for disadvantaged children. This is true whether children are in Head Start pro-grams, state-funded pre-K, subsidized child care, or parent-funded preschool.

Studies of programs around the country have shown that while teachers typically provide a warm and emotionally supportive climate, the quality of what they teach—and how it's taught—is mediocre at best.

We must address this disconnect between our high expectations for early childhood programs and the reality of what children are experiencing if we want to help poor children escape poverty. It's time for a change in mindset. For years, children have been treated to a social services model that emphasizes health, safety, socialization, and nutrition. The end result: Safe and nutritious holding tanks. This is obviously not good enough. Early childhood classrooms need to have the look and feel of the alligator lesson provided by the teacher in Classroom Two, with interactions that help develop children's language, cognitive and social skills. Although these programs should, of course, remain tightly coordinated with social services, our expectations can't end there. After all, if early education programs are going to enable poor children to compete with more affluent children, they must do *more*, not less, to level the playing field. A true anti-poverty system of education must start as soon as women are pregnant and continue until children are reading proficiently and are armed with the skills needed to learn on their own.

A progressive and proactive early education system for disadvantaged children should be built around two essential principles: 1) the use of pedagogy that promotes cognitive development, expanding children's use of language and providing a solid base of content knowledge, and 2) a seamless continuity of services—starting at birth and extending through the third grade—that buttresses learning and development.

This will take money. But some new investments are on the horizon. Despite the recession, most states with pre-K programs have so far avoided devastating cuts. A recent report from the advocacy group Pre-K Now showed that pre-K funding ticked up by 1 percent in the 2010 fiscal year. A one-time infusion of funding from the stimulus bill is now making its way to Head Start and child care centers, with enough funding to bring 55,000 additional families into Early Head Start, a program for babies, toddlers, and their mothers.

It would be a mistake to assume that more money and attention are magic bullets that make early education work for low-income youth. It is not as if children's readiness for school—and therefore their chances at academic and career success—will get an automatic lift once the fairy dust of more federal and state funding is sprinkled across the existing system. "This isn't just about keeping an eye on our children," President Obama said in a major education speech in March 2009. "It's about educating them." In asking how we might do just that, lets recap how we wound up with the present early childhood system.

THE EXISTING LANDSCAPE

The federal government started focusing on early childhood programs for poor children during President Johnson's War on Poverty, launching Head Start in 1965. The program provides free preschool to children in families at or below the federal poverty line. Approximately 920,000 three- and four-year-olds attend, and waiting lists are common in many cities. But while child advocates have always applauded Head Start, until recently there's been little proof that Head Start children make larger gains in their social and cognitive development than those who do not attend. So in 1998, Congress authorized a study comparing Head Start children with those who, though they were qualified, did not get into the program. The study analyzed how children are doing one year after Head Start as well as after kindergarten and first grade. It found that the Head Start children were more prepared for kindergarten than the control group, scoring higher on some, though not all, indicators of cognitive and social-emotional development. But it also found that by the end of first grade, there was little difference between the two groups.

These results have given pause to some policymakers who want more evidence that taxpayer dollars are being put to good use. Even the modest gains in kindergarten readiness have provided ammunition to some who believe that the government shouldn't be spending money on early learning experiences that they believe families should provide on their own....

Meanwhile, from the early- to mid-2000s, states launched their own sets of programs for preschoolers, most of which focused on getting them ready for school. Thirty-eight states now have what is called "state-funded pre-K" that provides a free half or full day of instruction in public schools or community-based centers. These programs vary greatly, but many serve families with incomes significantly higher than the poverty threshold, and some are available to every child, regardless of family income. Today, state pre-K programs serve more than 1.1 million children, according to the National Institute for Early Education Research.

The combination of Head Start, state-funded pre-K, and other subsidized child care centers has led to a system characterized by a hodgepodge of disconnected services. And the system is still far from being universal. Only about four-fifths of four-year-olds are in some kind of regular child care arrangement, according to the Census Bureau, and of those, it's unclear how many offer much more than babysitting. It has only been over the past few years that leaders of state pre-K and Head Start programs started to seriously consider integrating their services. Recently, advocates of child care subsidies have voiced a call for better coordination and quality of child care services as well. High-quality child care can become an important element of early education by providing wraparound services helping parents whose jobs do not allow them to pick up children at 3 or 4 p.m., when many full-day pre-K programs end.

THE WAY FORWARD

Even if Head Start, state-funded pre-K, and child care services were better connected, there is obviously no guarantee that they would provide anything like the experiences offered by Classroom Two. To the chagrin of many child development experts, children seem to be more likely to receive something like the thinner learning experience offered by Classroom One. Pre-literacy instruction in preschool is important, but introducing children to letters and print is only one component of preparing children to read. We need a system based on the principles of cognitive development and seamless integration. If these two reforms were taken truly seriously, early education could become a real poverty-killer.

IMPROVING PEDAGOGY

Research from leading reading experts has shown that children need frequent oral language interactions, coupled with frequent introduction to new vocabulary words, if they are going to have any luck in comprehending the books they'll be asked to read by second, third, and fourth grades.

To deliver something like that second alligator lesson, a teacher needs to be equipped with a rich knowledge base, a strong command of vocabulary and language, and a sound understanding of child development. The successful teacher will often, though not always, have a bachelor's degree and will receive training on how to engage children based on new findings in cognitive and social science.

Poor children do exceedingly well if they are fortunate enough to attend centers with such well-prepared teachers. High-profile studies have found that these children need fewer special education services, do better in school, and engage in less crime (as indexed by crime records), all of which lead to reduced costs to society.

In Head Start, analyses of data from the Congress-commissioned Impact Study show returns in line with or slightly greater than $1 for every dollar invested. Similar data does not exist for many state pre-K programs, and though some have shown that children arrive in kindergarten better prepared, quality varies greatly across the nation. A 2005 study of pre-K programs across 11 states showed classrooms to be, on average, of low-to-moderate quality. Researchers scored interactions between teachers and children, finding them to be in the mid-range for quality. And when it came to "instructional climate"—a measure of the quantity and quality of concepts taught, as well as how teachers provided feedback to spur more learning—scores dwelled around 2, the lower end of the 1-to-7 scale that researchers used.

And so we arrive at one of the hardest nuts to crack in early childhood policy. How do we improve this "instructional climate"? First, education schools and teacher preparation programs will need to greatly expand and improve their offerings, and policymakers must reward programs that hire teachers with strong content knowledge, language skills, and the know-how to introduce new concepts in ways that recognize children's stages of development.

Recruiting and retaining these teachers and caregivers are major challenges. The average salary of a Head Start teacher with a bachelor's degree is about $27,000 a year. It's no wonder that young adults with B.A. degrees decide to work in the elementary grades instead of in pre-K programs. To recruit better teachers, early learning centers will have to pay them what they would receive in the public schools. And yet only a handful of places—such as the state of Oklahoma and some districts in New Jersey—have mustered the political will (in Oklahoma's case) or the legal authority (as in the New Jersey Supreme Court's Abbott decision) to increase funding to that level. It's worth noting that a high-quality, random-assignment study of pre-K in Tulsa, Oklahoma, showed quite staggering improvements in Children's outcomes under its enriched program. There is good reason to believe that focusing on improved teaching could deliver much bang for the taxpayer's buck.

A SEAMLESS SYSTEM

But we get only halfway to a high-quality early education system by ramping up teaching. The history of Head Start gives us yet another lesson: starting children at age four is starting too late. Science has shown how much an unhealthy environment can negatively affect children's development, even in the womb. That's why in 1995, Early Head Start was established to provide support services to pregnant women and their babies, up to age three.

By the same token, halting interventions at age five is stopping too soon. When children move from high-quality learning environments to low-performing elementary schools, research shows that the pace of their social and cognitive development starts to slow. "It is magical thinking to expect that if we intervene in the early years, no further help will be needed by children in the elementary school years and beyond," wrote Jeanne Brooks-Gunn, a prominent psychologist at Columbia University, in a widely cited paper on early childhood education.

The good news is that experts in the field, including some federal policymakers, understand this. The new vision is to create a "birth-to-eight" network—a system of interlocking intervention services that build on existing programs serving pregnant women, babies, toddlers, preschool-age children, and elementary school students. This network will require data systems that share information on children's well-being and prior experiences, connecting them seamlessly to databases in public schools. It will force funding streams to be blended and eligibility parameters to be consistent across programs. It will require intense coordination between health departments and education departments—at both the state and local level—as well as between nonprofit organizations and public school systems. These requirements may seem daunting, but in fact we are already moving, if fitfully, toward just such a system.

The Obama administration has proposed a new competitive grant program that would reward states that have already taken steps to build these networks or that show a commitment to doing so. Called the Early Learning Challenge Fund, the program is part of a larger bill, the Student Aid and Fiscal Responsibility Act, that has been passed by the House and is expected to be taken up by the Senate this winter. It would distribute $1 billion a year to help states increase the number of disadvantaged children in high-quality early care and education programs, from birth to age five. An emphasis on high-quality environments pervades the legislation's language.

A remaining step—one that has not been fully articulated in many policies and needs more attention—is to stretch that quality network further into the primary grades. Studies by Robert C. Pianta, Dean of the Curry School of Education at the University of Virginia, show that elementary classrooms lack quality interactions as much as those for three- and four-year-olds. Research points to the need for what is called "the PreK-3rd approach," a strategy that provides high-quality early learning opportunities to every child before they arrive in kindergarten; that aligns standards, curricula, and assessments between the public schools and pre-K settings; and that provides continuous professional development and shared learning opportunities to well-qualified teachers. Paying pre-K teachers wages that are comparable to elementary school teachers would help ensure that all of these teachers feel like the critical professionals they are.

EARLY EDUCATION FROM A TO Z

This vision for early childhood intervention goes far beyond giving four-year-olds nutritious snacks and helping them identify the letters of the alphabet. It will not be easy. But if we could deliver a high-quality birth-to-eight system, just think about the potential for reducing poverty.

Imagine, for example, what might happen to a baby boy born to a mother who is poor, depressed, and on her own. She lives in a rough neighborhood. She is struggling to make ends meet. But now she receives free visits from a nurse who gives her tips on keeping her boy healthy and on controlling her temper on days when she's overwhelmed. She enrolls in Early Head Start. When her son turns three—full of "why" questions and fascinated by animals—he starts attending a high-quality pre-K/Head Start center, where he encounters Classroom Two's alligator lesson. In kindergarten, he receives the same caliber of instruction, and again in first grade, and again in second—each year building seamlessly on what he has learned the year before.

The little boy thrives. By third grade, he is reading chapter books and writing papers on veterinary science. His mother remains poor, struggling with family conflicts and on-and-off-again employment, but the boy's educational background has put him on a path toward college. By the time he is an adult, he will escape poverty. Not only that, but most of his neighborhood friends—all immersed in the same rich learning experiences from the day they were born—will too.

KEY CONCEPTS

Head Start program   educational reform

DISCUSSION QUESTIONS

1. What is the history of Head Start? Why was it started? What has the research shown regarding the effectiveness of the program?

2. Think back to your own experiences with pre-K schooling. Did you have a preschool that emphasized learning? Did it look more like Classroom One or Classroom Two from the article? How do you think your experience was different from that of other students from different backgrounds?

51

Charter Schools and the Public Good

LINDA A. RENZULLI AND VINCENT J. ROSCIGNO

The charter school movement has arisen in part because of national concerns that public schools are failing. Here Linda Renzulli and Vincent Roscigno evaluate whether charter schools are successful in promoting student achievement. They also assess the extent to which charter schools are held accountable for meeting educational standards.

According to the U.S Department of Education, "No Child Left Behind is designed to change the culture of America's schools by closing the achievement gap, offering more flexibility, giving parents more options, and teaching students based on what works." Charter schools—a recent innovation in U.S. education—are one of the most visible developments aimed at meeting these goals. Although they preceded the 2002 No Child Left Behind (NCLB) Act, charter schools are now supported politically and financially through NCLB. Charter schools are public schools set up and administered outside the traditional bureaucratic constraints of local school boards, with the goal of creating choice, autonomy, and accountability. Unlike regular public schools, charter schools are developed and managed by individuals, groups of parents, community members, teachers, or education-management organizations. In exchange for their independence from most state and local regulations (except those related to health,

SOURCE: Linda A. Renzulli and Vincent J. Roscigno. "Charter Schools and the Public Good." *Contexts* 6, no. 1 (Winter 2007): 31–36 © 2007 by the American Sociological Association.

safety, and nondiscrimination), they must uphold their contracts with the local or state school board or risk being closed. Each provides its own guidelines for establishing rules and procedures, including curriculum, subject to evaluation by the state in which it resides.

Charter schools are among the most rapidly growing educational institutions in the United States today. No charter schools existed before 1990, but such schools are now operating in 40 states and the District of Columbia. According to the Center for Educational Reform, 3,977 charter schools are now educating more than a million students.

Charter schools have received bipartisan support and media accolades. This, however, is surprising. The true academic value of the educational choices that charter schools provide to students, as well as their broader implications for the traditional system of public education, are simply unknown—a fact that became obvious in November 2004, when voters in the state of Washington rejected—for the third time—legislation allowing the creation of charter schools. Driven by an alliance of parents, teachers, and teacher unions against sponsorship by powerful figures such as Bill Gates, this rejection went squarely against a decade-long trend. Reflecting on Washington's rejection, a state Democrat told the *New York Times,* "Charter schools will never have a future here now until there is conclusive evidence, nationwide, that these schools really work. Until the issue of student achievement gets resolved, I'd not even attempt to start over again in the Legislature."

THE RATIONALE

Most justifications for charter schools argue that the traditional system of public schooling is ineffective and that the introduction of competition and choice can resolve any deficiencies. The leading rationale is that accountability standards (for educational outcomes and student progress), choice (in curriculum, structure, and discipline), and autonomy (for teachers and parents) will generate higher levels of student achievement. The result will be high-quality schools for all children, particularly those from poor and minority backgrounds, and higher levels of student achievement.

While wealthy families have always been able to send their children to private schools, other Americans have historically had fewer, if any, options. Proponents suggest that charter schools can address such inequality by allowing all families, regardless of wealth, to take advantage of these new public educational options. Opponents contend that charter schools cannot fix broader educational problems; if anything, they become instruments of segregation, deplete public school systems of their resources, and undermine the public good.

Given the rationales for charter schools before and after the NCLB Act, it is surprising how few assessments have been made of charter school functioning, impacts on achievement, or the implications of choice for school systems. Only a handful of studies have attempted to evaluate systematically the claims of charter school effectiveness, and few of these have used national data. The various justifications for charter schools—including the desire to increase achievement in the public school system—warrant attention, as do concrete research and evidence on whether such schools work. The debate, however, involves more than simply how to enhance student achievement. It also involves educational competition and accountability, individual choice and, most fundamentally, education's role in fostering the "public good."

IS THERE PROOF IN THE PUDDING?

Do students in charter schools do better than they would in traditional public schools? Unfortunately, the jury is still out, and the evidence is mixed. Profiles in the *New Yorker, Forbes, Time,* and *Newsweek,* for example, highlight the successes of individual charter schools in the inner cities of Washington, DC and New York, not to mention anecdotal examples offered by high-profile advocates like John Walton and Bill Gates. While anecdotes and single examples suggest that charter schools may work, they hardly constitute proof or even systematic evidence that they always do. In fact, broader empirical studies using representative and national data suggest that many charter schools have failed.

One noteworthy study, released by the American Federation of Teachers (AFT) in 2004, reports that charter schools are not providing a better education than traditional public schools. Moreover, they are not boosting student achievement. Using fourth- and eighth-grade test scores from the National Assessment of Educational Progress across all states with charter schools, the report finds that charter-school students perform *less* well, on average, in math and reading than their traditional-school counterparts. There appear to be no significant differences among eighth-graders and no discernable difference in black–white achievement gaps across school type.

Because the results reported in the AFT study—which have received considerable media attention—do not incorporate basic demographic, regional, or school characteristics simultaneously, they can only relate average differences across charter schools and public schools. But this ignores the huge effects of family background, above and beyond school environment. Without accounting for the background attributes of students themselves, not to mention other factors such as the race and social-class composition of the student body, estimates of the differences between charter schools and traditional public schools are overstated.

In response to the 2004 AFT report, economists Carolyn Hoxby and Jonah Rockoff compared charter schools to surrounding public schools. Their results contradict many of the AFT's findings. They examined students who applied to but did not attend charter schools because they lost lotteries for spots. Hoxby and Rockoff found that, compared to their lotteried-out fellow applicants, students who attended charter schools in Chicago scored higher in both math and reading. This is true especially in the early elementary grades compared to nearby public schools with similar racial compositions. Their work and that of others also shows that older charter schools perform better than newly formed ones—perhaps suggesting that school stability and effectiveness require time to take hold. Important weaknesses nevertheless remain in the research design. For example, Hoxby and Rockoff conducted their study in a single city—Chicago—and thus it does not represent the effect of charter schools in general.

As with the research conducted by the AFT, we should interpret selective case studies and school-level comparisons with caution. Individual student background is an important force in shaping student achievement, yet it rarely receives attention in this research or in the charter school achievement debate more generally. The positive influence of charter schools, where it is found, could easily be a function of more advantaged student populations drawn from families with significant educational resources at home. We know from prior research that parents of such children are more likely to understand schooling options and are motivated to ensure their children's academic success. Since family background, parental investments, and parental educational involvement typically trump school effects in student achievement, it is likely that positive charter school effects are simply spurious.

More recently, a report by the National Center for Education Statistics (NCES), using sophisticated models, appropriate demographic controls, and a national sample, has concurred with the AFT report—charter schools are not producing children who score better on standardized achievement tests. The NCES report showed that average achievement in math and reading in public schools and in charter schools that were linked to a school district did not differ statistically. Charter schools not associated with a public school district, however, scored significantly *less* well than their public school counterparts.

Nevertheless, neither side of the debate has shown conclusively, through rigorous, replicated, and representative research, whether charter schools boost student achievement. The NCES report mentioned above has, in our opinion, done the best job of examining the achievement issue and has shown that charter schools are not doing better than traditional public schools when it comes to improving achievement.

Clearly, in the case of charter schools, the legislative cart has been put before the empirical horse. Perhaps this is because the debate is about more than achievement. Charter school debates and legislation are rooted in more fundamental disagreements over competition, individualism, and, most fundamentally, education's role in the public good. This reflects an important and significant shift in the cultural evaluation of public education in the United States, at the crux of which is the application of a market-based economic model, complete with accompanying ideas of "competition" and "individualism."

COMPETITION AND ACCOUNTABILITY

To whom are charter schools accountable? Some say their clients, namely, the public. Others say the system, namely, their authorizers. If charter schools are accountable to the public, then competition between schools should ensure academic achievement and bureaucratic prudence. If charter schools are accountable to the system, policies and procedures should ensure academic achievement and bureaucratic prudence. In either case, the assumption is that charter schools will close when they are not successful. The successful application of these criteria, however, requires clear-cut standards, oversight, and accountability—which are currently lacking, according to many scholars. Indeed, despite the rhetoric of their advocates and legislators, charter schools are seldom held accountable in the market or by the political structures that create them.

In a "market" view of accountability, competition will ultimately breed excellence by "weeding out" ineffectual organizations. Through "ripple effects," all schools will be forced to improve their standards. Much like business organizations, schools that face competition will survive only by becoming more efficient and producing a better overall product (higher levels of achievement) than their private and public school counterparts.

Social scientists, including the authors of this article, question this simplistic, if intuitively appealing, application of neoliberal business principles to the complex nature of the educational system, children's learning, and parental choice for schools. If competition were leading to accountability, we would see parents pulling their children out of unsuccessful charter schools. But research shows that this seldom happens. Indeed, parents, particularly those with resources, typically choose schools for reasons of religion, culture, and social similarity rather than academic quality.

Nor are charter schools accountable to bureaucrats. Even though charter schools are not outperforming traditional public schools, relatively few (10 percent nationally) have actually been closed by their authorizers over the last decade. Although we might interpret a 10 percent closure rate as evidence of academic accountability at work, this would be misleading. Financial rather than academic issues are the principal reasons cited for these closures. By all indications, charter schools are not being held accountable to academic standards, either by their authorizers or by market forces.

In addition to measuring accountability through student performance, charter schools should also be held to standards of financial and educational quality. Here, some charter schools are faltering. From California to New York and Ohio, newspaper editorials question fiscal oversight. There are extreme cases such as the California Charter Academy, a publicly financed but privately run chain of 60 charter schools. Despite a budget of $100 million dollars, this chain became insolvent in August 2004, leaving thousands of children without a school to attend.

More direct accountability issues include educational quality and annual reports to state legislatures; here, charter-school performance is poor or mixed. In Ohio, where nearly 60,000 students now attend charter schools, approximately one-quarter of these schools are not following the state's mandate to report school-level test score results, and only 45 percent of the teachers at the state's 250 charter schools hold full teaching certification. Oversight is further complicated by the creation of "online" charter schools, which serve 16,000 of Ohio's public school students.

It is ironic that many charter schools are not held to the very standards of competition, quality, and accountability that legislators and advocates used to justify them in the first place. Perhaps this is why Fredrick Hess, a charter school researcher, recently referred to accountability as applied to charter schools as little more than a "toothless threat."

INDIVIDUALISM OR INEQUALITY?

The most obvious goal of education is student achievement. Public education in the United States, however, has also set itself several other goals that are not reducible to achievement or opportunity at an individual level but are important culturally and socially. Public education has traditionally managed diversity and integration, created common standards for the socialization of the next generation, and ensured some equality of opportunity and potential for meritocracy in the society at large. The focus of the charter school debate on achievement— rooted in purely economic rationales of competition and individual opportunism—has ignored these broader concerns.

Individual choice in the market is a key component of neoliberal and "free-market" theory—a freedom many Americans cherish. Therefore, it makes sense that parents might support choice in public schooling. Theoretically, school choice provides them market power to seek the best product for their children, to weigh alternatives, and to make changes in their child's interest. But this power is only available to informed consumers, so that educational institutions and policies that provide choice may be reinforcing the historical disadvantages faced by racial and ethnic minorities and the poor.

We might expect that students from advantaged class backgrounds whose parents are knowledgeable about educational options would be more likely to enroll in charter schools. White parents might also see charter schools as an educational escape route from integrated public schools that avoids the financial burden of private schooling. On the other hand, the justification for charter schools is often framed in terms of an "educational fix" for poor, minority-concentrated districts in urban areas. Here, charter schools may appear to be a better opportunity for aggrieved parents whose children are attending poorly funded, dilapidated public schools

National research, at first glance, offers encouraging evidence that charter schools are providing choices to those who previously had few options: 52 percent of those enrolled in charter schools are nonwhite compared to 41 percent of those in traditional public schools. These figures, however, tell us little about the local concentrations of whites and nonwhites in charter schools, or how the racial composition and distribution of charter schools compares to the racial composition and distribution of local, traditional public schools.

African-American students attend charter and noncharter schools in about the same proportion, yet a closer look at individual charter schools within districts reveals that they are often segregated. In Florida, for instance, charter schools are 82 percent white, whereas traditional public schools are only 51 percent white. Similar patterns are found across Arizona school districts, where charter school enrollment is 20 percent more white than traditional schools.

Amy Stuart Wells's recent research finds similar tendencies toward segregation among Latinos, who are underrepresented in California's charter schools. Linda Renzulli and Lorraine Evans's national analysis of racial composition within districts containing charter schools shows that charter school formation often results in greater levels of segregation in schools between whites and nonwhites. This is not to suggest that minority populations do not make use of charter schools. But, when they do, they do so in segregated contexts.

Historically, racial integration has been a key cause of white flight and it remains a key factor in the racial composition of charter schools and other schools of choice. Decades of research on school segregation have taught us that when public school districts become integrated, through either court mandates or simple population change, white parents may seek alternative schools for their children. Current research suggests that a similar trend exists with charter schools, which provide a public-school option for white flight without the drawbacks of moving (such as job changes and longer commutes). While those from less-privileged and minority backgrounds have charter schools at their disposal, the realities of poor urban districts and contemporary patterns of racial residential segregation may mean that the "choice" is between a racially and economically segregated charter school or an equally segregated traditional school, as Renzulli's research has shown. Individualism in the form of educational choice, although perhaps intuitively appealing, in reality may be magnifying some at the very inequalities that public education has been attempting to overcome since the *Brown v. Board of Education* decision in 1954.

Regarding equality of opportunity and its implications for the American ideal of meritocracy, there is also reason for concern. Opponents have pointed to the dilution of district resources where charter schools have emerged, especially as funds are diverted to charter schools. Advocates, in contrast, argue that charter schools have insufficient resources. More research on the funding consequences of charter school creation is clearly warranted. Why, within a system of public education, should some students receive more than others? And what of those left behind, particularly students from disadvantaged backgrounds whose parents may not be aware of their options? Although evidence on the funding question is sparse, research on public schools generally and charter school attendance specifically suggests that U.S. public education may be gravitating again toward a system of separate, but not equal, education.

THE FUTURE OF PUBLIC EDUCATION

Variation across charter schools prevents easy evaluation of their academic success or social consequences for public education. Case studies can point to a good school or a bad one. National studies can provide statistical averages and comparisons, yet they may be unable to reveal the best and worst effects of charter schools. Neither type of research has yet fully accounted for the influence of family background and school demographic composition. Although conclusions about charter school effectiveness or failure remain questionable, the most rigorous national analyses to date suggest that charter schools are doing no better than traditional public schools.

Certainly some charter schools are improving the educational quality and experience of some children. KIPP (Knowledge Is Power Program) schools, for example, are doing remarkable things for the students lucky enough to attend them. But for every KIPP school (of which there are only 45, and not all are charter schools), there are many more charter schools that do not provide the same educational opportunity to students, have closed their doors in the middle of the school year, and, in effect, isolate students from their peers of other races and social classes. Does this mean that we should prevent KIPP, for example, from educating students through the charter school option? Maybe. Or perhaps we should develop better program evaluations—of what works and what does not—and implement them as guideposts. To the dismay of some policymakers and "competition" advocates, however, such standardized evaluation and accountability would undercut significant charter school variations if not the very nature of the charter school innovation itself.

Student achievement is only part of the puzzle when it comes to the charter school debate; we need to consider social integration and equality as well. These broader issues, although neglected, warrant as much attention as potential effects on achievement. We suspect that such concerns, although seldom explicit, probably underlie the often contentious charter school and school choice debate itself. We believe it is time to question the logics pertaining to competition, choice, and accountability. Moreover, we should all scrutizine the existing empirical evidence, not to mention educational policy not firmly rooted in empirical reality and research. As Karl Alexander eloquently noted in his presidential address to the Southern Sociological Society, "The charter school movement, with its 'let 1,000 flowers bloom' philosophy, is certain to yield an occasional prize-winning rose. But is either of these approaches [to school choice] likely to prove a reliable guide for broad-based, systemic reform—the kind of reform that will carry the great mass of our children closer to where we want them to be? I hardly think so." Neither do we.

KEY CONCEPTS

charter school   educational attainment   No Child Left Behind

DISCUSSION QUESTIONS

1. What factors, other than the school itself, do Renzulli and Roscigno identify as affecting student achievement?

2. What do the authors mean, in the context of schooling, when they argue that "choice" is a racially and economically based concept?

D. Work

52

Nickel-and-Dimed: On (Not) Getting By in America

BARBARA EHRENREICH

Barbara Ehrenreich, a journalist and faculty member, "posed" for several weeks as an unskilled worker to find out how people survive doing low-wage work. Her account, later published in the best-selling book, Nickel and Dimed, is a compelling portrait of the conditions of work for millions of people in the United States.

At the beginning of June 1998 I leave behind everything that normally soothes the ego and sustains the body—home, career, companion, reputation, ATM card—for a plunge into the low-wage workforce. There, I become another, occupationally much diminished "Barbara Ehrenreich"—depicted on job-application forms as a divorced homemaker whose sole work experience consists of housekeeping in a few private homes. I am terrified, at the beginning, of being unmasked for what I am: a middle-class journalist setting out to explore the world that welfare mothers are entering, at the rate of approximately 50,000 a month, as welfare reform kicks in. Happily, though, my fears turn out to be entirely unwarranted: during a month of poverty and toil, my name goes unnoticed and for the most part unuttered. In this parallel universe where my father never got out of the mines and I never got through college, I am "baby," "honey," "blondie," and, most commonly, "girl."

SOURCE: Ehrenreich, Barbara. "Nickel-and-Dimed: On (Not) Getting By in America" *Harper's Magazine* 298 (January 1999): 37ff. Reprinted with permission.

My first task is to find a place to live. I figure that if I can earn $7 an hour—which, from the want ads, seems doable—I can afford to spend $500 on rent, or maybe, with severe economies, $600. In the Key West area, where I live, this pretty much confines me to flophouses and trailer homes—like the one, a pleasing fifteen-minute drive from town, that has no air-conditioning, no screens, no fans, no television, and, by way of diversion, only the challenge of evading the landlord's Doberman pinscher. The big problem with this place, though, is the rent, which at $675 a month is well beyond my reach. All right, Key West is expensive. But so is New York City, or the Bay Area, or Jackson Hole, or Telluride, or Boston, or any other place where tourists and the wealthy compete for living space with the people who clean their toilets and fry their hash browns. Still, it is a shock to realize that "trailer trash" has become, for me, a demographic category to aspire to.

So I decide to make the common trade-off between affordability and convenience, and go for a $500-a-month efficiency thirty miles up a two-lane highway from the employment opportunities of Key West, meaning forty-five minutes if there's no road construction and I don't get caught behind some sun-dazed Canadian tourists. I hate the drive, along a roadside studded with white crosses commemorating the more effective head-on collisions, but it's a sweet little place—a cabin, more or less, set in the swampy back yard of the converted mobile home where my landlord, an affable TV repairman, lives with his bartender girlfriend. Anthropologically speaking, a bustling trailer park would be preferable, but here I have a gleaming white floor and a firm mattress, and the few resident bugs are easily vanquished.

Besides, I am not doing this for the anthropology. My aim is nothing so mistily subjective as to "experience poverty" or find out how it "really feels" to be a long-term low-wage worker. I've had enough unchosen encounters with poverty and the world of low-wage work to know it's not a place you want to visit for touristic purposes; it just smells too much like fear. And with all my real-life assets—bank account, IRA, health insurance, multiroom home—waiting indulgently in the background, I am, of course, thoroughly insulated from the terrors that afflict the genuinely poor.

No, this is a purely objective, scientific sort of mission. The humanitarian rationale for welfare reform—as opposed to the more punitive and stingy impulses that may actually have motivated it—is that work will lift poor women out of poverty while simultaneously inflating their self-esteem and hence their future value in the labor market. Thus, whatever the hassles involved in finding child care, transportation, etc., the transition from welfare to work will end happily, in greater prosperity for all. Now there are many problems with this comforting prediction, such as the fact that the economy will inevitably undergo a downturn, eliminating many jobs. Even without a downturn, the influx of a million former welfare recipients into the low-wage labor market could depress wages by as much as 11.9 percent, according to the Economic Policy Institute (EPI) in Washington, D.C.

But is it really possible to make a living on the kinds of jobs currently available to unskilled people? Mathematically, the answer is no, as can be shown by taking $6 to $7 an hour, perhaps subtracting a dollar or two an hour for child care, multiplying by 160 hours a month, and comparing the result to the prevailing rents. According to the National Coalition for the Homeless, for example, in 1998 it took, on average nationwide, an hourly wage of $8.89 to afford a one-bedroom apartment, and the Preamble Center for Public Policy estimates that the odds against a typical welfare recipient's landing a job at such a "living wage" are about 97 to 1. If these numbers are right, low-wage work is not a solution to poverty and possibly not even to homelessness.

It may seem excessive to put this proposition to an experimental test. As certain family members keep unhelpfully reminding me, the viability of low-wage work could be tested, after a fashion, without ever leaving my study. I could just pay myself $7 an hour for eight hours a day, charge myself for room and board, and total up the numbers after a month. Why leave the people and work that I love? But I am an experimental scientist by training. In that business, you don't just sit at a desk and theorize; you plunge into the everyday chaos of nature, where surprises lurk in the most mundane measurements. Maybe, when I got into it, I would discover some hidden economies in the world of the low-wage worker. After all, if 30 percent of the workforce toils for less than $8 an hour, according to the EPI, they may have found some tricks as yet unknown to me. Maybe—who knows?—would even be able to detect in myself the bracing psychological effects of getting out of the house, as promised by the welfare wonks at places like the Heritage Foundation. Or, on the other hand, maybe there would be unexpected costs—physical, mental, or financial—to throw off all my calculations. Ideally, I should do this with two small children in tow, that being the welfare average, but mine are grown and no one is willing to lend me theirs for a month-long vacation in penury. So this is not the perfect experiment, just a test of the best possible case: an unencumbered woman, smart and even strong, attempting to live more or less off the land.

On the morning of my first full day of job searching, I take a red pen to the want ads, which are auspiciously numerous. Everyone in Key West's booming "hospitality industry" seems to be looking for someone like me—trainable, flexible, and with suitably humble expectations as to pay. I know I possess certain traits that might be advantageous—I'm white and, I like to think, well-spoken and poised—but I decide on two rules: One, I cannot use any skills derived from my education or usual work—not that there are a lot of want ads for satirical essayists anyway. Two, I have to take the best-paid job that is offered me and of course do my best to hold it; no Marxist rants or sneaking off to read novels in the ladies' room. In addition, I rule out various occupations for one reason or another: Hotel front-desk clerk, for example, which to my surprise is regarded as unskilled and pays around $7 an hour, gets eliminated because it involves standing in one spot for eight hours a day. Waitressing is similarly something I'd like to avoid, because I remember it leaving me bone tired when I was eighteen, and I'm decades of varicosities and back pain beyond that now. Telemarketing, one of the first refuges of the suddenly indigent, can be dismissed on grounds of personality. This leaves certain supermarket jobs, such as deli clerk, or housekeeping in Key West's thousands of hotel and guest rooms. Housekeeping is especially appealing, for reasons both atavistic and practical: it's what my mother did before I came along, and it can't be too different from what I've been doing part-time, in my own home, all my life.

So I put on what I take to be a respectful-looking outfit of ironed Bermuda shorts and scooped-neck T-shirt and set out for a tour of the local hotels and supermarkets. Best Western, Econo Lodge, and Hojo's all let me fill out application forms, and these are, to my relief, interested in little more than whether I am a legal resident of the United States and have committed any felonies. My next stop is Winn-Dixie, the supermarket, which turns out to have a particularly onerous application process, featuring a fifteen-minute "interview" by computer since, apparently, no human on the premises is deemed capable of representing the corporate point of view. I am conducted to a large room decorated with posters illustrating how to look "professional" (it helps to be white and, if female, permed) and warning of the slick promises that union organizers might try to tempt me with. The interview is multiple choice: Do I have anything, such as child-care problems, that might make it hard for me to get to work on time? Do I think safety on the job is the responsibility of management? Then, popping up cunningly out of the blue: How many dollars' worth of stolen goods have I purchased in the last year? Would I turn in a fellow employee if I caught him stealing? Finally, "Are you an honest person?"

Apparently, I ace the interview, because I am told that all I have to do is show up in some doctor's office tomorrow for a urine test. This seems to be a fairly general rule: if you want to stack Cheerio boxes or vacuum hotel rooms in chemically fascist America, you have to be willing to squat down and pee in front of some health worker (who has no doubt had to do the same thing herself). The wages Winn-Dixie is offering—$6 and a couple of dimes to start with—are not enough, I decide, to compensate for this indignity.

I lunch at Wendy's, where $4.99 gets you unlimited refills at the Mexican part of the Superbar, a comforting surfeit of refried beans and "cheese sauce." A teenage employee, seeing me studying the want ads, kindly offers me an application form, which I fill out, though here, too, the pay is just $6 and change an hour. Then it's off for a round of the locally owned inns and guest-houses. At "The Palms," let's call it, a bouncy manager actually takes me around to see the rooms and meet the existing housekeepers, who, I note with satisfaction, look pretty much like me—faded ex-hippie types in shorts with long hair pulled back in braids. Mostly, though, no one speaks to me or even looks at me except to proffer an application form. At my last stop, a palatial B&B, I wait twenty minutes to meet "Max," only to be told that there are no jobs now but there should be one soon, since "nobody lasts more than a couple weeks." (Because none of the people I talked to knew I was a reporter, I have changed their names to protect their privacy and, in some cases perhaps, their jobs.)

Three days go by like this, and, to my chagrin, no one out of the approximately twenty places I've applied calls me for an interview. I had been vain enough to worry about coming across as too educated for the jobs I sought, but no one even seems interested in finding out how overqualified I am. Only later will I realize that the want ads are not a reliable measure of the actual jobs available at any particular time. They are, as I should have guessed from Max's comment, the employers' insurance policy against the relentless turnover of the low-wage work-force. Most of the big hotels run ads almost continually, just to build a supply of applicants to replace the current workers as they drift away or are fired, so finding a job is just a matter of being at the right place at the right time and flexible enough to take whatever is being offered that day. This finally happens to me at one of the big discount hotel chains, where I go, as usual, for house-keeping and am sent, instead, to try out as a waitress at the attached "family res-taurant," a dismal spot with a counter and about thirty tables that looks out on a parking garage and features such tempting fare as "Polish [sic] sausage and BBQ sauce" on 95-degree days. Phillip, the dapper young West Indian who intro-duces himself as the manager, interviews me with about as much enthusiasm as if he were a clerk processing me for Medicare, the principal questions being what shifts can I work and when can I start. I mutter something about being woefully out of practice as a waitress, but he's already on to the uniform: I'm to show up tomorrow wearing black slacks and black shoes; he'll provide the rust-colored polo shirt with HEARTHSIDE embroidered on it, though I might want to wear my own shirt to get to work, ha ha. At the word "tomorrow," something between fear and indignation rises in my chest. I want to say, "Thank you for your time, sir, but this is just an experiment, you know, not my actual life."

So begins my career at the Hearthside, I shall call it, one small profit center within a global discount hotel chain, where for two weeks I work from 2:00 till 10:00 P.M. for $2.43 an hour plus tips. In some futile bid for gentility, the man-agement has barred employees from using the front door, so my first day I enter through the kitchen, where a red-faced man with shoulder-length blond hair is throwing frozen steaks against the wall and yelling, "Fuck this shit!" "That's just Jack," explains Gail, the wiry middle-aged waitress who is assigned to train me. "He's on the rag again"—a condition occasioned, in this instance, by the fact that the cook on the morning shift had forgotten to thaw out the steaks. For the next eight hours, I run after the agile Gail, absorbing bits of instruction along with fragments of personal tragedy. All food must be trayed, and the reason she's so tired today is that she woke up in a cold sweat thinking of her boyfriend, who killed himself recently in an upstate prison. No refills on lemonade. And the rea-son he was in prison is that a few DUIs caught up with him, that's all, could have happened to anyone. Carry the creamers to the table in a monkey bowl, never in your hand. And after he was gone she spent several months living in her truck, peeing in a plastic pee bottle and reading by candlelight at night, but you can't live in a truck in the summer, since you need to have the windows down, which means anything can get in, from mosquitoes on up.

At least Gail puts to rest any fears I had of appearing overqualified. From the first day on, I find that of all the things I have left behind, such as home and iden-tity, what I miss the most is competence. Not that I have ever felt utterly compe-tent in the writing business, in which one day's success augers nothing at all for the next. But in my writing life I at least have some notion of procedure: do the research, make the outline, rough out a draft, etc. As a server, though, I am beset by requests like bees: more iced tea here, ketchup over there, a to-go box for table fourteen, and where are the high chairs, anyway? Of the twenty-seven tables, up to six are usually mine at any time, though on slow afternoons or if Gail is off, I sometimes have the whole place to myself. There is the touch-screen computer-ordering system to master, which is, I suppose, meant to minimize server-cook contact, but in practice requires constant verbal fine-tuning: "That's gravy on the mashed, okay? None on the meatloaf," and so forth—while the cook scowls as if I were inventing these refinements just to torment him. Plus, something I had forgotten in the years since I was eighteen: about a third of a server's job is "side work" that's invisible to customers—sweeping, scrubbing, slicing, refilling, and re-stocking. If it isn't all done, every little bit of it, you're going to face the 6:00 P.M. dinner rush defenseless and probably go down in flames. I screw up dozens of times at the beginning, sustained in my shame entirely by Gail's support—"It's okay, baby, everyone does that sometime"—because, to my total surprise and despite the scientific detachment I am doing my best to maintain, I care.

The whole thing would be a lot easier if I could just skate through it as Lily Tomlin in one of her waitress skits, but I was raised by the absurd Booker T. Washingtonian precept that says: If you're going to do something, do it well. In fact, "well" isn't good enough by half. Do it better than anyone has ever done it before. Or so said my father, who must have known what he was talking about because he managed to pull himself, and us with him, up from the mile-deep copper mines of Butte to the leafy suburbs of the Northeast, ascending from boilermakers to martinis before booze beat out ambition. As in most endeavors I have encountered in my life, doing it "better than anyone" is not a reasonable goal. Still, when I wake up at 4:00 A.M. in my own cold sweat, I am not thinking about the writing deadlines I'm neglecting; I'm thinking about the table whose order I screwed up so that one of the boys didn't get his kiddie meal until the rest of the family had moved on to their Key Lime pies. That's the other powerful motivation I hadn't expected—the customers, or "patients," as I can't help thinking of them on account of the mysterious vulnerability that seems to have left them temporarily unable to feed themselves. After a few days at the Hearth-side, I feel the service ethic kick in like a shot of oxytocin, the nurturance hormone. The plurality of my customers are hard-working locals—truck drivers, construction workers, even housekeepers from the attached hotel—and I want them to have the closest to a "fine dining" experience that the grubby circumstances will allow. No "you guys" for me; everyone over twelve is "sir" or "ma'am." I ply them with iced tea and coffee refills; I return, mid-meal, to inquire how everything is; I doll up their salads with chopped raw mushrooms, summer squash slices, or whatever bits of produce I can find that have survived their sojourn in the cold-storage room mold-free....

Ten days into it, this is beginning to look like a livable lifestyle. I like Gail, who is "looking at fifty" but moves so fast she can alight in one place and then another without apparently being anywhere between them. I clown around with Lionel, the teenage Haitian busboy, and catch a few fragments of conversation with Joan, the svelte fortyish hostess and militant feminist who is the only one of us who dares to tell Jack to shut the fuck up. I even warm up to Jack when, on a slow night and to make up for a particularly unwarranted attack on my abilities, or so I imagine, he tells me about his glory days as a young man at "coronary school"—or do you say "culinary"?—in Brooklyn, where he dated a knock-out Puerto Rican chick and learned everything there is to know about food. I finish up at 10:00 or 10:30, depending on how much side work I've been able to get done during the shift, and cruise home to the tapes I snatched up at random when I left my real home—Marianne Faithfull, Tracy Chapman, Enigma, King Sunny Ade, the Violent Femmes—just drained enough for the music to set my cranium resonating but hardly dead. Midnight snack is Wheat Thins and Monterey Jack, accompanied by cheap white wine on ice and whatever AMC has to offer. To bed by 1:30 or 2:00, up at 9:00 or 10:00, read for an hour while my uniform whirls around in the landlord's washing machine, and then it's another eight hours spent following Mao's central instruction, as laid out in the Little Red Book, which was: Serve the people.

I could drift along like this, in some dreamy proletarian idyll, except for two things. One is management. If I have kept this subject on the margins thus far it is because I still flinch to think that I spent all those weeks under the surveillance of men (and later women) whose job it was to monitor my behavior for signs of sloth, theft, drug abuse, or worse. Not that managers and especially "assistant managers" in low-wage settings like this are exactly the class enemy. In the restaurant business, they are mostly former cooks or servers, still capable of pinch-hitting in the kitchen or on the floor, just as in hotels they are likely to be former clerks, and paid a salary of only about $400 a week. But everyone knows they have crossed over to the other side, which is, crudely put, corporate as opposed to human. Cooks want to prepare tasty meals; servers want to serve them graciously; but managers are there for only one reason—to make sure that money is made for some theoretical entity that exists far away in Chicago or New York, if a corporation can be said to have a physical existence at all. Reflecting on her career, Gail tells me ruefully that she had sworn, years ago, never to work for a corporation again. "They don't cut you no slack. You give and you give, and they take."

Managers can sit—for hours at a time if they want—but it's their job to see that no one else ever does, even when there's nothing to do, and this is why, for servers, slow times can be as exhausting as rushes. You start dragging out each little chore, because if the manager on duty catches you in an idle moment, he will give you something far nastier to do. So I wipe, I clean, I consolidate ketchup bottles and recheck the cheesecake supply, even tour the tables to make sure the customer evaluation forms are all standing perkily in their places—wondering all the time how many calories I burn in these strictly theatrical exercises. When, on a particularly dead afternoon, Stu finds me glancing at a USA Today a customer has left behind, he assigns me to vacuum the entire floor with the broken vacuum cleaner that has a handle only two feet long, and the only way to do that without incurring orthopedic damage is to proceed from spot to spot on your knees.

On my first Friday at the Hearthside there is a "mandatory meeting for all restaurant employees," which I attend, eager for insight into our overall marketing strategy and the niche (your basic Ohio cuisine with a tropical twist?) we aim to inhabit. But there is no "we" at this meeting. Phillip, our top manager except for an occasional "consultant" sent out by corporate headquarters, opens it with a sneer: "The break room—it's disgusting. Butts in the ashtrays, newspapers lying around, crumbs." This windowless little room, which also houses the time clock for the entire hotel, is where we stash our bags and civilian clothes and take our half-hour meal breaks. But a break room is not a right, he tells us. It can be taken away. We should also know that the lockers in the break room and whatever is in them can be searched at any time. Then comes gossip; there has been gossip; gossip (which seems to mean employees talking among themselves) must stop. Off-duty employees are henceforth barred from eating at the restaurant, because "other servers gather around them and gossip." When Phillip has exhausted his agenda of rebukes, Joan complains about the condition of the ladies' room and I throw in my two bits about the vacuum cleaner. But I don't see any backup coming from my fellow servers, each of whom has subsided into her own personal funk; Gail, my role model, stares sorrowfully at a point six inches from her nose. The meeting ends when Andy, one of the cooks, gets up, muttering about breaking up his day off for this almighty bullshit.

Just four days later we are suddenly summoned into the kitchen at 3:30 P.M., even though there are live tables on the floor. We all—about ten of us—stand around Phillip, who announces grimly that there has been a report of some "drug activity" on the night shift and that, as a result, we are now to be a "drug-free" workplace, meaning that all new hires will be tested, as will possibly current employees on a random basis. I am glad that this part of the kitchen is so dark, because I find myself blushing as hard as if I had been caught toking up in the ladies' room myself: I haven't been treated this way—lined up in the corridor, threatened with locker searches, peppered with carelessly aimed accusations—since junior high school. Back on the floor, Joan cracks, "Next they'll be telling us we can't have sex on the job." When I ask Stu what happened to inspire the crackdown, he just mutters about "management decisions" and takes the opportunity to upbraid Gail and me for being too generous with the rolls. From now on there's to be only one per customer, and it goes out with the dinner, not with the salad. He's also been riding the cooks, prompting Andy to come out of the kitchen and observe—with the serenity of a man whose customary implement is a butcher knife—that "Stu has a death wish today."

Later in the evening, the gossip crystallizes around the theory that Stu is himself the drug culprit, that he uses the restaurant phone to order up marijuana and sends one of the late servers out to fetch it for him. The server was caught, and she may have ratted Stu out or at least said enough to cast some suspicion on him, thus accounting for his pissy behavior. Who knows? Lionel, the busboy, entertains us for the rest of the shift by standing just behind Stu's back and sucking deliriously on an imaginary joint.

The other problem, in addition to the less-than-nurturing management style, is that this job shows no sign of being financially viable. You might imagine, from a comfortable distance, that people who live, year in and year out, on $6 to $10 an hour have discovered some survival stratagems unknown to the middle class. But no. It's not hard to get my co-workers to talk about their living situations, because housing, in almost every case, is the principal source of disruption in their lives, the first thing they fill you in on when they arrive for their shifts. After a week, I have compiled the following survey:

Gail is sharing a room in a well-known down-town flophouse for which she and a roommate pay about $250 a week. Her roommate, a male friend, has begun hitting on her, driving her nuts, but the rent would be impossible alone.

Claude, the Haitian cook, is desperate to get out of the two-room apartment he shares with his girlfriend and two other, unrelated, people. As far as I can determine, the other Haitian men (most of whom only speak Creole) live in similarly crowded situations.

Annette, a twenty-year-old server who is six months pregnant and has been abandoned by her boyfriend, lives with her mother, a postal clerk.

Marianne and her boyfriend are paying $170 a week for a one-person trailer.

Jack, who is, at $10 an hour, the wealthiest of us, lives in the trailer he owns, paying only the $400-a-month lot fee.

The other white cook, Andy, lives on his dry-docked boat, which, as far as I can tell from his loving descriptions, can't be more than twenty feet long. He offers to take me out on it, once it's repaired, but the offer comes with inquiries as to my marital status, so I do not follow up on it.

Tina and her husband are paying $60 a night for a double room in a Days Inn. This is because they have no car and the Days Inn is within walking distance of the Hearthside. When Marianne, one of the breakfast servers, is tossed out of her trailer for subletting (which is against the trailer-park rules), she leaves her boyfriend and moves in with Tina and her husband.

Joan, who had fooled me with her numerous and tasteful outfits (hostesses wear their own clothes), lives in a van she parks behind a shopping center at night and showers in Tina's motel room. The clothes are from thrift shops.

It strikes me, in my middle-class solipsism, that there is gross improvidence in some of these arrangements. When Gail and I are wrapping silverware in napkins—the only task for which we are permitted to sit—she tells me she is thinking of escaping from her roommate by moving into the Days Inn herself. I am astounded: How can she even think of paying between $40 and $60 a day?

But if I was afraid of sounding like a social worker, I come out just sounding like a fool. She squints at me in disbelief, "And where am I supposed to get a month's rent and a month's deposit for an apartment?" I'd been feeling pretty smug about my $500 efficiency, but of course it was made possible only by the $1,300 I had allotted myself for start-up costs when I began my low-wage life: $1,000 for the first month's rent and deposit, $100 for initial groceries and cash in my pocket, $200 stuffed away for emergencies. In poverty, as in certain propositions in physics, starting conditions are everything.

There are no secret economies that nourish the poor; on the contrary, there are a host of special costs. If you can't put up the two months' rent you need to secure an apartment, you end up paying through the nose for a room by the week. If you have only a room, with a hot plate at best, you can't save by cooking up huge lentil stews that can be frozen for the week ahead. You eat fast food, or the hot dogs and styrofoam cups of soup that can be microwaved in a convenience store. If you have no money for health insurance—and the Hearthside's niggardly plan kicks in only after three months—you go without routine care or prescription drugs and end up paying the price. Gail, for example, was fine until she ran out of money for estrogen pills. She is supposed to be on the company plan by now, but they claim to have lost her application form and need to begin the paperwork all over again. So she spends $9 per migraine pill to control the headaches she wouldn't have, she insists, if her estrogen supplements were covered. Similarly, Marianne's boyfriend lost his job as a roofer because he missed so much time after getting a cut on his foot for which he couldn't afford the prescribed antibiotic.

My own situation, when I sit down to assess it after two weeks of work, would not be much better if this were my actual life. The seductive thing about waitressing is that you don't have to wait for payday to feel a few bills in your pocket, and my tips usually cover meals and gas, plus something left over to stuff into the kitchen drawer I use as a bank. But as the tourist business slows in the summer heat, I sometimes leave work with only $20 in tips (the gross is higher, but servers share about 15 percent of their tips with the busboys and bartenders). With wages included, this amounts to about the minimum wage of $5.15 an hour. Although the sum in the drawer is piling up, at the present rate of accumulation it will be more than a hundred dollars short of my rent when the end of the month comes around. Nor can I see any expenses to cut. True, I haven't gone the lentil-stew route yet, but that's because I don't have a large cooking pot, pot holders, or a ladle to stir with (which cost about $30 at Kmart, less at thrift stores), not to mention onions, carrots, and the indispensable bay leaf. I do make my lunch almost every day—usually some slow-burning, high-protein combo like frozen chicken patties with melted cheese on top and canned pinto beans on the side. Dinner is at the Hearthside, which offers its employees a choice of BLT, fish sandwich, or hamburger for only $2. The burger lasts longest, especially if it's heaped with gut-puckering jalapenos, but by midnight my stomach is growling again.

So unless I want to start using my car as a residence, I have to find a second, or alternative, job. I call all the hotels where I filled out housekeeping applications weeks ago—the Hyatt, Holiday Inn, Econo Lodge, Hojo's, Best Western, plus a half dozen or so locally run guesthouses. Nothing. Then I start making the rounds again, wasting whole mornings waiting for some assistant manager to show up, even dipping into places so creepy that the front-desk clerk greets you from behind bulletproof glass and sells pints of liquor over the counter. But either someone has exposed my real-life housekeeping habits—which are, shall we say, mellow—or I am at the wrong end of some infallible ethnic equation: most, but by no means all, of the working housekeepers I see on my job searches are African Americans, Spanish-speaking, or immigrants from the Central European post-Communist world, whereas servers are almost invariably white and monolingually English-speaking. When I finally get a positive response, I have been identified once again as server material. Jerry's, which is part of a well-known national family restaurant chain and physically attached here to another budget hotel chain, is ready to use me at once. The prospect is both exciting and terrifying, because, with about the same number of tables and counter seats, Jerry's attracts three or four times the volume of customers as the gloomy old Hearthside....

I start out with the beautiful, heroic idea of handling the two jobs at once, and for two days I almost do it: the breakfast/lunch shift at Jerry's, which goes till 2:00, arriving at the Hearthside at 2:10, and attempting to hold out until 10:00. In the ten minutes between jobs, I pick up a spicy chicken sandwich at the Wendy's drive-through window, gobble it down in the car, and change from khaki slacks to black, from Hawaiian to rust polo. There is a problem, though. When during the 3:00 to 4:00 P.M. dead time I finally sit down to wrap silver, my flesh seems to bond to the seat. I try to refuel with a purloined cup of soup, as I've seen Gail and Joan do dozens of times, but a manager catches me and hisses "No eating!" though there's not a customer around to be offended by the sight of food making contact with a server's lips. So I tell Gail I'm going to quit, and she hugs me and says she might just follow me to Jerry's herself.

But the chances of this are minuscule. She has left the flophouse and her annoying roommate and is back to living in her beat-up old truck. But guess what? she reports to me excitedly later that evening: Phillip has given her permission to park overnight in the hotel parking lot, as long as she keeps out of sight, and the parking lot should be totally safe, since it's patrolled by a hotel security guard! With the Hearthside offering benefits like that, how could anyone think of leaving?

Gail would have triumphed at Jerry's, I'm sure, but for me it's a crash course in exhaustion management. Years ago, the kindly fry cook who trained me to waitress at a Los Angeles truck stop used to say: Never make an unnecessary trip; if you don't have to walk fast, walk slow; if you don't have to walk, stand. But at Jerry's the effort of distinguishing necessary from unnecessary and urgent from whenever would itself be too much of an energy drain. The only thing to do is to treat each shift as a one-time-only emergency: you've got fifty starving people out there, lying scattered on the battlefield, so get out there and feed them! Forget that you will have to do this again tomorrow, forget that you will have to be alert enough to dodge the drunks on the drive home tonight—just burn, burn, burn! Ideally, at some point you enter what servers call "a rhythm" and psychologists term a "flow state," in which signals pass from the sense organs directly to the muscles, bypassing the cerebral cortex, and a Zen-like emptiness sets in. A male server from the Hearthside's morning shift tells me about the time he "pulled a triple"—three shifts in a row, all the way around the clock—and then got off and had a drink and met this girl, and maybe he shouldn't tell me this, but they had sex right then and there, and it was like, beautiful.

But there's another capacity of the neuromuscular system, which is pain. I start tossing back drugstore-brand ibuprofen pills as if they were vitamin C, four before each shift, because an old mouse-related repetitive-stress injury in my upper back has come back to full-spasm strength, thanks to the tray carrying. In my ordinary life, this level of disability might justify a day of ice packs and stretching. Here I comfort myself with the Aleve commercial in which the cute blue-collar guy asks: If you quit after working four hours, what would your boss say? And the not-so-cute blue-collar guy, who's lugging a metal beam on his back, answers: He'd fire me, that's what. But fortunately, the commercial tells us, we workers can exert the same kind of authority over our painkillers that our bosses exert over us. If Tylenol doesn't want to work for more than four hours, you just fire its ass and switch to Aleve.

True, I take occasional breaks from this life, going home now and then to catch up on e-mail and for conjugal visits (though I am careful to "pay" for anything I eat there), seeing The Truman Show with friends and letting them buy my ticket. And I still have those what-am-I-doing-here moments at work, when I get so homesick for the printed word that I obsessively reread the six-page menu. But as the days go by, my old life is beginning to look exceedingly strange. The e-mails and phone messages addressed to my former self come from a distant race of people with exotic concerns and far too much time on their hands. The neighborly market I used to cruise for produce now looks forbiddingly like a Manhattan yuppie emporium. And when I sit down one morning in my real home to pay bills from my past life, I am dazzled at the two- and three-figure sums owed to outfits like Club BodyTech and Amazon.com....

I make the decision to move closer to Key West. First, because of the drive. Second and third, also because of the drive: gas is eating up $4 to $5 a day, and although Jerry's is as high-volume as you can get, the tips average only 10 percent, and not just for a newbie like me. Between the base pay of $2.15 an hour and the obligation to share tips with the busboys and dishwashers, we're averaging only about $7.50 an hour. Then there is the $30 I had to spend on the regulation tan slacks worn by Jerry's servers—a setback it could take weeks to absorb. (I had combed the town's two downscale department stores hoping for something cheaper but decided in the end that these marked-down Dockers, originally $49, were more likely to survive a daily washing.) Of my fellow servers, everyone who lacks a working husband or boyfriend seems to have a second job: Nita does something at a computer eight hours a day; another welds. Without the forty-five-minute commute, I can picture myself working two jobs and having the time to shower between them....

I can do this two-job thing, is my theory, if I can drink enough caffeine and avoid getting distracted by George's ever more obvious suffering....

Then it comes, the perfect storm. Four of my tables fill up at once. Four tables is nothing for me now, but only so long as they are obligingly staggered. As I bev table 27, tables 25, 28, and 24 are watching enviously. As I bev 25, 24 glowers because their bevs haven't even been ordered. Twenty-eight is four yuppyish types, meaning everything on the side and agonizing instructions as to the chicken Caesars. Twenty-five is a middle-aged black couple, who complain, with some justice, that the iced tea isn't fresh and the tabletop is sticky. But table 24 is the meteorological event of the century: ten British tourists who seem to have made the decision to absorb the American experience entirely by mouth. Here everyone has at least two drinks—iced tea and milk shake, Michelob and water (with lemon slice, please)—and a huge promiscuous orgy of breakfast specials, mozz sticks, chicken strips, quesadillas, burgers with cheese and without, sides of hash browns with cheddar, with onions, with gravy, seasoned fries, plain fries, banana splits. Poor Jesus! Poor me! Because when I arrive with their first tray of food—after three prior trips just to refill bevs—Princess Di refuses to eat her chicken strips with her pancake-and-sausage special, since, as she now reveals, the strips were meant to be an appetizer. Maybe the others would have accepted their meals, but Di, who is deep into her third Michelob, insists that everything else go back while they work on their "starters." Meanwhile, the yuppies are waving me down for more decaf and the black couple looks ready to summon the NAACP.

Much of what happened next is lost in the fog of war....

I leave. I don't walk out, I just leave. I don't finish my side work or pick up my credit-card tips, if any, at the cash register or, of course, ask Joy's permission to go. And the surprising thing is that you can walk out without permission, that the door opens, that the thick tropical night air parts to let me pass, that my car is still parked where I left it. There is no vindication in this exit, no fuck-you surge of relief, just an overwhelming, dank sense of failure pressing down on me and the entire parking lot. I had gone into this venture in the spirit of science, to test a mathematical proposition, but somewhere along the line, in the tunnel vision imposed by long shifts and relentless concentration, it became a test of myself, and clearly I have failed. Not only had I flamed out as a housekeeper/server, I had even forgotten to give George my tips, and, for reasons perhaps best known to hardworking, generous people like Gail and Ellen, this hurts. I don't cry, but I am in a position to realize, for the first time in many years, that the tear ducts are still there, and still capable of doing their job....

In one month, I had earned approximately $1,040 and spent $517 on food, gas, toiletries, laundry, phone, and utilities. If I had remained in my $500 efficiency, I would have been able to pay the rent and have $22 left over (which is $78 less than the cash I had in my pocket at the start of the month). During this time I bought no clothing except for the required slacks and no prescription drugs or medical care (I did finally buy some vitamin B to compensate for the lack of vegetables in my diet). Perhaps I could have saved a little on food if I had gotten to a supermarket more often, instead of convenience stores, but it should be noted that I lost almost four pounds in four weeks, on a diet weighted heavily toward burgers and fries.

How former welfare recipients and single mothers will (and do) survive in the low-wage workforce, I cannot imagine. Maybe they will figure out how to condense their lives—including child-raising, laundry, romance, and meals—into the couple of hours between full-time jobs. Maybe they will take up residence in their vehicles, if they have one. All I know is that I couldn't hold two jobs and I couldn't make enough money to live on with one. And I had advantages unthinkable to many of the long-term poor—health, stamina, a working car, and no children to care for and support. Certainly nothing in my experience contradicts the conclusion of Kathryn Edin and Laura Lein, in their recent book *Making Ends Meet: How Single Mothers Survive Welfare and Low-Wage Work,* that low-wage work actually involves more hardship and deprivation than life at the mercy of the welfare state. In the coming months and years, economic conditions for the working poor are bound to worsen, even without the almost inevitable recession. As mentioned earlier, the influx of former welfare recipients into the low-skilled workforce will have a depressing effect on both wages and the number of jobs available. A general economic downturn will only enhance these effects, and the working poor will of course be facing it without the slight, but nonetheless often saving, protection of welfare as a backup.

The thinking behind welfare reform was that even the humblest jobs are morally uplifting and psychologically buoying. In reality they are likely to be fraught with insult and stress. But I did discover one redeeming feature of the most abject low-wage work—the camaraderie of people who are, in almost all cases, far too smart and funny and caring for the work they do and the wages they're paid. The hope, of course, is that someday these people will come to know what they're worth, and take appropriate action.

KEY CONCEPTS

gender segregation   occupational segregation   working class

DISCUSSION QUESTIONS

1. What were the major lessons that Ehrenreich learned during her "experiment" as a low-wage worker, and how do they inform your understanding of social class?

2. Having read Ehrenreich's article, how do you understand the relationships between work, stress, and health? What workplace policies could be implemented to improve worker health (mental and physical)?

53

The Service Society and the Changing Experience of Work

CAMERON LYNNE MACDONALD AND CARMEN SIRIANNI

The U.S. economy has changed from being based primarily on manufacturing to being based on service industries. The transition to more "service work" has

SOURCE: Macdonald, Cameron Lynne, and Carmen Sinanni, eds. 1996, "The Service Society and the Changing Experience of Work." In *Working in the Service Society,* edited by Cameron Lynne Macdonald and Carmen Sirianni 1–24. Philadelphia: Temple University Press. Reprinted with permission.

changed the character of workplace control. The service economy is embedded in systems of race, gender, and class stratification that are revealed in patterns of employment and perceptions of who is most fit for particular jobs.

We live and work in a service society. Employment in the service sector currently accounts for 79 percent of nonagricultural jobs in the United States (U.S. Department of Labor 1994: 83). More important, 90 percent of the new jobs projected to be created by the year 2000 will be in service occupations, while the number of goods-producing jobs is projected to decline (Kutscher 1987: 5). Since the mid-nineteenth century the U.S. economy has been gradually transformed from an agriculture-based economy to a manufacturing-based economy to a service-based economy. Near the turn of the century, employment distribution among the three major economic sectors was equally divided at roughly one-third each. Since then, agriculture's labor market share has declined rapidly, now accounting for only about 3 percent of U.S. jobs, while the service sector provides over 70 percent and the goods-producing sector about 25 percent.

The decrease in proportion of manufacturing jobs occurred not because U.S. corporations manufacture fewer goods, but primarily because they use fewer workers to make the goods they produce (Albrecht and Zemke 1990). They use fewer workers due to increasing levels of automation and the exportation of manufacturing functions to low-wage job markets overseas. In addition, the feminization of the work force has created a self-fulfilling cycle in which the entrance of more women into the work force has led to increased demand for those consumer services once provided gratis by housewives (cleaning, cooking, child care, etc.), which in turn has produced more service jobs that are predominantly filled by women.

Still, these trends fail to account fully for the dominance of service work in the U.S. economy, since companies outside of the service sector also contain service occupations. For example, 13.2 percent of the employees in the manufacturing sector work in service occupations such as clerical work, customer service, telemarketing, and transportation (Kutscher 1987). Further, manufacturing and technical occupations are comprised increasingly of service components as U.S. firms adopt Total Quality Management (TQM) and other customer-focused strategies to generate a competitive edge in the global economy. When production efficiency and quality are maximized, the critical variable in the struggle for economic dominance is the quality of interactions with customers. As one business school professor remarks, "Sooner or later, new technology becomes available to everyone. Customer-oriented employees are a lot harder to copy or buy" (Schlesmger and Heskett 1991: 81). So whether one believes that U.S. manufacturing is going to Mexico, to automation, or to the dogs, it is clear that the United States is increasingly becoming a service society and that service work is here to stay.

What do we mean when we speak of "service work"? By definition, a service is intangible; it is produced and consumed simultaneously, and the customer generally participates in its production (Packham 1992).... Service work includes jobs in which face-to-face or voice-to-voice interaction is a fundamental element of the work. "Interactive service work" (Leidner 1993) generally requires some form of what Arlie Hochschild (1983) has termed "emotional labor," meaning the conscious manipulation of the workers' self-presentation either to display feeling states and/or to create feeling states in others. In addition, the guidelines, or "feeling rules," for this emotional labor are created by management and conveyed to the worker as a critical aspect of the job....

Much managerial and professional work also entails emotional labor. For example, doctors are expected to display an appropriate "bedside manner," lawyers are expert actors in and out of the courtroom, and managers, at the most fundamental level, try to instill feeling states and thus promote action in others. However, there remains a critical distinction between white-collar work and work in the emotional proletariat: in management and in the professions, guidelines for emotional labor are generated collegially and, to a great extent, are self-supervised. In front-line service jobs, workers are given very explicit instructions concerning what to say and how to act, and both consumers and managers watch to ensure that these instructions are carried out. However, one could argue that even those in higher ranking positions increasingly experience the kinds of monitoring of their interactive labor encountered by those lower on the occupational ladder, be it by customers, supervisors, or employees....

Given the rising dominance of service occupations in the labor force, what are the special difficulties and opportunities that workers encounter in a service society? A key problem seems to be how to inhabit the job. In the past there was a clear distinction between *careers,* which required a level of personalization, emotion management, authenticity in interaction, and general integration of personal and workplace identities, and *jobs,* which required the active engagement of the body and parts of the mind while the spirit and soul of the worker might be elsewhere. Workers in service occupations are asked to inhabit jobs in ways that were formerly limited to managers and professionals alone. They are required to bring some level of personal identity and self-expression into their work, even if it is only at the level of basic interactions, and even if the job itself is only temporary. The assembly-line worker could openly hate his job, despise his supervisor, and even dislike his co-workers, and while this might be an unpleasant state of affairs, if he completed his assigned tasks efficiently, his attitude was his own problem. For the service worker, inhabiting the job means, at the very least, pretending to like it, and, at most, actually bringing his whole self into the job, liking it, and genuinely caring about the people with whom he interacts.

This demand has several implications: who will be asked to fill what jobs, how they are expected to perform, and how they will respond to those demands. Because personal interaction is a primary component of all service occupations, managers continually strive to find ways to oversee and control those interactions, and worker responses to these attempts vary along a continuum from enthusiastic compliance to outright refusal. Hiring, control of the work process, and the stresses of bringing one's emotions to work are all shaped by the characteristics of the worker and the nature of the work. The self-presentation and other personal characteristics of the worker make up the work process and the work product, and are increasingly the domain of management-worker struggles (see Leidner 1993). In addition, because much of the labor itself is invisible, contests over control of the labor process are often more implicit than explicit....

There are three trends emanating from the rising dominance of service work. First, the need to supervise the production of an intangible, good service, has given rise to particularly invasive forms of workplace control and has led managers to attempt to oversee areas of workers' personal and psychic lives that have heretofore been considered off-limits. Second, the fact that workers' personal characteristics are so firmly linked to their "suitability" for certain service occupations continues to lead to increasing levels of stratification within the service *labor* force. Finally,.... how [do] workers respond to these and other aspects of working in the service society, and how might they build autonomy and dignity into their work, ensuring that service work does not equal servitude?...

GENDER, RACE, AND STRATIFICATION IN THE SERVICE SECTOR

Service industries tend to produce two kinds of jobs: large numbers of low-skill, low-pay jobs and a smaller number of high-skill, high-income jobs, with very few jobs that could be classified in the middle. As Joel Nelson (1994) notes, "Service workers are more likely than manufacturing workers to have lower incomes, fewer opportunities for full-time employment, and greater inequality in earnings" (p. 240). A typical example of this kind of highly stratified work force can be found in fast food industries. These firms tend to operate with a small core of managers and administrators and a large, predominantly part-time work force who possess few skills and therefore are considered expendable (Woody 1989).

As a result, service jobs fall into two broad categories: those likely to be production-line jobs and those likely to be empowered jobs. This distinction not only refers to the level of responsibility and autonomy expected of workers, but also to wages, benefits, job security, and potential for advancement. While empowered service jobs are associated with full-time work, decent wages and benefits, and internal job ladders, production-line jobs offer none of these. Some researchers have described the distinction between empowered and production-line service jobs as one between "core" and "periphery" jobs in the service economy (Hirschhorn 1988; Walsh 1990; Wood 1989)....

The core/periphery distinction may be a misleading characterization of functions in service industries, however. In many firms contingent workers perform functions essential to the operation of the firm and can comprise up to two-thirds of a firm's labor force while "core" workers perform nonessential functions (Walsh 1990). For example, a majority of key functions in industries such as hospitality, food service, and retail sales are performed by workers who, based on their level of benefits, pay, and job security, would be considered periphery workers. In low-skill service positions, job tenure has no relation to output or productivity. Therefore, employers can rely on contingent workers to provide high-quality service at low costs. As T.J. Walsh (1990: 527) points out, it is therefore likely that the poor compensation afforded these workers is due not to their productivity level but to the perceptions of their needs, level of commitment, and availability.

Given the proliferation of service jobs in the United States, key questions for labor analysts are what kinds of jobs are service industries producing, and who is likely to fill them? At the high end, service industries demand educated workers who can rapidly adapt to changing economic conditions. This means that employers may demand a college degree or better for occupations that formerly required only a high school diploma,

> even though many of the job-holders' activities have not changed, or appear relatively simple, because they want workers to be more responsive to the general situation in which they are working and the broader purposes of their work. (Hirschhorn 1988: 35)

In addition, core workers are frequently expected to take on more responsibilities, work longer hours, and intensify their output.

At the low end of the spectrum of service occupations are periphery workers who are frequently classified as part-time, temporary, contract, or contingent. These flexible-use workers act as a safety valve for service firms, allowing managers to redeploy labor costs in response to market conditions. In addition, they allow managers to minimize overhead because they rarely qualify for benefits and generally receive low wages. In 1988, 86 percent of all part-time workers worked in service sector industries, and this trend has continued since (Tilly 1992: p. 30). Labor analysts argue that the bulk of the expansion in part-time and contingent work is due to the expansion of the service sector, which has always used shift and part-time workers as cost-control mechanisms....

Service sector expansion has also sparked the rapid growth of the temporary help industry. Over the past decade, the number of temporary workers has tripled (Kilborn 1995). As the chairman of Manpower, Inc., the largest single employer in the United States, remarked, "The U.S. is going from just-in-time manufacturing to just-in-time employment. The employer tells us, 'I want them delivered exactly when I want them, as many as I need, and when I don't need them, I don't want them here'" (quoted in Castro 1993: 44). Like part-time workers, temporary workers carry no "overhead" costs in terms of taxes and benefits, and they are on call as needed. Since they experience little workplace continuity, they are less likely than full-time, continuous employees to organize or advocate changes in working conditions.

Women, youth, and minorities comprise the bulk of the part-time and contingent work force in the service sector. For example, Karen Brodkin Sacks (1990) has noted that the health care industry is "so stratified by race and gender that the uniforms worn to distinguish the jobs and statuses of health care workers are largely redundant" (p. 188). Patients respond to the signals implicitly transmitted via gender and race and act accordingly, offering deference to some workers and expecting it from others. Likewise in domestic service, race and gender determine who gets which jobs. White American and European women are most likely to be hired for domestic jobs defined primarily as child care, while women of color predominate in those defined as house cleaning (Rollins 1985), regardless of what the actual allocation of work might be. The same kinds of stratification can be found in secretarial work, food service, hotels, and sales occupations (Hochschild 1983; Leidner 1991). In all of these occupational groups, demographic characteristics of the worker determine the job title and thus other factors such as status, pay, benefits, and degree of autonomy on the job.

As a result, the shift to services has had a differential impact on various sectors of the labor market, increasing stratification between the well employed and the underemployed. The service sector work force is highly feminized, especially at the bottom. Within personal and business services, for example, Bette Woody (1989) found that "men are concentrated in high-ticket, high commission sales jobs, and women in retail and food service" (p. 57). And although the decline in manufacturing forced male workers to move into service industries, Jon Lorence (1992: 150) notes that within the service sector, occupations are highly stratified by gender. From 1950 to 1990, 60 percent of all new service sector employment and 74 percent of all new low-skill jobs were filled by women....

Service sector employment is equally stratified by race. For example, Woody (1989) finds that the shift to services has affected black women in two important ways. First, it has meant higher rates of unemployment and underemployment for black men, which increased pressure on black women to be the primary breadwinners for their families and contributed to the overall reduction in black family income. Second, although the increase in service sector jobs has meant greater opportunity for black women and has allowed them to move out of domestic service into the formal economy, as Evelyn Nakano Glenn (1996) also notes, they have remained at the lowest rung of the service employment ladder. Unlike some white women who "moved up" to male-intensive occupations with the shift to services, black women "moved over" to sectors traditionally employing white women (Woody 1989: 54). These traditionally male occupations are not only low-security, low-pay jobs, but they lack internal career ladders. As Ruth Needleman and Anne Nelson (1988) note, "There is no progression from nurse to physician, from secretary to manager" (p. 297).

Service work differs most radically from manufacturing, construction, or agricultural work in the relationship between worker characteristics and the job. Even though discrimination in hiring, differential treatment, differential pay, and other forms of stratification exist in all labor markets, service occupations are the only ones in which the producer in some sense equals the product. In no other area of wage labor are the personal characteristics of the workers so strongly associated with the nature of work. Because at least part of the job in all service occupations is to "manufacture social relations" (Filby 1992; 37), traits such as gender, race, age, and sexuality serve a signaling function, indicating to the customer/employer important cues about the tone of the interaction. Women are expected to be more nurturing and empathetic than men and to tolerate more offensive behavior from customers (Hochschild 1983; Leidner 1991; Pierce 1996, Sutton and Rafaeli 1989). Similarly, both women and men of color are expected to be deferential and to take on more demeaning tasks (Rollins 1985; Woody 1989). In addition, a given task may be viewed as more or less demeaning depending on who is doing it.

These occupations are so stratified that worker characteristics such as race and gender determine not only who is considered desirable or even eligible to fill certain jobs, but also who will want to fill certain jobs and how the job itself is performed. Worker characteristics shape what is expected of a worker by management and customers, how that worker adapts to the job, and what aspects of the job he or she will resist or embrace. The strategies workers use to adapt to the demands of service jobs are likely to differ according to gender and other characteristics. Women are more likely to embrace the emotional demands (e.g., nurturing, care giving) of certain types of service jobs because these demands generally fit their notion of gender-appropriate behavior. As Pierce notes, heterosexual men tend to resist these demands because they find "feminine" emotional labor demeaning; in response they either reframe the nature of the job to emphasize traditional masculine qualities or distance themselves by providing service by rote, making it clear that they are acting under duress.

All of these interconnections between worker, work, and product result in tendencies toward very specific types of labor market stratification. A long and heated debate has raged concerning the ultimate impact of deindustrialization on the structure of the labor market. On one side are those who argue that a shift to a service-based economy will produce skill upgrading and a leveling of job hierarchies as information and communications technologies reshape the labor market (see, for example, Bell 1973). Others take a more pessimistic view, arguing that the shift to services will give rise to two trends: "towards polarization and towards the proliferation of low-wage jobs" (Bluestone and Harrison 1988: 126). In a sense, both positions are correct but for different segments of the labor market.

Overall, the transition to a service-based economy will likely mean a more stratified work force in which more part-time and contingent jobs are filled predominantly by women, minorities, and workers without college degrees. In jobs lacking internal career ladders, these workers have little chance for upward mobility. Contingent workers are also less likely to organize successfully due to their tenuous attachment to specific employers and to the labor force in general. At the opposite end of the service sector occupational spectrum are highly educated managerial, professional, and para-professional workers, who will have equally weak attachments to employers, but who, due to their highly marketable skills, will move with relative ease from one well-paying job to another. Given this divided and economically segregated work force, what are the opportunities for workers to advocate, collectively or individually, for greater security, better working conditions, and a voice in shaping their work?...

REFERENCES

Albrecht, Karl, and Ron Zemke 1990 *Service America!: Doing Business in the New Economy.* New York: Warner Books.

Bell, Daniel. 1973. *The Coming of Post-Industrial Society: A Venture in Social Forecasting.* New York: Basic Books.

Bluestone, Barry, and Bennett Harrison. 1988. *The Great U-Turn: Corporate Restructuring and the Polarizing of America.* New York: Basic Books.

Castro, Janice. 1993. "Disposable Workers" *Time Magazine,* March 29, pp. 43–57.

Filby, M. P. 1992, "'The Figures, the Personality, and the Bums' : Service Work and Sexuality." *Work, Employment, and Society* 6 (March): 23–42.

Glenn, Evelyn Nakano. 1996. "From Servitude to Service Work: Historical Continuities in the Racial Division of Paid Reproductive Labor." In Cameron Lynne Macdonald and Carmen Sirianni, eds., *Working in the Service Society,* pp. 115–156 Philadelphia: Temple University Press.

Hirschhorn, Larry. 1988. "The Post-Industrial Economy: Labor, Skills, and the New Mode of Production." *The Service Industries Journal* 8: 19–38.

Hochschild, Arhe Russell. 1983. *The Managed Heart: Commercialization of Human Feeling.* Berkeley: University of California Press.

Kilborn, Peter T. 1995. "In New Work World, Employers Call All the Shots: Job Insecurity, a Special Report." *New York Times,* July 3, p. Al.

Kutscher, Ronald E. 1987. "Projections 2000: Overview and Implications of the Projections to 2000." *Monthly Labor Review* (September): 3–9.

Leidner, Robin. 1993. *Fast Food, Fast Talk: Service Work and the Routinization of Everyday Life.* Berkeley: University of California Press.

Leidner, Robin. 1991. "Serving Hamburgers and Selling Insurance: Gender, Work, and Identity in Interactive Service Jobs." *Gender & Society* 5: 154–77.

Lorence, Jon. 1992. "Service Sector Growth and Metropolitan Occupational Sex Segregation." *Work and Occupations* 19: 128–56.

Needleman, Ruth, and Anne Nelson. 1988. "Policy Implications: The Worth of Women's Work." In Anne Starham, Eleanor M. Miller, and Hans O. Mauksch, eds., *The Worth of Women's Work: A Qualitative Synthesis* (pp. 293–308). Albany: SUNY Press.

Nelson, Joel I. 1994. "Work and Benefits: The Multiple Problems of Service Sector Employment." *Social Problems* 41: 240–55.

Packham, John. 1992. "The Organization of Work on the Service-Sector Shop Floor." Unpublished paper.

Pierce, Jennifer L. 1996. *Gender Trials: Emotional Lives in Contemporary Law Firms.* Berkeley: University of California Press.

Rollins, Judith. 1985. *Between Women: Domestics and Their Employers*. Philadelphia: Temple University Press.

Sacks, Karen Brodkin. 1990. "Does It Pay to Care?" In Emily K. Abel and Margaret K. Nelson, eds., *Circles of Care: Work and Identity in Women's Lives* (pp. 188–206). Albany: SUNY Press.

Schlesinger, Leonard, and James Heskett. 1991. "The Service-Driven Service Company." *Harvard Business Review* (September–October): 71–81.

Sutton, Robert I., and Anat Rafaeli. 1989. "The Expression of Emotion in Organizational Life." *Research in Organizational Behavior* 11: 1–42.

Tilly, Chris. 1992. "Short Hours, Short Shrift: The Causes and Consequences of Part-Time Employment." In Virginia L. Du Rivage, ed., *New Policies for the Part-Time and Contingent Work-Force* (pp. 15–43). Armonk, NY: M. E. Sharpe, Economic Policy Institute Series.

U.S. Department of Labor, Bureau of Labor Statistics. 1994. *Monthly Labor Review* (July): 74–83.

Walsh, T. J. 1990. "Flexible Labor Utilization in the Private Service Sector." *Work, Employment, and Society* 4: 517–30.

Wood, Stephen, ed. 1989. *The Transformation of Work? Skill, Flexibility, and the Labor Process*. London: Unwin Hyman.

Woody, Bette. 1989. "Black Women in the Emerging Services Economy." *Sex Roles* 21: 45–67.

KEY CONCEPTS

division of labor   downsizing   service sector

DISCUSSION QUESTIONS

1. How has the transition to a service-based economy changed employment patterns in the contemporary economy? What implications does this have for the relationship between management and labor?

2. How do race and gender stratification influence the perceptions of some workers as being suitable for particular forms of service work? What evidence have you witnessed of this phenomenon in your experiences in a service-based society?

54

Harder Times: Undocumented Workers and the U.S. Informal Economy

RICHARD D. VOGEL

In this article Vogel describes the informal economy that employs people "off the book." He argues that these "underground" workers experience conditions much worse than those employed in the formal economic sector, including low pay and lack of benefits. He also argues that workers within the informal economy experience greater gender, ethnic, and racial discrimination. Undocumented workers also cost economies money in that the lack of taxes from these workers further burdens the public sector. He outlines the case of Los Angeles and also looks at other states' data on undocumented workers. His conclusion is that the persistence of a system that feels the need for undocumented workers further worsens conditions for the working class in America. For both native and immigrant workers in the informal economy, the historical patterns of exploitation continue.

Many of the informal economies operating in the world today are the off-spring of globalization and need to be understood as such. The economic and social prospects for people engaged in informal employment—sometimes referred to as "precarious" and "off-the-books employment"—as well as their families and communities, are substantially inferior to those associated with formal employment, and the current boom of informal economic activity bodes ill for all working people.

Referred to variously as "underground," "shadow," "invisible," and "black" (as in "black market") economies, many informal economies have developed around the economic survival activities of workers who have been excluded from the formal economy of their nation or region and exploited by entrepreneurs willing to take advantage of their desperation. Initially considered phenomena of the third world and developing nations, informal economies are now expanding rapidly in the free market nations of the western world, including the United States.

Work in the informal economy contrasts sharply with formal employment: wages and working conditions are substandard; there are no guaranteed

SOURCE: Vogel, Richard D. 2006. "Harder Times: Undocumented Workers and the U.S. Informal Economy." *Monthly Review* 58.

minimum or maximum hours of work, paid holidays, or vacations; gender, ethnic, and racial discrimination are uncontrolled and systematically exploited; no mandatory health or safety regulations protect workers from injury or death on the job; and workers are denied the traditional benefits of employment in the formal economy: workman's compensation; health, unemployment and life insurance; and pensions. Needless to say, few workers join the informal labor force voluntarily—the vast majority are recruited primarily through economic desperation unmitigated by even a minimal social safety net.

Worker participation in the informal economy is a complex phenomenon. While immigrant workers play a significant role in the informal U.S. economy, native workers also participate. These workers in the informal sector include widely divergent groups from professionals who do unreported jobs on the side, and craft workers who exchange work in kind, to marginalized native workers who, because of cutbacks in welfare programs, must accept any work they can find.

The informal sector of the U.S. economy is growing as the working population expands and employment opportunities in the formal economy do not keep pace. Not only are workers being displaced as companies move their operations offshore in search of lower labor costs, but an increasing number of U.S. corporate start-ups are overseas ventures. These current trends contribute to the exclusion of both new and veteran workers from the formal economy.

LOS ANGELES: A DEFINITIVE CASE STUDY

Although no comprehensive national studies of the informal U.S. economy have been published to date, a study of work in the informal economy of the City of Los Angeles by the Economic Roundtable, a nonprofit, public policy research organization, offers an in-depth look at a local informal economy. Since the recently reported case study of Los Angeles's informal economy presents the most refined estimates of informal employment and its impact available for any specific area of the country, it deserves special attention.

This [research] reveals a widening gap between worker and employer reported employment signaling an ongoing informalization of the Los Angeles County economy that has continued unabated through boom and bust for the last eighteen years.

The most remarkable feature … is in the period following the economic recession of 2001. The data indicates that as late as 2004 economic recovery was still out of sight. In spite of that, the informal economy held relatively steady during this period while the formal economy continued in serious decline. The 2005 data, not included in the Economic Roundtable study, indicates a continuation of the same trend. The conclusion of the Economic Roundtable researchers, that the economic stagnation of southern California that was triggered by the recession of 2001 would have been worse without the ameliorating effects of the informal economy, appears to be fully warranted.

In addition to documenting the general trends of informal economic activity, the Economic Roundtable study also offers credible estimates of the actual size of the informal economy in Los Angeles County. These researchers calculated low-range, mid-range, and high-range estimates for the number of county workers in the informal labor force and, after careful consideration, settled on the mid-range estimate of 679,000 workers in 2004. This number represented a substantial 15 percent of Los Angeles County's labor force in that same year.

The Economic Roundtable researchers used this number to determine the cost of informal employment to the public social safety net in Los Angeles County. Determining that the average annual wage for off-the-books jobs was slightly over $12,000 in 2004, they calculated that the informal economy produced an $8.1 billion payroll. This unreported and untaxed payroll shortchanged the public sector in Los Angeles County by the following amounts:

- $1 billion in Social Security taxes (paid by both employers and workers)
- $236 million in Medicare taxes (paid by both employers and workers)
- $96 million in California State Disability Insurance payments (paid by workers)
- $220 million in Unemployment Insurance payments (paid by employers)
- $513 million in Workers Compensation Insurance payments (paid by employers)

These losses added up to over $2 billion in unpaid payroll benefits and insurance that were needed to fund a minimal social safety net for workers. The payroll tax shortfall is continuing, and resistance on the part of California taxpayers to underwrite public relief measures has resulted in the widespread deterioration of social services for informal workers and their families across the state.

Though the Economic Roundtable study is focused primarily on work in the informal economy, researchers point out that informal workers in Los Angeles County spend an estimated $4.1 billion per year that should generate $440 million in sales tax revenue. However these workers purchase many goods and services from informal retailers and service providers who do not collect sales taxes and submit them to the state, further eroding support for the public sector.

The social and political crises of the region are being fueled by the fact that the expanding informal economy in southern California is based on the widespread exploitation of undocumented Mexican and Central American workers. This state of affairs has elevated the issue of illegal immigration in California (and, of course, the nation) to center stage.

THE EXPLOITATION OF UNDOCUMENTED
WORKERS

The issue of the exploitation of undocumented workers is fundamental to understanding the informal economy in Los Angeles. While it is true that many native workers and legal immigrants participate in Los Angeles's informal economy, undocumented immigrants represent the majority of the off-the-books workers. At the same time, while many individuals, both natives and immigrants, participate in both economies, these dual roles should not be allowed to obscure the dominant role of undocumented immigrant workers in Los Angeles's informal economy.

The Economic Roundtable study substantiates the dominant role of undocumented workers in Los Angeles County when it addresses the question of how many workers in the informal economy are undocumented immigrants and which industries employ them. Based on U.S. Immigration and Naturalization Service (INS) and Census 2000 data, the Economic Roundtable estimates that undocumented immigrant workers make up 61 percent of the informal labor force in Los Angeles County and 65 percent in the city. Census 2000 data also establishes that while there are sizable numbers of Asian and Middle Eastern immigrants in California, the vast majority of the immigrants residing and working in southern California are from Latin America.

While the proportion of undocumented workers in the informal economy of Los Angeles is truly remarkable, where these workers are employed is also surprising. The picture of the informal labor force ... contrasts sharply with popular perceptions of immigrant workers that are based on glimpses of day laborers soliciting on street corners, domestic workers waiting at bus stops, and landscape crews packed into pickup trucks. Derived from INS and Census 2000 data, this [research] indicates that a wide array of industries in Los Angeles County systematically exploit undocumented immigrant labor. The Economic Roundtable study reveals that, in fact, the majority of the informal workers labor regularly and out of public view in factories, mills, restaurants, warehouses, workshops, nursing homes, office buildings, and private homes. In actual numbers, the ERT researchers estimate that these top twenty industries in Los Angeles County alone employ over 180,000 undocumented workers.

Most of these undocumented workers are obviously not casual day laborers. The nature of the majority of the jobs ... entails stable relationships between employers and workers and requires extensive community networks to recruit and support those workers.

Although the earnings of immigrant workers in the informal economy are inferior to that of native workers across the board, they do not represent the rock bottom of exploitation—the Economic Roundtable's breakdown of earnings by gender reveals the severity of the super-exploitation of undocumented women workers in the Los Angeles informal economy.

THE SUPER-EXPLOITATION OF UNDOCUMENTED
IMMIGRANT WOMEN

Since this super-exploitation of undocumented women workers takes place almost exclusively out of sight, it all but escapes public attention. However, the

Economic Roundtable documents significant employment of undocumented women workers in the top twenty industries that employ the most informal workers. Generally, women tend to work in homes, personal service jobs, and light manufacturing, traditionally low-paying jobs, while men hold the majority of the jobs in transportation, construction, and other relatively higher-paying blue-collar jobs.

Economic Roundtable researchers highlight the super-exploitation of undocumented women workers in the Los Angeles informal economy when they compare wages by gender in specific job categories for the year 1999. For example, in the category of services to buildings, where the gender shares of jobs were exactly equal, men averaged $13,308 while women made an average of $6,869 (51 percent of the earnings of men). Even in private households, beauty salons, department stores, and health care services, jobs clearly dominated by women workers, men were paid more. While women held 75 percent of the jobs in these categories, they earned only 66 percent of the wages of men doing the same work. Overall, the average wage for undocumented women workers in the informal economy of Los Angeles County was $7,630 compared to $16,553 for men. That amounts to only 46 percent of undocumented men's annual earnings.

The economic basis of Los Angeles's informal economy can be summed up succinctly—undocumented women workers make less than one-half (46 percent) of what documented men workers make, and the average wage of all undocumented workers is less than one-half (41 percent) that of native workers in the same job categories in the formal economy. Even these glaring inequalities underestimate the super-exploitation of informal laborers because they do not account for the other employment benefits that workers in the formal economy receive, which range from 27 to 30 percent of their total employment compensation.

The Economic Roundtable study indicates that the informal sector of the Los Angeles economy, which depends on the exploitation of undocumented workers, has become an integral part of the area's economy. The implications of this extraordinary case study clearly suggest that the trends of the national informal economy deserve careful reconsideration.

NATIONAL TRENDS

Clearly the informal economies of Los Angeles and Los Angeles County are unique. The proximity of southern California to Mexico and the ready access to Mexican and Central American labor have historically shaped the economy, social structure, and culture of the area. The fact that Los Angeles remains the primary destination of Mexican immigrants, both legal and undocumented, is a matter of public record.

The Immigration and Naturalization Service statistics verify the concentration of undocumented immigrants in California but also reveal that Texas, New York, and Illinois (primarily New York City and Chicago) have considerable concentrations. All of these areas also have sizable established informal economies.

The big surprises in the INS study are the recent trends of worker immigration from Mexico and Central America to other destinations in the United States. This [research] shows that six of the top ten states that experienced the highest growth rate of undocumented immigrants during the last decade of the twentieth century are in the southeastern United States, with three states in mid-America not far behind. The phenomenal increase in immigration to the southeast is to the booming metropolitan areas where new construction and service-oriented businesses have created a huge demand for low-cost labor. During the period 1990–2000, the number of undocumented immigrants doubled from 3.5 to 7 million for the United States overall. Current estimates indicate that those immigration trends have accelerated and the nation's total is now between 11.5 and 12 million (http://www.pewhispanic.org).

The present patterns of illegal undocumented worker immigration in the United States signal a rapid expansion of the national informal economy. The practice of basing the informal economy on the exploitation of undocumented workers, well established in California and Texas, is rapidly spreading across the nation. The Pew Hispanic Center estimates that in March 2005, 7.2 million of the 12 million undocumented immigrants in the United States were working, making up about 4.9 percent of the total work force. Applying the same correction factor to this Pew assessment that the Economic Roundtable researchers applied to employment data in the Current Population Survey for L.A. County, to estimate total employment in the local informal economy, suggests that 8–10 percent of all workers might be employed in the informal economy nationwide.

This emerging informal economy signals perennial hard times for the undocumented workers and a continuing war of attrition against the U.S. working class at large.

HARDER TIMES, U.S. STYLE

Global economic theory discusses informal economies in terms of "restructuring." This neoclassical and neoliberal theory maintains that all the workers of the world are competing with each other in a zero-sum game. Informal economies arise when the workers in one economic region of the world lose out to the workers in another region, and the formal economy of the loser region contracts or collapses. According to this theory, informal economies arise to fill the resulting economic vacuum, and the mechanism of the free market will determine the outcome.

Neoliberal economists argue that displaced immigrant workers, primarily from Mexico and Central America, migrate to the United States and compete with native workers for scarce jobs in a post-industrial economy, as dictated by the "invisible hand" of the "free market." The ideological cover story for public consumption, meanwhile, is that immigrant workers take the jobs that native workers refuse.

But it is *political economics*, not free markets, that shape global, national, and local economies. The movement of industrial and service production offshore to take advantage of cheap labor requires the compliance, or outright cooperation, of both home and foreign governments and is therefore a quintessentially political act. Also political is the widespread subversion of immigration and labor law for the sake of profits. These political machinations—not some mythical "invisible hand"—are the engines of the informal economy in the United States.

The burgeoning informal economy in the United States is introducing new elements into familiar historical patterns of exploitation. While the U.S. economy has traditionally been fueled by immigrant labor, the current dependence on undocumented labor is unprecedented. Although the masses of Western and Eastern Europeans who immigrated to do America's dirty work in the past were rewarded with citizenship for their sacrifices, there is no indication that the current Mexican and Central American workers can expect the same. Naturalization is not part of the deal. Dirty and poorly paid work in the contemporary informal economy is not an initiation into the mainstream U.S. working class. The current wave of undocumented immigrants who work in the informal economy today are more likely than not to stay in the informal U.S. economy—as long as they are needed.

The U.S. economy is indebted to undocumented immigrant workers, but there is no indication that the debt is going to be paid. The proposed "guest worker" programs currently being debated in Congress are nothing more than reruns of the Bracero Program that lasted from 1942 to 1964, and the debts owed to workers from that program are still outstanding. The contemporary strategy is to use Mexican and Central American workers displaced by NAFTA as long as possible.

If, or when, they are no longer needed, they can be repatriated. It happened to their compatriots who were uprooted from their homes and communities and driven across the border at the onset of the Great Depression and again during the paramilitary repatriation campaign designated "Operation Wetback" in the recession that followed the Second World War.

The current boom in the informal economy bodes no better for native workers in both the informal and formal sectors of the U.S. economy. Real wages, benefits, and standards of living continue to decline for all workers, and the labor movement is stalled. Realizing the American Dream through hard work in a promising job is becoming a remote possibility rather than an accessible opportunity. And there's nothing like the naked specter of wage slavery in the informal economy and the ghettoization of the poor to keep the expectations of all workers in check. The revolution of rising expectations that gripped workers and minorities in the 1960s has been overshadowed by the prospect of pauperization.

Until the labor movement develops the solidarity necessary to confront the current assault of capitalism, the expanding informal economies of the world will continue to impoverish the lives of all working people. The call to all workers of the world to unite has never been more urgent than it is today.

KEY CONCEPTS

informal economy	labor force	undocumented worker

DISCUSSION QUESTIONS

1. Why do companies decide to employ undocumented workers? Outline the ways undocumented workers are exploited.
2. In your own work experience, have you been employed in the informal economy? Did you see this as an advantage? Can you understand how this is not ideal for immigrants and working-class people?

55

Diversity Management in Corporate America

FRANK DOBBIN, ALEXANDRA KALEV, AND ERIN KELLY

The research presented in this reading looks at the various strategies used by companies to increase diversity among company managers. The research shows that the most popular and most expensive programs to improve diversity are the least effective. Specifically, diversity training and diversity evaluations are not resulting in more women and minorities in positions of management. The authors find, however, that mentoring programs and diversity taskforces work much more effectively.

In the fall of 1996, a tape of Texaco executives joking about the "black jelly beans" working for the company hit the airwaves in the course of a discrimination suit. Texaco settled almost immediately for $176 million. Of that, $35 million was to fund diversity efforts. Texaco revamped training, expanded feedback to managers through diversity audits and diversity evaluations, and set up mentoring programs and affinity networks for women and minorities.

SOURCE: Dobbin, Frank, Alexandra Kalev, and Erin Kelly. 2007. "Diversity Management in Corporate America." *Contexts* 6: 21–27.

Texaco's report on its progress after five years championed the programs but did not make clear whether diversity had actually increased.

Coca-Cola's 2000 race-discrimination settlement promised similar improvements. Under its consent decree, Coca-Cola agreed to pay $192.5 million, $35 million of which was to fund half a dozen major diversity programs. In 2005, Coca-Cola reported to the court that women and minorities held 64 percent of salaried jobs, up from 61 percent in 2000.

Should Texaco and Coca-Cola be models for the rest of corporate America? How can we judge which of their programs should be copied elsewhere? Proponents of equal opportunity and diversity have faced this question for decades. Case studies and theories offer a variety of prescriptions for ending workplace inequality, but, as in medicine, discovering which remedies work best requires a large number of cases observed over time.

In the decades since Congress passed the Civil Rights Act of 1964, firms have experimented with dozens of diversity measures. Consultants have been pushing diversity training, diversity performance evaluations for managers, affinity networks, mentoring programs, diversity councils, and diversity managers, to name a few. But some experts question whether diversity programs are counterproductive, raising the hackles of white men who, after all, still do most of the hiring and firing. Certain programs even seem to get firms into trouble—Texaco executives were recorded talking about "black jelly beans" after hearing the term in diversity training seminars.

Until recently, no one had looked systematically at large numbers of companies to assess which kinds of programs work best, on average. Our research shows that certain programs do increase diversity in management jobs—the best test of whether a program works—but that others do little or nothing.

The good news is that companies that give diversity councils, or diversity managers, responsibility for getting more women and minorities into good jobs typically see significant increases in the diversity of managers. So do companies that create formal mentoring programs. Much less effective are diversity training sessions, diversity performance evaluations for managers, and affinity groups for women and minorities. There is no magic bullet for the problem of inequality. Programs that work in one firm may not work in another. Programs that fail on average may be just what the Acme Rocket Co. needs. But we are beginning to understand what works in general and what does not.

HOW DIVERSE ARE AMERICA'S MANAGERS?

The workforce as a whole has become dramatically more diverse since 1964. One reason is that mothers of young children are much more likely to be in the workforce than they were half a century ago. Another is that America has seen a new wave of immigration from Asia and Latin America. Yet white women, and African-American and Hispanic men and women, remain rare in management. Fifth-six percent of managers were white men. Only 3 percent were African American women. [W]e lump all managers together, but all management jobs are not created equal, and inequality is greatest at the top. While women made up more than one-third of all managers by 2002, they made up only 2 percent of Fortune 500 CEOs.

[Looking at] the percentage of people from each group who were managers, [w]hile 16 percent of white men were managers, only 9 percent of white women, 6 percent of black men, 4 percent of black women, 5 percent of Hispanic men, and 4 percent of Hispanic women were managers. Among Asian-Americans, 10 percent of men and 6 percent of women were managers.

FROM AFFIRMATIVE ACTION TO DIVERSITY MANAGEMENT

Until the 1960s, companies in both the North and the South practiced discrimination openly. King David Holmes had grown up in Connecticut and went off to college in the 1940s. When he returned with his degree at the end of the decade, he went to the local brass mill where family members had worked: "I came back, went in the employment office, said 'I want a job.' I filled out my form—'College graduate.'" Holmes wanted a job in sales, for which any college graduate should have been eligible. The personnel man made no bones about it: they were not hiring blacks in sales: "Oh, that's reserved," he said. They hired Holmes to make sheet metal in the old north mill, a job that required no reading or writing.

In his first year as president, John F. Kennedy decreed that companies wanting to do business with the federal government would have to take affirmative action to end discrimination of that sort, expanding orders from Roosevelt, Truman, and Eisenhower that applied to military contractors. The year after Kennedy's assassination, Lyndon Johnson signed the Civil Rights Act 1964, outlawing discrimination in education, housing, public accommodations, and employment.

In principle, Kennedy's affirmative action order and the Civil Rights Act made discrimination illegal in the private sector. In practice, most employers did not think the order would much affect them. Most thought they would have to take down "No Negroes" signs, and that would be the end of it.

Over time, however, Congress and the courts strengthened equal opportunity legislation. By the mid-1970s, the federal government was getting thousands of complaints a year, and personnel experts were setting up equal opportunity programs in the hope of fending off complaints and providing a defense in lawsuits. Because the law did not specify how employers should combat discrimination, firms experimented. Early on, they installed formal job tests, job ladders, and annual performance evaluations—measures designed to take the guesswork (and bias) out of hiring and promotion. The Supreme Court vetted many of these programs but overturned others, such as job tests that had been used as a subterfuge to exclude blacks from the workforce.

By the time of Ronald Reagan's election in 1980, most big employers had hired equal opportunity managers, if not entire departments, and were in the process of creating race-relations workshops, special recruitment systems, and a host of programs designed to improve opportunities for women and minorities. After Reagan's campaign vow to put the brakes on federal regulation, equal opportunity managers began to rebrand themselves as diversity managers. New programs emphasized business preparedness for an increase of women and minorities in the workforce. The forward-thinking executive would plan for change not because it was the cautious thing to do given the likelihood of lawsuits, but because it was the rational thing to do given growing numbers of African-Americans, Latinos, Asians, and women in the workforce.

Employers have invested in three broad approaches to increasing diversity: (1) changing the attitudes and behavior of managers; (2) improving the social ties of women and minorities; and (3) assigning responsibility for diversity to special managers and taskforces. We set out to discover which of these approaches works best.

THE STUDY

Texaco and Coca-Cola spent more than $7 million a year on diversity programs created under their consent decrees. Even Fortune 500 companies not under consent decrees can spend millions of dollars a year on trainers alone, without accounting for lost work time or travel expenses. Yet it is not easy to determine whether these programs actually improve opportunity for women and minorities.

Our challenge was to get accurate data on workforce diversity and on diversity programs for enough companies, over a long enough period of time, so that we could use sophisticated statistical techniques to isolate the effects of, say, diversity training on the percentage of black women in management. Our first break came when we obtained access to the EEOC's EEO-1 reports under a federal program requiring strict confidentiality. Every private firm with 100 or more employees must submit a form each year detailing workforce race, gender, and ethnicity in nine broad job categories. Those data are in good shape back to 1971. Because we were only interested in conducting statistical analyses, we were happy to pledge to maintain the confidentiality of records for individual companies. Our second break was that the EEOC had not changed job, race, or ethnicity categories, allowing us to compare trends over time.

We surveyed 829 of the companies covered in the EEOC's massive data file, putting together a life history of employment practices at each. We asked companies whether they had used dozens of different programs, and when those programs had been in place. Our research question was simple: If a company adopts a particular diversity program, what effect does it have on the share of women and minorities in management? To answer the question, we conducted statistical analyses on the 829 firms over 31 years. Some firms were not founded until the middle of our time frame, but we have more than 16,000 data points (each being a workplace in a particular year). Our analyses considered more than 60 work-place characteristics that might affect diversity so as to isolate the effects of diversity taskforces, training, and so on.

CHANGING MANAGERS' BEHAVIOR

One source of gender and racial inequality in the workplace is stereotyping and bias among managers who make hiring and promotion decisions. Research shows that educating people about members of other groups may reduce stereotyping. Such training first appeared in race-relations workshops conducted for government agencies and large federal contractors in the early 1970s. An industry of diversity trainers emerged in the 1980s to argue that people were unaware of their own racial, ethnic, and gender biases and that sensitivity training would help them to overcome stereotypes. Item number 1 in Texaco's 1997 settlement agreement was a "mandatory diversity learning experience," which included both regular diversity training and continuing education.

Some diversity experts dismissed training, arguing that attitudes are difficult to alter but that behavior can be changed with feedback. Instead they supported performance evaluations offering feedback on managers' diversity efforts.

...The current enthusiasm for training and evaluations has made them widely popular. Four in ten of the firms we surveyed offered training, and one in five had diversity evaluations. But these programs did not, on average, increase management diversity.

Our figure shows the percentage change in the proportion of managers from each group as a consequence of each program—holding the effects of other

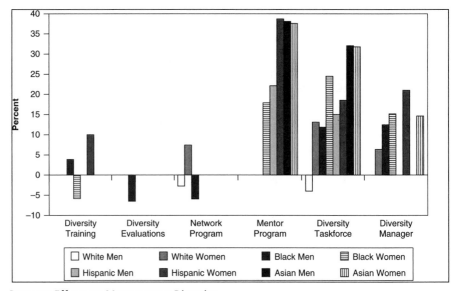

Program Effects on Management Diversity

factors constant. Because minorities are rare in management to begin with, for them the bars represent small numerical changes. African-American men held about 3 percent of management jobs in 2002, and so a 25 percent increase does not translate into big numerical gains. For white women, the numerical changes are more dramatic. But for all groups, an increase is an increase. Where the effect is not statistically significant, we show it in the figure as zero.

Training had small negative effects for African-American women and small positive effects for Hispanic women. Evaluations had only small negative effects on black men. In very large workplaces, with 1,500 or more workers, diversity training has a small positive effect on other groups, but nothing like the effects of the mentoring programs and taskforces discussed below. Evaluations were no more effective in large workplaces.

Some psychological research supports our finding that training may be ineffective. Laboratory experiments and field studies show that it is difficult to train away stereotypes, and that white men often respond negatively to training—particularly if they are concerned about their own careers. If training cannot eliminate stereotypes, and if it can elicit backlash, perhaps it is not surprising that, on average, it does not revolutionize the workplace.

We suspect that the potential of diversity performance evaluations is undermined by the complexity of rating systems. The diversity manager at a giant communications company told us that diversity performance makes up 5 percent of the total score. With so much of the score riding on other things, like sales performance, perhaps putting diversity into an employee's annual evaluation score is not enough to change behavior.

Overall, companies that try to change managers' behavior through training and evaluations have not seen much change. That is disappointing, because training is the single most popular program and, by most accounts, the most costly, and because many companies have put their money on diversity evaluations in recent years.

CREATING SOCIAL CONNECTIONS

Another way to view the problem of inequality in management jobs is from the supply side. Do women and minorities have the social resources needed to succeed? Many firms have well-educated white women and minorities in their ranks who fail to move up, and some sociologists suggest this is because they lack the kinds of social connections that ambitious white men develop easily with coworkers and bosses. Strong social networks and mentors have long been thought crucial to career success. Mark Granovetter's 1974 classic book, *Getting a Job*, showed that people often find jobs through their social networks. *A Harvard Business Review* article from 1978 titled "Everyone Who Makes It Has a Mentor" suggested that mentors are crucial. The idea that women and minorities have trouble moving up because they lack good network contacts and mentoring relationships has caught on among diversity experts.

Consultants argued that formal programs could help women and minorities develop social networks at work. Company affinity networks became popular. Under these affinity networks, people gather regularly to hear speakers and talk about their experiences.

Many companies also put in formal mentoring programs that match aspiring managers with seasoned higher-ups for regular career-advising sessions. These programs can target women and minorities but are often open to all employees. The Texaco and Coca-Cola settlements both called for new mentoring programs.

When it comes to improving the position of women and minorities in management, network programs do little on average, but mentor programs show strong positive effects. In the average workplace, network programs lead to slight increases for white women and decreases for black men. Perhaps that is not surprising, as studies of these network programs suggest that while they sometimes give people a place to share their experiences, they often bring together people on the lowest rungs of the corporate ladder. They may not put people in touch with what they need to know, or whom they need to know, to move up.

Mentor programs, by contrast, appear to help women and minorities. They show positive effects for black women, Latinos, and Asians. Moreover, in industries with many highly educated workers who are eligible for management jobs, they also help white women and black men. Mentor programs put aspiring managers in contact with people who can help them move up, both by offering advice and by finding them jobs. This strategy appears to work.

MAKING SOMEONE RESPONSIBLE

... Today, despite new university rules about advertising open positions, many departments still rely on word of mouth to find new recruits.

The best way to bring about change, management theorists argue, is to make new programs the responsibility of a person or a committee. Instead of just sending around a memo asking managers to practice equal opportunity, military contractors, fearful of losing their contracts in the 1960s, put someone in charge of affirmative action programs. That person's sole job was to make sure the firm hired, and kept, women and minorities. Now that is the job of the diversity manager, or chief diversity officer. Recently firms have also put in taskforces, or diversity councils, comprising managers from different departments and charged with finding ways to increase diversity.

Our analyses show that making a person or a committee responsible for diversity is very effective. Companies that establish taskforces typically see small decreases in the number of white men in management, and large increases for every other group (white men dominate management to begin with, so a small percentage decrease for white men can make space for big percentage gains for the other groups). Firms that put in diversity managers see increases for all groups of women, and for black men.

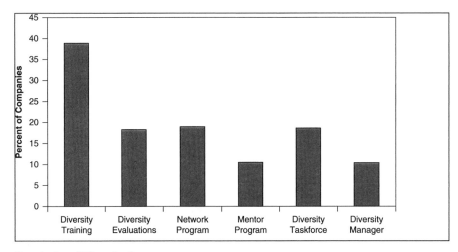

Popularity of Diversity Programs

In interviews, executives tell us that diversity managers and taskforces are effective because they identify specific problems and remedies. If the taskforce sees that the company has not been recruiting African-American engineers, it will suggest sending recruiters to historically black colleges. If a company has trouble retaining women, the diversity manager may talk to women at risk of leaving and try to work out arrangements that will keep them on the job. Managers and taskforces feel accountable for change, and they monitor quarterly employment data to see if their efforts are paying off. If not, it is back to the drawing board to sketch new diversity strategies. Taskforces may be so widely effective, some tell us, because they cause managers from different departments to "buy into" the goal of diversity.

LESSONS FOR EXECUTIVES

What are the most cost-effective strategies for increasing diversity? Our last figure shows what companies were doing in 2002. Many companies were not investing in the strategies that have proven most effective. Three of the four most popular programs—diversity training, evaluations, and network programs—have no positive effects in the average workplace. The two least popular initiatives, mentoring and diversity managers, were among the most effective.

On average, programs designed to reduce bias among managers responsible for hiring and promotion have not worked. Neither diversity training to extinguish stereotypes, nor diversity performance evaluations to provide feedback and oversight to people making hiring and promotion decisions, have accomplished much. This is not surprising in the light of research showing that stereotypes are difficult to extinguish.

There are two caveats about training. First, it does show small positive effects in the largest of workplaces, although diversity councils, diversity managers, and mentoring programs are significantly more effective. Second, optional (not mandatory) training programs and those that focus on cultural awareness (not the threat of the law) can have positive effects. In firms where training is mandatory or emphasizes the threat of lawsuits, training actually has negative effects on management diversity. Psychological studies showing backlash in response to diversity training suggest a reason: managers respond negatively when they feel that someone is pointing a finger at them. Most managers are still white and male, so forced training focusing on the law may backfire. Unfortunately, among firms with training, about three-quarters make it mandatory and about three-quarters cover the dangers posed by lawsuits.

Our findings suggest that firms should look into how they can make training and evaluations more effective. Some experts suggest that training should focus on hiring and promotion routines that can quash subjectivity and bias. Others suggest that diversity performance evaluations based on objective indicators (minority hiring and retention, for instance) and tied to significant incentives (attractive bonuses, or promotions) would work. But few companies follow that model.

One piece of good news is that mentor programs appear to be quite effective. Such programs can provide women and minorities with career advice and vital connections to higher-ups. While mentoring can be costly—involving release time for mentors and mentees, travel to meetings, and training for both groups—the expense generally pays off. One reason mentoring may not elicit backlash, as training often does, is that many companies make it available to men and women and to majority and minority workers. The theory is that women and minorities less often find mentors on their own (if they did, there would be less of a problem to start with), and thus benefit more from formal programs.

Another piece of good news is that companies that assign responsibility for diversity to a diversity manager, or to a taskforce made up of managers from different departments, typically see significant gains in diversity. Management experts have long argued that if a firm wants to achieve a new goal, it must make someone responsible for that goal. To hire a diversity manager or appoint a taskforce or council is to make someone responsible. Both managers and taskforces scrutinize workforce data to see if their efforts are paying off, and both propose specific solutions to the company's problems in finding, hiring, keeping, and promoting women and minorities. Taskforces have the added advantage of eliciting buy-in—they focus the attention of department heads from across the firm who sit together with a collective mission.

If companies regrouped and put their time and energy into programs known to work elsewhere, they would likely see small but steady increases in the representation of women and minorities in management jobs. Programs that assign responsibility for change and that connect promising management talent with mentors (another, less formal way to assign responsibility), seem to hold the best hope. Managers might also spend more time assessing programs that don't seem to be working and trying to figure out how to make them effective.

KEY CONCEPTS

diversity training   equal employment opportunity   mentoring

DISCUSSION QUESTIONS

1. What does the reading tell us about diversity training? Is there a difference between mandatory and voluntary diversity training?

2. Have you ever attended a diversity training workshop? What was the experience like? Do you see value in these programs? What about mentoring? How do you see mentoring applying to diversity in the workplace? What about mentoring for students on campus?

E. Government and Politics

56

The Power Elite

C. WRIGHT MILLS

C. Wright Mills's classic book, The Power Elite, first published in 1956, remains an important analysis of the system of power in the United States. He argues that national power is located in three particular institutions: the economy, politics, and the military. An important point in his article is that the power of elites is derived from their institutional location, not their individual attributes.

The powers of ordinary men are circumscribed by the everyday worlds in which they live, yet even in these rounds of job, family, and neighborhood they often seem driven by forces they can neither understand nor govern. "Great changes" are beyond their control, but affect their conduct and outlook none the less. The very framework of modern society confines them to projects not their own, but from every side, such changes now press upon the men and women of the mass society, who accordingly feel that they are without purpose in an epoch in which they are without power.

But not all men are in this sense ordinary. As the means of information and of power are centralized, some men come to occupy positions in American society from which they can look down upon, so to speak, and by their decisions mightily affect, the everyday worlds of ordinary men and women. They are not made by their jobs; they set up and break down jobs for thousands of others; they are not confined by simple family responsibilities; they can escape. They may live in many hotels and houses, but they are bound by no one community. They need not merely "meet the demands of the day and hour"; in some part, they create these demands, and cause others to meet them. Whether or not they profess their power, their technical and political experience of it far transcends that of the underlying population. What Jacob Burckhardt said of "great men," most Americans might well say of their elite: "They are all that we are not."

The power elite is composed of men whose positions enable them to transcend the ordinary environments of ordinary men and women; they are in positions to make decisions having major consequences. Whether they do or do not make such decisions is less important than the fact that they do occupy such pivotal positions: their failure to act, their failure to make decisions, is itself an act that is often of greater consequence than the decisions they do make. For they are in command of the major hierarchies and organizations of modern society. They rule the big corporations. They run the machinery of the state and claim its prerogatives. They direct the military establishment. They occupy the strategic command posts of the social structure, in which are now centered the effective means of the power and the wealth and the celebrity which they enjoy.

The power elite are not solitary rulers. Advisers and consultants, spokesmen and opinion-makers are often the captains of their higher thought and decision. Immediately below the elite are the professional politicians of the middle levels of power, in the Congress and in the pressure groups, as well as among the new and old upper classes of town and city and region. Mingling with them, in curious ways which we shall explore, are those professional celebrities who live by being continually displayed but are never, so long as they remain celebrities, displayed enough. If such celebrities are not at the head of any dominating hierarchy, they do often have the power to distract the attention of the public or afford sensations to the masses, or, more directly, to gain the ear of those who do occupy positions of direct power. More or less unattached, as critics of morality and technicians of power, as spokesmen of God and creators of mass sensibility, such celebrities and consultants are part of the immediate scene in which the drama of the elite is enacted. But that drama itself is centered in the command posts of the major institutional hierarchies.

The truth about the nature and the power of the elite is not some secret which men of affairs know but will not tell. Such men hold quite various theories about their own roles in the sequence of event and decision. Often they are uncertain about their roles, and even more often they allow their fears and their hopes to affect their assessment of their own power. No matter how great their actual power, they tend to be less acutely aware of it than of the resistances of others to its use. Moreover, most American men of affairs have learned well the rhetoric of public relations, in some cases even to the point of using it when they are alone, and thus coming to believe it. The personal awareness of the actors is only one of the several sources one must examine in order to understand the higher circles. Yet many who believe that there is no elite, or at any rate none of any consequence, rest their argument upon what men of affairs believe about themselves, or at least assert in public.

There is, however, another view: those who feel, even if vaguely, that a compact and powerful elite of great importance does now prevail in America often base that feeling upon the historical trend of our time. They have felt, for example, the domination of the military event, and from this they infer that generals and admirals, as well as other men of decision influenced by them, must be enormously powerful. They hear that the Congress has again abdicated to a handful of men decisions clearly related to the issue of war or peace. They know that the bomb was dropped over Japan in the name of the United States of America, although they were at no time consulted about the matter. They feel that they live in a time of big decisions; they know that they are not making any.

Accordingly, as they consider the present as history, they infer that at its center, making decisions or failing to make them, there must be an elite of power.

On the one hand, those who share this feeling about big historical events assume that there is an elite and that its power is great. On the other hand, those who listen carefully to the reports of men apparently involved in the great decisions often do not believe that there is an elite whose powers are of decisive consequence.

Both views must be taken into account, but neither is adequate. The way to understand the power of the American elite lies neither solely in recognizing the historic scale of events nor in accepting the personal awareness reported by men of apparent decision. Behind such men and behind the events of history, linking the two, are the major institutions of modern society. These hierarchies of state and corporation and army constitute the means of power; as such they are now of a consequence not before equaled in human history—and at their summits, there are now those command posts of modern society which offer us the sociological key to an understanding of the role of the higher circles in America.

Within American society, major national power now resides in the economic, the political, and the military domains. Other institutions seem off to the side of modern history, and, on occasion, duly subordinated to these. No family is as directly powerful in national affairs as any major corporation; no church is as directly powerful in the external biographies of young men in America today as the military establishment; no college is as powerful in the shaping of momentous events as the National Security Council. Religious, educational, and family institutions are not autonomous centers of national power; on the contrary, these decentralized areas are increasingly shaped by the big three, in which developments of decisive and immediate consequence now occur.

Families and churches and schools adapt to modern life; governments and armies and corporations shape it; and, as they do so, they turn these lesser institutions into means for their ends. Religious institutions provide chaplains to the armed forces where they are used as a means of increasing the effectiveness of its morale to kill. Schools select and train men for their jobs in corporations and their specialized tasks in the armed forces. The extended family has, of course, long been broken up by the industrial revolution, and now the son and the father are removed from the family, by compulsion if need be, whenever the army of the state sends out the call. And the symbols of all these lesser institutions are used to legitimate the power and the decisions of the big three.

The life-fate of the modern individual depends not only upon the family into which he was born or which he enters by marriage, but increasingly upon the corporation in which he spends the most alert hours of his best years; not only upon the school where he is educated as a child and adolescent, but also upon the state which touches him throughout his life; not only upon the church in which on occasion he hears the word of God, but also upon the army in which he is disciplined.

If the centralized state could not rely upon the inculcation of nationalist loyalties in public and private schools, its leaders would promptly seek to modify the decentralized educational system. If the bankruptcy rate among the top five hundred corporations were as high as the general divorce rate among the thirty-seven million married couples, there would be economic catastrophe on an international scale. If members of armies gave to them no more of their lives than do believers to the churches to which they belong, there would be a military crisis.

Within each of the big three, the typical institutional unit has become enlarged, has become administrative, and, in the power of its decisions, has become centralized. Behind these developments there is a fabulous technology, for as institutions, they have incorporated this technology and guide it, even as it shapes and paces their developments.

The economy—once a great scatter of small productive units in autonomous balance—has become dominated by two or three hundred giant corporations, administratively and politically interrelated, which together hold the keys to economic decisions.

The political order, once a decentralized set of several dozen states with a weak spinal cord, has become a centralized, executive establishment which has taken up into itself many powers previously scattered, and now enters into each and every cranny of the social structure.

The military order, once a slim establishment in a context of distrust fed by state militia, has become the largest and most expensive feature of government, and, although well versed in smiling public relations, now has all the grim and clumsy efficiency of a sprawling bureaucratic domain.

In each of these institutional areas, the means of power at the disposal of decision makers have increased enormously; their central executive powers have been enhanced; within each of them modern administrative routines have been elaborated and tightened up.

As each of these domains becomes enlarged and centralized, the consequences of its activities become greater, and its traffic with the others increases. The decisions of a handful of corporations bear upon military and political as well as upon economic developments around the world. The decisions of the military establishment rest upon and grievously affect political life as well as the very level of economic activity. The decisions made within the political domain determine economic activities and military programs. There is no longer, on the one hand, an economy, and, on the other hand, a political order containing a military establishment unimportant to politics and to moneymaking. There is a political economy linked, in a thousand ways, with military institutions and decisions. On each side of the world-split running through central Europe and around the Asiatic rimlands, there is an ever-increasing interlocking of economic, military, and political structures. If there is government intervention in the corporate economy, so is there corporate intervention in the governmental process. In the structural sense, this triangle of power is the source of the *interlocking directorate* that is most important for the historical structure of the present....

At the pinnacle of each of the three enlarged and centralized domains, there have arisen those higher circles which make up the economic, the political, and the military elites. At the top of the economy, among the corporate rich, there are the chief executives; at the top of the political order, the members of the political directorate; at the top of the military establishment, the elite of soldier-statesmen clustered in and around the Joint Chiefs of Staff and the upper echelon. As each of these domains has coincided with the others, as decisions tend to become total in their consequence, the leading men in each of the three domains of power—the warlords, the corporation chieftains, the political directorate—tend to come together, to form the power elite of America. ...

By the powerful we mean, of course, those who are able to realize their will, even if others resist it. No one, accordingly, can be truly powerful unless he has access to the command of major institutions, for it is over these institutional means of power that the truly powerful are, in the first instance, powerful. Higher politicians and key officials of government command such institutional power; so do admirals and generals, and so do the major owners and executives of the larger corporations. Not all power, it is true, is anchored in and exercised by means of such institutions, but only within and through them can power be more or less continuous and important.

Wealth also is acquired and held in and through institutions. The pyramid of wealth cannot be understood merely in terms of the very rich; for the great inheriting families, as we shall see, are now supplemented by the corporate institutions of modern society: every one of the very rich families has been and is closely connected—always legally and frequently managerially as well—with one of the multi-million dollar corporations.

The modern corporation is the prime source of wealth, but, in latter-day capitalism, the political apparatus also opens and closes many avenues to wealth. The amount as well as the source of income, the power over consumer's goods as well as over productive capital, are determined by position within the political economy. If our interest in the very rich goes beyond their lavish or their miserly consumption, we must examine their relations to modern forms of corporate property as well as to the state; for such relations now determine the chances of men to secure big property and to receive high income....

If we took the one hundred most powerful men in America, the one hundred wealthiest, and the one hundred most celebrated away from the institutional positions they now occupy, away from their resources of men and women and money, away from the media of mass communication that are now focused upon them— then they would be powerless and poor and uncelebrated. For power is not of a man. Wealth does not center in the person of the wealthy. Celebrity is not inherent in any personality. To be celebrated, to be wealthy, to have power requires access to major institutions, for the institutional positions men occupy determine in large part their chances to have and to hold these valued experiences....

KEY CONCEPTS

interlocking directorate   power   power elite model

DISCUSSION QUESTIONS

1. What evidence do you see of the presence of the power elite in today's economic, political, and military institutions? Suppose that Mills were writing his book today; what might he change about his essay?

2. Mills argues that the power elite use institutions such as religion, education, and the family as the means to their ends. Find an example of this from the daily news, and explain how Mills would see this institution as being shaped by the power elite.

57

Has the Power Elite Become Diverse?

RICHARD L. ZWEIGENHAFT AND G. WILLIAM DOMHOFF

Richard Zweigenhaft and G. William Domhoff ask whether the power elite has changed by incorporating more diverse groups into the power structure of the United States. Although there is now more diversity in these circles of power, the "newcomers" tend to share the values of those already in power.

In justices based on race, gender, ethnicity, and sexual orientation have been the most emotionally charged and contested issues in American society since the end of the 1960s, far exceeding concerns about social class and rivaled only by conflicts over abortion. These issues are now subsumed under the umbrella term *diversity,* which has been discussed extensively from the perspectives of both the aggrieved and those at the middle levels of the social ladder who resist any changes.

... [W]e look at diversity from a new angle: we examine its impact on the small group at the top of American society that we call the *power elite,* those

SOURCE: Richard L. Zweigenhaft and G. William Domhoff. 2006. *Diversity in the Power Elite: How It Happened, Why It Matters.* Lanham, MD: Rowman & Littlefield.

who own and manage large banks and corporations, finance the political campaigns of conservative Democrats and virtually all Republicans at the state and national levels, and serve in government as appointed officials and military leaders. We ask whether the decades of civil disobedience, protest, and litigation by civil rights groups, feminists, and gay and lesbian rights activists have resulted in a more diverse power elite. If they have, what effects has this new diversity had on the functioning of the power elite and on its relation to the rest of society?

We also compare our findings on the power elite with those from our parallel study of Congress to see whether there are differences in social background, education, and party affiliation for women and other underrepresented groups in these two realms of power. We explore the possibility that elected officials come from a wider range of occupational and income backgrounds than members of the power elite. In addition, we ask whether either of the major political parties has been more active than the other in advancing the careers of women and minorities.

According to many popular commentators, the composition of the higher circles in the United States had indeed changed by the late 1980s and early 1990s....

Since the 1870s, the refrain about the new diversity of the governing circles has been closely intertwined with a staple of American culture created by Horatio Alger Jr., whose name has become synonymous with upward mobility in America. Far from being a Horatio Alger himself, the man who gave his name to upward mobility was born into a patrician family in 1832; his father was a Harvard graduate, a Unitarian minister, and a Massachusetts state senator. Horatio Jr. graduated from Harvard at the age of nineteen, after which he pursued a series of unsuccessful efforts to establish himself in various careers. Finally, in 1864, Alger was hired as a Unitarian minister in Brewster, Massachusetts. Fifteen months later, he was dismissed from this position for homosexual acts with boys in the congregation.

Alger returned to New York, where he soon began to spend a great deal of time at the Newsboys' Lodging House, founded in 1853 for footloose youngsters between the ages of twelve and sixteen and home to many youths who had been mustered out of the Union Army after serving as drummer boys. At the Newsboys' Lodging House, Alger found his literary niche and his subsequent claim to fame: writing books in which poor boys make good. His books sold by the hundreds of thousands in the last third of the nineteenth century, and by 1910, they were enjoying annual paperback sales of more than one million.[1]

The deck is not stacked against the poor, according to Alger. When they simply show a bit of gumption, work hard, and thereby catch a break or two, they can become part of the American elite. The persistence of this theme, reinforced by the annual Horatio Alger Award given to such well-known personalities as Ronald Reagan, Bob Hope, and Billy Graham (who might not have been so eager to accept the award had they known that Alger did not fit their fantasy of a straight, white, patriarchal, American male), suggests that we may be dealing once again with a cultural myth.

In its early versions, of course, the story concerned the great opportunities available for poor white boys willing to work their way to the top. More recently, the story has featured black Horatio Algers who started in the ghetto, Latino Horatio Algers who started in the barrio, Asian American Horatio Algers whose parents were immigrants, and female Horatio Algers who seem to have no class backgrounds, all of whom now sit on the boards of the country's largest corporations.[2]

Few people read Horatio Alger today, but they still believe in upward mobility in an era when real wages have been stagnant for the bottom 80 percent since the 1970s, the percentage of low-income people finishing college is decreasing, and the rate of upward mobility is declining.[3]...

But is any of the talk about Horatio Alger and upward mobility true? Can anecdotes, dime novels, and self-serving autobiographical accounts about diversity, meritocracy, and upward social mobility survive a more systematic analysis? Have very many women or members of other previously excluded groups made it to the top? Has class lost its importance in shaping life chances?

... [We] address these and related questions within the framework provided by the iconoclastic sociologist C. Wright Mills in his classic *The Power Elite,* published half a century ago in 1956 when the media were in the midst of what Mills called the "Great American Celebration," and still accurate today in terms of many of the issues he addressed. In spite of the Great Depression of the 1930s, Americans had pulled together to win World War II, and the country was both prosperous at home and influential abroad. Most of all, according to enthusiasts, the United States had become a relatively classless and pluralistic society, where power belonged to the people through their political parties and public opinion. Some groups certainly had more power than others, but no group or class had too much. The New Deal and World War II had forever transformed the corporate-based power structure of earlier decades.

Mills challenged this celebration of pluralism by studying the social backgrounds and career paths of the people who occupied the highest positions in what he saw as the three major institutional hierarchies in postwar America, the corporations, the executive branch of the federal government, and the military. He found that almost all members of this leadership group, which he called the power elite, were white, Christian males who came from "at most, the upper third of the income and occupational pyramids," despite the many Horatio Algeresque claims to the contrary.[4] A majority came from an even narrower stratum, the 11 percent of U.S. families headed by busniesspeople or highly educated professionals like physicians and lawyers. Mills concluded that power in the United States in the 1950s was just about as concentrated as it had been since the rise of the large corporations, although he stressed that the New Deal and World War II had given political appointees and military chieftains more authority than they had exercised previously.

It is our purpose, therefore, to take a detailed look at the social, educational, and occupational backgrounds of the leaders of these three institutional hierarchies to see whether they have become more diverse in terms of gender, race, ethnicity, and sexual orientation, and also in terms of socioeconomic origins. Unlike Mills, however, we think the power elite is more than a set of institutional leaders; it is also the leadership group for the small upper class of owners and managers of large, income-producing properties, the 1 percent of American households that owned 44.1 percent of all privately held stock, 58.0 percent of financial securities, and 57.3 percent of business equity in 2001, the last year for which systematic figures are available. (By way of comparison, the bottom 90 percent, those who work for hourly wages or monthly salaries, have a mere 15.5 percent of the stock, 11.3 percent of financial securities, and 10.4 percent of business equity.) Not surprisingly, we think the primary concern of the power elite is to support the kind of policies, regulations, and political leaders that maintain this structure of privilege for the very rich.[5]

We first study the directors and chief executive officers (CEOs) of the largest banks and corporations, as determined by annual rankings compiled by *Fortune* magazine. The use of *Fortune* rankings is now standard practice in studies of corporate size and power. Over the years, *Fortune* has changed the number of corporations on its annual list and the way it groups them. For example, the separate listings by business sector in the past, like "life insurance companies," "diversified financial companies," and "service companies," have been combined into one overall list, primarily because many large businesses are now involved in more than one of the traditional sectors. Generally speaking, we use the *Fortune* list or lists available for the time period under consideration.

Second, again following Mills, we focus on the appointees to the president's cabinet when we turn to the "political directorate," Mills's general term for top-level officials in the executive branch of the federal government. We also have included the director of the CIA... because of the increased importance of that agency since Mills wrote. Third, and rounding out our portrait of the power elite, we examine the same two top positions in the military, generals and admirals, that formed the basis for Mills's look at the military elite.

As we have noted, we also study Congress. In the case of senators, we do the same kind of background studies that we do for members of the power elite. For members of the House of Representatives, we concern ourselves only with party affiliation for most groups. We include findings on senators and representatives from underrepresented groups for two reasons. First, this allows us to see whether there is more diversity in the electoral system than in the power elite. Second, we do not think, as Mills did, that Congress should be relegated to the "middle level" of power. To the contrary, we believe that Congress is an integral part of the power structure in America. Similarly, unlike Mills, because we think that the Supreme Court is a key institution within the power elite, we have added information on Supreme Court appointments....

In addition to studying the extent to which women and other previously excluded groups have risen in the system, we focus on whether they have followed different avenues to the top than their predecessors did, as well as on any special roles they may play. Are they in the innermost circles of the power elite, or are they more likely to serve as buffers and go-betweens? Do they go just so far and no farther? What obstacles does each group face?

We also examine whether or not their presence affects the power elite itself. Do they influence the power elite in a more liberal direction, or do they end up endorsing traditional conservative positions, such as opposition to trade unions, taxes, and government regulation of business?… We argue that the diversity forced on the power elite has had the ironic effect of strengthening it by providing it with a handful of people who can reach out to the previously excluded groups and by showing that the American system can deliver on its most important promise, an equal opportunity for every individual. It is as if the diversity efforts in the final thirty-five years of the twentieth century were scripted for the arrival of a George W. Bush, the most conservative and uncompromising president since Herbert Hoover, a president with the most diverse cabinet in the history of the country; in the first five years, its members have included six women, four African Americans, three Latinos, and two Asian Americans who also held a total of ten corporate directorships and two corporate law partnerships before joining the cabinet.

The issues we address are not simple. They involve both the nature of the American power structure and the way in which people's need for self-esteem and a coherent belief system mesh with the hierarchical social structure they face. Moreover, the answers to some of the questions we ask vary greatly depending on which previously disadvantaged group we are talking about. Nonetheless, in the course of our research, a few general patterns have emerged….

The power elite now shows considerable diversity, at least as compared with its state in the 1950s, but its core group continues to consist of wealthy, white, Christian males, most of whom are still from the upper third of the social ladder. Like the white, Christian males, those who are newly arrived to the power elite have been filtered through the same small set of elite schools in law, business, public policy, and international relations.

1. In spite of the increased diversity of the power elite, high social origins continue to be the most important factor in making it to the top. There are relatively few rags-to-riches stories in the groups we studied, and those we did find tended to have received scholarships to elite schools or to have been elected to office, usually within the Democratic Party. In general, it still takes the few who make it at least three generations to rise from the bottom to the top in the United States.

2. The new diversity within the power elite is transcended by common values and a subjective sense of hard-earned and richly deserved class privilege. The newcomers to the power elite have found ways to signal that they are willing to join the game as it has always been played, assuring the old guard that they will call for no more than relatively minor adjustments, if that. There are few liberals and fewer crusaders in the power elite, despite its newly acquired diversity. Class backgrounds, current roles, and future aspirations are more powerful in shaping behavior in the power elite than gender, ethnicity, race, or sexual orientation.

3. Not all the groups we studied have been equally successful in contributing to the new diversity in the power elite. Women, African Americans, Latinos,

Asian Americans, and openly homosexual men and women are all under-represented, but to varying degrees and with different rates of increasing representation. There is a real possibility that Americans are replacing the old black-versus-white distinction with a black-versus-non-black dividing line, with "white" coming to be just another word for the in-group....

4. There is greater diversity in Congress than in the power elite, especially in terms of class origins, and the majority of these more diverse elected officials are Democrats, whose presence has forced the Republicans to play catch-up by including some candidates from the previously excluded groups.

5. Although the corporate, political, and military elites have accepted diversity only in response to pressure from social-movement activists and feminists, the power elite has benefited from the presence of new members. Some serve either a buffer or a liaison function with such groups and institutions as consumers, angry neighborhoods, government agencies, and wealthy foreign entrepreneurs. More generally, their simple presence at the top serves a legitimating function. Tokenism does work in terms of reassuring the general population that the system is fair....

In what may be the greatest and most important irony of them all, the diversity forced upon the power elite may have helped to strengthen it. Diversity has given the power elite buffers, ambassadors, tokens, and legitimacy. This is an unintended consequence that few insurgents or social scientists foresaw. As recent social psychology experiments show and experience confirms, it often takes only a small number of upwardly mobile members of previously excluded groups, perhaps as few as 2 percent, to undermine an excluded group's definition of who is "us" and who is "them," which contributes to a decline in collective protest and disruption and increases striving for individual mobility. That is, those who make it are not only "role models" for individuals, but they are safety valves against collective action by aggrieved groups.

Tokens at the top create ambiguity and internal doubt for members of the subordinated group. Maybe "the system" is not as unfair to their group as they thought it was. Maybe there is something about them personally that keeps them from advancing. Once people begin to ponder such possibilities, the likelihood of any sustained group action declines greatly. Because a few people have made it, the general human tendency to think of the world as just and fair reasserts itself: since the world is fair, and some members of my group are advancing, then it may be my fault that I have been left behind....

DO MEMBERS OF PREVIOUSLY EXCLUDED GROUPS ACT DIFFERENTLY?

Perhaps it is not surprising that when we look at the business practices of the members of previously excluded groups who have risen to the top of the corporate world, we find that their perspectives and values do not differ markedly from those of their white male counterparts. When Linda Wachner, one of the first women to become CEO of a *Fortune-level* company, the Warnaco Group, concluded that one of Warnaco's many holdings, the Hathaway Shirt Company, was unprofitable, she decided to stop making Hathaway shirts and to sell or close down the factory. It did not matter to Wachner that Hathaway, which started making shirts in 1837, was one of the oldest companies in Maine, that almost all of the five hundred employees at the factory were working-class women, or even that the workers had given up a pay raise to hire consultants to teach them to work more effectively and, as a result, had doubled their productivity. The bottom-line issue was that the company was considered unprofitable, and the average wage of the Hathaway workers, $7.50 an hour, was thought to be too high. (In 1995, Wachner was paid $10 million in salary and stock, and Warnaco had a net income of $46.5 million.) "We did need to do the right thing for the company and the stockholders," explained Wachner.[6]

Nor did ethnic background matter to Thomas Fuentes, a senior vice president at a consulting firm in Orange County, California, a director of Fleetwood Enterprises, and chairman of the Orange County Republican Party. Fuentes targeted fellow Latinos who happened to be Democrats when he sent uniformed security guards to twenty polling places in 1988 "carrying signs in Spanish and English warning people not to vote if they were not U.S. citizens." The security firm ended up paying $60,000 in damages when it lost a lawsuit stemming from this intimidation.[7] We also can recall that the Fanjuls, the Cuban American sugar barons, have no problem ignoring labor laws in dealing with their migrant labor force, and that Sue Ling Gin, one of the Asian Americans on our list of corporate directors, explained to an interviewer that, at one point in her career, she had hired an all-female staff, not out of feminist principles but "because women would work for lower wages." Linda Wachner, Thomas Fuentes, the Fanjuls, and Sue Ling Gin acted as employers, not as members of disadvantaged groups. That is, members of the power elite of both genders and of all ethnicities practice class politics.[8]

CONCLUSION

The black and white liberals and progressives who challenged Christian, white, male homogeneity in the power structure starting in the 1950s and 1960s sought to do more than create civil rights and new job opportunities for men and women who had previously been mistreated and excluded, important though these goals were. They also hoped that new perspectives in the boardrooms and the halls of government would spread greater openness throughout the society. The idea was both to diversify the power elite and to shift some of its power to underrepresented groups and social classes. The social movements of the 1960s were strikingly successful in increasing the individual rights and freedoms available to all Americans, especially African Americans. As we have shown, they also created pressures that led to openings at the top for individuals from groups that had previously been ignored.

But as some individuals made it, and as the concerns of social movements, political leaders, and the courts gradually came to focus more and more on individual rights and individuals advancement, the focus on "distributive justice," general racial exclusion, and social class was lost. The age-old American commitment to individualism, reinforced by tokenism and reassurances from members of the power elite, won out over the commitment to greater equality of income and wealth that had been one strand of New Deal liberalism and a major emphasis of left-wing activism in the 1960s.

We therefore conclude that the increased diversity in the power elite has not generated any changes in an underlying class system in which the top 1 percent of households (the upper class) own 33.4 percent of all marketable wealth, and the next 19 percent (the managerial, professional, and small business stratum) have 51 percent, which means that just 20 percent of the people own a remarkable 84 percent of the privately owned wealth in the United States, leaving a mere 16 percent of the wealth for the bottom 80 percent (wage and salary workers).[9] In fact, the wealth and income distributions became even more skewed starting in the 1970s as the majority of whites, especially in the South and Great Plains, switched their allegiance to the Republican Party and thereby paved the way for a conservative resurgence that is as antiunion, antitax, and antigovernment as it is determined to impose ultraconservative social values on all Americans.

The values of liberal individualism embedded in the Declaration of Independence, the Bill of Rights, and American civic culture were renewed by vigorous and courageous activists in the years between 1955 and 1975, but the class structure remains a major obstacle to individual fulfillment for the overwhelming majority of Americans. The conservative backlash that claims to speak for individual rights has strengthened this class structure, one that thwarts advancement for most individuals from families in the bottom 80 percent of the wealth distribution. This solidification of class divisions in the name of individualism is more than an irony. It is a dilemma.

Furthermore, this dilemma combines with the dilemma of race to obscure further the impact of class and to limit individual mobility, simply because the majority of middle-American whites cannot bring themselves to make common cause with African Americans in the name of greater individual opportunity and economic equality through a progressive income tax and the kind of government programs that lifted past generations out of poverty. These intertwined dilemmas of class and race lead to a nation that celebrates individualism, equal opportunity, and diversity but is, in reality, a bastion of class privilege, African American exclusion, and conservatism.

NOTES

1. See Richard M. Huber, *The American Idea of Success* (New York: McGraw-Hill, 1971), 44–46; Gary Scharnhorst, *Horatio Alger, Jr.* (Boston: Twayne, 1980), 24, 29, 141. For a discussion of the general pattern by which the media eulogize tycoons as "self-made," see Richard L. Zweigenhaft, "Making Rags out of Riches: Horatio Alger and the Tycoon's Obituary," *Extra!,* January/February 2004, 27–28.

2. As C. Wright Mills wrote half a century ago, "Horatio Alger dies hard." C. Wright Mills, *The Power Elite* (Oxford University New York Press, 1956), 91.

3. See Harold R. Kerbo, *Social Stratification and Inequality* (New York: McGraw-Hill, 2006), ch. 12. See, also, the series on class in the *New York Times,* "A Portrait of Class in America," spring 2005.

4. Mills, *The Power Elite.* 279 For Mills's specific findings, see 104–105, 128–29, 180–81, 393–94, and 400–401.

5. Edward N. Wolff, "Changes in Household Wealth in the 1980s and 1990s in the U.S." (working paper 407, Levy Economics Institute, Bard College, 2004), at www.levy.org. See table 2, p. 30 , and table 6, p. 34.

6. Sara Rimer, "Fall of a Shirtmaking Legend Shakes Its Maine Hometown," *New York Times*, May 15, 1996. See, also, Floyd Norris, "Market Place," *New York Times,* June 7, 1996; Stephanie Strom, "Double Trouble at Linda Wachner's Twin Companies," *New York Times,* August 4, 1996. Strom's article reveals that Hathaway Shirts "got a reprieve" when an investor group stepped in to save it.

7. Claudia Luther and Steven Churm, "GOP Official Says He OK'd Observers at Polls," *Los Angeles Times,* November 12, 1988; Jeffrey Perlman, "Firm Will Pay $60,000 in Suit over Guards at Polls," *Los Angeles Times,* May 31, 1989.

8. Edward N. Wolff, "Changes in Household Wealth in the 1980s and 1990s in the U.S." (working paper 407, Levy Economics Institute, Bard College, 2004), at www.levy.org.

KEY CONCEPTS

interlocking directorate   power

DISCUSSION QUESTIONS

1. To what extent do Zweigenhaft and Domhoff see diverse groups as becoming a part of the power elite? What factors do they identify as important in gaining entrance to the power elite, and how does this challenge the myth of upward mobility typically symbolized by the Horatio Alger story?

2. At the heart of Zweigenhaft and Domhoff's analysis is the question, "Does having more women and people of color in positions of power change institutions?" Using empirical evidence, how would you answer this question?

58

Why You Voted

BY ANDREW J. PERRIN

This reading examines the characteristics that predict voting behavior. Past research has successfully been able to identify those people who are more likely to go to the polls and generally predict who they will vote for. Perrin argues that the ritual of participating in elections represents more than our individual political preferences. Instead, he argues that the very act of voting represents our political individuality and our only political involvement with community. Our vote represents our expression of personal beliefs.

On November 4, 2008, probably 140 million Americans cast votes in the election for president of the United States. Nearly as many citizens, although eligible, chose not to vote, whether out of inertia, disgust, or apathy.

From one point of view, not voting is a rational thing to do. Political scientist Anthony Downs showed decades ago that voting costs time, energy, transportation, and more, and the chances one's own vote will actually change the election's outcome are vanishingly small. It makes sense to stay home.

And yet 140 million of us do it. We take time away from our responsibilities, travel to a place we might never otherwise go, wait in line, and emerge with nothing more than a tiny lapel sticker proclaiming "I Voted" and a feeling of superiority over our non-voting fellow citizens.

Moreover, roughly 20 percent of the eligible population will lie about voting. In an enduring puzzle of public opinion research, more people will tell survey researchers they voted in any given election than actually did so.

This happens because voting is more than straightforward choice-making. Voting is never just the educated, emotion-free weighing of issues and the subsequent casting of a ballot. Indeed, it is a ritual in which lone citizens express personal beliefs that reflect the core of who they are and what they want for their countrymen, balancing strategic behavior with the opportunity to express their inner selves to the world.

In other words, voting in America has two faces: the first, a ritualistic expression of personal belief without regard to strategy; the second, a cold, calculating form of citizenship where what anthropologist Julia Paley calls the

SOURCE: Perrin, Andrew J. 2008. "Why You Voted." *Contexts* 7:22–25.

"choice-making citizen" weighs the costs and benefits of particular policies and votes accordingly.

We can't understand who votes, and how, without understanding the two faces of voting that come together in citizens' minds and activities.

WE DON'T VOTE OVER THE PHONE

The standard approach to studying voting decisions generally ignores the ritualistic face of democratic decision-making.

The modern study of voting dates most prominently to *The American Voter*, a 1960 study by political scientists Angus Campbell, Philip E. Converse, Warren E. Miller, and Donald E. Stokes. Building upon earlier studies that had considered voting patterns in specific cities or regions, *The American Voter* mobilized newly available techniques of scientific public opinion research to understand how Americans made such decisions.

The authors explained voting decisions as a "funnel of causality" pushing in on individuals, a hierarchy of influences on their decisions that grew progressively stronger as the act of voting drew near. To predict whether any given individual would vote, and for whom, one only needed a few known criteria—essentially, what kind of person you were, how much you knew, what you believed, and whom you knew.

This approach was to become the gold standard of voting studies. But it's important to understand some of the decisions these innovative researchers made. First and foremost, they conceived of the voting decision as essentially an individual activity.

They mounted an extraordinary polling effort, asking a national sample of voters a series of well-crafted questions that have become the staples of the American National Election Studies (NES). These surveys, funded by the National Science Foundation and taken every two years, form an immense proportion of what we know about voting and political participation. They are, in fact, the basis of more than 5,000 articles and books published between 1960 and 2008.

Surveys of individual voters, however, as we all know from the barrage of polls we witness each election cycle, rely on contacting individual citizens by telephone and asking them a standard set of multiple choice questions. But this is a pretty poor approximation of how people actually vote, because it takes voters outside their normal social contexts—the neighborhoods, workplaces, schools, unions, clubs, and religious groups in which they actually live their lives and form their views.

As productive as the *American Voter* model of studying voting has been, conceiving of voting as decisions made by individual citizens who understand the issues, weigh them, and dispassionately select a candidate has put limits on how scholars have understood voting and how Americans have decided whether, and how, to vote.

THE EVOLUTION OF THE MODERN VOTER

Not so long before the 1950s, voting was altogether different. In the late 1800s, citizens voted in the open, their choices available for all to see. Political parties mobilized their supporters, told them whom to vote for, got them to the polls, and even printed the ballots themselves. Voting was social, collective, exciting, and fraught with corruption (at least to our modern sensibilities).

The Progressive movement of the early twentieth century changed this, applying the then-overwhelming faith in scientific rationality to reforms in the political arena. Progressive reforms included the "Australian Ballot," the secret government-provided ballot voters now see and consider utterly obvious. They prohibited personal rewards from being handed out by elected officials, substituting objective rules and expertise for the personal networks and influence of the prior era.

At the same time, progressive reforms made voting less exciting, harder to figure out, more dependent on individual rather than collective knowledge, and, certainly, more isolated. These are the characteristics voting maintains today. This was a classic case of "rationalization," the double-edged sword sociological giant Max Weber considered the centerpiece of modern life.

By the 1980s, political pundits were increasingly worried that low turnout—typically around half of registered voters in presidential elections, and much lower in local elections—was a bad omen for American democracy. The so-called "me generation" was chastised for caring little about the concerns of community or society, and the decline in voter turnout was a prominent symptom.

A wide variety of answers to why this happened surfaced, including the complexity of ballots, the influence of big money and special interests in politics, perceived lack of difference among the candidates, and the logistical hassles of registering and voting. A prominent set of studies found that voters were likely to be white as well as more educated and wealthier than the average population. The effects of this inequality were exacerbated by the fact that people were more likely to vote if they were contacted personally by a campaign or party—and the people most likely to be contacted were also white, wealthy, and educated.

Convinced that registration barriers were keeping particularly low-income and African-American citizens from the polls, social scientists Frances Fox Piven and Richard Cloward argued for making registration easier in their influential 1988 book *Why Americans Don't Vote*. In 1993 their campaign paid off with passage of the Motor-Voter Bill, which required states to allow citizens to register to vote when they applied for driver's licenses, thereby substantially reducing the burden of registration. The same impulse is behind recent trends to allow "no excuses" early voting, "vote by mail," "one-stop voting," and other reforms designed to make it easier for citizens to vote. Interestingly, most studies have found little evidence these reforms have, indeed, increased voter turnout.

VOTING AND EXPRESSION

Voting in America is among the most cherished ways of expressing political individuality, and in many cases it's the only way citizens actually participate in their political communities. To take part in this ritual, citizens must often decipher complicated ballots in carefully created and guarded isolation. This isolation is not just physical, it's also psychological. We work hard to give citizens the idea that the vote they will cast is *their own*, that the vote says something important about what they truly believe, who they are, and that it is among the most important things they can do as citizens.

And it works. Consider, for example, the candidates who periodically run for president as independents or nominees of minor parties. In virtually all such cases, the independent candidate stands no real likelihood of winning and is often accused of being a "spoiler"—a candidate who, by virtue of being in the race, distorts the outcome from what it otherwise should be. Voters are regularly implored not to "waste votes" on such candidates, since their votes would be ineffective or even counterproductive.

Yet voters continue to cast ballots for such candidates in substantial numbers, and both the 1992 and 2000 elections were probably significantly affected by these votes. In recent elections with a significant third-party candidate, 4 million voters (about 4 percent of the total) in 2000 and a striking 20.4 million (nearly 20 percent) in 1992 "wasted" these votes. Even in the closely contested 2004 election, in which there was no serious third-party candidate, more than 1.2 million voters (about 1 percent) voted for a candidate other than George W. Bush or John Kerry. Why?

If we understand a vote as a strategic resource, something like a purchase—exchanging something of symbolic value for one selection among several—it's impossible to figure out why citizens would "throw away" their votes by casting them for a candidate with no possibility of winning. But if we instead consider voting as individuals' opportunity to express their own private, core beliefs, it is priceless.

… [O]ur votes are understood as expressions of who we are, our deepest ideals and values. But this presents a strange paradox. Why should such a thoroughly social behavior, a practice that expresses our core values about how society should be structured, be practiced in enforced privacy? The answer lies not only in the history of voting, but in the importance of ritual. As political theorist Danielle S. Allen writes, the ritual of voting *simultaneously* allows us to imagine ourselves as members of an abstract national community and as effective, thinking, competent individuals.

THE RITUAL

Rituals like voting are the practices we use to hold society together—to help us, in the words of the anthropologist Benedict Anderson, imagine ourselves as a community. We carry with us the memories of elections past, refracted through the collective imagination provided by the news media and everyday conversations. Voting connects citizens to these memories, making us a part of them and infusing them with meaning.

Nearly 50 years after *The American Voter,* a team of political scientists analyzed the 2000 version of the NES, which the original book had launched. The basic model, they found, remains unchanged. Interviewing thousands of voters in isolation (and also separate from their voting booths and their feelings on voting day), they found that the most important elements of the voting decision remained individual in character.

It still holds true that if you tell me who you are, what you know, whom you know, and what you believe, chances are good I can tell you (and the world) whether you will vote, and for whom. Who we are as citizens—our class, race, sex, region of the country, and education—does say a lot about whether we are likely to vote and for whom. Stable political identifications—particularly identification with a political party, the importance of which *The American Voter* first demonstrated and which remains crucial—tell us yet more.

But voting and citizenship are about more than who you are and whom you know (the bread-and-butter concepts of studies like these). They're about what you believe, what you can imagine, and what communities you are part of.

As we move farther away from the narrow end of *The American Voter*'s funnel, it becomes increasingly important to understand how people imagine these communities and their own interests within those communities. In essence, it becomes important to understand how we become who we are, how we learn what we know, how we meet those we know, and how we come to believe what we do.

RITUAL AND REFORM

There are often calls for major reforms to fix some of the perceived problems with voting. Two of the most common are adding direct or deliberative features to our democratic practice and making it easier to vote by encouraging early voting, voting by mail, and easing registration requirements.

Americans have long been excited by the ideal of direct democracy, whether by town meeting, electronic plebiscite, or ballot initiative. What could be more democratic than bringing an issue directly to the demos—the people—to decide for themselves instead of relying on a clumsy, hierarchical system of representation? Similarly, the idea of some sort of national conversation—whether by town meetings, public forums, or electronic debates—sparks Americans' ambitions to improve democratic practice.

But the ritual aspect of voting complicates all these pictures.

If direct democracy allows citizens to answer questions more directly, who gets to ask the questions? How do citizens sort themselves into groups when deliberating? How do these groups help determine the outcome of deliberation?

None of this means voting reforms—whether institutional, "direct," or deliberative—should be off the table. But none will be successful unless it takes into account both faces of the curious practice of voting in America. The ritual face of American democracy is every bit as important as its procedural face.

KEY CONCEPTS

democracy   ritual

DISCUSSION QUESTIONS

1. Explain Perrin's argument for how voting represents more than just a list of characteristics about ourselves. What meaning does the act of voting have for individuals?

2. Have you voted in a major election (local or national)? Do you see value in the voting process? Why or why not?

F. Health Care

59

The Social Meanings of Illness

ROSE WEITZ

In this article, Rose Weitz outlines the sociological study of health and illness, summarizes the medical model of illness, and compares the two models. She explains that sociologists largely see illness as socially constructed and that race, class, and gender inequalities play a part in how we define health and sickness.

All Marco Oriti has ever wanted, ever imagined, is to be taller. At his fifth birthday party at a McDonald's in Los Angeles, he became sullen and withdrawn because he had not suddenly grown as big as his friends who were already five: in his simple child's calculus, age equaled height, and Marco had awakened that morning still small. In the six years since then, he has grown, but slowly, achingly, unlike other children. "Everybody at school calls me shrimp and stuff like that," he says.

"They think they're so rad. I feel like a loser. I feel like I'm nothing." At age 11, Marco stands 4 feet 1 inch— 4 inches below average—and weighs 49 pounds. And he dreams, as all aggrieved kids do, of a sudden, miraculous turnaround: "One day I want to, like, surprise them. Just come in and be taller than them."

Marco, a serious student and standout soccer player, more than imagines redress. Every night but Sunday, after a dinner he seldom has any appetite for, his mother injects him with a hormone known to stimulate bone growth. The drug, a synthetic form of naturally occurring human growth hormone (HGH) produced by the pituitary, has been credited with adding up to 18 inches to the predicted adult height of children who produce insufficient quantities of the

SOURCE: Rose Weitz. 2007. *The Sociology of Health, Illness, and Health Care,* 4th ed. Belmont, CA: Thomson Wadsworth, pp. 125–151.

hormone on their own—pituitary dwarfs. But there is no clinical proof that it works for children like Marco, with no such deficiency. Marco's rate of growth has improved since he began taking the drug, but his doctor has no way of knowing if his adult height will be affected. Without HGH, Marco's predicted height was 5 feet 4 inches, about the same as the Nobel Prize-winning economist Milton Friedman and ... Masters golf champion, Ian Woosnam, and an inch taller than the basketball guard Muggsy Bogues of the Charlotte Hornets. Marco has been taking the shots for six years, at a cost to his family and their insurance company of more than $15,000 a year [$21,000 in 2005 dollars]....

A Cleveland Browns cap splays Marco Oriti's ears and shadows his sparrow-ish face. Like many boys his age, Marco imagines himself someday in the NFL. He also says he'd like to be a jockey—making a painful incongruity that mirrors the wild uncertainty over his eventual size. But he is unequivocal about his shots, which his mother rotates nightly between his thighs and upper arms. "I hate them," he says.

He hates being short far more. Concord, the small Northern California city where the Oriti family now lives, is a high-achievement community where competition begins early. So Luisa Oriti and her husband, Anthony, a bank vice president, rationalize the harshness of his treatment. "You want to give your child that edge no matter what," she says, "I think you'd do just about anything." (Werth, 1991)

Does Marco have an illness? According to his doctors, who have recommended that he take an extremely expensive, essentially experimental, and potentially dangerous drug, it would seem that he does. To most people, however, Marco simply seems short....

MODELS OF ILLNESS

The Medical and Sociological Models of Illness

What do we mean when we say something is an illness? As Marco's story suggests, the answer is far from obvious. Most Americans are fairly confident that someone who has a cold or cancer is ill. But what about the many postmenopausal women whose bones have become brittle with age, and the many older men who have bald spots, enlarged prostates, and urinary problems? Or the many young boys who have trouble learning, drink excessively, or enjoy fighting? Depending on who you ask, these conditions may be defined as normal human variations, as illnesses, or as evidence of bad character. As these questions suggest, defining what is and is not an illness is far from a simple task. In this section we explore the medical model of illness: what doctors typically mean when they say something is an illness. This medical model is not accepted in its entirety by all doctors—those in public health, pediatrics, and family practice are especially likely to question it—and is not rejected by all sociologists, but it is the dominant conception of illness in the medical world. The sociological model of illness summarizes critical sociologists' retort to the medical model of illness.

This sociological model reflects sociologists' view of how the world currently operates, not how it ideally should operate....

The medical model of illness begins with the assumption that illness is an *objective* label given to anything that deviates from normal biological functioning (Mishler, 1981). Most doctors, if asked, would explain that polio is caused by a virus that disrupts the normal functioning of the neurological system, that menopause is a "hormone deficiency disease" that, among other things, impairs the body's normal ability to regenerate bone, and that men develop urinary problems when their prostates grow excessively large and unnaturally compress the urinary tract. Doctors might further explain that, because of scientific progress, all educated doctors can now recognize these problems as illnesses, even though they were not considered as such in earlier eras.

In contrast, the sociological model of illness begins with the statement that illness (as the term is actually used) is a *subjective* label, which reflects personal and social ideas about what is normal as much as scientific reasoning (Weitz, 1991). Sociologists point out that ideas about normality differ widely across both individuals and social groups. A height of 4 feet 6 inches would be normal for a Pygmy man but not for an American man. Drinking three glasses of wine a day is normal for Italian women but could lead to a diagnosis of alcoholism in American medical circles. In defining normality, therefore, we need to look not only at individual bodies but also at the broader social context. Moreover, even within a given group, "normality" is a range and not an absolute. The median height of American men, for example, is 5 feet 9 inches, but most people would consider someone several inches taller or shorter than that as still normal. Similarly, individual Italians routinely and without social difficulties drink more or less alcohol than the average Italian. Yet medical authorities routinely make decisions about what is normal and what is illness based not on absolute, objective markers of health and illness but on arbitrary, statistical cutoff points—deciding, for example, that anyone in the fifth percentile for height or the fiftieth percentile for cholesterol level is ill. Culture, too, plays a role: Whereas the American Society of Plastic and Reconstructive Surgeons recommends breast enlargement for small breasts, which it considers a disease ("micromastia") and believes "results in feelings of inadequacy, lack of self-confidence, distortion of body image and a total lack of well-being due to a lack of self-perceived femininity" (1989: 4—5), in Brazil large breasts are denigrated as a sign of African heritage and breast *reduction* is the most popular cosmetic surgery (Gilman, 1999).

Because the medical model assumes illness is an objective, scientifically determined category, it also assumes there is no *moral* element in labeling a condition or behavior as an illness. Sociologists, on the other hand, argue that illness is inherently a moral category, for deciding what is illness always means deciding what is good or bad. When, for example, doctors label menopause a "hormonal deficiency disease," they label it an undesirable deviation from normal. In contrast, many women consider menopause both normal and desirable and enjoy the freedom from fear of pregnancy that menopause brings (E. Martin, 1987). In the same manner, when we define cancer, polio, or diabetes as illnesses, we judge the bodily changes these conditions produce to be both abnormal and undesirable, rather than simply normal variations in functioning, abilities, and life expectancies. (Conversely, when we define a condition as healthy, we judge it normal and desirable.)

Similarly, when we label an individual as ill, we also suggest that there is something undesirable about that *person*. By definition, an ill person is one whose actions, ability, or appearance do not meet social norms or expectations within a given culture regarding proper behavior or appearance. Such a person will typically be considered less whole and less socially worthy than those deemed healthy. Illness, then, like virginity or laziness, is a moral status: a social condition that we believe indicates the goodness or badness, worthiness or unworthiness, of a person.

From a sociological stand, illness is not only a moral status but (like crime or sin) a form of deviance (Parsons, 1951). To sociologists, labeling something deviant does not necessarily mean that it is immoral. Rather, deviance refers to behaviors or conditions that socially powerful persons within a given culture *perceive,* whether accurately or inaccurately, as immoral or as violating social norms. We can tell whether behavior violates norms (and, therefore, whether it is deviant) by seeing if it results in *negative social sanctions*. This term refers to any punishment, from ridicule to execution. (Conversely, positive social sanctions refers to rewards, ranging from token gifts to knighthood.) These social sanctions are enforced by social control agents including parents, police, teachers, peers, and doctors. Later in this [article] we will look at some of the negative social sanctions imposed against those who are ill.

For the same reasons that the medical model does not recognize the *moral* aspects of illness labeling, it does not recognize the *political* aspects of that process. Although some doctors at some times are deeply immersed in these political processes—arguing, for example, that insurance companies should cover treatment for newly labeled conditions such as fibromyalgia or multiple chemical sensitivity—they rarely consider the ways that politics underlie the illness-labeling process in general. In contrast, sociologists point out that any time a condition or behavior is labeled as an illness, some groups will benefit more than others, and some groups will have more power than others to enforce the definitions that benefit them. As a result, there are often open political struggles over illness definitions (a topic we will return to later in this [article]). For example, vermiculite miners and their families who were constantly exposed to asbestos dust and who now have strikingly high rates of cancer have fought with insurance companies and doctors, in clinics, hospitals, and the courts, to have "asbestosis" labeled an illness; meanwhile, the mining companies and the doctors they employed have argued that there is no such disease and that the high rates of health problems in mining communities are merely coincidences (A. Schneder and McCumber, 2004).

In sum, from the sociological perspective, illness is a *social construction,* something that exists in the world not as an objective condition but *because we have defined it as existing*. This does not mean that the virus causing measles does not exist, or that it does not cause a fever and rash. It does mean, though, that when we talk about measles as an illness, we have organized our ideas about that virus, fever, and rash in only one of the many possible ways. In another place or time, people might conceptualize those same conditions as manifestations of witchcraft, as a healthy response to the presence of microbes, or as some other illness altogether. To sociologists, then, *illness,* like *crime* or *sin,* refers to biological, psychological, or social conditions subjectively defined as undesirable by those within a given culture who have the power to enforce such definitions.

In contrast, and as we have seen, the medical model of illness assumes that illness is an objective category. Based on this assumption, the medical model of health care assumes that each illness has specific features, universally recognizable in all populations by all trained doctors, that differentiate it both from other illnesses and from health (Dubos, 1961; Mishler, 1981). The medical model thus assumes that diagnosis is an objective, scientific process.

Sociologists, on the other hand, argue that diagnosis is a subjective process. The subjective nature of diagnosis expresses itself in three ways. First, patients with the same symptoms may receive different diagnoses depending on various social factors. Women who seek medical care for chronic pain, for example, are more likely to receive psychiatric diagnoses than are men who report the same symptoms. Similarly, African Americans (whether male or female) are more likely than whites are to have their chest pain diagnosed as indigestion rather than as heart disease (Hoffman and Tarzian, 2001; Nelson, Smedley, and Stith, 2002). Second, patients with the same underlying illness may experience different symptoms, resulting in different diagnoses. For example, the polio virus typically causes paralysis in adults but only flu-like symptoms in very young children, who often go undiagnosed. Third, different cultures identify a different range of symptoms and categorize those symptoms into different illnesses. For example, U.S. doctors assign the label of attention deficit disorder (ADD) to children who in Europe would be considered lazy troublemakers. And French doctors often attribute headaches to liver problems, whereas U.S. doctors seek psychiatric or neurological explanations (Payer, 1996). In practice, the American medical model of illness assumes that illnesses manifest themselves in other cultures in the same way as in American culture and, by extension, that American doctors can readily transfer their knowledge of illness to the treatment and prevention of illness elsewhere.

Finally, the medical model of illness assumes that each illness has not only unique symptoms but also a unique *etiology,* or cause (Mishler, 1981). Modern medicine assumes, for example, that *tuberculosis,* polio, HIV disease, and so on, are each caused by a unique microorganism. Similarly, doctors continue to search for limited and unique causes of heart disease and cancer, such as high-cholesterol diets and exposure to asbestos. Yet even though illness-causing microorganisms exist everywhere and environmental health dangers are common, relatively few people become ill as a result. By the same token, although cholesterol levels and heart disease are strongly correlated among middle-aged men, many men eat high-cholesterol diets without developing heart disease, and others eat low-cholesterol diets but die of heart disease anyway. The doctrine of unique ecology discourages medical researchers from asking why individuals respond in such different ways to the same health risks and encourages researchers to search for magic bullets—a term first used by Paul Ehrlich, discoverer of the first effective treatment for syphilis, in referring to drugs that almost miraculously prevent or cure illness by attacking one specific etiological factor....

MEDICINE AS SOCIAL CONTROL

Creating Illness: Medicalization

The process through which a condition or behavior becomes defined as a medical problem requiring a medical solution is known as medicalization (Conrad and Schneider, 1992; Conrad, 2005). For example, as social conditions have changed, activities formerly considered sin or crime, such as masturbation, homosexual activity, or heavy drinking, have become defined as illnesses. The same has happened to various natural conditions and processes such as uncircumcised penises, limited sexual desire, aging, pregnancy, and menopause (e.g., F. Armstrong, 2000; Barker, 1998; Figert, 1996; Rosenfeld and Faircloth, 2005). The term *medicalization* also refers to the process through which the definition of an illness is *broadened*. For example, when the World Health Organization (WHO) in 1999 lowered the blood sugar level required for diagnosis with diabetes, the number of persons eligible for this diagnosis increased in some populations by as much as 30 percent (Shaw, de Courten, Boyko, and Zimmet, 1999).

For medicalization to occur, one or more organized social groups must have both a vested interest in it and sufficient power to convince others (including doctors, the public, and insurance companies) to accept their new definition of the situation. Not surprisingly, doctors often play a major role in medicalization, for medicalization can increase their power, the scope of their practices, and their incomes. For example, during the first half of the twentieth century, improvements in the standard of living coupled with the adoption of numerous public health measures substantially reduced the number of seriously ill children. As a result, the market for pediatricians declined, and their focus shifted from treating serious illnesses to treating minor childhood illnesses and offering well-baby care. Pediatrics thus became less well-paid, interesting, and prestigious. To increase their market while obtaining more satisfying and prestigious work, some pediatricians have expanded their practices to include children whose behavior concerns their parents or teachers and who are now defined as having medical conditions such as attention deficit disorder or antisocial personality disorder (Halpern, 1990). Doctors have played similar roles in medicalizing premenstrual syndrome (Figert, 1996), drinking during pregnancy (E. Armstrong, 1998), impotence (Loe, 2004; Tiefer, 1994), and numerous other conditions....

The Consequences of Medicalization In some circumstances, medicalization can be a boon, leading to social awareness of a problem, sympathy toward its sufferers, and development of beneficial therapies. Persons with epilepsy, for example, lead far happier and more productive lives now that drugs usually can control their seizures, and few people view epilepsy as a sign of demonic possession. But defining a condition as an illness does not necessarily improve the social status of those who have that condition. Those who use alcohol excessively, for example, continue to experience social rejection even when alcoholism is labeled a disease. Moreover, medicalization also can lead to new problems, known by sociologists as unintended negative consequences (Conrad and Schneider, 1992; Zola, 1972)....

The Rise of Demedicalization   The dangers of medicalization have fostered a counter movement of demedicalization (R. Fox, 1977). A quick look at medical textbooks from the late 1800s reveals many "diseases" that no longer exist. For example, nineteenth-century medical textbooks often included several pages on the health risks of masturbation. One popular textbook from the late nineteenth century asserted that masturbation caused "extreme emaciation, sallow or blotched skin, sunken eyes,... general weakness, dullness, weak back, stupidity, laziness,... wandering and illy defined pains," as well as infertility, impotence, consumption, epilepsy, heart disease, blindness, paralysis, and insanity (Kellogg, 1880: 365). Today, however, medical textbooks describe masturbation as a normal part of human sexuality.

Like medicalization, demedicalization often begins with lobbying by consumer groups. For example, medical ideology now defines childbirth as an inherently dangerous process, requiring intensive technological, medical assistance. Since the 1940s, however, growing numbers of American women have attempted to redefine childbirth as a generally safe, simple, and natural process and have promoted alternatives ranging from natural childbirth classes, to hospital birthing centers, to home births assisted only by midwives (Sullivan and Weitz, 1988). Similarly ... gay and lesbian activists have at least partially succeeded in redefining homosexuality from a pathological condition to a normal human variation. More broadly, in recent years, books, magazines, television shows, and popular organizations devoted to teaching people to care for their own health rather than relying on medical care have proliferated. For example, in the early 1970s, the Boston Women's Health Book Collective published a 35-cent mimeographed booklet on women's health. From this, they have built a virtual publishing empire that has sold to consumers worldwide millions of books (including the best-selling *Our Bodies, Ourselves)* on the topics of childhood, adolescence, aging, and women's health....

Social Control and the Sick Role

... Medicine also can work as an institution of social control by pressuring individuals to *abandon* sickness, a process first recognized by Talcott Parsons (1951).

Parsons was one of the first and most influential sociologists to recognize that illness is deviance. From his perspective, when people are ill, they cannot perform the social tasks normally expected of them. Workers stay home, homemakers tell their children to make their own meals, students ask to be excused from exams. Because of this, either consciously or unconsciously, people can use illness to evade their social responsibilities. To Parsons, therefore, illness threatened social stability.

Parsons also recognized, however, that allowing some illness can *increase* social stability. Imagine a world in which no one could ever "call in sick." Over time, production levels would fall as individuals, denied needed recuperation time, succumbed to physical ailments. Morale, too, would fall while resentment would rise among those forced to perform their social duties day after day without relief. Illness, then, acts as a kind of pressure valve for society— something we recognize when we speak of taking time off work for "mental health days."

From Parsons's perspective, then, the important question was how did society control illness so that it would increase rather than decrease social stability? The author's emphasis on social stability reflected his belief in the broad social perspective known as functionalism. Underlying functionalism is an image of society as a smoothly working, integrated whole, much like the biological concept of the human body as a homeostatic environment. In this model, social order is maintained because individuals learn to accept society's norms and because society's needs and individuals' needs match closely, making rebellion unnecessary. Within this model, deviance—including illness—is usually considered dysfunctional because it threatens to undermine social stability.

Defining the Sick Role   Parsons's interest in how society manages to allow illness while minimizing its impact led him to develop the concept of the sick role. The sick role refers to social expectations regarding how society should view sick people and how sick people should behave. According to Parsons, the sick role as it currently exists in Western society has four parts. First, the sick person is considered to have a legitimate reason for not fulfilling his or her normal social role. For this reason, we allow people to take time off from work when sick rather than firing them for malingering. Second, sickness is considered beyond individual control, something for which the individual is not held responsible. This is why, according to Parsons, we bring chicken soup to people who have colds rather than jailing them for stupidly exposing themselves to germs. Third, the sick person must recognize that sickness is undesirable and work to get well. So, for example, we sympathize with people who obviously hate being ill and strive to get well and question the motives of those who seem to revel in the attention their illness brings. Finally, the sick person should seek and follow medical advice. Typically, we expect sick people to follow their doctors' recommendations regarding drugs and surgery, and we question the wisdom of those who do not.

Parsons's analysis of the sick role moved the study of illness forward by highlighting the social dimensions of illness, including identifying illness as deviance and doctors as agents of social control. It remains important partly because it was the first truly sociological theory of illness. Parsons's research also has proved important because it stimulated later research on interactions between ill people and others. In turn, however, that research has illuminated the analytical weaknesses of the sick role model.

Critiquing the Sick Role Model   Many recent sociological writings on illness—including this [article]—have adopted a conflict perspective rather than a functionalist perspective. Whereas functionalists envision society as a harmonious whole held together largely by socialization, mutual consent, and mutual interests, those who hold a conflict perspective argue that society is held together largely by power and coercion, as dominant groups impose their will on others. Consequently, whereas functionalists view deviance as a dysfunctional element to be controlled, conflict theorists view deviance as a necessary force for social change and as the conscious or unconscious expression of individuals who refuse to conform to an oppressive society. Conflict theorists therefore have stressed the need to study social control agents as well as, if not more than, the need to study deviants....

In sum, the sick role model is based on a series of assumptions about both the nature of society and the nature of illness. In addition, the sick role model confuses the experience of *patienthood* with the experience of *illness* (Conrad, 1987). The sick role model focuses on the interaction between the ill person and the mainstream health care system. Yet interactions with the medical world form only a small part of the experience of living with illness or disability, as the next chapter will show. For these among other reasons, research on the sick role has declined precipitously; whereas *Sociological Abstracts* listed 71 articles on the sick role between 1970 and 1979, it listed only 7 articles between 1990 and 1999, even though overall far more academic articles were published during the 1990s than during the 1970s.

CONCLUSION

The language of illness and disease permeates our everyday lives. We routinely talk about living in a "sick" society or about the "disease" of violence infecting our world, offhandedly labeling anyone who behaves in a way we don't understand or don't condone as "sick."

This metaphoric use of language reveals the true nature of illness: behaviors, conditions, or situations that powerful groups find disturbing and believe stem from internal biological or psychological roots. In other times or places, the same behaviors, conditions, or situations might have been ignored, condemned as sin, or labeled crime. In other words, illness is both a social construction and a moral status.

In many instances, using the language of medicine and placing control in the hands of doctors offers a more humanistic option than the alternatives. Yet, as this [article] has demonstrated, medical social control also carries a price. The same surgical skills and technology for cesarean sections that have saved the lives of so many women and children now endanger the lives of those who have cesarean sections unnecessarily. At the same time, forcing cesarean sections on women potentially threatens women's legal and social status. Similarly, the development of tools for genetic testing has saved many individuals from the anguish of rearing children doomed to die young and painfully, but has cost others their jobs or health insurance.

In the same way, then, that automobiles have increased our personal mobility in exchange for higher rates of accidental death and disability, adopting the language of illness and increasing medical social control bring both benefits and costs. These benefits and costs will need to be weighed carefully as medicine's technological abilities grow.

REFERENCES

Armstrong, Elizabeth M. 1998. "Diagnosing moral disorder: The discovery and evolution of fetal alcohol syndrome." *Social Science and Medicine* 47: 2025–2042.

———. 2000. "Lessons in control: Prenatal education in the hospital." *Social Problems* 47: 583–605.

Barker, K. K. 1998. "A ship upon a stormy sea: The medicalization of pregnancy." *Social Science and Medicine* 47: 1067–1076.

Conrad, Peter. 1987. "The experience of illness: Recent and new directions." *Research in the Sociology of Health Care* 6: 1–32.

———. 2005. "The shifting engines of medicalization." *Journal of Health and Social Behavior* 46: 3–14.

Conrad, Peter, and Joseph W. Schneider. 1992. *Deviance and Medicalization: From Badness to Sickness.* Philadelphia: Temple University Press.

Dubos, Rene. 1961. *Mirage of Health.* New York: Anchor.

Figert, Anne E. 1996. *Women and the Ownership of PMS: The Structuring of a Psychiatric Disorder.* New York: Aldine DeGruyter.

Fox, Renee. 1977. "The medicalization and demedicalization of American society." *Daedalus* 106: 9–22.

Gilman, Sander L. 1999. *Making the Body Beautiful: A Cultural History of Aesthetic Surgery.* Princeton, NJ: Princeton University Press.

Halpern, S. A. 1990. "Medicalization as a professional process: Postwar trends in pediatrics." *Journal of Health and Social Behavior* 31: 28–42.

Hoffman, Diane E., and Anita J. Tarzian. 2001. "The girls who cried pain: A bias against women in the treatment of pain." *Journal of Law, Medicine, and Ethics* 29: 13–27.

Kellogg, J. H. 1880. *Plain Facts for Young and Old.* Burlington, IA: Segner and Condit.

Loe, Meika. 2004. *The Rise of Viagra: How the Little Blue Pill Changed Sex in America.* New York: New York University Press.

Martin, Emily. 1987. *The Woman in the Body.* Boston: Beacon.

Mishler, Elliot G. 1981. "Viewpoint: Critical perspectives on the biomedical model." Pp. 1–23 in *Social Contexts of Health, Illness, and Patient Care,* edited by Elliot G. Mishler. Cambridge, UK: Cambridge University Press.

Nelson, Alan R., Brian D. Smedley, and Adreinne Y. Stith. 2002. *Unequal Treatment: Confronting Racial and Ethnic Disparities in Health Care.* Washington, DC: Institute of Medicine, National Academy Press.

Parsons, Talcott. 1951. *The Social System.* New York: Free Press.

Payer, Lynn. 1996. *Medicine and Culture.* Rev. ed. New York: Holt.

Rosenfeld, Dana, and Christopher A. Faircloth (eds.). 2005. *Medicalized Masculinities.* Philadelphia: Temple University Press.

Schneider, Andrew, and David McCumber. 2004. *An Air That Kills: How the Asbestos Poisoning of Libby, Montana Uncovered a National Scandal.* New York: Putnam's Sons.

Shaw, Jonathan E., Maximilian de Courten, Edward J. Boyko, and Paul Z. Zimmet. 1999. "Impact of new diagnostic criteria for diabetes on different populations." *Diabetes Care* 22: 762–766.

Sullivan, Deborah A., and Rose Weitz. 1988. *Labor Pains: Modern Midwives and Home Birth*. New Haven, CT: Yale University Press.

Tiefer, Leonore. 1994. "The medicalization of impotence: Normalizing phallocentrism." *Gender & Society* 8: 363–377.

Weitz, Rose. 1991. Life with AIDS. New Brunswick, NJ: Rutgers University Press.

Werth, Barry. 1991. "How short is too short? Marketing human growth hormone." *New York Times Magazine* June 16: 14+.

Zola, Irving K. 1972. "Medicine as an institution of social control." *Sociological Review* 20: 487–504.

KEY CONCEPTS

conflict theory   functionalism   medicalization of illness

DISCUSSION QUESTIONS

1. Compare and contrast the sociological and the medical models of illness. What are the pros and cons of each model?

2. What does the author mean by saying that "illness is a moral status"?

60

The Paradox

T. R. REID

This reading is taken from a book that compares the American health care system with other first world countries. T.R. Reid finds that Americans pay more for care that is of lower quality. He outlines the main reasons for this problem and suggests that we can learn from other models of health care.

We start with a paradox: At the beginning of the twenty-first century, the United States of America is the most powerful, most innovative, and richest nation the planet has ever known. But while this great nation is strong, smart,

SOURCE: Reid, T.R. 2009. "The Paradox." Chapter Three from *The Healing of America: A Global Quest for Better, Cheaper, and Fairer Health Care*. New York: The Penguin Press.

and wealthy, it is not particularly healthy compared to other developed nations. When it comes to the essential task of providing health care for people, the mighty USA is a fourth-rate power.

This is particularly paradoxical because the American medical establishment boasts many assets that no other country can match. The United States has the best-educated doctors, nurses, and medical technicians of any nation. We have the best-equipped hospitals. American laboratories lead the world in medical research; American companies set the global standard in developing miracle drugs and advanced medical technology. If you walk along Main Street or through the mall in any American city, you will almost certainly pass people who would be dead if it weren't for the skill and dedication of some physician. This is the picture of American medicine conveyed by TV shows like *House* and *Grey's Anatomy*, where dedicated, highly trained professionals save a half-dozen lives each week before the first commercial break. For anyone with the money—or the insurance policy—to pay for it, American medical treatment ranks with the best on earth. That's why seriously rich people all over the world tend to board their private jets and race to some famous American clinic when they face a medical emergency. That's why, when I visited a sparkling new state-of-the-art hospital in Singapore, the sign out-side said the facility was run by Duke University Medical School. The government of Singapore—an island nation floating off the Malay Peninsula in the South China Sea, about as far from North Carolina as you can get—decided that the best possi-ble place to find medical expertise was in Durham, North Carolina, USA.

But the sad fact is, we've squandered this treasure. We've wasted our shining medical assets because of a health care payment system—or, more precisely, a crazy quilt of several overlapping and often conflicting systems—that prevents millions from receiving the treatment they need and that undermines the quality of care for millions more. The shortcomings of our system can be grouped into three basic problems: coverage, quality, and cost.

Our system doesn't cover everybody. All the other developed countries see to it that every person has a right to health care when necessary. We don't.

There are tens of millions of Americans who can't go to the doctor when they're sick, or don't take the pills that could keep them well, because they can't pay for the office visit or the prescription. Some Americans get world-class, state-of-the-art treatment for a chronic disease, while other Americans die from the same disease for lack of treatment. In the richest country on earth, there are chil-dren going to bed at night with an earache, with a toothache, with an asthma attack that leaves them gasping for the next breath, because their parents don't dare face a doctor bill. In other developed countries, those sick children would see a doctor and get the medicine they need regardless of the family's income. In comparative studies of health system performance in twenty-three developed na-tions, the Commonwealth Fund, a private U.S. foundation dedicated to promot-ing a better U.S. health care system, ranked the USA last when it comes to providing universal access to medical care. When the World Health Organiza-tion rated the national health care systems of 191 countries in terms of "fairness," the United States ranked fifty-fourth. That put us slightly ahead of Chad and Rwanda, but just behind Bangladesh and the Maldives.

This gap in coverage is not just a moral issue; it has severe practical impacts. As we'll see shortly, the United States does poorly on common benchmarks like curing people who have curable diseases. A key reason is that millions of us can't get to the doctor for a cure. Americans die every day from medical problems like lupus, cervical cancer, and diabetes, which could have been cured, or at least controlled, with medical care. Of course American doctors know how to treat those ailments—but only if the sick person has access to treatment. The cohort of Americans who don't have health insurance on any given day numbers over 45 million (about 15 percent of the population). Americans who don't have enough money or enough insurance to buy medical care can sometimes go to the emergency room—but only if they're on the verge of death or in active labor. For the vast majority of sick people, the emergency room is not an option. Beyond that, you can't go down to the emergency room for the physical exam or the blood test or the breast palpation that could head off some disease before it threatens your life. You can't go back to the ER to refill the prescription for the pills required to keep you alive.

In addition to those who have no health insurance coverage, tens of millions of Americans have coverage so limited that they are not protected against any serious bill from a doctor or a hospital. For those Americans who are uninsured or under-insured, any bout with illness can be terrifying on two levels. In addition to the risk of disability or death due to the disease, there's the risk of financial ruin due to the medical and pharmaceutical bills. This is a uniquely American problem. When I was traveling the world on my quest, I asked the health ministry of each country how many citizens had declared bankruptcy in the past year because of medical bills. Generally, the officials responded to this question with a look of astonishment, as if I had asked how many flying saucers from Mars landed in the ministry's parking lot last week. How many people go bankrupt because of medical bills? In Britain, zero. In France, zero. In Japan, Germany, the Netherlands, Canada, Switzerland: zero. In the United States, according to a joint study by Harvard Law School and Harvard Medical School, the annual figure is around 700,000.

For all the money America spends on health care, our health outcomes are worse on many basic measures than those in countries that spend much less.

Some Americans get the world's best medical care (some Singaporeans do, too, by buying it from us). Overall, though, the quality of care provided by the U.S. medical system is mediocre by global standards. For all its organizational problems, I had thought that hands-on American medicine was top-notch. But comparative studies repeatedly demonstrate that this is not so.

One classic benchmark for a national medical system is "avoidable mortality"—that is, how well a country does at curing diseases that are curable. A 2008 report by the Commonwealth Fund, "Deaths Before Age 75 from Conditions That Are at Least Partially Modifiable with Effective Medical Care," concluded that the United States is the worst of the developed countries on this measure. Among nineteen wealthy countries, the United States ranked nineteenth in curing people who could be cured with decent care. (However, we did better than any of the world's poor countries.) The number of people under seventy-five who die from curable illness was almost twice as high in the United States as in the countries that do the best on this measure: France, Japan, and Spain.

Another way to measure the quality of medical treatment is to compare the survival rate from major diseases. On this score, too, the United States generally comes out badly in comparison to other rich countries. A Commonwealth Fund study of nine developed countries between 2001 and 2004 showed that Americans diagnosed with asthma die sooner than their counterparts in seven of the countries. (British asthmatics fared even worse than Americans did.) Americans with diabetes die younger than diabetics in any of the other countries. After kidney transplants, Americans have the worst survival rate. And if you've been thinking about having major surgery in the United States, here's a statistic to ponder: Among those nine rich nations, the per-capita rate of "Deaths Due to Surgical or Medical Mishaps" was the highest by far in the USA. For some particular ailments, U.S. medicine tops the world. America's five-year survival rate for women diagnosed with breast cancer was the best of the nine countries in this study. But overall, we lag the other rich countries in treating many of the diseases that medicine knows how to treat.

In terms of life expectancy—how many years the average newborn baby is likely to live—the United States ranks below most European countries and rich East Asian countries like Japan, Taiwan, and Singapore. But our country's score here is skewed somewhat, because more people die young in America than in other rich countries. For those of us who are adults, and wondering how many good years we've got left, the medical researchers have a more relevant benchmark: "healthy life expectancy at age sixty." That predicts not just how many more years a sixty-year-old can expect to live, but how long she can expect to feel pretty good; that is, how long she can expect to live before the onset of predictable ailments of the aged, such as Alzheimer's disease and rheumatoid arthritis. Since this is largely a function of medical care, any country's score on "healthy life expectancy at age sixty" turns out to be a pretty good measure of that country's medical system. And on this one, too, we're in the basement. Among twenty-three countries in a 2006 survey by the Commonwealth Fund, the United States was tied for last. (Japan came in first.)

[T]he United States is by far the world's biggest spender on health care. Whether measured as a percentage of the nation's GDP or as per-capita spending, we pour roughly twice as much into medicine as other rich countries do. Given what we've just read about coverage and quality, that raises an obvious question: If we're getting only fair-to-middling performance from a system that leaves tens of millions of people without reliable health care, why are we paying more than anybody else? Why does American health care cost so much? ...

... [T]he major reasons our national medical bill is so much higher than any other country's are two things that the United States does differently from every other country: the way we manage health insurance and the complexity of our health care system.

The United States is the only developed country that relies on profit-making health insurance companies to pay for essential and elective care. About 80 percent of non-elderly Americans have health insurance; generally they get it through the job, with the employer paying part of the premium as well. The monthly premium goes toward paying the worker's medical bills, but the insurance firms also soak up a significant share of the premium dollar to cover the costs of marketing, underwriting, and administration, as well as their profit. Economists agree that this is about the most expensive possible way to pay for a nation's health care. That's why all the other developed countries have decided that basic health insurance must be a nonprofit operation. In those countries, the insurance plans—sometimes run by government, sometimes private entities—exist only to pay people's medical bills, not to provide dividends for investors.

… Still, when American-style private health insurance works, it can definitely help the insured. For almost everyone, it's better to have health insurance than not to. But unlike people in other rich countries, Americans under sixty-five can't get health insurance that is permanent. If you leave your job, voluntarily or otherwise, you lose your insurance. No other country uses that model, because it hits the victim with a double whammy. She not only loses most of her income, but she loses her family's insurance coverage precisely at the moment when she is most economically vulnerable. In the rest of the world, this is considered unbelievably cruel. "Excuse me, Mr. Reid, but I don't understand your approach to health care," a junior minister in Sweden's health department said to me. "It seems to me that your country takes away the insurance when people most need it." Of all the mysterious nooks and crannies of U.S. health policy, the rule that takes away insurance coverage from those most in need may be the hardest to justify. In France, Germany, Japan, etc., people get health insurance as a benefit of employment, but the coverage continues if the job ends. Government pays the premium until the unlucky employee can get back to work. She may not have a job, but she can still afford to take her sick child to the doctor.

The second major anomaly of the U.S. system—the flaw that forces us to spend more than any other country on health—is sheer complexity. We have developed, more or less by accident, the most fragmented health care system in the developed world, with "providers" sending bills to a vast array of different payers.

All the other developed countries have settled on one health care system for everybody; that means every patient is treated equally, and there's one set of rules governing treatment and payment. (Some countries have a separate military medical system for soldiers overseas, but as soon as those soldiers get home, they go into the national system with everybody else.) The United States, in contrast, is a crazy quilt of different payment systems. There's one system for Americans over sixty-five. There's one for military personnel, and a different one for veterans. There's a separate system for Native Americans and yet another for people with end-stage renal failure. There's one system for Americans under sixteen living in poor families and a different one for people over sixteen in poor families. And there are scores, perhaps hundreds, of different private insurance plans. Each paying entity has its own distinct rules about what care it will pay for and how much it will pay. Quite often, neither the buyer (the patient) nor the seller (the doctor) knows how much a particular treatment costs.

…The presence of countless different payers and fee schedules drives another unique feature of American health care: the cost shift. Medical providers—doctors, hospitals, labs—naturally try to shift costs toward the highest payer. If Medicare, with its recurrent budget problems, cuts the fee it pays a hospital for a particular procedure, the hospital will raise the price for other payers to make up the difference. That's another reason why the same operation in the same hospital on the same day can have ten different prices, depending on who is paying.

The administrative patchwork makes everything about American medicine more complex and more expensive than it needs to be. A British hospital, a Taiwanese hospice, or a Canadian clinic will deal with one paying entity and one standard payment schedule. When you go to the doctor in France, the standard fee schedule for each potential treatment is posted on the wall, showing exactly what the bill will be and how much of it the insurance plan will cover. U.S. hospitals, in contrast, routinely deal with twenty, fifty, or a hundred different public and private payers. Even a neighborhood doctor's office, with three or four family physicians and four nurses, will have a corps of four to eight people in the back room just to handle the billing. Not surprisingly, one of the fastest-growing aspects of the American health care industry is the booming business for "compilers," middlemen who compile the bills that doctors submit and then shuttle them through the payment system. This makes life easier for doctors, but at a price: It adds an extra level of complexity and yet another layer of bills to the overall cost of American medicine.

The U.S. health care system's troubles with quality, coverage, and cost control are well-known in the rest of the developed world. When health industry executives, public health officials, or health care economists gather for international meetings, bashing the U.S. system is a standard agenda item. "You get used to it after a while," says Princeton professor Uwe E. Reinhardt, one of the most distinguished American economists in the field. "Economists love to disagree, and they argue about everything. But at these conferences, the one thing they all agree on is that the American system is a huge mess. In health care, the United States has become the bogeyman of the world."

I don't think that Americans are any more willing to ditch our own health care system and replace it wholesale with a British or German or Canadian model. But there are useful approaches, ideas, and techniques we could learn from health care systems that are fairer, cheaper, and more effective than ours.

KEY CONCEPTS

health care   health insurance   life expectancy

DISCUSSION QUESTIONS

1. How is the insurance system we presently have in America contributing to the problem of poor health care in America?

2. Why do medical procedures vary so much in cost? What is the argument made in this reading about how the American system differs with other countries with regard to how medical billing works?

61

Neighborhood Racial Composition, Neighborhood Poverty, and the Spatial Accessibility of Supermarkets in Metropolitan Detroit

SHANNON N. ZENK, AMY J. SCHULZ, BARBARA A. ISRAEL, SHERMAN A. JAMES, SHUMING BAO, AND MARK L. WILSON

This reading summarizes research that systematically plotted the location of supermarkets in Detroit and the surrounding areas. Looking at both racial composition of the neighborhoods and the poverty level of neighborhoods, the researchers found that there are considerably fewer supermarkets available to those that live in predominantly black, disadvantaged communities. This has severe consequences for diet-related health problems among the residents of these neighborhoods.

Four of the 10 leading causes of death in the United States are chronic diseases for which diet is a major risk factor.[1] Racial disparities in the burden of these chronic, diet-related diseases are well documented, with African Americans often having the highest morbidity and mortality.[2–5] Because health risks and resources are spatially and socially structured and African Americans disproportionately live in economically disadvantaged neighborhoods, increased attention has been focused on how residential environments shape health and contribute to racial disparities in health.[6–9] An extensive body of literature now associates residence in economically disadvantaged neighborhoods, after control for individual socioeconomic status, with a variety of adverse diet-related health outcomes.[10–12]

Despite numerous research efforts that have examined neighborhood variations in health, relatively little is known about the mechanisms by which neighborhood environments affect health.[13–15] One hypothesis is that economically

SOURCE: Zenk, Shannon N., Amy J. Schulz, Barbara A. Israel, Sherman A. James, Shuming Bao, and Mark L. Wilson. 2005. "Neighborhood Racial Composition, Neighborhood Poverty, and the Spatial Accessibility of Supermarkets in Metropolitan Detroit." *American Journal of Public Health* 95: 660–667.

and socially disadvantaged neighborhoods have inadequate access to healthy foods, thus negatively affecting dietary quality and health. Although the presence of supermarkets may not always be beneficial for neighborhood residents (e.g., if supermarkets displace smaller stores with owners who had positively contributed to and invested in the neighborhood), such large stores can be neighborhood health resources providing generally better availability and selection, higher quality, and lower cost of foods compared with smaller food stores.[16] These food resource factors influence dietary patterns.[17–24] Previous studies have found that fewer supermarkets are located in African American neighborhoods compared with White neighborhoods[25,26] and are located in economically disadvantaged neighborhoods compared with affluent neighborhoods.[16,26–29] Other studies have found no differences in the accessibility of supermarkets according to racial or socioeconomic characteristics of neighborhoods.[30–32] This discrepancy could reflect either differences in the definition of supermarkets or true variability in results across time and place that may be caused by differences in the degree of racial or economic segregation. Lower purchasing power is an often-cited but disputed explanation for the relative scarcity of supermarkets in economically disadvantaged neighborhoods.[33–35]

Analysis of the role of race without regard to poverty and of poverty without regard to race offers an incomplete picture of the potential importance of these factors in shaping the spatial accessibility of supermarkets. Understanding these relationships is critical for informing intervention and policy efforts. Such an understanding is particularly important, given the roles of racial residential segregation and economic restructuring in concentrating poverty in African American neighborhoods of older industrial cities of the Northeast and Upper Midwest.[36–40] Therefore, we sought to determine whether supermarkets are located at farther distances from the center of African American neighborhoods compared with White neighborhoods regardless of neighborhood economic conditions or if racial disparities in supermarket accessibility occur only in higher-poverty contexts.

METHODS

Setting and Sample

The setting for this study was the Detroit metropolitan area in Michigan. Metropolitan Detroit is characterized by extreme economic inequalities across neighborhoods.[36] Economic inequalities can be traced to the period just after World War II, when highway construction and cheap land outside the city led many industries to relocate to the suburbs.[39,41–43] Between 1950 and 1990, the city of Detroit lost approximately 350000 jobs,[41,43,44] largely owing to the relocation of industries to the suburbs, deindustrialization, and other facets of economic restructuring. At the same time, discriminatory federal housing policies and lending practices, racial steering (the act of real estate agents systematically showing African Americans to different neighborhoods than Whites), restrictive covenants, and violence created and reinforced racial residential segregation in metropolitan Detroit. In effect, African Americans were confined to the least desirable, older residential neighborhoods of the city, whereas Whites were able to move to more desirable, newer suburban locations.[35,37,41,44–47] The city of Detroit shifted from 16.2% African American in 1950 to 81.2% African American in 2000,[43,48] a sharp contrast with the 84.8% of metropolitan Detroit residents outside the city limits who identified as non-Hispanic White in 2000.[48] Metropolitan Detroit remains one of the most racially segregated areas in the United States—ranked second overall in residential segregation of African Americans in 2000.[49] The sample for this study was 869 neighborhoods (we used census tracts as proxies) in the tricounty Detroit metropolitan area. These neighborhoods are located in the city of Detroit and in the tricounty Detroit metropolitan area (Wayne, Oakland, Macomb counties) within a 10-mile buffer of Detroit.

Measures

We used 2000 decennial census data to characterize the neighborhoods. Population density was computed as the total population per square mile (median = 5367.44). Racial composition was defined as the percentage of non-Hispanic African American residents (median = 6.06%). Neighborhood poverty was defined as the percentage of residents below the poverty line (median = 8.21%). Tertiles for percentage of African American residents (0%–1.98%, 1.99%–62.63%, and 63.11%–98.43%) and percentage of residents in poverty (0%–5.03%, 5.10%–17.20%, and 17.23%–81.96%) were used in statistical analyses. Given that 92% of residents in tricounty metropolitan Detroit were either non-Hispanic White (67.3%) or non-Hispanic African American (24.9%), neighborhoods with low proportions of African Americans generally correspond with predominately White neighborhoods.

Supermarkets were defined as supercenters (e.g., Meijer, Super Kmart) and full-line grocery stores (e.g., Farmer Jack, Kroger) associated with a national or regional grocery chain, i.e., a chain with 11 or more retail stores.[50] To identify supermarkets, we obtained a 2001 list of grocery stores from the Michigan Department of Agriculture. We used 2001–2002 paper telephone directories, as well as online telephone directories and company Web sites, in the fall of 2002 to verify the addresses of these supermarkets and to identify additional supermarkets. We confirmed the address of any supermarket not on the Michigan Department of Agriculture list by telephoning the store. One hundred and sixty supermarkets were identified in Detroit and in the metropolitan area within a 15-mile buffer of Detroit (Table 1). The additional 5-mile buffer of supermarkets around the sampled neighborhoods helped to ensure that we could calculate supermarket accessibility for neighborhoods at the periphery.

Thus, supermarket accessibility represents the distance to the nearest supermarket for a resident positioned in the middle of the neighborhood.[51] The median Manhattan block distance to the nearest supermarket was 1.43 miles (range: 0.05 to 5.05 miles).

T A B L E 1   **No. of Chain Supermarkets, by Type, in Detroit and Within 15 Miles: Tricounty Detroit Metropolitan Area, 2002**

	No. of Chain Supermarkets
City of Detroit	
Total supermarkets	9
Full-line grocery stores	7
Supercenters	2
Tricounty Detroit metropolitan area (excluding city of Detroit)	
Total supermarkets	151
Full-line grocery stores	123
Supercenters	28
Total supermarkets	160

Note: We did not include wholesale clubs (e.g., Sam's Club) or limited-assortments stores (e.g., Save-A-Lot), which are generally smaller and offer a more limited variety of foods, in the sample.

RESULTS

As shown in model 1, the nearest supermarket was significantly further away in neighborhoods with a high proportion of African Americans (tertile 3) and in the most impoverished (tertile 3) neighborhoods compared with neighborhoods with a low proportion of African Americans (tertile 1) and the least impoverished (tertile 1) neighborhoods, respectively. (These results were adjusted for population density.) In model 2, 2 of the 4 terms to capture the effects of interaction between percentage African American and percentage poor were statistically significant. Moreover, the addition of interaction terms significantly improved the fit of the spatial model.

To examine the interaction between tertiles of percentage African American and percentage poor, we calculated and plotted predicted values for distance to the nearest supermarket by tertiles of percentage African American and percentage poor (Figure 1). Mean distance to the nearest supermarket was similar in the least impoverished (tertile 1) neighborhoods across all tertiles of percentage African American residents (Figure 1). Mean distance to the nearest supermarket increased with each successive tertile of percentage poor for neighborhoods with a high proportion of African Americans but remained approximately the same across all tertiles of percentage poor for neighborhoods with a low proportion of African Americans (predominately White). Among the most impoverished neighborhoods, distance to the nearest supermarket varied considerably by percentage African American, with the nearest supermarket averaging 1.10 to 1.15 miles farther in neighborhoods with medium (tertile 2) and high (tertile 3) proportions of African Americans, respectively, than in neighborhoods with low proportions of African Americans (tertile 1).

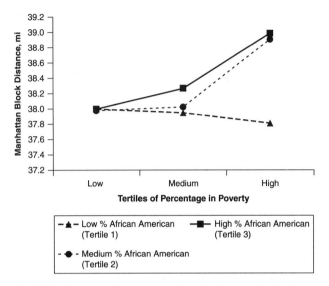

FIGURE 1 Predicted values for Manhattan block distance (in miles) to the nearest supermarket, by tertiles of percentage African American and percentage poor.

Note: Because the spatial linear trend was removed from distance to the nearest supermarket, the absolute values of the predicted values along the y-axis are not meaningful. The relative differences among groups are meaningful.

DISCUSSION

Disparities in Supermarket Accessibility

The relationship between neighborhood racial composition and supermarket accessibility varied according to neighborhood poverty level in metropolitan Detroit. The distance to the nearest supermarket was similar among the least impoverished neighborhoods across the 3 tertiles of percentage African American. However, disparities in supermarket accessibility on the basis of race were evident among the most impoverished neighborhoods: the most impoverished neighborhoods in which African Americans resided, on average, were 1.1 miles farther from the nearest supermarket than were the most impoverished White neighborhoods. Most African Americans in tricounty metropolitan Detroit reside in neighborhoods that are in the upper tertile for percentage poor and that have either a high proportion of African Americans (60%) or a medium proportion of African Americans (20%), as defined in this study.

Inadequate accessibility to supermarkets may contribute to less-nutritious diets and hence to greater risk for chronic, diet-related diseases. In a recent qualitative study, Detroit residents reported that lack of access to supermarkets was a barrier to healthy eating.[61] At least 3 previous quantitative studies, all of which examined chain supermarkets, have suggested that closer proximity to supermarkets is associated with better-quality diets.[62–64] The observation that the nearest supermarket averaged 1.1 miles further in the most impoverished neighborhoods in which African Americans resided (tertile 3 poverty, tertiles 2 and 3 African American) compared with the most impoverished White neighborhoods (tertile 3 poverty, tertile 1 African American) is particularly salient, given that 23% and 28% (tertiles 2 and 3 African American) of households in the most impoverished neighborhoods in which African Americans resided did not own a car in 2000.

The most impoverished neighborhoods in which African Americans resided (tertiles 2 and 3 African American) averaged 2.3 and 2.7 (respectively) fewer supermarkets within a 3-mile radius and had lower potential supermarket accessibility relative to the most impoverished White neighborhoods.

Race appears to be an important factor with respect to supermarket accessibility in the context of more impoverished neighborhoods; 76% of neighborhoods with a high proportion of African Americans were among the most impoverished. The disproportionate representation of African Americans in more impoverished neighborhoods in Detroit can be traced historically. Until the 1940s—a decade in which Detroit's African American population doubled from 149119 to 300506 because of the influx of African Americans from the South for manufacturing jobs—African Americans generally resided in central Detroit and east central Detroit.[66,67] Facing overcrowded and substandard housing, African Americans began moving to other parts of the city.[39,66,68] Nevertheless, institutional racism—specifically racial residential segregation—confined African Americans to Detroit neighborhoods that began losing employment opportunities, particularly in the manufacturing industry, in the 1950s.[39,41,44] Between 1948 and 1967, Detroit lost nearly 130,000 manufacturing jobs.[39]

Often confronting strong resistance from Whites, African Americans first moved to nearby neighborhoods in central Detroit.[46,69] Hence, many neighborhoods located in central Detroit[68] transitioned from White to African American in the 1950s and 1960s when African Americans moved in and Whites moved out to newly constructed housing in northeast and northwest Detroit and the suburbs. Businesses closed soon thereafter, particularly after the 1967 racial discord, further compounding the adverse economic impact of the loss of manufacturing jobs.[39,44,70] Hence, the number of abandoned homes and businesses, including grocery stores, grew, and poverty increased substantially.

This pattern of White flight and economic divestment was repeated during several decades across Detroit neighborhoods. Residential patterns of African Americans generally expanded outward in a stepwise progression from central and east central Detroit toward the northern city boundaries and eventually to Southfield, a suburb adjacent to northwest Detroit.[39,44,69] Neighborhoods located in the northernmost portion of Detroit near Eight Mile Road, the infamous African American–White racial dividing line of metropolitan Detroit, and particularly in northwest and northeast Detroit were among the most recent neighborhoods to transition from White to African American. Some neighborhoods, such as those in the far northwest Detroit community of Redford, shifted to African American as late as the 1990s.[44,69,71,72] Similarly, in the suburb of Southfield, the African American population has grown tremendously, from 102 people (0.1% of the population) in 1970 to 42 259 people (54%) in 2000.[69] The number of

African Americans in Southfield increased by 48% between 1990 and 2000 alone, despite an increase of only 3% in the city's total population.

This social history of metropolitan Detroit neighborhoods is relevant to our study because, among the least impoverished neighborhoods, all but 1 of the predominately African American neighborhoods with supermarket accessibility equivalent to that of their predominately White counterparts were located in northwest and north central Detroit and in Southfield. An optimistic interpretation of our findings is that supermarkets have newly opened or have remained open in or nearby these middle-income, yet racially transitioning, neighborhoods. This interpretation provides hope that supermarkets will invest or stay invested in African American neighborhoods as long as the residents have sufficient purchasing power to make these outlets profitable. An alternative interpretation of the findings is that among the least impoverished neighborhoods, African American neighborhoods have supermarket accessibility equivalent to that of predominantly White neighborhoods only because the supermarkets located in and nearby are remnants of historically White neighborhoods. Longitudinal data are needed to empirically test these different theories. If these areas remain African American, if they maintain a middle-income population, and if supermarkets remain open in or near these neighborhoods, then economic development may be a key intervention strategy to improve supermarket accessibility in African American neighborhoods. If, conversely, supermarkets close or do not open new sites in these economically stable African American neighborhoods, then factors associated with race are a more likely cause of disparities in supermarket accessibility. Indeed, our finding of disparate supermarket accessibility among the most impoverished neighborhoods by neighborhood racial composition warrants further investigation to identify contributing factors.

Practice and Policy Implications

The results of this study have several practice and policy implications. Pursuit of these strategies would benefit from a partnership approach between public health professionals and community members to ensure the local relevance of intervention strategies and to enhance community capacity for future intervention efforts.[81] In the first of these implications, the results suggest the critical importance of working to redress fundamental inequalities between African Americans and Whites in order to reduce chronic, diet-related diseases among African Americans.[82] For example, the economic development of African American neighborhoods could be enhanced by policies of creating jobs that pay a fair wage, to improve educational quality and opportunities for adults to increase job skills, to subsidize child care, and to attract new businesses. Second, working to attract supermarkets to economically disadvantaged African American neighborhoods in Detroit is a specific economic development strategy that may directly improve food access. Supermarket development can enhance local economic vitality by (1) providing jobs for residents, (2) increasing the local tax base, (3) making foods available at lower prices, thereby increasing the spending power of residents, and (4) attracting other forms of retail.[34,35,83] Supporting African American ownership of and employment at these supermarkets may be critical to their acceptance and success.[56]

Third, a metropolitanwide planning approach to the food system needs to be pursued.[84] Ideally, the food system would be evaluated holistically to ensure that all communities are served equitably. Fourth, in the short term, inadequate transportation is a significant barrier for residents of economically disadvantaged African American neighborhoods' gaining access to supermarkets.[61] Affordable public transportation needs to be improved by integrating transportation routes with supermarket locations.[85,86]

Fifth, on the basis of our findings of disparate access to supermarkets among the most impoverished neighborhoods by percentage African American residents, efforts to expand the Community Reinvestment Act, a law designed to combat discrimination in commercial real estate lending, may be warranted. Finally, given that racial ideologies are likely to shape the political will to pursue these intervention strategies, public health researchers and practitioners need to work to challenge racial stereotypes in public discourse.[42,43,87]

Helping to contextualize the plight of African Americans historically and spatially and to identify its ramifications for health were the primary intents of this study. We found that the historically influenced concentration of African Americans in higher-poverty neighborhoods in Detroit adversely affects spatial access to supermarkets, a resource of potential great importance in promoting the health of African Americans.

NOTES

1. National Center for Health Statistics, Centers for Disease Control and Prevention. Deaths—leading causes. Available at: http://www.cdc.gov/nchs/fastats/lcod.htm. Accessed February 13, 2004.

2. Fried, VM, Prager K, MacKay AP, Xia H. Chart-book on trends in the health of Americans. In: *Health, United States, 2003*. Hyattsville, Md: National Center for Health Statistics; 2003.

3. Jemal A, Thomas A, Murray T, Thun M. Cancer statistics, 2002. *CA Cancer J Clin.* 2002;52:23–47.

4. Ries LAG, Eisner MP, Kosary CL, et al., eds. SEER cancer statistics review, 1973–1999. Bethesda, Md: National Cancer Institute; 2002. Available at: http://seer.cancer.gov/csr/1973_1999. Accessed February 13, 2004.

5. *Healthy People 2010: Understanding and Improving Health*. Washington, DC: US Dept of Health and Human Services; 2001. Also available at: http://web.health.gov/healthypeople/document. Accessed January 1, 2005.

6. Fitzpatrick K, LaGory M. *Unhealthy Places: The Ecology of Risk in the Urban Landscape*. New York, NY: Routledge; 2000.

7. Lillie-Blanton M, LaVeist T. Race/ethnicity, the social environment, and health. *Soc Sci Med.* 1996;43: 83–91.

8. Macintyre S, Maciver S, Sooman A. Area, class, and health: should we be focusing on places or people? *J Soc Policy.* 1993;22:213–234.

9. Yen I, Syme SL. The social environment and health: a discussion of the epidemiologic literature. *Annu Rev Public Health.* 1999;20:287–308.

10. Ellen I, Mijanovich GT, Dillman K. Neighborhood effects on health: exploring the links and assessing the evidence. *J Urban Aff.* 2001;23:391–408.

11. Pickett KE, Pearl M. Multilevel analyses of neighborhood socioeconomic context and health outcomes: a critical review. *J Epidemiol Community Health.* 2000; 55:111–122.

12. Robert SA. Socioeconomic position and health: the independent contribution of community socioeconomic context. *Annu Rev Sociol.* 1999;25:489–516.

13. Diez Roux AV. Investigating neighborhood and area effects on health. *Am J Public Health.* 2001;91: 1783–1789.

14. Hillemeier MM, Lynch J, Harper S, Casper M. Measuring contextual characteristics for community health. *Health Serv Res.* 2004;38:1645–1717.

15. Macintyre S, Ellaway A, Cummins S. Place effects on health: how can we conceptualize, operationalize, and measure them? *Soc Sci Med.* 2002;55:125–139.

16. Mantovani RE, Daft L, Macaluso TF, Welsh J, Hoffman K. *Authorized Food Retailers' Characteristics and Access Study.* Alexandria, Va: US Dept of Agriculture; 1997.

17. Cheadle A, Psaty BM, Curry S, et al. Community-level comparisons between the grocery store environment and individual dietary practices. *Prev Med.* 1991; 20:250–261.

18. Cohen NL, Stoddard AM, Sarouhkhanians S, Sorensen G. Barriers toward fruit and vegetable consumption in a multiethnic worksite population. *J Nutr Educ.* 1998;30:381–386.

19. Drewnowski A, Specter SE. Poverty and obesity: the role of energy density and energy costs. *Am J Clin Nutr.* 2004;79:6–16.

20. French SA, Story M, Jeffery RW. Environmental influences on eating and physical activity. *Annu Rev Public Health.* 2001;22:309–335.

21. Furst T, Connors M, Bisogni CA, Sobal J, Falk LW. Food choice: a conceptual model of the process. *Appetite.* 1996;26:247–266.

22. Glanz K, Basil M, Mailbach E, Goldberg J, Snyder D. Why Americans eat what they do: taste, nutrition, cost, convenience, and weight control concerns as influences on food consumption. *J Am Diet Assoc.* 1998;98: 1118–1126.

23. Huang K. Role of national income and prices. In: Frazao E, ed. *America's Eating Habits: Changes and Consequences.* US Dept of Agriculture; 1999:161–171. Agriculture Information Bulletin No. 750.

24. Reicks M, Randall J, Haynes B. Factors affecting vegetable consumption in low-income households. *J Am Diet Assoc.* 1994;94:1309–1311.

25. Race and place matter for major Chicago area grocers. Chicago, Ill: Metro Chicago Information Center. Available at: http://www.mcic.org. Accessed January 3, 2005.

26. Morland K, Wing S, Diez Roux A, Poole C. Neighborhood characteristics associated with the location of food stores and food service places. *Am J Prev Med.* 2002;22:23–29.

27. Alwitt LF, Donley TD. Retail stores in poor urban neighborhoods. *J Consum Aff.* 1997;31:139–164.

28. Chung C, Myers SL. Do the poor pay more for food? An analysis of grocery store availability and food price disparities. *J Consum Aff.* 1999;33:276–296.

29. Cotterill RW, Franklin AW. *The Urban Grocery Store Gap.* Storrs, Conn: Food Marketing Policy Center, University of Connecticut; 1995.

30. Cummins S, Macintyre S. A systematic study of an urban foodscape: the price and availability of food in Greater Glasgow. *Urban Stud.* 2002;39:2115–2130.

31. Cummins S, Macintyre S. "Food deserts"—Evidence and assumption in health policy making. *BMJ.* 2002; 325:436–438.

32. Jones SJ. *The Measurement of Food Security at the Community Level: Geographic Information Systems and Participatory Ethnographic Methods* [dissertation]. Chapel Hill, NC: University of North Carolina at Chapel Hill; 2002.

33. Donohue RM. *Abandonment and Revitalization of Central City Retailing: The Case of Grocery Stores* [dissertation]. Ann Arbor, Mich: University of Michigan; 1997.

34. The business case for pursuing retail opportunities in the inner city. Boston, Mass: The Boston Consulting Group and The Initiative for a Competitive Inner City; 1998. Available at: http://www.icic.org. Accessed January 3, 2005.

35. The changing models of inner city grocery retailing. Boston, Mass: The Initiative for a Competitive Inner City; 2002. Available at: http://www.icic.org. Accessed January 3, 2005.

36. Jargowsky PA. *Poverty and Place: Ghettos, Barrios, and the American City.* New York, NY: Russell Sage Foundation; 1997.

37. Massey DS, Denton NA. *American Apartheid: Segregation and the Making of the Underclass.* Cambridge, Mass: Harvard University Press; 1993.

38. Massey DS, Fischer MJ. How segregation concentrates poverty. *Ethn Racial Stud.* 2000;23:670–691.

39. Sugrue TJ. *The Origins of the Urban Crisis: Race and Inequality in Postwar Detroit.* Princeton, NJ: Princeton University Press; 1996.

40. Wilson WJ. *The Truly Disadvantaged: The Inner City, the Underclass, and Public Policy.* Chicago, Ill: University of Chicago Press; 1987.

41. Farley R, Danziger S, Holzer HJ. *Detroit Divided.* New York, NY: Russell Sage Foundation; 2000.

42. Geronimus AT. To mitigate, resist, or undo: addressing structural influences on the health of urban populations. *Am J Public Health.* 2000;90:867–872.

43. Schulz A, Williams DR, Israel B, Lempert LB. Racial and spatial relations as fundamental determinants of health in Detroit. *Milbank Q.* 2002;80:677–707.

44. Darden JT, Hill RC, Thomas J, Thomas R. *Detroit: Race and Uneven Development.* Philadelphia, Pa: Temple University Press; 1987.

45. Collins CA, Williams DR. Segregation and mortality: the deadly effects of racism. *Sociol Forum.* 1999; 14:495–523.

46. Thomas JM. *Redevelopment and Race: Planning a Finer City in Postwar Detroit.* Baltimore, Md: The Johns Hopkins University Press; 1997.

47. Williams DR, Collins CA. Racial residential segregation: a fundamental cause of racial disparities in health. *Public Health Rep.* 2001;116:404–416.

48. *Michigan Metropolitan Information Center 2000 Census Demographic Characteristics.* Detroit, Mich: The Center for Urban Studies, Wayne State University. Available at: http://www.cus.wayne.edu/census/censuspubs.aspx. Accessed January 3, 2005.

49. Iceland J, Weinberg DH, Steinmetz E. *Racial and Ethnic Residential Segregation in the United States: 1980–2000.* Washington, DC: US Government Printing Office; 2002. US Census Bureau. Series CENSR-3.

50. *Supermarket Facts Industry Overview 2002.* Washington, DC: Food Marketing Institute; 2002. Available at: http://www.fmi.org/facts_figs/superfact.htm. Accessed April 21, 2004.

51. Gimpel JG, Schuknecht JE. Political participation and the accessibility of the ballot box. *Polit Geogr.* 2003;22:471–488.

52. Anselin L. Interactive techniques and exploratory spatial data analysis. In: Longley PA, Goodchild MF, Maguire DJ, Rhind DW, eds. *Geographic Information Systems: Principles, Techniques, Management, and Applications.* New York, NY: John Wiley & Sons; 1999: 253–266.

53. Anselin L. *Spatial Econometrics: Methods and Models.* Boston, Mass: Kluwer Academic; 1988.

54. Legendre P, Fortin MJ. Spatial pattern and ecological analysis. *Vegetatio.* 1989;80: 107–138.

55. Stralberg D, Bao S. Identifying the spatial structure in error terms with spatial covariance models: a case study on urbanization influence in chaparral bird species. *Geographic Information Sciences.* 1999;5: 106–20.

56. *The African American Grocery Shopper 2000.* Washington DC: Food Marketing Institute; 2000.

57. Kaluzny SP, Vega SC, Cardoso TP, Shelly AA. *S+ SpatialStats User's Manual for Windows and UNIX.* New York, NY: Springer; 1998.

58. Cressie NAC. *Statistics for Spatial Data.* New York, NY: John Wiley & Sons; 1993.

59. Bailey TC, Gatrell AC. *Interactive Spatial Data Analysis.* Essex, UK: Addison Wesley Longman; 1995.

60. Haining R. *Spatial Data Analysis: Theory and Practice.* Cambridge, Mass: Cambridge University Press; 2003.

61. Kieffer E, Willis S, Odoms-Young A, et al. Reducing disparities in diabetes among African American and Latino residents of Detroit: the essential role of community planning focus groups. *Ethn Dis.* 2004; 14:S1-27–S1-37.

62. Laraia BA, Siega-Riz AM, Kaufman JS, Jones SJ. Proximity to supermarkets is positively associated with diet quality index for pregnancy. *Prev Med.* 2004;39: 869–875.

63. Morland K, Wing S, Diez Roux A. The contextual effect of the local food environment on residents' diets: the atherosclerosis risk in communities study. *Am J Public Health.* 2002;92:1761–1767.

64. Wrigley N, Warm D, Margetts B, Whelan A. Assessing the impact of improved retail access on diet in a "food desert": a preliminary report. *Urban Stud.* 2002;39:2061–2082.

65. Zenk SN. *Neighborhood Racial Composition, Neighborhood Poverty, and Food Access in Metropolitan Detroit: Geographic Information Systems and Spatial Analysis* [dissertation]. Ann Arbor, Mich: University of Michigan; 2004.

66. Hartigan J. *Racial Situations: Class Predicaments of Whiteness in Detroit.* Princeton, NJ: Princeton University Press; 1999.

67. Zunz O. *The Changing Face of Inequality: Urbanization, Industrial Development, and Immigrants in Detroit, 1880–1920.* Chicago, Ill: University of Chicago Press; 1982.

68. McWhirter C. Life of one street mirrors city's fall: racial fears trigger white flight in '50s. *The Detroit News.* June 17, 2001. Available at: http://www.detnews.com/specialreports/2001/elmhurst. Accessed January 5, 2005.

69. Metzger K, Booza J. *African Americans in the United States, Michigan, and Metropolitan Detroit.* Detroit, Mich: Center for Urban Studies, Wayne State University; 2002. Center for Urban Studies Working Paper Series, No. 8.

70. McWhirter C. 1967 riot sent street into wrenching spiral: once stable block withers as property owners desert. *The Detroit News.* June 18, 2001. Available at: http://www.detnews.com/specialreports/2001/elmhurst. Accessed January 5, 2005.

71. *1990 Census Subcommunity Profiles for the City of Detroit.* Detroit, Mich: Southeast Michigan Census Council; 1993.

72. *2000 Census Subcommunity Profiles for the City of Detroit.* Detroit, Mich: United Way Community Services; 2001.

73. *Key Facts: Supermarket Sales.* Washington, DC: Food Marketing Institute; 2004. Available at: http://www.fmi.org/facts_figs/keyfacts/grocery.htm. Accessed June 9, 2004.

74. Fortney J, Rost K, Warren J. Comparing alternative methods of measuring geographic access to health services. *Health Services and Outcomes Research Methodology.* 2000;1:173–184.

75. Martin D, Wrigley H, Barnett S, Roderick P. Increasing the sophistication of access measurement in a rural healthcare study. *Health Place.* 2002;8:3–13.

76. Kwan M. Gender and individual access to urban opportunities: a study using space-time measures. *Prof Geogr.* 1999;51:210–227.

77. Bott J. Reaction in Detroit: residents feel angry, betrayed by decision. *The Detroit Free Press.* January 15, 2003. Available at: http://www.freep.com/money/business/kdet15_20030115.htm. Accessed January 3, 2005.

78. Dixon J. Farmer Jack drops 4 stores. *The Detroit Free Press.* March 28, 2003. Available at: http://www.freep.com/money/business/fj28_20030328.htm. Accessed January 3, 2005.

79. Guest G. Farmer Jack will close 13 metro stores. *The Detroit Free Press.* January 7, 2004. Available at: http://www.freep.com/money/business/fjack7_20040107.htm. Accessed January 3, 2005.

80. Haber G. A nearby convenience: new Farmer Jack store gives city residents a grocery shopping option closer to home. *The Detroit Free Press.* June 26, 2003. Available at: http://www.freep.com/money/business/farmer26_20030626.htm. Accessed January 3, 2005.

81. Israel BA, Schulz AJ, Parker EA, Becker AB. Review of community-based research: assessing partnership approaches to improve public health. *Annu Rev Public Health.* 1998;19:173–201.

82. James SA. Primordial prevention of cardiovascular disease among African Americans: a social epidemiological perspective. *Prev Med.* 1999;29:S84–S89.

83. Pothukuchi K. Attracting supermarkets to inner city neighborhoods: economic development out of the box. *Econ Dev Q.* In press.

84. Pothukuchi K, Kaufman JL. Placing the food system on the urban agenda: the role of municipal institutions in food systems planning. *Agric Human Values*. 1999;16:213–224.

85. Ashman L, Vega J, Dohan M, Fisher A, Hippler R, Romain B. *Seeds of Change: Strategies for Food Security for the Inner City*. Los Angeles, Calif: University of California Los Angeles; 1993.

86. Gottlieb R, Fisher A, Dohan M, O'Connor L, Parks V. *Homeward Bound: Food-Related Transportation Strategies for Low Income and Transit Dependent Communities*. Los Angeles, Calif: University of California Transportation Center; 1996. Available at: http://www.foodsecurity.org/homewardbound.pdf. Accessed February 13, 2004.

87. Cohen HW, Northridge ME. Getting political: racism and urban health. *Am J Public Health*. 2000;90: 841–842.

KEY CONCEPTS

segregation   White flight

DISCUSSION QUESTIONS

1. Why is lack of supermarkets a health problem for poor communities? What are the consequences of this for the health of residents?

2. Think of your own community, both at home and at school. What types of stores are available to you? How easy is it to find affordable and healthy food? Do you think this is true everywhere? How is it different from one community to another?

62

Health and Wealth

Our Appalling Health Inequality Reflects and Reinforces Society's Other Gaps.

LAWRENCE R. JACOBS AND JAMES A. MORONE

This reading summarizes how there is unequal access to health care and unequal quality of health care among Americans. The authors argue that America can and should be reformed with regard to access to quality, affordable health care. The American system currently disadvantages too many, which leads to unfair disparities in disability, life-expectancy, and overall health.

A look at Americans' health reveals the astonishing inequalities in our society. American girls are born with a life expectancy that ranks 19th in the world (in another survey they fall to 28th). Male babies rank 31st—in a dead tie with Brunei. Among the 13 wealthiest countries, the United States ranks last or nearly so in almost every way we measure health: infant mortality, low birth weight, life expectancy at birth, life expectancy for infants. The average American boy lives three and a half fewer years than the average Japanese baby, despite higher rates of cigarette smoking in Japan. The American adolescent death rate is twice as high as, say, England's.

These dismal American averages mask vast differences across our population. A male born in some sections of Washington, D.C., for example, has a life expectancy 40 years lower than a woman born in many wealthy neighborhoods. In short, great differences in wealth match up to—indeed, they create—terrible differences in health.

Why do Americans come out so badly in the cross-national health statistics? Why are our infants more likely to die than those in, say, Croatia? Our health troubles have three interrelated causes: inequality, poverty, and the way we organize our health-care system.

Let's start with inequality. A famous study of the British civil service found that with each rung up the ladder of success, people suffered fewer fatal heart attacks—the clerks and messengers at the bottom were four times more likely

SOURCE: Jacobs, Lawrence R. and James A. Morone. 2004. "Health and Wealth: Our Applying Health Inequality Reflects and Reinforces Society's Other Gaps." *The American Prospect* 15: A20–A21.

to die than the executives at the top. Researchers following up this study reached a surprising conclusion that seems to hold up in one nation after another: The wider the inequality, the worse the nation's overall health.

Why should this be so? For one thing, falling behind in the race to make ends meet generates stress and physiological harm—the results are depression, hypertension, other illnesses, and high mortality rates. In addition, the middle-class scramble to get ahead erodes neighborly feelings, frays our communities, and lowers trust in institutions like churches and governments. All of these are factors in other countries. But most industrial nations buffer their citizens against economic uncertainty and lost jobs. In the United States, only the market winners get security.

Of course, American health problems go beyond inequality and are closely correlated with the poverty in which more than one in 10 Americans now live. Of our 34.6 million "poor" citizens, according to the U.S. Census Bureau, more than 14 million are "severely poor," meaning they don't even make it halfway to the federal poverty line. The numbers are worse for minorities, with nearly a quarter of blacks and more than a fifth of Hispanics living in poverty.

And poverty brings troubles like hunger (33 million Americans live with "food insecurity," as defined by the Department of Agriculture) and homelessness (perhaps as many as 3.5 million a year), which disproportionately fall on kids. Poor neighborhoods face high crime, inferior schools, few good jobs, and inadequate health-care facilities. Instead, poverty attracts danger—too much alcohol and tobacco, illegal drugs, and fast foods. One observer after another has gone off to study poor communities and come back with the same report. The lives of the poor are full of stress and the struggle to get by.

People die younger in Harlem than in Bangladesh. Why? It is not what most people think—homicide, drug abuse, and AIDS are far down the list. Rather, as *The New England Journal of Medicine* reports, the leading causes of death in poor black neighborhoods are "unrelenting stress," "cardiovascular disease," "cancer," and "untreated medical conditions."

Finally, beyond the fundamentals—inequality and poverty—there is that stubborn American policy dilemma: No other industrial nation tolerates such yawning gaps in health insurance. According to the Congressional Budget Office, 43.6 million people were uninsured in 2002, with 19.9 million coming from the ranks of full-time workers; 74.7 million Americans under 65 were without health insurance for all or part of 2001 and 2002. Part of the problem is that workplace coverage is unraveling as more employers shift costs like premiums, co-payments, and coverage limitations onto their workers. Meanwhile, medical costs are rising faster than personal-income growth.

Simple medical care—annual check-ups, screenings, vaccinations, eyeglasses, dentistry—saves lives, improve well-being, and is shockingly uneven. Well-insured people get assigned hospital beds; the uninsured get patched up and sent back to the streets. From diagnostic procedures—prostate screenings, mammograms, and Pap smears—to treatment for asthma, the uninsured get less care, they get it later in their illnesses, and they are roughly three times more likely to have an adverse health outcome. The Institute of Medicine recently blamed gaps in insurance coverage for 17,000 preventable deaths a year.

Even middle-class parents worry about the next medical emergency or, in many cases, the routine trip to the doctor's office. Life without health insurance means constantly measuring aches and fevers against the next payday. Changing jobs brings a new set of anxieties about shifts in medical coverage. Health bills are the largest cause of personal bankruptcy in the United States.

Of course, no health-care system treats everyone the same way. But in America, our disparities are unusually wide and deep.

How can we reverse these trends and begin to build the good society? Recent experience counsels incremental reform that builds on past successes while pushing bold new proposals for the future.

As recent history shows, even half steps—like adding amendments to bipartisan legislation—can add up to something important. Back when the Reagan administration was attacking poverty programs while cutting taxes and running up enormous deficits, California Congressman Henry Waxman oversaw bipartisan support for a series of minor expansions in Medicaid eligibility. The result: In the late 1980s, the program grew to cover an additional 5 million children and 500,000 pregnant women.

While Bill Clinton's failure to pass national health insurance got most of the press, his administration quietly enacted the Children's Health Insurance Program for states in 1997. Using federal matching funds as a prod, the program pushed states to widen coverage to uninsured children, helping Medicaid reach 20 million kids by 2000 and funding non-Medicaid programs to cover an additional 2 million.

Even further below the national radar screen, the Robert Wood Johnson Foundation induced state governments to place health-care clinics directly in schools. Families in underserved neighborhoods suddenly—and usually for the first time—found it easy for their kids to get into a physician's office. Despite strong initial opposition from the cultural right over birth control, teachers, public-health advocates, parents, and community organizers have managed to open 1,498 school centers from Maine to California.

Reforms beyond medical care can also improve general living conditions and boost American health. The Earned Income Tax Credit, for example, has lifted millions of low-income workers and their children out of poverty. To be sure, making Americans healthy means addressing the economic insecurity that threatens these struggling families, forcing middle-class Americans to work double shifts and the poor to confront hunger and homelessness.

Making Americans healthy also means casting off the political torpor of this new Gilded Age and reclaiming a long-standing commitment to our neighbors and communities. Only great aspirations will galvanize a new populist politics and leverage our reluctant state.

There is not much mystery about what works. Other industrial countries rely on three familiar paths to good health. First, government plays an important role through such policies as family and housing allowances, universal health care, pensions, and tax credits. The generous welfare states of northern Europe and nations with more modest programs like France, Germany, and Canada all have poor, middle-class, and wealthy populations. However, all these nations achieve much narrower income gaps among groups than now exist in the United States.

A second type of policy fosters opportunity. Governments invest in education to expand the supply of skilled labor and help workers help themselves. Lowering the barriers to college education and worker retraining reduces the high premium for skilled labor. In addition, European governments collaborate with businesses by regularly adjusting the minimum wage and overseeing the negotiations between business and labor.

Finally, most wealthy nations maintain taxes. The new global economy was expected to spark dramatic tax cuts as governments competed with one another to create an attractive business climate and lure investment and skilled labor. In Europe and Canada, international pressures did not eviscerate the government's capacity to raise revenues. Instead, domestic support to maintain programs (and international pressure to limit deficits) barred governments from plunging into tax-cut wars.

In short, America's allies have tried to defend all their citizens from the worst effects of a global economy. The results across the industrial world are powerful: Policies that moderate income disparities turn out to be good for your health.

American public policy has, on balance, gone the other way: Tax cuts, deregulation, and unmediated markets sabotage our incremental stabs at fostering real opportunity. Some individuals have grown fantastically wealthy; most struggle to make ends meet. The dirty policy secret lies in the health consequences: Our population suffers more illness and dies younger.

Our call to reform is simple: A civilized society should not accept gaping disparities in life and death, health and disability. Americans are too generous and fair-minded to tolerate so much preventable suffering. This moral vision undergirds a hardheaded analysis of the rapidly changing global economy that has reshuffled the distribution of money in American society and unsettled the life circumstances that nurture and protect the health of the country. The solutions are no mystery. Other nations successfully protect their people. So can we.

KEY CONCEPTS

industrial nations   global economy   life expectancy   welfare state

DISCUSSION QUESTIONS

1. What explanation do Jacobs and Morone suggest for why there is such a gap between who is insured and who is not? What are their suggestions for how this could be fixed?

2. Look up the current health care reform bill on the Internet. Does it adequately address these issues? How does the current presidential administration attempt to provide health insurance for all Americans?

Applying Sociological Knowledge: An Exercise for Students

Investigate insurance company policies regarding standard and preventative care and treatment for illness. One group of students should research the Medicare program, including prescription coverage. Then report back to the class, presenting information about ease of accessing policy particulars and coverage details. Think specifically about how easy or difficult it is for elderly and chronically ill Americans to obtain the coverage and information they need. What happens to the uninsured? Find the most recent statistics on how many Americans remain uninsured.

✳

Population, Urbanization, and the Environment

63

American Apartheid

DOUGLAS S. MASSEY AND NANCY A. DENTON

Douglas S. Massey and Nancy A. Denton argue that segregation, particularly residential segregation, is a fundamental dimension of race relations in the United States and is all too often ignored by policymakers and even scholars. It is a major cause of many of the ills of race relations in this country. They argue that it is the "missing link" in past attempts to understand the urban poor.

> It is quite simple. As soon as there is a group area then all your uncertainties are removed and that is, after all, the primary purpose of this Bill [requiring racial segregation in housing].
>
> Minister of the Interior,
> Union of South Africa legislative debate on the Group Areas
> Act of 1950

During the 1970s and 1980s a word disappeared from the American vocabulary. It was not in the speeches of politicians decrying the multiple ills besetting American cities. It was not spoken by government officials responsible for administering the nation's social programs. It was not mentioned by journalists reporting on the rising tide of homelessness, drugs, and violence in urban America. It was not discussed by foundation executives and think-tank experts proposing new programs for unemployed parents and unwed mothers. It was not articulated by civil rights leaders speaking out against the persistence of racial inequality; and it was nowhere to be found in the thousands of pages written by social scientists on the urban underclass. The word was segregation.

Most Americans vaguely realize that urban America is still a residentially segregated society, but few appreciate the depth of black segregation or the degree to which it is maintained by ongoing institutional arrangements and contemporary individual actions. They view segregation as an unfortunate holdover from a racist past, one that is fading progressively over time. If racial residential segregation persists, they reason, it is only because civil rights laws passed during the 1960s have not had enough time to work or because many blacks still prefer to live in black neighborhoods. The residential segregation of blacks is viewed charitably as a "natural" outcome of impersonal social and economic forces, the same forces that produced Italian and Polish neighborhoods in the past and that yield Mexican and Korean areas today.

But black segregation is not comparable to the limited and transient segregation experienced by other racial and ethnic groups, now or in the past. No group in the history of the United States has ever experienced the sustained high level of residential segregation that has been imposed on blacks in large American cities for the past fifty years. This extreme racial isolation did not just happen; it was manufactured by whites through a series of self-conscious actions and purposeful institutional arrangements that continue today. Not only is the depth of black segregation unprecedented and utterly unique compared with that of other groups, but it shows little sign of change with the passage of time or improvements in socioeconomic status.

If policymakers, scholars, and the public have been reluctant to acknowledge segregation's persistence, they have likewise been blind to its consequences for American blacks. Residential segregation is not a neutral fact; it systematically undermines the social and economic well-being of blacks in the United States. Because of racial segregation, a significant share of black America is condemned to experience a social environment where poverty and joblessness are the norm, where a majority of children are born out of wedlock, where most families are on welfare, where educational failure prevails, and where social and physical deterioration abound. Through prolonged exposure to such an environment, black chances for social and economic success are drastically reduced.

Deleterious neighborhood conditions are built into the structure of the black community. They occur because segregation concentrates poverty to build a set of mutually reinforcing and self-feeding spirals of decline into black neighborhoods. When economic dislocations deprive a segregated group of employment and increase its rate of poverty, socioeconomic deprivation inevitably becomes more concentrated in neighborhoods where that group lives. The damaging social consequences that follow from increased poverty are spatially concentrated as well, creating uniquely disadvantaged environments that become progressively isolated—geographically, socially, and economically—from the rest of society.

The effect of segregation on black well-being is structural, not individual. Residential segregation lies beyond the ability of any individual to change; it constrains black life chances irrespective of personal traits, individual motivations, or private achievements. For the past twenty years this fundamental fact has been swept under the rug by policymakers, scholars, and theorists of the urban underclass. Segregation is the missing link in prior attempts to understand the plight of the urban poor. As long as blacks continue to be segregated in American cities, the United States cannot be called a race-blind society.

THE FORGOTTEN FACTOR

The present myopia regarding segregation is all the more startling because it once figured prominently in theories of racial inequality. Indeed, the ghetto was once seen as central to black subjugation in the United States. In 1944 Gunnar Myrdal wrote in *An American Dilemma* that residential segregation "is basic in a mechanical sense. It exerts its influence in an indirect and impersonal way: because Negro people do not live near white people, they cannot... associate with each other in the many activities founded on common neighborhood. Residential segregation... becomes reflected in uni-racial schools, hospitals, and other institutions" and creates "an artificial city... that permits any prejudice on the part of public officials to be freely vented on Negroes without hurting whites."

Kenneth B. Clark, who worked with Gunnar Myrdal as a student and later applied his research skills in the landmark *Brown v. Topeka* school integration case, placed residential segregation at the heart of the U.S. system of racial oppression. In *Dark Ghetto,* written in 1965, he argued that "the dark ghetto's invisible walls have been erected by the white society, by those who have power, both to confine those who have no power and to perpetuate their powerlessness. The dark ghettos are social, political, educational, and—above all—economic colonies. Their inhabitants are subject peoples, victims of the greed, cruelty, insensitivity, guilt, and fear of their masters."

Public recognition of segregation's role in perpetuating racial inequality was galvanized in the late 1960s by the riots that erupted in the nation's ghettos. In their aftermath, President Lyndon B. Johnson appointed a commission chaired by Governor Otto Kerner of Illinois to identify the causes of the violence and to propose policies to prevent its recurrence. The Kerner Commission released its report in March 1968 with the shocking admonition that the United States was "moving toward two societies, one black, one white—separate and unequal." Prominent among the causes that the commission identified for this growing racial inequality was residential segregation.

In stark, blunt language, the Kerner Commission informed white Americans that "discrimination and segregation have long permeated much of Amencan life; they now threaten the future of every American." "Segregation and poverty have created in the racial ghetto a destructive environment totally unknown to most white Americans. What white Americans have never fully understood—but what the Negro can never forget—is that white society is deeply implicated in the ghetto. White institutions created it, white institutions maintain it, and white society condones it."

The report argued that to continue present policies was "to make permanent the division of our country into two societies; one, largely Negro and poor, located in the central cities; the other, predominantly white and affluent, located in the suburbs." Commission members rejected a strategy of ghetto enrichment coupled with abandonment of efforts to integrate, an approach they saw "as another way of choosing a permanently divided country." Rather, they insisted that the only reasonable choice for America was "a policy which combines ghetto enrichment with programs designed to encourage integration of substantial numbers of Negroes into the society outside the ghetto."

America chose differently. Following the passage of the Fair Housing Act in 1968, the problem of housing discrimination was declared solved, and residential segregation dropped off the national agenda. Civil rights leaders stopped pressing for the enforcement of open housing, political leaders increasingly debated employment and educational policies rather than housing integration, and academicians focused their theoretical scrutiny on everything from culture to family structure, to institutional racism, to federal welfare systems. Few people spoke of racial segregation as a problem or acknowledged its persisting consequences. By the end of the 1970s residential segregation became the forgotten factor in American race relations.

While public discourse on race and poverty became more acrimonious and more focused on divisive issues such as school busing, racial quotas, welfare, and affirmative action, conditions in the nation's ghettos steadily deteriorated. By the end of the 1970s, the image of poor minority families mired in an endless cycle of unemployment, unwed childbearing, illiteracy, and dependency had coalesced into a compelling and powerful concept: the urban underclass. In the view of many middle-class whites, inner cities had come to house a large population of poorly educated single mothers and jobless men—mostly black and Puerto Rican—who were unlikely to exit poverty and become self-sufficient. In the ensuing national debate on the causes for this persistent poverty, four theoretical explanations gradually emerged: culture, racism, economics, and welfare.

Cultural explanations for the underclass can be traced to the work of Oscar Lewis, who identified a "culture of poverty" that he felt promoted patterns of behavior inconsistent with socioeconomic advancement. According to Lewis, this culture originated in endemic unemployment and chronic social immobility, and provided an ideology that allowed poor people to cope with feelings of hopelessness and despair that arose because their chances for socioeconomic success were remote. In individuals, this culture was typified by a lack of impulse control, a strong present-time orientation, and little ability to defer gratification. Among families, it yielded an absence of childhood, an early initiation into sex, a prevalence of free marital unions, and a high incidence of abandonment of mothers and children.

Although Lewis explicitly connected the emergence of these cultural patterns to structural conditions in society, he argued that once the culture of poverty was established, it became an independent cause of persistent poverty. This idea was further elaborated in 1965 by the Harvard sociologist and then Assistant Secretary of Labor Daniel Patrick Moynihan, who in a confidential report to the President focused on the relationship between male unemployment, family instability, and the inter-generational transmission of poverty, a process he labeled a "tangle of pathology." He warned that because of the structural absence of employment in the ghetto, the black family was disintegrating in a way that threatened the fabric of community life.

When these ideas were transmitted through the press, both popular and scholarly, the connection between culture and economic structure was somehow lost, and the argument was popularly perceived to be that "people were poor because they had a defective culture." This position was later explicitly adopted by the conservative theorist Edward Banfield, who argued that lower-class culture—with its limited time horizon, impulsive need for gratification, and psychological self-doubt—was primarily responsible for persistent urban poverty. He believed that these cultural traits were largely imported, arising primarily because cities attracted lower-class migrants.

The culture-of-poverty argument was strongly criticized by liberal theorists as a self-serving ideology that "blamed the victim." In the ensuing wave of re-action, black families were viewed not as weak but, on the contrary, as resilient and well adapted survivors in an oppressive and racially prejudiced society. Black disadvantages were attributed not to a defective culture but to the persistence of institutional racism in the United States. According to theorists of the underclass such as Douglas Glasgow and Alphonso Pinkney, the black urban underclass came about because deeply imbedded racist practices within American institutions—particularly schools and the economy—effectively kept blacks poor and dependent.

As the debate on culture versus racism ground to a halt during the late 1970s, conservative theorists increasingly captured public attention by focusing on a third possible cause of poverty: government welfare policy. According to Charles Murray, the creation of the underclass was rooted in the liberal welfare state. Federal antipoverty programs altered the incentives governing the behavior of poor men and women, reducing the desirability of marriage, increasing the benefits of unwed childbearing, lowering the attractiveness of menial labor, and ultimately resulting in greater poverty.

A slightly different attack on the welfare state was launched by Lawrence Mead, who argued that it was not the generosity but the permissiveness of the U.S. welfare system that was at fault. Jobless men and unwed mothers should be required to display "good citizenship" before being supported by the state. By not requiring anything of the poor, Mead argued, the welfare state undermined their independence and competence, thereby perpetuating their poverty.

This conservative reasoning was subsequently attacked by liberal social scientists, led principally by the sociologist William Julius Wilson, who had long been arguing for the increasing importance of class over race in understanding the social and economic problems facing blacks. In his 1987 book *The Truly Disadvantaged,* Wilson argued that persistent urban poverty stemmed primarily from the structural transformation of the inner-city economy. The decline of manufacturing, the suburbanization of employment, and the rise of a low-wage service sector dramatically reduced the number of city jobs that paid wages sufficient to support a family, which led to high rates of joblessness among minorities and a shrinking pool of "marriageable" men (those financially able to support a family). Marriage thus became less attractive to poor women, unwed childbearing increased, and female-headed families proliferated. Blacks suffered disproportionately from these trends because, owing to past discrimination, they were concentrated in locations and occupations particularly affected by economic restructuring.

Wilson argued that these economic changes were accompanied by an increase in the spatial concentration of poverty within black neighborhoods. This new geography of poverty, he felt, was enabled by the civil rights revolution of the 1960s, which provided middle-class blacks with new opportunities outside the ghetto. The out-migration of middle-class families from ghetto areas left behind a destitute community lacking the institutions, resources, and values necessary for success in postindustrial society. The urban underclass thus arose from a complex interplay of civil rights policy, economic restructuring, and a historical legacy of discrimination.

Theoretical concepts such as the culture of poverty, institutional racism, welfare disincentives, and structural economic change have all been widely debated. None of these explanations, however, considers residential segregation to be an important contributing cause of urban poverty and the underclass. In their principal works, Murray and Mead do not mention segregation at all, and Wilson refers to racial segregation only as a historical legacy from the past, not as an outcome that is institutionally supported and actively created today. Although Lewis mentions segregation sporadically in his writings, it is not assigned a central role in the set of structural factors responsible for the culture of poverty, and Banfield ignores it entirely. Glasgow, Pinkney, and other theorists of institutional racism mention the ghetto frequently, but generally call not for residential desegregation but for race-specific policies to combat the effects of discrimination in the schools and labor markets. In general, then, contemporary theorists of urban poverty do not see high levels of black–white segregation as particularly relevant to understanding the underclass or alleviating urban poverty.

The purpose of this [argument] is to redirect the focus of public debate back to issues of race and racial segregation and to suggest that they should be fundamental to thinking about the status of black Americans and the origins of the urban underclass. Our quarrel is less with any of the prevailing theories of urban poverty than with their systematic failure to consider the important role that segregation has played in mediating, exacerbating, and ultimately amplifying the harmful social and economic processes they treat.

We join earlier scholars in rejecting the view that poor urban blacks have an autonomous "culture of poverty" that explains their failure to achieve socioeconomic success in American society. We argue instead that residential segregation has been instrumental in creating a structural niche within which a deleterious set of attitudes and behaviors—a culture of segregation—has arisen and flourished. Segregation created the structural conditions for the emergence of an oppositional culture that devalues work, schooling, and marriage and that stresses attitudes and behaviors that are antithetical and often hostile to success in the larger economy. Although poor black neighborhoods still contain many people who lead conventional, productive lives, their example has been overshadowed in recent years by a growing concentration of poor, welfare-dependent families that is an inevitable result of residential segregation.

We readily agree with Douglas, Pinkney, and others that racial discrimination is widespread and may even be institutionalized within large sectors of American society, including the labor market, the educational system, and the welfare bureaucracy. We argue, however, that this view of black subjugation is incomplete without understanding the special role that residential segregation plays in enabling all other forms of racial oppression. Residential segregation is the institutional apparatus that supports other racially discriminatory processes and binds them together into a coherent and uniquely effective system of racial subordination. Until the black ghetto is dismantled as a basic institution of American urban life, progress ameliorating racial inequality in other arenas will be slow, fitful, and incomplete.

We also agree with William Wilson's basic argument that the structural transformation of the urban economy undermined economic supports for the black community during the 1970s and 1980s. We argue, however, that in the absence of segregation, these structural changes would not have produced the disastrous social and economic outcomes observed in inner cities during these decades. Although rates of black poverty were driven up by the economic dislocations Wilson identifies, it was segregation that confined the increased deprivation to a small number of densely settled, tightly packed, and geographically isolated areas.

Wilson also argues that concentrated poverty arose because the civil rights revolution allowed middle-class blacks to move out of the ghetto. Although we remain open to the possibility that class-selective migration did occur, we argue that concentrated poverty would have happened during the 1970s with or without black middle-class migration. Our principal objection to Wilson's focus on middle-class out-migration is not that it did not occur, but that it is misdirected: focusing on the flight of the black middle class deflects attention from the real issue, which is the limitation of black residential options through segregation.

Middle-class households—whether they are black, Mexican, Italian, Jewish, or Polish—always try to escape the poor. But only blacks must attempt their escape within a highly segregated, racially segmented housing market. Because of segregation, middle-class blacks are less able to escape than other groups, and as a result are exposed to more poverty. At the same time, because of segregation no one will move into a poor black neighborhood except other poor blacks. Thus both middle-class blacks and poor blacks lose compared with the poor and middle class of other groups: poor blacks live under unrivaled concentrations of poverty, and affluent blacks live in neighborhoods that are far less advantageous than those experienced by the middle class of other groups.

Finally, we concede Murray's general point that federal welfare policies are linked to the rise of the urban underclass, but we disagree with his specific hypothesis that generous welfare payments, by themselves, discouraged employment, encouraged unwed childbearing, undermined the strength of the family, and thereby caused persistent poverty. We argue instead that welfare payments were only harmful to the socioeconomic well-being of groups that were residentially segregated. As poverty rates rose among blacks in response to the economic dislocations of the 1970s and 1980s, so did the use of welfare programs. Because of racial segregation, however, the higher levels of welfare receipt were confined to a small number of isolated, all-black neighborhoods. By promoting the spatial concentration of welfare use, therefore, segregation created a residential environment within which welfare dependency was the norm, leading to the intergenerational transmission and broader perpetuation of urban poverty....

Our fundamental argument is that racial segregation—and its characteristic institutional form, the black ghetto—are the key structural factors responsible for the perpetuation of black poverty in the United States. Residential segregation is the principal organizational feature of American society that is responsible for the creation of the urban underclass....

KEY CONCEPTS

apartheid   hypersegregation   segregation
culture of poverty

DISCUSSION QUESTIONS

1. Regardless of your race or ethnicity, did you grow up in a racially segregated environment? How central in your life was this fact? What consequences did it have? Think of other people you know who did grow up in a racially segregated environment. What, in your estimation, were the effects on their life?

2. What is the "culture of poverty" view? Do you agree with it? What do Massey and Denton have to say about it?

64

Environmental Justice in the 21st Century: Race Still Matters

ROBERT D. BULLARD

This reading outlines the grassroots movements to address the way environmental hazards are disproportionally located near minority communities. Bullard summarizes some of the research that shows how minority residents are hardest hit by the health consequences of toxic dumping grounds and industries that create poor air quality. His conclusion is that environmental justice has not been obtained; rather, environmental racism is still prevalent.

SOURCE: Bullard, Robert D. 2001. "Environmental Justice in the 21st Century: Race Still Matters." *Phylon* 49: 151–171.

Hardly a day passes without the media discovering some community or neighborhood fighting a landfill, incinerator, chemical plant, or some other polluting industry. This was not always the case. Just three decades ago, the concept of environmental justice had not registered on the radar screens of environmental, civil rights, or social justice groups.[1] Nevertheless, it should not be forgotten that Dr. Martin Luther King, Jr., went to Memphis in 1968 on an environmental and economic justice mission for the striking black garbage workers. The strikers were demanding equal pay and better work conditions. Of course, Dr. King was assassinated before he could complete his mission.

Another landmark garbage dispute took place a decade later in Houston, when African-American homeowners in 1979 began a bitter fight to keep a sanitary landfill out of their suburban middle-income neighborhood.[2] Residents formed the Northeast Community Action Group or NECAG. NECAG and their attorney, Linda McKeever Bullard, filed a class-action lawsuit to block the facility from being built. The 1979 lawsuit, *Bean v. Southwestern Waste Management, Inc.*, was the first of its kind to challenge the siting of a waste facility under civil rights law.

The landmark Houston case occurred three years before the environmental justice movement was catapulted into the national limelight in the rural and mostly African-American Warren County, North Carolina. The environmental justice movement has come a long way since its humble beginning in Warren County, North Carolina, where a PCB landfill ignited protests and over 500 arrests. The Warren County protests provided the impetus for a U.S. General Accounting Office study, *Siting of Hazardous Waste Landfills and Their Correlation with Racial and Economic Status of Surrounding Communities*.[3] That study revealed that three out of four of the off-site, commercial hazardous waste landfills in Region 4 (which comprises eight states in the South) happen to be located in predominantly African-American communities, although African-Americans made up only 20% of the region's population. More important, the protesters put "environmental racism" on the map. Fifteen years later, the state of North Carolina is required to spend over $25 million to clean up and detoxify the Warren County PCB landfill.

The Warren County protests also led the Commission for Racial Justice to produce *Toxic Waste and Race*,[4] the first national study to correlate waste facility sites and demographic characteristics. Race was found to be the most potent variable in predicting where these facilities were located—more powerful than poverty, land values, and home ownership. In 1990, *Dumping in Dixie: Race, Class, and Environmental Quality* chronicled the convergence of two social movements—social justice and environmental movements—into the environmental justice movement. This book highlighted African-Americans, environmental activism in the South, the same region that gave birth to the modern civil rights movement. What started out as local and often isolated community-based struggles against toxics and facility siting blossomed into a multiissue, multiethnic, and multiregional movement.

The 1991 First National People of Color Environmental Leadership Summit was probably the most important single event in the movement's history. The

Summit broadened the environmental justice movement beyond its early anti-toxics focus to include issues of public health, worker safety, land use, transportation, housing, resource allocation, and community empowerment.[5]

The meeting also demonstrated that it is possible to build a multiracial grass-roots movement around environmental and economic justice.[6]

Held in Washington, DC, the four-day Summit was attended by over 650 grassroots and national leaders from around the world. Delegates came from all fifty states including Alaska and Hawaii, Puerto Rico, Chile, Mexico, and as far away as the Marshall Islands. People attended the Summit to share their action strategies, redefine the environmental movement, and develop common plans for addressing environmental problems affecting people of color in the United States and around the world.

On September 27, 1991, Summit delegates adopted 17 "Principles of Environmental Justice." These principles were developed as a guide for organizing, networking, and relating to government and nongovernmental organizations (NGOs). By June 1992, Spanish and Portuguese translations of the Principles were being used and circulated by NGOs and environmental justice groups at the Earth Summit in Rio de Janeiro.

In response to growing public concern and mounting scientific evidence, President Clinton on February 11, 1994 (the second day of the national health symposium), issued Executive Order 12898, "Federal Actions to Address Environmental Justice in Minority Populations and Low-Income Populations." This Order attempts to address environmental injustice within existing federal laws and regulations.

Executive Order 12898 reinforces the 35-year old Civil Rights Act of 1964, Title VI, which prohibits discriminatory practices in programs receiving federal funds. The Order also focuses the spotlight back on the National Environmental Policy Act (NEPA), a twenty-five-year-old law that set policy goals for the protection, maintenance, and enhancement of the environment. NEPA's goal is to ensure for all Americans a safe, healthful, productive, and aesthetically and culturally pleasing environment. NEPA requires federal agencies to prepare a detailed statement on the environmental effects of proposed federal actions that significantly effect the quality of human health.

The Executive Order calls for improved methodologies for assessing and mitigating impacts, health effect from multiple and cumulative exposure, collection of data on low-income and minority populations who may be disproportionately at risk, and impacts on subsistence fishers and wildlife consumers. It also encourages participation of the impacted populations in the various phases of assessing impacts—including scoping, data gathering, alternatives, analysis, mitigation, and monitoring.

The Executive Order focuses on "subsistence" fishers and wildlife consumers. Everybody does not buy fish at the supermarket. There are many people who are subsistence fishers, who fish for protein, who basically subsidize their budgets, and their diets by fishing from rivers, streams, and lakes that happen to be polluted. These subpopulations may be underprotected when basic assumptions are made using the dominant risk paradigm.

Many grassroots activists are convinced that waiting for the government to act has endangered the health and welfare of their communities. Unlike the federal EPA, communities of color did not first discover environmental inequities in 1990. The federal EPA only took action on environmental justice concerns in 1990 after extensive prodding from grassroots environmental justice activists, educators, and academics.[7]

People of color have known about and have been living with inequitable environmental quality for decades—most without the protection of the federal, state, and local governmental agencies. Environmental justice advocates continue to challenge the current environmental protection apparatus and offer their own framework for addressing environmental inequities, disparate impact, and unequal protection.

AN ENVIRONMENTAL JUSTICE FRAMEWORK

The question of environmental justice is not anchored in a debate about whether or not decision makers should tinker with risk management. The framework seeks to prevent environmental threats before they occur.[8] The environmental justice framework incorporates other social movements that seek to eliminate harmful practices (discrimination harms the victim) in housing, land use, industrial planning, health care, and sanitation services. The impact of redlining, economic disinvestment, infrastructure decline, deteriorating housing, lead poisoning, industrial pollution, poverty, and unemployment are not unrelated problems if one lives in an urban ghetto or barrio, rural hamlet, or reservation.

The environmental justice framework attempts to uncover the underlying assumptions that may contribute to and produce unequal protection. This framework brings to the surface the ethical and political questions of "who gets what, why, and how much." Some general characteristics of the framework include:

(1) *The environmental justice framework incorporates the principle of the "right" of all individuals to be protected from environmental degradation.* The precedents for this framework are the Civil Rights Act of 1964, Fair Housing Act of 1968 and as amended in 1988, and Voting Rights Act of 1965.

(2) *The environmental justice framework adopts a public health model of prevention (elimination of the threat before harm occurs) as the preferred strategy.* Impacted communities should not have to wait until causation or conclusive "proof" is established before preventive action is taken. For example, the framework offers a solution to the lead problem by shifting the primary focus from *treatment* (after children have been poisoned) to *prevention* (elimination of the threat via abating lead in houses).

Overwhelming scientific evidence exists on the ill effects of lead on the human body. However, very little action has been taken to rid the nation of childhood lead poisoning in urban areas. Former Health and Human Secretary Louis Sullivan tagged this among the "number one environmental health threats to children."[9]

The Natural Resources Defense Council, NAACP Legal Defense and Educational Fund, ACLU, and Legal Aid Society of Alameda County joined forces in

1991 and won an out-of-court settlement worth $15–20 million for a blood-lead testing program in California. The *Matthews v. Coye* lawsuit involved the State of California not living up to the federally mandated testing of some 557,000 poor children for lead who receive Medicaid. This historic agreement triggered similar actions in other states that failed to live up to federally mandated screening.[10]

Lead screening is an important element in this problem. However, screening is not the solution. Prevention is the solution. Surely, if termite inspections can be mandated to protect individual home investment, a lead-free home can be mandated to protect public health. Ultimately, the lead abatement debate, public health (who is affected) vs. property rights (who pays for cleanup), is a value conflict that will not be resolved by the scientific community.

(3) *The environmental justice framework shifts the burden of proof to polluters/dischargers who do harm, discriminate, or who do not give equal protection to racial and ethnic minorities, and other "protected" classes.* Under the current system, individuals who challenge polluters must "prove" that they have been harmed, discriminated against, or disproportionately impacted. Few impacted communities have the resources to hire lawyers, expert witnesses, and doctors needed to sustain such a challenge.

The environmental justice framework would require the parties that are applying for operating permits (landfills, incinerators, smelters, refineries, chemical plants, etc.) to "prove" that their operations are not harmful to human health, will not disproportionately impact racial and ethnic minorities and other protected groups, and are nondiscriminatory.

(4) *The environmental justice framework would allow disparate impact and statistical weight, as opposed to "intent," to infer discrimination.* Proving intentional or purposeful discrimination in a court of law is next to impossible, as demonstrated in Bean v. Southwestern Waste. It took nearly a decade after *Bean v. Southwestern Waste* for environmental discrimination to resurface in the courts.

(5) *The environmental justice framework redresses disproportionate impact through "targeted" action and resources.* This strategy would target resources where environmental and health problems are greatest (as determined by some ranking scheme but not limited to risk assessment). Reliance solely on "objective" science disguises the exploitative way the polluting industries have operated in some communities and condones a passive acceptance of the status quo.

Human *values* are involved in determining *which* geographic areas are worth public investments. In the 1992 EPA report, *Securing Our Legacy*, the agency describes geographic initiatives as "protecting what we love."[11]

The strategy emphasizes "pollution prevention, multimedia enforcement, research into causes and cures of environmental stress, stopping habitat loss, education, and constituency building."[12] Geographic initiatives are underway in the Chesapeake Bay, Great Lakes, Gulf of Mexico programs, and the U.S.-Mexican Border program. Environmental justice targeting would channel resources to "hot spots," communities that are overburdened with more than their "fair" share of environmental and health problems.

The dominant environmental protection paradigm reinforces instead of challenges the stratification of people (race, ethnicity, status, power, etc.), place (central cities, suburbs, rural areas, unincorporated areas, Native American reservations, etc.), and work (i.e., office workers are afforded greater protection than farm workers). The dominant paradigm exists to manage, regulate, and distribute risks. As a result, the current system has (1) institutionalized unequal enforcement, (2) traded human health for profit, (3) placed the burden of proof on the "victims" and not the polluting industry, (4) legitimated human exposure to harmful chemicals, pesticides, and hazardous substances, (5) promoted "risky" technologies such as incinerators, (6) exploited the vulnerability of economically and politically disenfranchised communities, (7) subsidized ecological destruction, (8) created an industry around risk assessment, (9) delayed cleanup actions, and (10) failed to develop pollution prevention as the overarching and dominant strategy.[13]

The mission of the federal EPA was never designed to address environmental policies and practices that result in unfair, unjust, and inequitable outcomes. EPA and other government officials are not likely to ask the questions that go to the heart of environmental injustice: What groups are most affected? Why are they affected? Who did it? What can be done to remedy the problem? How can the problem be prevented? Vulnerable communities, populations, and individuals often fall between the regulatory cracks.

IMPETUS FOR A PARADIGM SHIFT

The environmental justice movement has changed the way scientists, researchers, policy makers, and educators go about their daily work. This bottom–up movement has redefined environment to include where people live, work, play, go to school, as well as how these things interact with the physical and natural world. The impetus for changing the dominant environmental protection paradigm did *not* come from within regulatory agencies, the polluting industry, academia, or the "industry" that has been built around risk management. The environmental justice movement is led by a loose alliance of grassroots and national environmental and civil rights leaders who question the foundation of the current environmental protection paradigm.

Despite significant improvements in environmental protection over the past several decades, millions of Americans continue to live, work, play, and go to school in unsafe and unhealthy physical environments.[14] During its 30-year history, the U.S. EPA has not always recognized that many of our government and industry practices (whether intended or unintended) have adverse impact on poor people and people of color. Growing grassroots community resistance emerged in response to practices, policies, and conditions that residents judged to be unjust, unfair, and illegal. Discrimination is a fact of life in America. Racial discrimination is also illegal.

THE IMPACT OF RACIAL APARTHEID

Apartheid-type housing, development, and environmental policies limit mobility, reduce neighborhood options, diminish job opportunities, and decrease choices for millions of Americans.[15] The infrastructure conditions in urban areas are a result of a host of factors including the distribution of wealth, patterns of racial and economic discrimination, redlining, housing and real estate practices, location decisions of industry, and differential enforcement of land use and environmental regulations. Apartheid-type housing and development policies have resulted in limited mobility, reduced neighborhood options, decreased environmental choices, and diminished job opportunities for African Americans.

Race still plays a significant part in distributing public "benefits" and public "burdens" associated with economic growth. The roots of discrimination are deep and have been difficult to eliminate. Housing discrimination contributes to the physical decay of inner-city neighborhoods and denies a substantial segment of the African-American community a basic form of wealth accumulation and investment through home ownership.[16] The number of African-American homeowners would probably be higher in the absence of discrimination by lending institutions.[17] Only about 59 percent of the nation's middle-class African Americans own their homes, compared with 74 percent of whites.

Eight out of every ten African Americans live in neighborhoods where they are in the majority. Residential segregation decreases for most racial and ethnic groups with additional education, income, and occupational status. However, this scenario does not hold true for African Americans. African Americans, no matter what their educational or occupational achievement or income level, are exposed to higher crime rates, less effective educational systems, high mortality risks, more dilapidated surroundings, and greater environmental threats because of their race. For example, in the heavily populated South Coast air basin of the Los Angeles area, it is estimated that over 71 percent of African Americans and 50 percent of Latinos reside in areas with the most polluted air, while only 34 percent of whites live in highly polluted areas.[18]

Institutional racism continues to influence housing and mobility options available to African Americans of all income levels—and is a major factor that influences quality of neighborhoods they have available to them. The "web of discrimination" in the housing market is a result of action and inaction of local and federal government officials, financial institutions, insurance companies, real estate marketing firms, and zoning boards. More stringent enforcement mechanisms and penalties are needed to combat all forms of discrimination.

Racial and ethnic inequality is perpetuated and reinforced by local governments in conjunction with urban-based corporations. Race continues to be a potent variable in explaining urban land use, streets and highway configuration, commercial and industrial development, and industrial facility siting. Moreover, the question of "who gets what, where, and why" often pits one community against another.[19]

ENVIRONMENTAL RACISM

Many of the differences in environmental quality between black and white communities result from institutional racism. Institutional racism influences local land use, enforcement of environmental regulations, industrial facility siting, and where people of color live, work, and play. The roots of institutional racism are deep and have been difficult to eliminate. Discrimination is a manifestation of institutional racism and causes life to be very different for whites and blacks. Historically, racism has been and continues to be a major part of the American sociopolitical system, and as a result, people of color find themselves at a disadvantage in contemporary society.

Environmental racism is real. It is just as real as the racism found in the housing industry, educational institutions, employment arena, and judicial system. What is environmental racism and how does one recognize it? *Environmental racism refers to any policy, practice, or directive that differentially affects or disadvantages (whether intended or unintended) individuals, groups, or communities based on race or color.* Environmental racism combines with public policies and industry practices to provide benefits for whites while shifting costs to people of color.[20] Environmental racism is reinforced by government, legal, economic, political, and military institutions.

Environmental decision making and policies often mirror the power arrangements of the dominant society and its institutions. Environmental racism disadvantages people of color while providing advantages or privileges for whites. A form of illegal "exaction" forces people of color to pay costs of environmental benefits for the public at large. The question of who *pays* and who *benefits* from the current environmental and industrial policies is central to this analysis of environmental racism and other systems of domination and exploitation.

Racism influences the likelihood of exposure to environmental and health risks as well as accessibility to health care.[21] Many of the nation's environmental policies distribute the costs in a regressive pattern while providing disproportionate benefits for whites and individuals who fall at the upper end of the education and income scale. Numerous studies, dating back to the seventies, reveal that people of color have borne greater health and environmental risk burdens than the society at large.[22]

Elevated public health risks are found in some populations even when social class is held constant. For example, race has been found to be independent of class in the distribution of air pollution,[23] contaminated fish consumption,[24] location of municipal landfills and incinerators,[25] toxic waste dumps,[26] cleanup of superfund sites,[27] and lead poisoning in children.[28]

Lead poisoning is a classic example of an environmental health problem that disproportionately impacts children of color at every class level. Lead affects between 3 to 4 million children in the United States—most of whom are African American and Latinos who live in urban areas. Among children 5 years old and younger, the percentage of African-American children who have excessive levels of lead in their blood far exceeds the percentage of whites at all income levels.

THE RIGHT TO BREATHE CLEAN AIR

Urban air pollution problems have been with us for some time now. Before the federal government stepped in, issues related to air pollution were handled primarily by states and local governments. Because states and local governments did such a poor job, the federal government set out to establish national clean air standards. Congress enacted the Clean Air Act (CAA) in 1970 and mandated the U.S. Environmental Protection Agency (EPA) to carry out this law. Subsequent amendments (1977 and 1990) were made to the CAA that form the current federal program. The CAA was a response to states' unwillingness to protect air quality. Many states used their lax enforcement of environmental laws as lures for business and economic development.[29]

Central cities and suburbs do not operate on a level playing field. They often compete for scarce resources. One need not be a rocket scientist to predict the outcome between affluent suburbs and their less affluent central city competitors.[30] Freeways are the lifeline for suburban commuters, while millions of central-city residents are dependent on public transportation as their primary mode of travel. But recent cuts in mass transit subsidies and fare hikes have reduced access to essential social services and economic activities. Nevertheless, road construction programs are booming—even in areas choked with automobiles and air pollution.[31]

The air quality impacts of transportation are especially significant to people of color who are more likely than whites to live in urban areas with reduced air quality. National Argonne Laboratory researchers discovered that 437 of the 3,109 counties and independent cities failed to meet at least one of the EPA ambient air quality standards.[32] Specifically, 57 percent of whites, 65 percent of African Americans, and 80 percent of Hispanics live in 437 counties with substandard air quality. Nationwide, 33 percent of whites, 50 percent of African Americans, and 60 percent of Hispanics live in the 136 counties in which two or more air pollutants exceed standards. Similar patterns were found for the 29 counties designated as nonattainment areas for three or more pollutants. Again, 12 percent of whites, 20 percent of African Americans, and 31 percent of Hispanics resided in the worse nonattainment areas.

Asthma is an emerging epidemic in the United States. The annual age-adjusted death rate from asthma increased by 40% between 1982 through 1991, from 1.34 to 1.88 per 100,000 population,[33] with the highest rates being consistently reported among blacks aged 15–24 years of age during the period 1980–1993.[34] Poverty and minority status are important risk factors for asthma mortality.

Children are at special risk from ozone.[35] Children also represent a considerable share of the asthma burden. It is the most common chronic disease of childhood. Asthma affects almost 5 million children under 18 years. Although the overall annual age-adjusted hospital discharge rate for asthma among children under 15 years old decreased slightly from 184 to 179 per 100,000 between 1982 and 1992, the decrease was slower compared to other childhood diseases[36] resulting in a 70% increase in the proportion of hospital admissions related to asthma during the 1980s.[37] Inner-city children have the highest rates for asthma prevalence, hospitalization, and mortality.[38] In the United States, asthma is the fourth leading cause of disability among children aged less than 18 years.[39]

...Nationally, African Americans and Latino Americans have significantly higher prevalence of asthma than the general population. A 1996 report from the federal Centers for Disease Control shows hospitalization and deaths rates from asthma increasing for persons twenty-five years or less. The greatest increases occurred among African Americans. African Americans are two to six times more likely than whites to die from asthma. Similarly, the hospitalization rate for African Americans is 3 to 4 times the rate for whites.

GLOBAL DUMPING GROUNDS

There is a direct correlation between exploitation of land and exploitation of people. It should not be a surprise to anyone to discover that Native Americans have to contend with some of the worst pollution in the United States.[40] Native American nations have become prime targets for waste trading.[41] More than three dozen Indian reservations have been targeted for landfills, incinerators, and other waste facilities.[42] The vast majority of these waste proposals were defeated by grassroots groups on the reservations. However, "radioactive colonialism" is alive and well.[43] The legacy of institutional racism has left many sovereign Indian nations without an economic infrastructure to address poverty, unemployment, inadequate education and health care, and a host of other social problems. In 1999, Eastern Navajo reservation residents filed suit against the Nuclear Regulatory Commission to block uranium mining in Church Rock and Crown Point communities.

In the real world, all people, communities, and nations are not created equal. Some populations and interests are more equal than others. Unequal interests and power arrangements have allowed poisons of the rich to be offered as short term remedies for poverty of the poor.

SETTING THE RECORD STRAIGHT

The environmental protection apparatus is broken and needs to be fixed. The environmental justice movement has set out clear goals of eliminating unequal enforcement of environmental, civil rights, and public health laws. Environmental justice leaders have made a difference in the lives of people and the physical environment. They have assisted public decision makers in identifying "at-risk" populations, toxic "hot spots," research gaps, and action models to correct existing imbalances and prevent future threats. However, impacted communities are not waiting for the government or industry to get their acts together. Grassroots groups have taken the offensive to ensure that government and industry do the right thing.

NOTES

1. Robert D. Bullard, 1994, *Dumping in Dixie: Race, Class and Environmental Quality.* Boulder, CO: Westview Press.

2. Robert D. Bullard, "Solid Waste Sites and the Black Houston Community," *Sociological Inquiry* 53 (Spring 1983): 273–288.

3. U.S. General Accounting Office (1983), *Siting of Hazardous Waste Landfills and Their Correlation with Racial and Economic Status of Surrounding Communities*, Washington, DC: Government Printing Office.

4. Commission for Racial Justice (1987), *Toxic Wastes and Race in the United States*, New York: United Church of Christ.

5. Charles Lee, 1992, *Proceedings: The First National People of Color Environmental Leadership Summit.* New York: United Church of Christ Commission for Racial Justice.

6. Dana Alston, "Transforming a Movement: People of Color Unite at Summit against Environmental Racism," *Sojourner* 21 (1992), pp. 30–31.

7. William K. Reilly, "Environmental Equity: EPA's Position," *EPA Journal* 18 (March/April 1992): 18–19.

8. R.D. Bullard and B.H. Wright, "The Politics of Pollution: Implications for the Black Community," *Phylon* 47 (March 1986): 71–78.

9. Robert D. Bullard, "Race and Environmental Justice in the United States," *Yale Journal of International Law* 18 (Winter 1993): 319–335; Robert D. Bullard, "The Threat of Environmental Racism," *Natural Resources & Environment* 7 (Winter 1993): 23–26, 55–56.

10. Louis Sullivan, "Remarks at the First Annual Conference on Childhood Lead Poisoning," in Alliance to End Childhood Lead Poisoning, *Preventing Child Lead Poisoning: Final Report.* Washington, DC: Alliance to End Childhood Lead Poisoning, October, 1991, p. A–2.

11. Bill Lann Lee, "Environmental Litigation on Behalf of Poor, Minority Children, Mathews v. Coye: A Case Study." Paper presented at the Annual Meeting of the American Association for the Advancement of Science, Chicago (February 9, 1992).

12. Ibid., p. 32.

13. Ibid.

14. Robert D. Bollard, "The Environmental Justice Framework: A Strategy for Addressing Unequal Protection." Paper presented at Resources for the Future Conference on Risk Management, Annapolis, MD (November 1992).

15. Marianne Lavelle and Marcia Coyle, "Unequal Protection," *National Law Journal* (September 21, 1992): S1–S2.

16. Robert D. Bullard, ed., *Confronting Environmental Racism: Voices from the Grassroots.* Boston: South End Press, 1993, chapter 1; Robert D. Bullard, "Waste and Racism: A Stacked Deck?" *Forum for Applied Research and Public Policy* 8 (Spring, 1993): 29–35; Robert D. Bollard (ed.), *In Search of the New South – The Black Urban Experience in the 1970s and 1980s* (Tuscaloosa. AL: University of Alabama Press, 1991).

17. Florence Wagman Roisman, "The Lessons of American Apartheid: The Necessity and Means of Promoting Residential Racial Integration," *Iowa Law Review* 81 (December 1995): 479–525.

18. Joe R. Feagin, "A House Is Not a Home: White Racism and U.S. Housing Practices," in R.D. Bullard, J.E. Grigsby, and Charles Lee, eds., *Residential Apartheid: The American Legacy.* Los Angeles: UCLA Center for Afro-American Studies Publication, 1994, pp. 17–48.

19. Bunyan Bryant and Paul Mohai, *Race and the Incidence of Environmental Hazards* (Boulder, CO: Westview Press, 1992); Bunyan Bryant, ed., *Environmental Justice,* pp. 8–34.

20. U.S. General Accounting Office, *Siting of Hazardous Waste Landfills and Their Correlation with Racial and Economic Status of Surrounding Communities.* Washington, DC: U.S. General Accounting Office, 1983, p. 1.

21. Robert D. Bullard, ed., *Confronting Environmental Racism: Voices from the Grassroots* Boston: South End, 1993; Robert D. Bullard. "The Threat of Environmental Racism," *Natural Resources & Environment* 7 (Winter 1993): 23–26; Bunyan Bryant and Paul Mohai, eds., *Race and the Incidence of Environmental Hazards.* Boulder, CO: Westview Press, 1992; Regina Austin and Michael Schill, "Black, Brown, Poor and Poisoned: Minority Grassroots Environmentalism and the Quest for Eco-Justice." *The Kansas Journal of Law and Public Policy* 1 (1991): 69–82; Kelly C. Colquette and Elizabeth A. Henry Robertson, "Environmental Racism: The Causes, Consequences, and Commendations." *Tulane Environmental Law Journal* 5 (1991): 153–207; Rachel D. Godsil, "Remedying Environmental Racism." *Michigan Law Review* 90 (1991): 394–427.

22. Bullard and Feagin, "Racism and the City," pp. 55–76; Robert D. Bullard, "Dismantling Environmental Racism in the USA," *Local Environment* 4 (1999): 5–19.

23. W. J. Kruvant, "People, Energy and Pollution." pp. 125–167 in D. K. Newman and Dawn Day, eds., *The American Energy Consumer.* Cambridge, Mass.: Ballinger, 1975; Robert D. Bullard, "Solid Waste Sites and the Black Houston Community." *Sociological Inquiry* 53 (Spring 1983): 273–288; United Church of Christ Commission for Racial Justice, *Toxic Wastes and Race in the United States.* New York: Commission for Racial Justice, 1987; Dick Russell, "Environmental Racism." *The Amicus Journal* 11 (Spring 1989): 22–32; Eric Mann, *L.A.'s Lethal Air: New Strategies for Policy, Organizing, and Action.* Los Angeles: Labor/Community Strategy Center, 1991; D. R. Wernette and L. A. Nieves, "Breathing Polluted Air: Minorities are Disproportionately Exposed." *EPA Journal* 18 (March/April 1992): 16–17; Bryant and Mohai, Race and the Incidence of Environmental Hazards; Benjamin Goldman and Laura J. Fitton, *Toxic Wastes and Race Revisited.* Washington, DC: Center for Policy Alternatives, NAACP, and United Church of Christ, 1994.

24. Myrick A. Freedman, "The Distribution of Environmental Quality." in Allen V. Kneese and Blair T. Bower (eds.), *Environmental Quality Analysis.* Baltimore: Johns Hopkins University Press for Resources for the Future, 1971; Michel Gelobter, "The Distribution of Air Pollution by Income and Race." Paper presented at the Second Symposium on Social Science in Resource Management, Urbana, Illinois (June 1988); Gianessi et al., "The Distributional Effects of Uniform Air Pollution Policy in the U.S." *Quarterly Journal of Economics* (May 1979): 281–301.

25. Patrick C. West, J. Mark Fly, and Robert Marans, "Minority Anglers and Toxic Fish Consumption: Evidence from a State-Wide Survey in Michigan." In Bryant and Mohai, *Race and the Incidence of Environmental Hazards,* pp. 100–113.

26. Robert D. Bullard, "Solid Waste Sites and the Black Houston Community." *Sociological Inquiry* 53 (Spring 1983): 273–288; Robert D. Bullard, *Invisible Houston: The Black Experience in Boom and Bust.* College Station, TX: Texas A&M University Press, 1987, chapter 6; Robert D. Bollard, "Environmental Racism and Land Use." *Land Use Forum: A Journal of Law, Policy & Practice* 2 (Spring 1993): 6–11.

27. United Church of Christ Commission for Racial Justice, *Toxic Wastes and Race*; Paul Mohai and Bunyan Bryant, "Environmental Racism: Reviewing the Evidence." in Bryant and Mohai, *Race and the Incidence of Environmental Hazards*; Paul Stretesky and Michael J. Hogan, "Environmental Justice: An Analysis of Superfund Sites in Florida," *Social Problems* 45 (May 1998): 268–287.

28. Marianne Lavelle and Marcia Coyle, "Unequal Protection: The Racial Divide in Environmental Law." *National Law Journal*, September 21, 1992.

29. James L. Pirkle, D.J. Brody, E.W. Gunter, R.A. Kramer, D.C. Paschal, K.M. Glegal, and T.D. Matte, "The Decline in Blood Lead Levels in the United States: The National Health and Nutrition Examination Survey (NHANES III)," *Journal of the American Medical Association* 272 (1994): 284–291.

30. Arnold W. Reitze, Jr., "A Century of Air Pollution Control Law: What Worked; What Failed; What Might Work," *Environmental Law* 21 (1991): 1549.

31. For an in-depth discussion of transportation investments and social equity issues, see R.D. Bullard and C.S. Johnson, eds., *Just Transportation: Dismantling Race and Class Barriers to Mobility.* Gabriola Island, BC: New Society Publishers, 1997.

32. Sid Davis, "Race and the Politics of Transportation in Atlanta," in R.D. Bullard and G.S. Johnson, Just Transportation, pp. 84–96; Environmental Justice Resource Center, *Sprawl Atlanta: Social Equity Dimensions of Uneven Growth and Development.* A Report prepared for the Turner Foundation, Atlanta: Clark Atlanta University (January 1999).

33. D.R. Wernette and L.A. Nieves, "Breathing Polluted Air: Minorities Are Disproportionately Exposed," *EPA Journal* 18 (March 1992): 16–17.

34. CDC, "Asthma—United States, 1982–1992." *MMWR* 43 (1995): 952–955.

35. CDC, "Asthma mortality and hospitalization among children and young adults – United States, 1980–1993." *MMWR* 45(1996): 350–353.

36. Anna E. Pribitkin, "The Need for Revision of Ozone Standards: Why Has the EPA Failed to Respond?" *Temple Environmental Law & Technology Journal* 13 (1994): 104.

37. CDC/NCHS, *Health United States* 1994. DHHS Pub.No.(PHS) 95-1232; Tables 83, 84, 86, & 87.

38. CDC, "Asthma—United States, 1982–1992." *MMWR* 43 (1995): 952–955.

39. CDC, "Disabilities among children aged less than or equal to 17 years–United States, 1991–1992." *MMWR* 44(1995): 609–613.

40. Conger Beasley, "Of Pollution and Poverty: Deadly Threat on Native Lands," *Buzzworm,* 2(5) (1990): 39–45; Robert Tomsho, "Dumping Grounds: Indian Tribes Contend with Some of the Worst of America's Pollution," *The Wall Street Journal* (November 29, 1990); Jane Kay, "Indian Lands Targeted for Waste Disposal Sites," *San Francisco Examiner* (April 10, 1991); Valerie Taliman, "Stuck Holding the Nation's Nuclear Waste," *Race, Poverty & Environment Newsletter* (Fall 1992): 6–9.

41. Bradley Angel, *The Toxic Threat to Indian Lands: A Greenpeace Report.* San Francisco: Greenpeace, 1992; Al Geddicks, *The New Resource Wars: Native and Environmental Struggles Against Multinational Corporations.* Boston: South End Press, 1993.

42. Jane Kay, "Indian Lands Targeted for Waste Disposal Sites," *San Francisco Examiner* (April 10, 1991).

43. Ward Churchill and Winona LaDuke, "Native America: The Political Economy of Radioactive Colonialism." *Insurgent Sociologist* 13(1) (1983): 61–63.

KEY CONCEPTS

environmental justice   environmental racism

DISCUSSION QUESTIONS

1. Can you find a newspaper or Internet news story about a community fighting against a polluting industry? What are the key issues facing a neighborhood with an environmental concern?

2. What about in your own community? Do you see environmental hazards near your school or your home? Do you see a difference in the racial composition of neighborhoods that are close to environmental hazards versus ones that are not?

65

Global Warming and Sociology

CONSTANCE LEVER-TRACY

Lever-Tracy argues that, even with the need for scientific studies of the effects of climate change, sociology has a particular role to play in this growing environmental crisis. Her argument stresses the need for multidisciplinary approaches to the consequences of global warming.

SOURCE: Lever-Tracy. Constance. 2008. "Global Warming and Sociology". *Current Sociology*, 56: 445–466.

... If any of the scenarios of probable serious effects or possible escalating feed-backs or runaway climate change materialize beyond a point of no return, then global society as a whole, its structure, culture and trajectory, will surely be completely changed, as will the nature and relations of states to society and to each other. Similarly, fundamental changes would result from any of a number of alternative responses to such a crisis.

Evidence of a dramatic speeding up of the rate of change in natural processes has revolutionary implications for society and for sociology.

SCIENTISTS AND GLOBAL WARMING

...Numerous reputable, peer-reviewed studies have contributed to establish-ing an almost complete scientific consensus about the reality, causation and seriousness of the threat, although with a range of possible scenarios. This culminated in July 2005, in a letter from the heads of 11 influential national science academies (of Brazil, Canada, China, France, Germany, India, Italy, Japan, Russia, the UK and the US), who wrote to the G8 leaders warning that global climate change was 'a clear and increasing threat' and that they must act immediately. They outlined strong and long-term evidence 'from direct measurements of rising surface air temperatures and sub-surface ocean temperatures and from phenomena such as increases in average global sea levels, retreating glaciers and changes to many physical and biological systems' (*New Scientist*, 2005).

It is true that the popular media often give equal time to others who have questioned the methodology of each study, stressing uncertainty and complex-ity. However, the same few 'climate sceptics' (often business economists) ap-pear repeatedly, and may be dubiously funded. Alternative explanations may be given for each of the findings and none will be immune to later correction. However, their diversity and cumulating mutual reinforcement, and the pro-gressive resolution of anomalies, have left few climatologists with any serious doubts.

The most powerful argument of the sceptics was that measurements from satellites and balloons in the lower troposphere indicated cooling, contradicting those from the surface and the upper troposphere. On 20 August 2005, *New Scientist* reported the publication in *Science* of the findings of three independent studies, described as 'nails in the coffin' of the scep-tics' case. These showed that the data that lay behind the anomaly were faulty. 'The debate will linger. But there is [*sic*] no longer any data contra-dicting the predictions of global warming', said one of these researchers (Merali, 2005: 10).

While sociologists may not be trained to evaluate the scientific evidence, they should know to be wary of the powerful corporate interests motivated to deny global warming, and respectful of the choice of 'speaking truth to power' that many scientists have now adopted.

THE EVENTS OF 2005

… Global catastrophe may be relatively unlikely, but few scientists would now consider it inconceivable, especially if major Third World countries such as China, India or Indonesia achieve western emissions levels, or if the Amazon rainforest is destroyed or positive feedbacks cascade unpredictably. While most models now project a maximum warming of between 5 and 6°C by the end of the century, some new scenarios suggest much worse if feedback loops are triggered (Leggett, 2005: 118–21)….

SOCIOLOGY AND GLOBAL WARMING

The Strange Silence of Mainstream Sociology

The dangers that scientists have been warning about pose major threats to society or call for a rapid, total reorientation of its most important activities, structures and values. Yet mainstream sociology and its major journals continue to be oblivious of this elephant in the room….

Why this Silence?

There is a mystery in this lack of interest in developments that could conceivably open the door to chaos and barbarism later this century, or whose prevention might require a transformation in the core processes of industrial society.

… Arguably, it derives from the interaction of two factors. The first is our recently acquired suspicion of teleology and our mirroring of an indifference we find in contemporary society towards the future. The second factor is our continuing foundational suspicion of naturalistic explanations for social facts, which has often led us to question or ignore the authority of natural scientists, even in their own field of study. Together, these two have often blinded us to the predicted, fateful convergence of social and natural time, in a new teleological countdown to possible disaster, coming towards us from the future.

… Sociologists have thus described at length how contemporary society has turned its eyes away from the future, its people focusing on immediate consumption and ephemeral fashions, its politicians on the next election and its industrial leaders on the next annual report. To take global warming seriously involves asking the kinds of questions about future directions that most sociologists believe they have now put behind them. Preoccupied with analysing these 'social facts', sociologists are unwilling to be disturbed by the voices of natural scientists, reporting from inaccessible upper atmospheres, ancient ice cores or deep oceans, where no social facts exist. Unable themselves to judge the validity of the evidence, and increasingly uncomfortable with predictions and teleologies, they prefer to avoid the subject.

… Social processes that impact on nature in unintended ways, such as emissions caused by economic growth and the destruction of carbon sink forests, have been speeding up exponentially since the industrial revolution. The result has been an un-expected and unprecedented speeding up also of changes in natural processes. Natural change is usually very slow. It used to be believed, for example, that it would take 10,000 years to melt an ice sheet, but we can no longer assume that, for practical purposes, changes in natural processes are not relevant to social analysis. Global climate changes are now likely to impact within our own lives or those of our children. The urgency for remedial action is now measured in decades, not able to be postponed to some indefinite future. But even decades have now receded out of sight.

... Daniel Bell, in his influential *Post Industrial Society*, proposed a three-part schema, comprising pre-industrial (or traditional), industrial and post-industrial stages. The third would be based on information technology, rather than on the use of energy and raw materials, and on the displacement of the secondary, manufacturing sector by what we now call 'services'. In his schema, the 'game against nature' was relegated to the 'pre-industrial stage' (with no hint that it might return), and the 'game against fabricated nature' of the industrial stage was now also about to be dis-placed by the post-industrial 'game between persons' (Bell, 1974: 117). Others later added theories of 'information society' and of 'dematerialized production' (Stehr, 2001: 77) to the concept of a post-industrial society – often ignoring the fact that energy-intensive material production has been globalized rather than displaced, and continues to grow absolutely despite large increases in efficiency.

CONCLUSION: QUESTIONS FOR SOCIOLOGY

The year 2005 was also a social 'tipping point', with global warming perhaps irre-versibly on the public agenda. Although emissions continue to rise, with Kyoto's limited targets unmet, the old debates on the existence or seriousness of anthropo-genic causes of global warming appear to be effectively over. The voices of scien-tists are regularly penetrating the media, saying we would need to start now on a path that would cut emissions by 50 percent or more by mid-century, if we are to avoid possible escalating feedback effects (Hogan, 2005; Leggett, 2005: 161, 283). Al Gore's film on global warming is breaking box office records for a documen-tary. George Bush and John Howard have ceased their denial, and become parti-sans of carbon sequestration and nuclear power. Business, politicians and environmentalists are now internally split in novel ways between advocates of re-duced consumption, of efficiency and of a range of alternative energy sources, in-cluding gas, solar, wind, nuclear, hydro, biomass, geothermal, carbon sequestration, etc. The advocates of each accuse the others of grandstanding without serious intent, or of ignoring costs, needed time scales and technical feasibility or the negative impacts and dangers for society or the environment.

Most sociologists have not yet taken stock of these changes. We have already wasted too much time, and may awaken too late to have any impact. It is now essential that the discipline as a whole reorients itself, to overcome the two in-hibitions described earlier—the old one against listening seriously to what scien-tists say about nature and the more recent one against thinking about the future.

Twenty years ago, Piore and Sabel (1984) proposed that society periodically came to a crossroads indicating alternative technological futures, but a direction once chosen would become increasingly hard to reverse. Past turning points have included the development of steam power, the internal combustion engine, microelectronics and information and communications technology. The choice of which technology to develop, and especially of how and by whom, has shaped not just the physical and economic but also the social, political and cultural world. The decision of how much, how and by whom global warming should be confronted is perhaps the biggest of such moments of choice.

The first choice will be about priority and resources. If this is too little, too late, we will face growing dangers and probable economic and political breakdowns. If amplifying feedback loops are triggered, improbable catastrophes become more likely. The Gulf Stream flow could be halted, freezing Britain and Northern Europe. Droughts could wipe out the agriculture of hundreds of millions in Africa and Australia and in Bangladesh and China, where they depend on Himalayan melt water and monsoon rains. Will this accentuate global inequality or produce new winners and losers? Will it bring the world together, as in responses to the Tsunami, or will it provoke closing borders, xenophobia and national or communal conflict? How well will different societies be prepared? Can we avoid repeating the discriminatory and incompetent response to Hurricane Katrina?

If effective action is chosen, it will have to be on a massive scale, involving a rapid retooling of production and distribution systems, particularly agriculture, energy, transport and urban structure. Will this be achieved by a market-based regime (requiring huge public subsidies) or by a shift to a national or a global regulatory regime? What agents would have the power and interest to achieve each of these outcomes? Who would benefit and who would lose from such changes? What impact would such a shift of resources, from consumption to investment in producer goods, have on industrial structure, skills and above all on culture? Is it likely, for example, that the practices and values of free markets, individualism, diversity and choice will not be significantly modified?

Other questions are posed by the measure or mix of measures chosen. Would a strategy of reducing or halting economic growth or consumption require draconian enforcement? Which kind of alternative energy development could be effective in time and at a realistic cost? How would the choice of alternative affect centralization or decentralization of economic and political power? Tackling all these questions requires the cooperation of social and natural science and for awareness of the issue to penetrate all branches of sociology and be incorporated into their research.

REFERENCES

Bell, Daniel (1974) *The Coming of Post Industrial Society: A Venture in Social Forecasting.* London: Heinemann.

Hogan, Jenny (2005) 'Only Huge Emissions Cuts Will Curb Climate Change', *New Scientist* 3 February; at: www.newscientist.com/channel/earth/dn6964

Leggett, Jeremy (2005) *Half Gone: Oil, Gas, Hot Air and the Global Energy Crisis*. London: Portobello Books.

Merali, Zeeya (2005) 'Sceptics Forced into Climate Climb-Down', *New Scientist* 20 August: 10.

New Scientist (2005) 'Breaking News. Global Warming is a Clear and Increasing Threat', 1 July, at: www.newscientist.com (accessed 19 December 2005).

Piore, Michael and Sabel, Charles (1984) *The Second Industrial Divide*. New York: Basic Books.

Stehr, Nico (2001) 'Economy and Ecology in an Era of Knowledge-Based Economies', *Current Sociology* 49(1): 67–90.

KEY CONCEPTS

climate change   environmental sociology   global warming
                                                                    teleology

DISCUSSION QUESTIONS

1. What consequences for your community might there be with climate change? What groups would be most affected and how?

2. Lever-Tracy suggests that physical sciences alone cannot address the needs of our society, given the potential consequences of climate change. What is the role of sociology in thinking about such changes?

Applying Sociological Knowledge: An Exercise for Students

The BP oil spill in the Gulf of Mexico in the year 2010 is one of the largest environmental disasters in history. Using various sources of evidence, explore the social consequences of this disaster for the different communities and groups of people affected by it. Analyze what you have found by examining such social facts as the race, class, and gender consequences for different communities and groups of this environmental catastrophe.

✳

PART XV

Social Movements and Social Change

66

Generations X, Y, and Z

Are They Changing America?

DUANE F. ALWIN

What creates social change? Is it the values of young people who bring new perspectives and new issues to society? Or is it the influence of major historical events? In this article, Duane Alwin examines generational sources of social change, arguing that historical events and shifts in individual lives due to aging are both sources of social change.

The Greatest Generation saved the world from fascism. The Dr. Spock Generation gave us rebellion and free love. Generation X made cynicism and slacking off the hallmarks of the end of the 20th century. In the media, generation is a popular and all-purpose explanation for change in America. Each new generation replaces an older one's Zeitgeist with its own.

Generational succession is increasingly a popular explanation among scholars, too. Recently, political scientist Robert Putnam argued in *Bowling Alone* that civic engagement has declined in America even though individual Americans have not necessarily become less civic minded. Instead, he argues, older engaged citizens are dying off and being replaced by younger, more alienated Americans who are less tied to institutions such as the church, lodge, political party and bowling league.

Next to characteristics like social class, race, and religion, generation is probably the most common explanatory tool used by social scientists to account for differences among people. The difficulties in proving such explanations, however, are not always apparent and are often overshadowed by the seductiveness of the idea. Generational arguments do not always hold the same allure once they are given closer scrutiny.

Changes in the worldviews of Americans result not only from the progression of generations but also from historical events and patterns of aging. For example, generational replacement seems to explain why fewer Americans now than 30 years ago say they trust other people, but historical events seem to explain why fewer say they trust government. Similarly, historical events in

SOURCE: Alwin, Duane F. 2002. "Generations X,Y, and Z: Are They Changing America?" *Contexts* 1 (Winter) 42–50.

interaction with aging (or life cycle change) may explain lifetime changes in church attendance and political partisanship better than generational shifts.

EXPLAINING SOCIAL CHANGE

Some rather massive changes over the past 50 years in Americans' attitudes need explaining. Consider this short list of examples:

- In 1977, 66 percent said that it is better if the man works and the woman stays home; in 2000, only 35 percent did.
- In 1972, 48 percent said that sex before marriage is wrong; in 2000, 36 percent did.
- In 1972, 39 percent said that there should be a law against interracial marriage; in 2000, 12 percent did.
- In 1958, 78 percent said that one could trust the government in Washington to do right; in 2000 only 44 percent did.

Do changes in beliefs and behaviors reflect the experiences of specific generations, do they occur when Americans of all ages change their orientations, or do they result from something else? Although the idea of generational succession is promising and useful, it also has problems that may limit it as an all-purpose explanation of social change.

SOME PRELIMINARIES

Before we begin to deconstruct the idea of generational replacement we need to clarify a few issues. The first is that when sociologists use the term *generation*, it can refer to one of three quite different things:

1. All people born at the same time.
2. A unique position within a family's line of descent (as in the second generation of Bush presidents).
3. A group of people self-consciously defined, by themselves and by others, as part of an historically based social movement (as in the "hippie" generation).

There are many examples in the social science literature of all three uses, and this can create a great deal of confusion. Here I refer mainly to the first use, measuring generation by year of birth. Demographers prefer to use the term cohort. Either way the reference is to the historical period in which people grow to maturity. I use the terms cohort and generation more or less interchangeably.

When sociologists are discussing social change in less precise terms, they may refer to generations in a somewhat more nuanced, cultural sense. Generations in this usage do not necessarily map neatly to birth years. Rather, the distinction between generations is a matter of quality, not degree, and their exact time boundaries cannot always be easily identified. It is also clear that statistically there is no way to identify cohort or generation effects unequivocally. The interpretation of generational differences depends entirely on one's ability or willingness to make some rather hefty assumptions about other processes, such as how aging affects attitudes, but as we shall see, we can nonetheless develop reasonable conclusions.

COHORT REPLACEMENT

Cohort replacement is a fact of social life. Earlier-born cohorts die off and are replaced by those born more recently. The question is: Do the unique formative experiences of cohorts become distinctively imprinted onto members' worldviews, making them distinct generations over the course of their lifetimes, or do people of all cohorts adapt to change, remaining pliable in their beliefs throughout their lives?

When historical events mainly affect the young, we have the makings of a generation. Such an effect—labeled a cohort effect—refers to the outcomes attributable to having been born in a particular historical period. When, for example, people describe the Depression generation as particularly thrifty, they imply that the experience of growing up under privation permanently changed the economic beliefs and style of life of people who grew to maturity during that time.

Unique events that happen during youth are no doubt powerful. Certainly, some eras and social movements, like the Civil Rights era and the women's movement, or some new ideologies (e.g. Roosevelt's New Deal) provide distinctive experiences for youth during particular times. As Norman Ryder put it, "the potential for change is concentrated in the cohorts of young adults who are old enough to participate directly in the movements impelled by change, but not old enough to have become committed to an occupation, a residence, a family of procreation or a way of life."

To some observers, today's younger generations—Generation X and its younger counterpart—display a distinctive lack of social commitment. The goals of individualism and the good life have replaced an earlier generation's involvement in social movements and organizations. Is this outlook simply part of being young, or is it characteristic of a particular generation?

Each generation resolves issues of identity in its own way. In the words of analyst Erik Erikson, "No longer is it merely for the old to teach the young the meaning of life … it is the young who, by their responses and actions, tell the old whether life as represented by the old and presented to the young has meaning; and it is the young who carry in them the power to confirm those who confirm them and, joining the issues, to renew and to regenerate or to reform and to rebel." …

Before we accept this way of understanding change, however, we should consider other possibilities. One is that people change as they get older, which we call an effect of aging. The older people get, for example, the more intensely they may hold to their views. America, as a whole, may be becoming more politically partisan because the population is getting older—an age effect. Another possibility is that people change in response to specific historical events, what sociologists call period effects. The Civil Rights movement, for example, may have changed many Americans' ideas about race, not just the views of the generation growing up in the 1960s. The events of September 11, 2001, likely had an effect on the entire nation, not just those in the most impressionable years of youth.

A third possibility is that the change is located in only one segment of society. Members of the Roman Catholic faith, for example, may be the most responsive to the current turmoil over the sexual exploitation of youth by some priests in ways that hardly touch the lives of Protestants. Let us weigh these possibilities more closely, looking at the issues raised by Putnam in *Bowling Alone*.

CHANGES IN SOCIAL CONNECTEDNESS AND TRUST

It is often relatively easy to construct a picture of generational differences by comparing data from different age groups in social surveys and polls, but determining what produced the data is considerably more complex.

Take, for example, one of the key empirical findings of Putnam's analysis: the responses people give to the question of whether they trust their fellow human beings. The General Social Survey (administered regularly to a nationwide, representative sample of American residents since 1972) asks the following question: "Do you think most people would try to take advantage of you if they got the chance, or would they try to be fair?" Figure 1 ... presents the

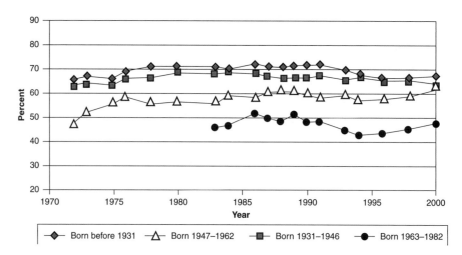

FIGURE 1   Percent of U.S. Population Who Believe "Most People Try to Be Fair"

percentage of respondents in each set of cohorts who responded that people would try to be fair. The results show that birth cohorts were consistently different from one another, the recent ones being more cynical about human nature. There has been little change in this outcome over the years except insofar as new generations replaced older ones. These results reinforce the Putnam thesis, that the degree of social connectedness in the formative years of people's generation shapes their sense of trust.

Still, I would note some problems with these conclusions. First, generational experiences are not the only factors that differentiate these four groups; they also differ by age. Second, these data do not depict the young lives of the cohorts born before 1930 (who were 42 years of age or older in 1972), so we have little purchase on their beliefs before 1972. Third, there is remarkable growth in trust among the Baby Boom cohorts—those born from 1947 to 1962—over their midlife period, and in 2000 they had achieved a level of trust on a par with earlier cohorts. Finally, even the most recent cohorts (the lowest line in the figure) show some tendency to gain trust in recent years. The point is that while the data appear to show a pattern of generational differences—less trust among more recent cohorts—age might be just as plausible an explanation of the differences: trust goes up as people mature.

There may be more than one way to explain changes in Americans' trust of people, but generations do not explain changes in Americans' trust of government. In 1958 the National Election Studies (NES) began using the following question: "How much of the time do you think you can trust the government in Washington to do what is right—just about always, most of the time, or only some of the time?"

There are two important things to note about Figure 2. First, there are hardly any differences among birth cohorts who say most of the time or always in their responses to this question; the lines are virtually identical. Thus,

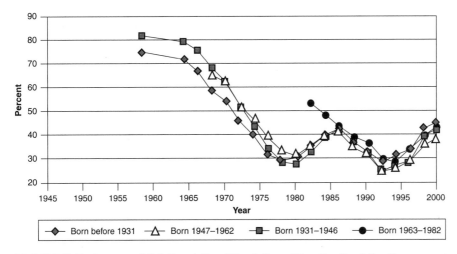

FIGURE 2 Percent of U.S. Population Who Believe "You Can Trust the Government in Washington to Do What Is Right"

generational replacement explains none of the very dramatic decline of trust in government. That decline may be better explained by historical events that affected all cohorts—the Vietnam War, the feminist movement, or the Watergate and Whitewater scandals—and there is little basis for arguing that more recent cohorts are more alienated from government than those born earlier. (Note that affirmations of trust in government rose dramatically right after 9/11.)...

GENERATIONS AND SOCIAL CHANGE

Society reflects, at any given time, the sum of its generations. Where one set of cohorts is especially large—like the Baby Boomers—its lifestyle dominates the society as it passes through the life course. Baby Boomers' taste in music and clothes, for example, disproportionately influence the whole culture. However, in cases where there are no major differences among generations (as in the example of trust in government), then generational succession cannot explain social change.

Where generations persistently differ, however, their succession will produce social change. Certainly, if the more recent generations have less affiliation and involvement with traditional religious groups, this will lead to social change, at least until they develop their own form of religiosity.

Because of the Baby Boomer generation's sheer size, its liberal positions on political and social issues will probably shape beliefs and behavior well into the new century, as Boomers replace the generations that came before. But even here, the Baby Boomers' distinctiveness may wane under the influence of historical events and processes of aging. Baby Boomers, for example, may be growing more conservative with age. This argues in favor of an alternative to the generational view: Generations do not necessarily differ in the same ways over time; individuals are not particularly consistent over their lives; and social change results as much from shifts in individual lives due either to aging or historical events....

The existence of generation effects may depend very much on when one takes the snapshot of generational differences, and how generations differ may depend on which groups in society one examines. All fair warnings for the next essay you read on Generations X,Y or Z.

KEY CONCEPTS

| age cohort | cohort effect | social change |

DISCUSSION QUESTIONS

1. What is a cohort effect, and how is it significant to the process of social change? What social changes are the result of cohort effects among your age generation?

2. As you imagine the future, what social changes do you think will result from the various sources of generational change that Alwin identifies?

67

Is the Economic Crisis Driving Wedges Between Young and Old? Rich and Poor?

ANDREA LOUISE CAMPBELL

This reading considers whether or not the economic crisis has created greater animosity between the young and the old. Young people are working and contributing to governmental programs designed to support the aging population. Campbell's research, however, points to no evidence of a generational difference in attitudes about the programs or any animosity between the generations. Instead, she finds that class differences pervade. Privileged Americans are less positive about supporting social insurance programs; disadvantaged Americans are more reliant on these programs because of the economic crisis.

The current economic crisis, with its myriad threats to the financial well-being of Americans from all walks of life, raises the specter of conflict between generations and between classes. Perhaps working-age people facing high unemployment and underemployment, large declines in home values, and an uncertain economic future will come to resent as onerous the payroll taxes that support a public pension and healthcare system for older people that might be greatly curtailed when their turn for retirement comes. Alternatively, class divisions might be heightened, with affluent Americans, who hold the lion's share of

SOURCE: Campbell, Andrea Louise. 2009. "Is the Economic Crisis Driving Wedges Between Young and Old? Rich and Poor?" *Generations* 33: 47–53.

wealth and who therefore have experienced the greatest losses in the recent fall, resisting policy changes focused on raising taxes or cutting benefits for them.

Using survey data where possible and informed speculation when necessary, I argue that the economic crisis has not spurred intergenerational cleavages. The 1980s literature on intergenerational conflict failed to find emergent cleavages then; I find little evidence that such conflict is increasing now. The more relevant cleavage is a class divide. Affluent Americans are more resistant to expanding social insurance programs than are their lower-income counterparts. While age differences in policy preferences are muted—indeed, younger Americans are often more supportive of social insurance expansion than are older beneficiaries—longstanding class differences in preferences about the role and size of government remain. And it is these class differences that may have significant consequences going forward, as solutions to both the current crisis and the long-term fiscal challenges facing the United States fall exactly along these lines.

EARLIER DECADES: LITTLE EVIDENCE OF AN INTERGENERATIONAL DIVIDE

During the 1980s, a literature on intergenerational conflict argued that cleavages would emerge between "greedy geezers" and resentful workers—the former sucking the economy dry with their generous benefits and the latter resentful of the high and regressive payroll taxes that undermined their own ability to make ends meet and that they were forced to pay, all for a social insurance system from which they might not even benefit in retirement (e.g., Fairlie, 1988; Peterson and Howe, 1989). These tensions were purportedly exacerbated by trends in political activity, in which the increased participation of older adults (age 65 and older) virtually forced elected politicians to meet their demands even at the expense of workers, further fueling such resentments.

In earlier work (Campbell, 2003), I showed that while older adults (age 65 and above) have in fact outstripped younger people, or "the nonelderly," (from age 18 to age 64) in their rates of political participation, the hypothesized intergenerational resentments never materialized. Data from the 1980s and 1990s revealed that the nonelderly were actually more likely than the elderly to desire increased spending on Social Security; that the elderly scored second only to teachers as the most highly regarded societal group in National Election Study "thermometer ratings"; that two-thirds of respondents in a 1999 survey thought older people were getting too little attention from Washington; and that payroll taxes were among the least disliked of all the levies Americans face. Cynicism among younger people about the likelihood that they themselves would ever receive Social Security or Medicare benefits did not translate into a desire to trim their parents' or grandparents' current benefits (Jacobs and Shapiro, 1998).

The sanguine attitudes of the past notwithstanding, perhaps the extreme conditions of the current recession could alter these warm attitudes toward older Americans and their programs. With workers in many sectors saddled with furlough days, reduced hours, and salary freezes, perhaps those mandatory payroll taxes are becoming more painful. With the rolls of the uninsured increasing as employers drop health insurance coverage after years of galloping premium increases, perhaps expensive Medicare coverage for elders will spawn resentment. An examination of more recent public opinion data suggests, however, that intergenerational conflict remains muted.

CONTINUED WARMTH TOWARD ELDERS AND THEIR PROGRAMS

Even during the current economic downturn, support for the major social insurance programs for older people remains high, with people ages 18 to 64 actually more likely than older people to support increases in spending, just as they were in earlier years. For example, a December 2008 Kaiser Family Foundation poll asked respondents whether they wanted to see federal spending increased, decreased, or kept about the same on a variety of programs. Despite the fact that the question lead-in mentioned the federal government's "substantial budget deficit," people under age 65 were more likely than older people to say spending on Medicare should be increased, 45 percent to 36 percent. In the same Kaiser Family Foundation poll, respondents were asked about spending more federal money to expand the Medicare prescription drug benefit to fill in the "donut hole" coverage gap, in which older people have to pay the full costs of their medications out of pocket. Again, younger people were more supportive of closing the gap than were older people, 81 percent versus 69 percent.

Beyond preferences on the major social insurance programs, I found that other recent and current attitudes mirrored those observed in earlier decades. In a 2004 AARP poll, younger and older people had almost identical responses to a question about how much influence "retired older Americans have in this country today." Half of each group thought older people have too little influence, and two-fifths of each group thought they had the right amount; very few members of either age group said they had too much influence. It is true that the percentage of younger people saying that older people have too little influence has declined over time—two-thirds thought so in a comparable 1981 poll—but it is certainly not the case that resentments are boiling over: the percentage of those under 65 saying older people have "too much influence" crept up just 6 points, from 3 percent to 9 percent, between 1981 and 2004 (AARP, 2006, p. 56). Nor have much-predicted antagonisms toward the payroll tax emerged. In the 2009 NASI/Rockefeller Foundation poll, large percentages of Americans said they didn't "mind paying Social Security taxes," 72 percent because they knew they would be getting retirement benefits themselves; 76 percent because otherwise they'd have to support family members financially; and 87 percent because the program provides "security and stability" to millions of retirees, disabled people, and survivors. More than three-quarters agreed that "it is critical that we preserve Social Security for future generations" even if it means increasing taxes. Some observers have wondered whether a younger Latino population would be willing to continue funding the benefits of an older Anglo population; in a June 2008 Rockefeller Foundation/*Time Magazine* poll, 84 percent of self-described Hispanics favored a payroll tax increase "to guarantee every worker at least 70 percent of their preretirement income in retirement," compared to 68 percent of whites. (Favorable attitudes toward the payroll tax may be facilitated by the Earned Income Tax Credit, which began in 1975 and was expanded multiple times through 2001 and which mitigates the effects of the payroll tax for many low-wage earners.)

WHY SO LITTLE INTERGENERATIONAL CONFLICT?

We might imagine several explanations for these continuing warm feelings. One factor could be ignorance or misinformation about current policy or about the older population....

It is unlikely that ignorance is the only factor at play, however. Some of the continuing positive feelings are probably due to the function that Social Security and Medicare serve in our society. They provide benefits for a population that continues to be viewed as deserving. The programs represent emancipation for younger people, very few of whom support their older parents financially. Indeed, respondents ages 65 and over were three times more likely to report having given financial support to their adult children than having received it (Pew Research Center, 2005). The programs have enabled older people to retire, created steady incomes and a sense of financial certainty, and have facilitated independence, something quite valued in the United States. And while Americans may overstate the degree to which older people live in poverty, this overstatement probably derives from an intuitive sense that variation in incomes is quite high among older people and that there are subgroups within that population who face far higher poverty levels than the mean would indicate. Everyone can think of a very low-income widow, for example; many of us have them in our own families.

The economic slump has, if anything, heightened the value of these programs in many Americans' eyes. Indeed, 88 percent of respondents to the 2009 NASI/Rockefeller poll agreed that "with the economy and the stock market as bad as it is right now, Social Security benefits are more important than ever to ensure that retirees have a dependable income when they retire" (Reno and Lavery 2009, p. 6).

Both objective measures and behavioral patterns among elders and younger people demonstrate the impact of Social Security and Medicare on the financial stability of the older group. The incomes of older people are low on average—just over half those of middle-aged people (U.S. Census Bureau, 2009, Table 676)—but the very stability and predictability of their incomes, combined with other aspects of their financial situations, have allowed older people to meet the economic crisis with fewer behavioral changes than their children. Eighty percent of the elderly own homes, and over two-thirds of them own free and clear (U.S. Census Bureau, 2008, Table 17; U.S. Department of Housing and Urban Development, 2008). They are far less likely than younger people to hold financial debt of other kinds as well (U.S. Census Bureau, 2009, Table 1134). The 2007 personal bankruptcy filing rate of older people was less than one-third of that of people ages 35 to 44 (Thorne, Warren, and Sullivan, 2008, p. 5). Of course older people have been affected by the decline of the stock market, and many wonder how they will make up for their losses. But with the valuable Social Security annuity making up the majority of income for the majority of older people (Federal Interagency Forum on Aging-Related Statistics, 2008, p. 15), they have experienced far fewer financial or psychological hits from the economic crisis.

Predicted antagonisms across generations have not cropped up so far. Indeed, the current economic crisis, rather than making such cleavages more likely, seems, if anything, to mute them, as the value of the major programs for older people is apparent to many Americans.

THE MORE POTENT DIVIDE: CROSS-CLASS

The more potent and relevant cleavage in American politics is socioeconomic rather than generational. As noted above, there are differences in the preferences on age-related policies across different age groups, but mostly it is younger people who are more generous than older people themselves in their orientations. Socio-economic differences in preferences are fundamentally different: high-income Americans are generally less positively disposed toward the big social insurance programs than are lower income Americans. Recall the items from the December 2008 Kaiser Family Foundation poll asking about Medicare spending and filling in the prescription drug "donut hole." While younger people were more supportive of increased federal spending in these areas than were older people, high-income respondents were more supportive than lower-income respondents were. Forty-nine percent of those with incomes under $40,000 per year wanted federal spending on Medicare increased, compared to just 38 percent of those with incomes over $40,000. Still larger differences obtain when we look at the ends of the income spectrum: Some 52 percent of those with incomes of less than $25,000 wanted federal Medicare spending boosted, versus just 30 percent among those earning $100,000 or more per year. Similarly, filling in the donut was strongly favored by 64 percent of those earning under $40,000 in income but just 50 percent among those earning more. Strong support was found among 69 percent of those with incomes under $25,000 and only 44 percent among those with incomes over $100,000 (Kaiser Family Foundation, 2008).

The implication of these public opinion patterns across income groups is that we have to be quite careful in how we execute entitlement reform. As valuable as the major social insurance programs have proven to be during the current economic crisis, reform is necessary to sustain them in the long term....

At lower income levels, the concern is trimming benefits too much. Although Medicare pays for most of the big-ticket items, older people nonetheless currently pay for about half of their healthcare out of pocket. Social Security provides 80 percent of the income of the bottom two-fifths of elders, and more than half for the majority of them. We can't trim these benefits too much given many Americans' low incomes, low savings rates, and inability to self-insure: adequacy concerns remain acute. And there are political consequences as well. Social Security over time has enhanced the political participation rates of older people, but the effects are largest for poorer elders, to whom the program redistributes income. The steep gradients we see in the U.S. between income and political participation are much less steep for older people because of this democratizing effect. In a society that lacks European-style labor unions or political parties of the left, these universal entitlements are virtually the only institutions that mobilize low-income people. Reducing the benefits of the modestly endowed could undermine their political participation, muting the voices of one of the few politically active low-income groups.

We must be wary of cutting benefits or raising taxes too much at the top of the income spectrum as well. The other piece of political magic that Social Security and Medicare have wrought is a cross-class coalition in support of these programs. Universal programs elicit the support of the affluent with large absolute benefits.

Of course the temptation is to place the financial burdens of reform on the affluent: That's where the money is. But driving them out of the supportive coalition could leave these programs even more vulnerable in the future and could mean going down the slippery slope from universalism to means-testing, precipitating the political unraveling of the programs. We already know that imposing net costs on the affluent reduces their support for social insurance programs. The 1983 Social Security amendments made the program a worse deal for the affluent by taxing their benefits, raising their payroll taxes, and increasing their payback times. And the gap in support for Social Security between the affluent and the poor, which had been ten points or so before the reform, grew to twenty or thirty points (Campbell and Morgan, 2005).... The point is that given past experience and continuing attitudinal differences across income groups, not just the economic consequences of entitlement reform, but the political consequences as well, must be taken into account.

CONCLUSION

In the politics of aging, there is more conflict between rich and poor than across generations. The effect of economic crisis has not been to heighten the latter. Indeed, the recession serves to highlight the value of the social safety net. Not only have older people been able to meet the economic downturn with fewer lifestyle changes than have their children, but also these programs serve as cross-cohort life preservers for many families. The more significant cleavage is an economic one, with the more affluent less enthusiastic about these programs than lower-income Americans. Sustaining our major age-related policies into the future will take some major readjustments, reforms that unfortunately threaten to exacerbate these class differences.

REFERENCES

AARP. 2006. *Images of Aging in America 2004.* Washington, D.C.

Campbell, A. L. 2003. *How Policies Make Citizens: Senior Citizen Activism and the American Welfare State.* Princeton, N.J.: Princeton University Press.

Campbell, A. L., and Morgan, K. J. 2005. "Financing the Welfare State: Elite Politics and the Decline of the Social Insurance Model in America." *Studies in American Political Development* 19: 173–95.

Fairlie, H. 1988. "Talkin' 'Bout My Generation." *New Republic*, March 28.

Federal Interagency Forum on Aging-Related Statistics. 2008. *Older Americans 2008: Key Indicators of Well-Being.* Washington, D.C.: U.S. Government Printing Office.

Jacobs, L. R., and Shapiro, R. Y. 1998. "Myths and Misunderstandings About Public Opinion Toward Social Security." In R. D. Arnold, M. J. Graetz, and A. H. Munnell, eds., *Framing the Social Security Debate.* Washington, D.C.: National Academy of Social Insurance.

Kaiser Family Foundation. 2008. *The Public's Health Care Agenda for the New President and Congress.* Poll #2008-POL019. Conducted on December 4–14.

Peterson, P. G., and Howe, N. 1989. *On Borrowed Time: How the Growth in Entitlement Spending Threatens America's Future.* New York: Simon & Schuster.

Pew Research Center. 2005. *Pew Social Trends: Family Bonds.* Poll #2005-SDT01. Conducted from October 5 to November 6.

Reno, V. P., and Lavery, J. 2009. *Economic Crisis Fuels Support for Social Security: Americans' Views on Social Security.* Washington, D.C.: National Academy of Social Insurance.

Rockefeller Foundation/*Time Magazine*. 2008. *National American Worker Survey.* Conducted on June 19–29.

Thorne, D., Warren, E., and Sullivan, T. A. 2008. *Generations of Struggle.* AARP Public Policy Institute #2008-11. Washington, D.C.: AARP.

U.S. Census Bureau. 2008. *Housing Vacancies and Homeownership Survey.* www.census.gov/hhes/www/housing/hvs/annual08/ann08tl7.xls. Washington, D.C.: Government Printing Office.

U.S. Census Bureau. 2009. *Statistical Abstract of the United States.* Washington, D.C.: Government Printing office.

KEY CONCEPTS

AARP	intergenerational conflict	Medicare
		Social Security

DISCUSSION QUESTIONS

1. What are some of the governmental programs that workers contribute to in the formal labor sector? What are your taxes used for?

2. What happens in an economic recession that puts these programs at risk? What does that mean for those disadvantaged members of society? What possible solutions are there during an economic downturn?

68

Children of the Great Recession

A Tour of the Generational Landscape, from Struggles to Successes, Coast to Coast.

RONALD BROWNSTEIN

This reading summarizes the problems facing young people today as they try to gain employment after school. The recession created a problem of too many workers and too few jobs. Brownstein finds that young people are not necessarily discouraged. They turn, instead, to finding education bargains, starting their own companies, or deciding on public service work.

The April sun shines brightly over the career development center at California State University (Los Angeles). Yet, despite the perfect weather, the center's conference room is crammed with students taking notes intently. Judy Narcisse, a counselor at the center, is running a workshop on résumé writing.

Gesturing toward a PowerPoint presentation on a large screen, Narcisse marches her charges through the basics. Her tone is casual, and her humor dead-pan, but her advice is hard-headed. "If you're going to leave your cellphone number, don't have rap music on your voice mail," she explains. "You can't say, 'Hey, this is Naomi! Give me a holla!' "

As Narcisse speaks, heads bob and pens scribble. Her students at this heavily Hispanic and Asian, working-class university of about 20,000 located east of

SOURCE: Brownstein, Ronald. 2010. "Children of the Great Recession: A Tour of the Generational Landscape, from Struggles to Successes, Coast to Coast." *National Journal*, May 8, 2010.

downtown Los Angeles are attentive to a fault. "What if I don't have enough experience to fill a piece of paper?" one young woman asks. Print your résumé with wider margins or a larger font size, Narcisse replies. She is full of such practical suggestions: Find internships, she recommends, attend career fairs, volunteer, proofread your résumé. At times, Narcisse sounds like a coach delivering a halftime pep talk. 'We're in a tough economic market," she says bluntly. "You're going to have to hustle."

The message is not lost on these children of the Great Recession, or on the massive Millennial Generation they embody. The Millennials, best defined as people born between 1981 and 2002, are now the largest generation in the United States, with nearly 92 million members according to the latest Census Bureau figures. They are the most diverse generation in American history (more than two-fifths are nonwhite, according to the census), and they are en route to possibly becoming the best-educated: About half of Millennial men and 60 percent of Millennial women older than 18 have at least some college education. They are frequently described as a civic generation, more oriented toward public service and volunteerism and more instinctively optimistic than any cohort of young people since the fabled GI Generation that fought World War II and built postwar America.

And now their great expectations are colliding with diminished circumstances. Millions in the generation are navigating the transition to adulthood through the howling gale of the worst economic downturn since the Depression. Federal statistics tell a bracing story. The unemployment rate among 20-to-24-year-olds stood at nearly 16 percent in March, more than double the level of three years earlier, before the recession began. For those 25 to 29, unemployment stood at 11.5 percent, also more than double the level of three years ago. For teenagers, 20-something African-Americans and Hispanics, and young people without a college degree, the unemployment rates spike as high as 28 percent—debilitating levels of the kind that Americans haven't seen in decades.

EMPLOYMENT

Adult Millennials have been hit hard by the recession, with their rate of unemployment half again as high as that of older workers. But the situation looks still worse when you factor in those who have stopped looking for work or opted to remain in school. The percentage of Millennials with jobs—called the employment-population ratio—has dropped much more sharply than that for other age groups.

Even in this dire climate, some Millennials are finding new ways to weather the storm, displaying the instinct for pragmatic problem-solving that generational theorists consider one of their defining characteristics. They are flocking into service-oriented alternatives such as Teach for America or partnering with peers to launch innovative entrepreneurial ventures.

But for many others, the working world has become an inscrutable maze of part-time jobs, temporary gigs, and full-time positions that abruptly dissolve into layoffs and start the entire disorienting cycle again. Such widespread uncertainty could impose lasting costs, measured not only in diminished opportunity and earnings for the Millennials themselves, but lost productivity and multiplying social challenges for American society at large. "If the jobless recovery... drags on for three or four years, then I think we will face some large problems that we haven't faced since the Great Depression," says Robert Reich, who served as Labor secretary in the Clinton administration. "Young people are just not going to form the habits and attachments to work that their older siblings or parents have had, and that could conceivably create a whole variety of social problems."

The Broken Escalator

When the job market works well, it functions like an escalator. Young people finish their schooling, whether high school or higher education, and move onto the escalator's bottom steps with entry-level jobs. As they acquire more skills and experience, they rise into better-paying and more-senior positions, clearing space on the first steps for the next crop of new graduates. Meanwhile, on the top steps, older workers step off into retirement, creating room that allows everyone below them to ascend through their own careers.

But now the escalator is jammed at every level. With jobs scarce, many young people are stuck at the bottom, unable to take that first step. Those who have been lucky or skillful enough to get on the escalator in the past few years are often not rising smoothly. They might gain a job, lose it, and fall back several steps or off the escalator altogether. There they must jostle with each successive class of graduates trying to squeeze on at the bottom. Meanwhile, at the other end, with the stock market's collapse decimating 401(k) plans, fewer older workers are moving briskly off the escalator into retirement. (Federal statistics show that while participation in the labor force is declining slightly for middle-aged workers, and sharply for younger workers, it has actually increased since 2007 among workers ages 55 to 64.) The continued presence of this older generation on the escalator makes it more difficult for everyone behind them to ascend, intensifying the pileup at the lower levels. Compounding the pressure, more middle-aged workers are being toppled from the upper steps by layoffs, which force them to compete for space lower down that junior colleagues might once have occupied. Rather than advancing in smooth procession, everyone is stepping on everybody else.

The pileup is being felt keenly this spring on college campuses. Last year, the bottom fell out of the job market for the class of 2009. The Collegiate Employment Research Institute at Michigan State University, which tracks these trends nationwide, found that last spring large employers hired 42 percent fewer graduating students than they had originally targeted when the school year began. It was a "rout," the group reported in its latest annual survey. This year, the job market for graduates appears to have stabilized and may even be rebounding slightly—albeit from that sharply reduced new starting point—says Phil Gardner, the institute's director. "We see sporadic little shoots grow up here and there," he explains, "but... no surge."

Career counselors at a wide variety of colleges and universities echo Gardner's equivocal verdict. 'Things are starting to thaw," says Paula Klempay, the director of M.B.A. career services at the University of Washington's Foster School of Business. "We're not as robust as we were two years ago, but we're better than last year." Jaime Velasquez, assistant director of career services at the University of Illinois (Chicago), says that the number of employers at the school's job fair jumped from 50 last year to 62 this spring, "and they all had positions available." Still, he notes, "five years ago, we had 110 [firms]."

School Daze

The climate has improved enough—modestly but perceptibly—that success stories on campus are again growing more common. Kesha McLaren, a senior at the University of Missouri's College of Agriculture, struck out when she tried submitting her résumé directly to employers, but she was hired by an agricultural finance company in St. Louis after a professor steered a recruiter in her direction. "It was the only position I was offered, but that doesn't bother me, because it was kind of my dream job," she says. Halfway across the country, the undergraduate business school at Howard University, a historically black college in Washington, D.C., this year has placed 90 percent of its honors students in graduate school or full-time work. Ann Jackson, a 21-year-old graduating this spring, is moving to Portland, Ore., after winning a job at Intel's human relations department. "I got lucky, honestly," she says. "I'd been in a lot of interviews, and there were quite a few I didn't get."

Despite these flickers of improvement, the challenges remain substantial, even for graduates of elite institutions. In 2009, the unemployment rate among 20-to-24-year-olds with four-year college degrees stood at 8.8 percent, twice what it was just two years earlier. Many college officials concur with John Noble, the director of career counseling at Williams College in Massachusetts, who says that even with some recent revival, "this is the worst market I've seen since I started in this line of work in 1982."

Complicating matters for the graduating class of 2010 are the substantial number of students from previous classes who are still competing for entry-level positions. Sean Sposito, for instance, graduated from the University of Missouri's journalism school in 2009 but is looking for work again after completing a one-year internship with the Newark, N.J., *Star-Ledger*—which he supplemented with jobs as a bar bouncer. "To me this is regular," he says of the uncertainty he faces. "I haven't known a professional life outside of this."

Economists say that young people who enter the workforce during recessions never entirely recover from the delays they face at the foot of the escalator. Over the course of their working lives, their wages remain lower than those of young people who enter the workforce when times are flush. The social cost for these detours and reversals is high too. Gardner says the economy will be dealing for years with a mismatch between the skills that new graduates have acquired and the work that is available for them. "When this all gets said and done ... we are going to have a pile of highly skilled people underutilized in the economy, frustrated because they don't know how to get out of these positions, and maybe just giving up," he says.

The challenges facing Millennials only rise as their education levels decline. In 2009, 20 percent of 20-to-24-year-olds with only a high school degree were unemployed, according to federal statistics. For those in the age group who did not finish high school, the unemployment rate soared to almost 28 percent. Carlotta Workman feels the weight of those numbers from her position as guidance counselor at the high school in blue-collar Zanesville, Ohio. "Kids are thinking, 'My mom just got laid off, so how am I going to get a job?'" she says.

Somewhat surprisingly, the federal data show that the unemployment rate was no greater last year for graduates of two-year community colleges than for graduates with four-year degrees. But that statistic says nothing about the quality of jobs available to the community college graduates. Roseanne Buckley, director of career services at Burlington County College in New Jersey, finds graduates still working in pizza parlors or holding down the other unskilled jobs that they used to help pay for school in the first place. "Before," she says, "you just expected that once you graduated and got your degree, you'd be able to land a job in that field." Instead, her graduates are often jostling for positions with older workers who have been laid off from more-lucrative positions. "They have no experience," Buckley says of her graduates. "How are they supposed to compete with somebody that's a little bit older, that has a lot of work experience?"

Coping Mechanisms

Social analysts say that as a "civic generation," Millennials are more inclined to light candles than to curse the darkness. They tend to be pragmatic, team-oriented, resilient, and optimistic. In a recent Pew Research Center survey, for instance, only about one-third of Millennials said they were earning enough money now to live the kind of life they wanted. But among the two-thirds who said they weren't earning enough money now, nearly nine in 10 said they expected to earn enough someday. Morley Winograd, co-author of the 2008 book *Millennial Makeover: My Space, YouTube, and the Future of American Politics*, says that although this generation is as likely as its elders to believe that big institutions such as government and business are broken, its reaction to that conclusion is different. "It isn't like the Baby Boomers who said, 'It's broken, let's tear down everything,'" says Winograd, a fellow at the Democratic advocacy group NDN. "In this generation, they say, 'These institutions are broken, we can fix them.'"

It is in that spirit that many Millennials are blazing innovative paths through the Great Recession. Alexander Develov, a senior finishing this spring at business-oriented Babson College in Wellesley, Mass., for instance, has already started a business that provides online video marketing services. "The recession is fantastic for somebody trying to start a business," he says. "You can't raise as much money, but you can get the best people."

Winograd notes that polls have found that Millennials list among their top economic priorities the opportunity to collaborate with peers and the chance to have an impact on the institutions where they work. Such attitudes could lead them to develop new ways of organizing their work lives, especially if their opportunities for stable long-term employment at a single company continue to narrow.

Other Millennials are choosing to ride out the labor market's squalls by pursuing more education. Community college applications have soared 17 percent in the past two years—although the increase is driven in part by older people seeking new skills after layoffs. Demand at community colleges is so high that Bunker Hill Community College in Boston has been forced to add classes on a graveyard shift that begins at 11 p.m. Many young people with four-year degrees are seeking further education as well, although this reliance on graduate school as a harbor could prove a mixed blessing: While some experts say that acquiring more skills almost always pays off, others worry that students will simply accumulate more debt without obtaining the experience they need to begin moving up the workplace escalator. "Graduate school never solved a labor market problem," says Gardner of Michigan State.

But arguably this generation's most distinctive response to hard times is not entrepreneurship or extended education but rather a pursuit of public service, in all of its variations—a trend that was evident even before the recession struck. Edna Medford, who has spent more than two decades teaching history at Howard, sees the impulse toward service as a defining characteristic of this generation. "In the last couple of years, there is a return to that whole idea of being willing to commit to the larger group," she says. "It's truly refreshing."

Closed doors in the private workforce have only reinforced this inclination. Applications are surging at virtually every form of service-oriented institution. AmeriCorps, the national service program created by President Clinton, received nearly 250,000 applications last year—more than two and a half times the number it received in 2008. And so far this year, applications are running almost 60 percent higher than even that peak. Applications for the Peace Corps jumped almost 20 percent last year, to their highest level in the program's nearly 50-year history. At Teach for America, which recruits young college graduates to teach for two years in high-need urban and rural public schools, applications have nearly doubled since 2008, to more than 46,000. "I think the silver lining of the economic downturn is that it's caused people to re-examine what's truly important to them, and that's led folks to choose careers where they can make a difference," says Kerci Marcello Stroud, the communications director of Teach for America.

A similar wave is washing over the military, which is now able to choose from a growing pool of better-prepared applicants. The share of Army recruits with a high school degree jumped to 95 percent in 2009, from 79 percent as recently as 2007, and the Marines have a backlog of young people trying to join. "Continued high unemployment is one reason for interest in the military," acknowledges Curtis Gilroy, director of accession policy in the Office of the Defense Undersecretary for Personnel and Readiness. "But another is the patriotism exhibited by young people today." Howard University senior Frank Bonner, who enlisted to fly aircraft in the Navy after he graduates this spring, exemplifies these intertwined motivations. "A lot of it comes down to job security," he says, "and dedication to my country."

The Big Fixes

In many respects, the economic challenges facing young people are simply a more acute version of the threats confronting all Americans in this sharp downturn. And, as a result, many of the questions about how to create more opportunity for young Americans inevitably become tangled in the larger debates between the political parties about how to promote more prosperity for society overall—that is, whether the economy is more likely to respond to an agenda focused mostly on tax cuts and deregulation or one centered on government investment in areas such as education and alternative energy.

Yet in some ways the problems confronting the Millennial Generation are unique. Although unemployment has increased among all age groups, younger workers are falling out of the workforce much more rapidly than older workers—and more rapidly than they did in previous recessions. In fact, federal statistics show that the share of young people either working or actively looking for work was declining through the first part of the past decade, even before the recession hit. Such data suggest a growing problem in connecting young people to the world of work—in effect, a breakdown in the process of getting them to step onto the escalator. "They are just simply delaying their careers and their labor force connections," laments Reich, who is now a public policy professor at the University of California (Berkeley). "It's a social problem if young people, fresh out of school… can't get into the labor force and have to sit around for a year or two or more."

What can be done to ease the pathway into work for more young people? Obviously, nothing will help more than an overall acceleration in job creation. But smaller, targeted steps could make a difference as well. Among them:

Expanding access to community college. These two-year institutions are a powerful means of providing young people with skills that improve their marketability. Last year, President Obama set a goal of graduating an additional 5 million people from community colleges by 2020 and to that end has sought $12 billion in increased funding; Congress recently approved $2 billion in grants as a down payment on that figure. Experts caution, however, that for community colleges to fulfill expectations, they must ensure that their programs are tied to hiring needs in the local workforce—and they must also improve their record on guiding students through to completion. Federal data show that fewer than three in 10 community college students who enrolled in 2004 had received degrees by the summer of 2007. "We believe [community colleges] are one of the smartest investments we can make" says Melody Barnes, the chief White House domestic policy adviser. "But in doing so, we also feel that it is important to make sure that they are matching their programs to the needs of students today."

Helping students manage the burden of debt. With two-thirds of students now graduating with debts, Obama recently won congressional approval for a significant expansion in federal Pell Grants, as well as reforms in student loan programs that will limit repayments to 10 percent of a student's discretionary income after 2014. But policy makers also need to explore more carefully ways to break the link between expanded student aid and rising tuition—such as proposals from congressional Republicans to limit federal subsidies for schools that raise costs too rapidly.

Broadening internship opportunities. College placement officials agree that employers increasingly expect graduating students to have acquired experience through internships. But because most internships offer little or no pay, low-income students who must earn money either to help meet their family's bills or fund their educations often can't afford to pursue them, which leaves these students disadvantaged in the job hunt after graduation. In a recent study, the liberal think-tank Demos argued that to even the playing field, Washington should provide grants to subsidize internships for low-income students. House Education and Labor Committee Chairman George Miller, D-Calif., recently proposed $500 million in government-funded apprenticeships that could serve a similar function, although with federal deficits gaping, the idea's prospects are uncertain.

Providing more chances to serve. Michigan State's Gardner says that his strongest advice to young people unable to find work is "to find an issue and get involved in it" through volunteering. Nationally, he says, one of the best ways to respond to youth unemployment would be "to expand opportunities for service." To this end, last year Obama proposed and Congress approved legislation that will more than triple the size of AmeriCorps by 2017. Other pending proposals would enlarge public service alternatives such as the Peace Corps.

Updating the safety net for freelance workers. The American social safety net delivers most of its benefits, such as pensions and health care, through the workplace. But that model may be increasingly obsolete for Millennials, given a job environment where so many move back and forth between contract and conventional employment, notes Sara Horowitz, founder of the Freelancers Union, which advocates for freelance workers. Health care reform moved the nation toward accommodating this new reality by creating exchanges and federal subsidies that will make it easier for the self-employed to obtain coverage. But Washington has only just begun rethinking how to provide security to a more fluid workforce.

Perhaps the one sure thing is that the Millennials themselves, as they assume growing responsibility in public and private institutions, will increasingly shape the debate over how to provide opportunity and security for America's rising generations. Few of them seem daunted by that prospect; indeed, many in this enormous, civically oriented, politically engaged generation appear impatient to place their stamp on society. The most farsighted and ambitious among them see the recession that has raged across the economy as something akin to a forest fire that leaves terrible damage in its wake but cultivates new growth by clearing away obstructions. "Recessions are moments of realignment where everyone realizes the way people have been producing work and making money has broken down somehow and it's time to retool," says Julienne Alexander, another partner in Washington's Steadfast Associates. "We want to be a part of that retooling by giving smart, creative youngsters a place to do both commercial and creative work." With luck, this generation's efforts may someday be seen as seedlings on a charred forest floor—the first step toward regeneration after a season of fearsome loss.

KEY CONCEPTS

labor force participation   millennial generation   recession

DISCUSSION QUESTIONS

1. Based on your reading of this article, do you think the value of a college education has gone down? Is the cost of your higher education going to pay off with better work and better pay?

2. What consequences of the recession have you observed or experienced? How are your plans for the future influenced by the economic situation?

Applying Sociological Knowledge: An Exercise for Students

Previous generations in society have all been remembered for specific changes they brought about as they came of age. Ask those in your generation what types of things they think make your generation distinctive compared to the previous one. How do you think your generation will be remembered in years to come?

Glossary

A

AARP the American Association of Retired Persons is one of the most powerful lobbying groups in America; organization that protects rights of older Americans

achievement gap the differences between groups of students, typically based on socioeconomic status and race-ethnicity, in their academic performance

adolescence the period of time between childhood and adulthood, often thought of as a period of many adjustments

age cohort group of people born during the same time period

age prejudice negative attitude about an age group that is generalized to all people in that group

age stereotypes oversimplified set of beliefs about older people

age stratification hierarchical ranking of age groups in society

ageism institutionalized practice of age prejudice and discrimination

alienation feelings of powerlessness and separation from one's group or society

American dream a belief system that the United States is an open society where anyone can be economically and socially successful with hard work

anticipatory socialization the learning of social roles that precedes the actual holding of a given role

antimiscegenation laws laws outlawing the mixing of races through marriage

apartheid system in which different groups (typically racial groups) are completely segregated from each other in all aspects of public life

assimilation process by which a minority becomes socially, economically, and culturally absorbed within the dominant society

B

blaming the victim pattern whereby individuals are blamed for their own misfortune

bourgeoisie term used to loosely describe either the ruling or middle class in a capitalist society

Brown versus Board of Education Supreme Court decision in 1954 that declared the principle of "separate but equal" in public facilities to be unconstitutional

bureaucracy type of formal organization characterized by an authority hierarchy, with a clear division of labor, explicit rules, and impersonality

C

capitalism economic system based on the principles of market competition, private property, and the pursuit of profit

capitalist class those persons who own the means of production in a society

caste system system of stratification (characterized by low social mobility) in which one's place in the stratification system is determined and fixed by birth

charter school public school that has been granted exemption from some local or state regulations

class see social class

class consciousness the understanding of one's position in the class system and an awareness of the exploitation of groups because of their class status

class privilege the typically unrecognized benefit received from higher class status

climate change the environmental problem associated with global warming

clique narrow exclusive circle or group of persons, especially one held together by common interests, views, or purposes

cohabitation living together as an intimate couple while not married

cohort effect the concept that similar occurrences happen to members of a research sample because they share a particular even in common (such as birth year)

collective consciousness body of beliefs that are common to a community or society and that give people a sense of belonging

color-blind racism believing that racism no longer defines people's status in society, including being oblivious to continuing racial inequities

coming-out process process of defining oneself as gay or lesbian

communism economic system in which the state is the sole owner of the systems of production

community a group of people, often thought to be in rather close proximity, who share a common perception of themselves as linked through social contact

conflict theory perspective that emphasizes the role of power and coercion in producing social order

consumerism preoccupation with and inclination toward the buying of consumer goods

contingent worker temporary worker

controlling images powerful, but demeaning, depictions of minority groups, usually found in the media

counterterrorism efforts and social movements to combat terrorism

cultural hegemony pervasive and excessive influence of one culture throughout society

cultural imperialism forcing all groups in a society to accept the dominant culture as their own

cultural relativism idea that something can be understood and judged only in relationship to the cultural context in which it appears

culture of poverty argument that poverty is a way of life and, like other cultures, is passed on from generation to generation

culture complex system of meaning and behavior that defines the way of life for a given group or society

cyberspace interaction sharing by two or more persons of a a virtual reality experience via communication and interaction with each other

D

data systematic information that sociologists use to investigate research questions

de facto by practice or in fact, even if not in law

de jure by rule, policy, or law

debunking the sociological process of challenging taken-for-granted assumptions about social reality

deindustrialization structural transformation of a society from a manufacturing-based economy to a service-based economy

democracy system of government based on the principle of representing all people through the right to vote

desegregation the process of integrating previously segregated groups, such as by race or gender

deviance behavior that is recognized as violating expected rules and norms

discrimination overt negative and unequal treatment of members of some social group or stratum solely because of their membership in that group or stratum, typically involves the powerful group keeping the less powerful group oppressed

diversity variety of group experiences that result from the social structure of society

diversity training programs typically used in corporations to help workplaces be more accepting of a diverse work force

division of labor systematic interrelation of different tasks that develops in complex societies

divorce rate number of divorces per some number (usually 1,000) in a given year

doing gender analytical framework that interprets gender as an activity accomplished through everyday interaction

dominant culture culture of the most powerful group in society

double consciousness realizing that one is being viewed as the "other" in a social situation and simultaneously perceiving and understanding the dominant group

downsizing action by companies to eliminate job positions in order to cut the firm's operating costs

dramaturgical model perspective used to suggest that social interaction has a likeness to dramas presented on a stage

dual labor market theoretical description of the occupational system as divided into two major segments: the primary and secondary labor markets

E

economic crisis the phenomenon beginning around 2009 by which banks failed, people defaulted on their mortgages, and unemployment and general economic malaise was pervasive

educational attainment total years of formal education

educational reform policies and programs designed to improve the educational system

education debt the concept that society is creating greater and greater weakness in our educational system by not addressing the needs of all students

egalitarian societies or groups in which men and women share power, such as in families and marriages

elites high status groups in society

emotional labor (or management) work intended to produce a desired emotional effect on a client

environmental justice social movements that challenge the presence of greater pollution and toxic waste dumping in poor and minority communities

environmental racism disproportionate location of sources of toxic pollution in or very near communities of color

equal employment opportunity civil rights legislation that guarantees equality in access to paid work in larger companies

ethnic enclave regional or neighborhood location that contains people of distinct cultural origins

ethnic group social category of people who share a common culture, such as a common language or dialect, a common religion, and common norms, practices, and customs ethnography descriptive account of social life and culture in a particular social system based on observations of what people actually do

extreme poverty situation in which people live on less than $1 (or its equivalent) per day

F

family wage the system that assumes men as the major earner in families

fatherhood a social status of male parenting that carries with it certain roles, ideals, and responsibilities

feminism beliefs, actions and theories that attempt to bring justice, fairness, and equity to all women, regardless of their race, age, class, sexual orientation, or other characteristics

fictive fatherhood the idea that men can be active parents, even if not the biological parent of a child

field research process of gathering data in a naturally occurring social setting

functionalism theoretical perspective that interprets each part of society in terms of how it contributes to the stability of the whole society

fundamentalism a form of religious extremism that takes a highly conservative view of a given faith

G

gender socially learned expectations and behaviors associated with members of each sex

gender identity one's definition of self as a woman or man

gender role leaned expectations associated with being a man or a woman

gender segregation distribution of men and women in different positions in a social system

gender socialization process by which men and women learn the expectations associated with their sex

gender theory a perspective of society that relies on an understanding of gender relations, gender power differences, and the structural arrangements influenced by gender

gender trouble conflict that emerges from uncertain or changing gender identities (*see gender identity*)

gendered institution total pattern of gender relationships embedded in social institutions

gendered revolution the social change by which women's and men's roles both in work and family have been re-formulated in major ways

gerontology one who studies the branch of knowledge dealing with aging and the problems of the aged

global care chain social network in which a series of personal links between mothers across the globe transfer caring (paid or unpaid) to children other than their own

global culture diffusion of a single culture throughout the world

global economy term acknowledging that all dimensions of the economy now cross national borders

global warming concept that the planet is experiencing a slow and steady change in temperature, resulting in various detrimental environmental consequences

globalization increased economic, political, and social interconnectedness and interdependence among societies in the world

H

health care a system of caring for illness and disease and preventative care for individuals

health insurance a product that gives individuals monetary aid in paying fees associated with health care

hegemonic masculinity pervasiveness of culturally supported attributes associated with the dominant ideas of masculinity

heteronormativity the commonality of heterosexual relations and heterosexual assumptions

heterosexism institutionalization of heterosexuality as the only socially legitimate sexual orientation

heterosexual privilege benefit obtained from a society that values heterosexual relations over other types of sexual relationships; the typically unrecognized benefit gained from being a heterosexual

homophobia fear and hatred of homosexuality

hooking up the act of casual sexual relations between two people

hypersegregation very high levels of segregation

hypothesis an "if, then" statement that is tested through rigorous research

hypothesis statement about what one expects to find when one does research

I

identity one's self-definition

identity lockbox

identity management (see impression management)

immigrant incorporation the process by which immigrant groups are incorporated into the values, norms, and institutions of the host society

immigration migration of people into one society from another society

impression management process by which people control how others perceive them

income the amount of money earned in a given period of time

individualism doctrine that states that the interests of the individual are and ought to be paramount

industrialization sustained economic growth following the application of raw materials and other, more intellectual resources to mechanized production

industrial nations nations marked by sustained economic growth from raw materials, intellectual resources, and other mechanized production

informal economy (or informal sector) part of the labor force and economic system that is not included in the tax structure and other formal ways of being counted in a review of national economic picture

in-group group with which one feels a sense of solidarity or community of interests

institution see social institution

intergenerational conflict the trouble created when two different age groups have opposing needs and ways to address those needs

interlocking directorate organizational linkages created when the same people sit on the boards of directors of a number of different corporations

issues problems that affect large numbers of people and have their origins in the institutional arrangements and history of a society

L

labor force participation the rate at which people stay employed in the formal sector of the economy

labor market available supply of jobs

life expectancy average number of years individuals in a particular group can expect to live

longitudinal study research design dealing with change within a specific group over a period of time

M

managed care use of collective bargaining on the part of large collections of HMOs

marriage rate number of marriages per some number (usually 1,000) in a given year

masculinity a set of social expectations associated with men's gender roles

mass media channels of communication that arc available to very wide segments of the population

McDonaldization process by which increasing numbers of services share the bureaucratic and rationalized processes associated with this food chain

means of production system by which goods are produced and distributed

Medicaid government assistance program that provides health-care assistance for the poor

medicalization of deviance social process through which a norm-violating behavior is culturally defined as a disease and is treated as a medical condition

Medicare government assistance program that provides health-care assistance for the elderly

mentoring the act of one person offering guidance to someone in a Tower position of power as they adjust to a new institution, such as a school or a workplace

meritocracy system (as an educational system) presuming that the most talented are chosen and moved ahead on the basis of their talent and achievement

methodology practices and techniques used to gather, process, and interpret theories about social life

middle class an amorphous group generally considered to share a common economic and social status

millennial generation those individuals born between 1981 and 2002; the older set of millenials are now in their 20-something years and emerging into adulthood

myth of the absent father the misleading idea that African American men are unlikely to assume strong roles as parents

N

neoliberalism the belief and policy system based on the idea that economic markets should be left to function without intervention by the government

nonrepresentative sample sample not similar to the population from which it was drawn

norms specific cultural expectations for how to act in given situations

nuclear family social unit comprised of a man and a woman living together with their children

O

occupational segregation pattern by which workers are separated into different occupations on the basis of social characteristics such as race and gender

open-ended interviews process of asking research questions that allow respondents flexibility in their responses, as compared to closed-ended interviews where the answers are pre-determined and selected from a list of multiple choices

organizations large scale units in society with a complex and hierarchical form

out-group group that is distinct from one's own and is usually an object of hostility or dislike

P

participant observation method whereby the sociologist is both a participant in the group being studied and a scientific observer of the group

patriarchy society or group in which men have power over women

pluralism state of society in which members of different ethnic, racial, religious, and social groups maintain their distinctive cultural traditions, respect the traditions of other groups, and share a common civilization

popular culture beliefs, practices, and objects that arc part of everyday traditions

post-racial society the idea that race no longer determines people's social location in society

poverty line figure established by the government to indicate the amount of money needed to support the basic needs of a household

power elite model theoretical model of power positing a strong link between government and business

power ability of a person or group to exercise influence and control over others

prejudice negative evaluation of a social group and individuals within that group based on misconceptions about that group

presentation of self Erving Goffman's phrase referring to the way people display themselves during social interaction

prestige subjective value with which different groups or people are judged

proletariat lowest social or economic class of a community; the laboring class, especially the class of industrial workers who lack their own means of production and hence sell their labor to live

Protestant ethic belief that hard work and self-denial lead to salvation and success

Q

queer theory a conceptual understanding of society from the perspective of homosexual relations and the marginalization of homosexuals

R

race social category or social construction based on certain characteristics, some biological, that have been assigned social importance in the society racial inequality the pattern whereby racial groups have differing degrees of advantage and disadvantage

racism perception and treatment of a racial or ethnic group or member of that group as intellectually, socially, and culturally inferior to one's own

recession a period of economic decline, usually defined by economists as a time when gross domestic product declines for two consecutive economic quarters; sociologically, it is a time when most economic indicators are low, leading financial struggles across class boundaries

recidivism relapse into a former pattern of criminal behavior

redlining discriminating against racial groups in housing, home-loan funds, or insurance

religion institutionalized system of symbols, beliefs, values, and practices by which a group of people interprets and responds to what they believe is sacred and that provides answers to questions of ultimate meaning

religiosity intensity and consistency of practice of a person's (or group's) faith

representative sample a sample drawn from a larger population to be used in a research project that mirrors very closely the key characteristics of that larger population; typically a sample drawn using random methods that eliminate selection bias

research design the plan for a research investigation that is derived from the particular question being asked

resegregation the process by which schools (or other places) become less integrated than in the past

rite of passage ceremonial events that mark the transition from one status to another; especially common as part of major life events

ritual symbolic activities that express a group's spiritual convictions

role expected behavior associated with a given status in society

S

sample any subset of units from a population that a researcher studies

secular ordinary beliefs of daily life that are specifically not religious

segregation pattern by which groups based on some social characteristic (such as gender, age, or race_are separated from one another

service sector part of the labor market composed of the nonmanual, nonagricultural jobs

sex trafficking international pattern of selling sex, often involving moving sex workers from one country to another

sex work paid labor involving sexual services

sex biological identity as male or female

sexism system of practices and beliefs through which women are controlled and exploited because of the significance given to differences between the sexes

sexual orientation manner in which individuals experience sexual arousal and pleasure

sexuality sexual desire and behavior

sexualization of culture process by which sexuality has come to permeate images and behaviors in popular culture

social change process by which societies are transformed

social class (or class) social structural position that groups hold relative to the economic, social, political, and cultural resources of society

social control process by which groups and individuals within those groups are brought into conformity with dominant social expectations

social institution established, organized system of social behavior with a recognized purpose

social interaction behavior between two or more people that is given meaning

social mobility movement over time by a person from one social class to another

social movement group that acts with some continuity and organization to promote or resist change in society

social myths false beliefs that nonetheless can guide people's understanding of social issues

Social Security government program that supports former workers in their older years, based on the amount of earnings over a lifetime

social stratification relatively fixed hierarchical arrangement in society by which groups have different access to resources, power, and perceived social worth; a system of structured social inequality

social structure patterns of social relationships and social institutions that comprise society

society system of social interactions that includes both culture and social organization

sociological imagination ability to see the societal patterns that influence individual and group life

sociology study of human behavior in society

spirituality an internalized feeling of faith, not necessarily associated with a given religious denomination

stereotype oversimplified set of beliefs about the members of a social group or social stratum that is used to categorize individuals of that group

stigma attribute that is socially devalued and discredited

subculture culture of groups whose values and norms of behavior are somewhat different from those of the dominant culture

subprime mortgage mortgages that carry an interest rate that is higher than the prime rate

survey to query (someone or a group or a sample) in order to collect data for the analysis of some aspect of a group or area

T

teleology the study of purposes in natural phenomena; or, an argument where phenomena are explained by their purposes theory of displacement the idea from Freud that emotion can be redirected from one person or object to another

tracking grouping, or stratifying, students in school on the basis of ability test scores

transsexual a person born with one biological sex, but in the process of, or having completed, the transition to the other sex

troubles privately felt problems that come from events or feelings in one individual's life

U

undocumented worker an employed person who receives pay "off the books" and is not typically counted among full- or part-time workers in national counting of labor force participation rates

urban underclass grouping of people, largely minority and poor, who live at the absolute bottom of the socioeconomic ladder in urban areas

V

victimization experiences of people who have been victims of crimes or systematic oppression

virtual communities communities that exist via computer interactions voter turnout percentage of a given population that participates in elections

W

wealth monetary value of what someone owns, calculated by adding all financial assets (stocks, bonds, property, insurance, investments, etc.) and subtracting debts; also called net worth

welfare state a system within a nation of public benefits designed to provide for those who cannot fully provide for themselves and their families

white flight the concept that as neighborhoods become more racially diverse, White people will move out to other neighborhoods

white privilege system in which white people benefit collectively from the social and economic history of racism

work productive human activity that produces something of value, either goods or services

working class those persons who do not own the means of production and therefore must sell their labor in order to earn a living

world economy refers to the increasingly interlocking nature of national economies in a global system

Name Index

Subject Index

Dropout factories, 407
Dubious data, 20
'Dude, You're a Fag': Adolescent Masculinity and the Fag Discourse, (Pascoe), 300
Dumping in Dixie: Race, Class, and Environmental Quality, 527
Dungeons and Dragons Online (game), 37

E
Early Learning Challenge Fund, 415
Earned Income Tax Credit, 515, 556
Earth Summit, 528
Econo Lodge, 428, 434
Economic crisis
 cleavages, intergenerational, 553–554
 discussed, 553–559
 reform, entitlement, 557–558
Economic disparity, statistics, 233
Economic Policy Institute (EPI), 426–427
Economic Roundtable, 448–451
The Economist, 234
Economy
 agriculture, dependence on, 201
 Americans, effect on, 181
 bankruptcy, causes of, 192
 Big One, The, 191
 boom, 1960s, 181
 cards, credit, 193
 core expenses, 181
 crisis, subprime mortgage, 192
 deindustrialization, 132–133
 development, economic, 506
 development, supermarket, 506
 downturn, global, 199
 families, effect on, 364
 family, effects on, 350
 financial industry, 181
 global, 516
 global, Freud in, 206
 informal, 447–453
 inner-city, poverty link to, 523
 institution, power, 464–468
 low-wage work, effects of, 426
 middle class, loss of, 180–182
 national debt, 394, 400
 national deficit, 394

net, global safety, 199–204
opportunism, 192
Order, New Economic, 202–203
preschool, effects on, 412
recession, impact of, 190–195
retail jobs, change to, 133
service industries, transition to, 438–445
taxes, effect on, 516
ultimatum game, 161
workers, displaced, 201–202
Edo, ethnic group, 217
Education, practical credentialists, 387
Education debt
 addressing, reasons for, 400
 contributing factors, 395
 discussed, 391–402
 economic, 396–397
 historical, 395–396
 moral, 398–399
 sociopolitical, 397–398
Education research, 392
Education Trust, 401
Educational system. *See Aslo Colleges/Universities*
 achievement, factors of, 393
 birth to eight network, 415–416
 Brown v. Board of Education, 392, 409, 423
 Center for Educational Reform, 418
 charter schools, 417–424
 debt of, 394
 desegregation, 406
 disparities, funding, 396
 disparities, funding, statistics, 397
 dropout factories, 407
 Early Learning Challenge Fund, 415
 gap, achievement, 392–393
 Goals 2000 project, 406
 Head Start, 410–413
 Head Start, Early, 415–416
 instructional climate, 414
 Knowledge Is Power Program (KIPP), 424
 meritocracy, 422
 minority, largest, Latino, 406
 No Child Left Behind Act (NCLB), 406–409

pedagogy, 414
poverty, escape from, 412
power elite, controlled by, 467
Pre-K Now, 412
PreK-3rd approach, 416
preschool, importance of, 410–416
principles, essential, 412
principles, neoliberal business, 421
schools, apartheid, 401
segregation, 401
segregation, increasing, 405–409
services, continuity of, 412
Student Aid and Fiscal Responsibility Act, 415
students, minority,statistics, 406
trends of, 391–402
trust, relational, 400
War on Poverty, 412
EEOC. *See* Equal Employment Opportunity Commission (EEOC)
Emancipation, U.S., 224
eMarketer, 100
Emergency rooms, 496
Eminem exception, faggot, epithet, 307
Emoticons, 40
Emotional labor
 defined, 440
 men, view of, 444
 occupations, service vs professional, 440
Empathy, strategy, counterterrorism, 163
Employee turnover, 140
Employment. *See Aslo* Labor
 African Americans, 364–365
 America, low-wage work in, 425–438
 China, unemployed, statistics, 202
 core/periphery jobs, 441
 degree, college, required, 442
 freelance, 567
 garbage pickers, 200
 gender expectations, 261
 gender transitions, 260–274
 identity, personal, 355
 industries, fast food, 441
 industries, service, 438–445